WASTE MATERIALS USED IN CONCRETE MANUFACTURING

WASTE MATERIALS USED IN CONCRETE MANUFACTURING

Edited by

Satish Chandra

Division of Concrete Structures
Chalmers University of Technology
Göteborg, Sweden

np

NOYES PUBLICATIONS
Westwood, New Jersey, U.S.A.

Library of Congress Catalog Card Number: 96-12577
ISBN: 0-8155-1393-3
Printed in the United States

Published in the United States of America by
Noyes Publications
369 Fairview Avenue, Westwood, New Jersey 07675

10 9 8 7 6 5 4 3 2 1

Library of Congress Cataloging-in-Publication Data

Waste materials used in concrete manufacturing / edited by Satish Chandra
p. cm.
Includes bibliographical references and index.
ISBN 0-8155-1393-3
1. Concrete--Additives. 2. Waste products. I. Chandra, Satish.
TP884.A3W38 1996
666'.893--dc20 96-12577
CIP

BUILDING MATERIALS SCIENCE SERIES

Editor

V. S. Ramachandran, National Research Council Canada

CONCRETE ADMIXTURES HANDBOOK; Properties, Science and Technology, Second Edition: edited by V. S. Ramachandran

HANDBOOK OF CONCRETE AGGREGATES; A Petrographic and Technological Evaluation: by Ludmila Dolar-Mantuani

CONCRETE ALKALI-AGGREGATE REACTIONS: edited by P. E. Grattan-Bellew

HANDBOOK OF FIBER-REINFORCED CONCRETE; Principles, Properties, Developments, and Applications: by James J. Beaudoin

CONCRETE MATERIALS; Properties, Specifications and Testing, Second Edition: by Sandor Popovics

HANDBOOK OF POLYMER MODIFIED CONCRETE AND MORTARS; Properties and Process Technology: by Yoshihiko Ohama

WASTE MATERIALS USED IN CONCRETE MANUFACTURING: edited by Satish Chandra

Related Titles

CORROSION AND CHEMICAL RESISTANT MASONRY MATERIALS HANDBOOK: by W. L. Sheppard, Jr.

To My Beloved Parents

Preface

One of the most critical problems we face today is acid rain. One of the main causes is the burning of fossil fuels and, as a result, sulfuric acid and carbon dioxide are added to the atmosphere. These pollutant gases have detrimental effect on building materials. Thus there are two problems: 1) to decrease the gas emission, and 2) to produce construction material which is more durable to aggressive pollutant gases and acid rain.

The environmental aspects involved in the production and use of cement, concrete and other building materials are of growing importance. CO_2 emission, for example, is 0.8–1.3 ton per ton of cement production in dry process. SO_2 emission is also very high, depending upon the type of fuel used. Energy consumption is also very high at 100–150 kWT/ton of cement produced. It is costly to erect new cement plants. It costs approximately $230 US per ton of installed capacity in Europe and $230 US in developing countries. Substitution of waste materials will conserve dwindling resources and will avoid the environmental and ecological damages caused by quarrying and exploitation of the raw materials for making cement. To some extent, it will help to solve the problem otherwise encountered in disposing of the wastes. Partial replacement of clinker or Portland cement by slag, fly ash, silica fume, and natural rock minerals illustrates these aspects. Partial replacement by natural materials that require little or no processing, such as pozzolans, calcined clays, etc., saves energy and decreases emission of gases. The annual production of Portland cement is more than 1000 M tons, while the output of waste materials suitable as cement replacement (slags, fly ashes, silica fumes, rice husk ash, etc.) is more than double this amount.

These waste materials can partly be used, or processed, to produce materials suitable as aggregates or fillers in concrete. These can also be used as clinker raw materials or processed into cementing systems. New grinding and mixing technology will make the use of these secondary materials simpler. Developments in chemical admixtures (superplasticizers, air-entraining agents, etc.) help in controlling production techniques and in achieving the desired properties in concrete.

The use of waste products is not only a partial solution to environmental and ecological problems. It significantly improves the microstructure, and consequently the durability properties of concrete, which are difficult to achieve by the use of pure Portland cement. The aim is not only to make the cements and concrete less expensive, but to provide a blend of tailored properties of waste materials and Portland cements suitable for a specified purpose. This however requires a better understanding of chemistry and materials science.

There is also an increasing demand for better understanding of material properties, as well as better control of the microstructure developing in the construction material, to increase durability. The combination of different binders and modifiers to produce cheaper and more durable building materials will be a factor in solving the ecological and environmental problems.

Göteborg, Sweden
January, 1997

Satish Chandra

Contributors

A. A. Abang Abdullah
University of Pertanian Malaysia
Serdang, Malaysia

Leif Berntsson
Chalmers University of Technology
Göteborg, Sweden

Jiri Brandstetr
Technical University of Brno
Department of Materials Chemistry
Brno, Czech Republic

Satish Chandra
Chalmers University of Technology
Göteborg, Sweden

Marek Gawlicki
University of Mining and Metallurgy
Cracow, Poland

Shunsuke Hanehara
Chichibu Onoda Cement Corporation
Sakura, Chiba, Japan

Jaromir Havlica
Technical University of Brno
Department of Materials Chemistry
Brno, Czech Republic

Chao Lung Hwang
National Taiwan Institute of Technology
Taipei, Taiwan

Pavel V. Krivenko
State Technical University of Construction and Architecture
Kiev, Ukraine

Wieslaw Kurdowski
University of Mining and Metallurgy
Cracow, Poland

Ajoy Kumar Mullick
National Council for Cement and Building Materials
New Delhi, India

Wieslawa Nocuń-Wczelik
University of Mining and Metallurgy
Cracow, Poland

Ivan Odler
Technical University of Clausthal
Institute for Non-metallic Materials
Clausthal-Zellerfeld, Germany

Wojciech Roszczynialski
University of Mining and Metallurgy
Cracow, Poland

Francois Sorrentino
LAFARGE
Central Research Laboratory
Saint-Quentin-Fallavier, France

Bob Talling
Renotech Ltd.
Turku, Finland

Hiroshi Uchikawa
Chichibu Onoda Cement Corporation
Sakura, Chiba, Japan

Aimin Xu
STK Inter TEST AB
Helsingborg, Sweden

About the Editor

Satish Chandra is presently an assistant professor at Chalmers University of Technology, Department of Concrete Structures, Göteborg, Sweden. A graduate of Lucknow University, Lucknow, India, he then was awarded his Doctor of Technology from the University of Mining and Metallurgy, Cracow, Poland. His area of special interest is in materials development, durability and deterioration mechanisms of building materials. He also is interested in the use of mineral admixtures in concrete.

He is a member of ACI Committees 547 and 548; RILEM Committees TC 105 and 113; a Life member of the Indian Concrete Institute; a Member of the Swedish Chemical Society and Swedish Silicate Society. He has also organized and chaired numerous meetings and seminars in the Cement and Concrete area. Dr. Chandra has contributed to three books and has published sixty-five technical papers in journals of international repute.

NOTICE

To the best of our knowledge the information in this publication is accurate; however the Publisher does not assume any responsibility or liability for the accuracy or completeness of, or consequences arising from, such information. This book is intended for informational purposes only. Mention of trade names or commercial products does not constitute endorsement or recommendation for use by the Publisher. Final determination of the suitability of any information or product for use contemplated by any user, and the manner of that use, is the sole responsibility of the user. We recommend that anyone intending to rely on any recommendation of materials or procedures mentioned in this publication should satisfy himself as to such suitability, and that he can meet all applicable safety and health standards.

CONTENTS

5 BLAST FURNACE SLAG - THE ULTIMATE BINDER 235

Bob Talling and Pavel Krivenko

1

PROPERTIES AND USE OF SOLID RESIDUE FROM FLUIDIZED BED COAL COMBUSTION

Jiri Brandštetr, Jaromir Havlica and Ivan Odler

INTRODUCTION

The combustion of coal is expected to be the most widely used source of energy for the production of electricity at least in the foreseeable future. In conventional power plants, the combustion of the fuel takes place at temperatures between about 1200-1700°C and the process is associated with the liberation of large amounts of sulfur oxide in addition to CO_2. Worldwide, millions of tons of nSO_2 are generated in this way annually [1]. Thus, the necessity exists to reduce the emission of SO_2 and to clean the produced flue gases.

In recent years a variety of "clean" technologies for coal combustion were developed [2]. Out of these, technologies based on fluidized bed combustion found greatest acceptance [4-6]. Here, finely ground coal is burnt together with ground limestone, acting as a sorbent of sulfur dioxide, in a fluidized bed at temperatures of approximately 850°C. Under these conditions, the SO_2 produced in the combustion process remains absorbed in the ash and additional cleaning of the flue gas is not necessary. It may be expected that large fraction of newly built power plants, especially those with a smaller or medium capacity will be based on this technology. Instead of limestone also dolomite may be employed as a sorbent which, however, is less suitable should the ash formed in the process be utilized, rather than disposed.

The solid residue of the fluidized combustion of bituminous or sub-bituminous coals differ distinctly in their chemical and phase composition from ashes produced in conventional power plants and due to their generally favorable properties the potential exists for their large scale utilization [7-18]. In fact, at a site in the Netherlands, solid mining residues are processed in a fluidized bed reactor to produce a good quality inorganic binder, while the electricity, produced simultaneously, is considered to be a by-product [19].

A factor supporting the general trend to utilize industrial by-products [1,15,17,19-26] is the gradual depletion of high quality raw materials in the world and the steadily increasing volume of waste of all kinds which must be disposed, unless it is utilized in a meaningful way. This necessitates the development of technologies aimed at the utilization and/or recycling of secondary raw materials.

The permanently increasing number of power plants employing fluidized bed combustion results in an increasing interest in the utilization of the formed ashes whose composition and properties may vary in a wide range, depending on the composition of the fuel used, on the quality and amount of sorbent and on the combustion technology employed.

Fluidized bed combustion may be done either under atmospheric pressure (Atmospheric Fluidized Bed Combustion, AFBC) or at elevated pressure (Pressurized Fluidized Bed Combustion, PFBC). The latter approach, developed only recently, is used only in three power plants of the world and there are no published data on the properties of formed ashes so far.

New materials require new testing procedures to assess their quality and possible utilization. Should they be used under industrial conditions, these methods must also be sufficiently simple to make the testing procedure acceptable [27].

To attain a wide acceptance of new materials, it is essential to publish information about their properties and potential uses. In the case of fluidized bed ashes produced in flue gas de-sulfurization, the steadily increasing costs of their disposal may also stimulate efforts to utilize them in different applications.

The prevailing majority of papers on fluidized bed combustion of coal published so far deals with the optimization of the combustion process and on the equipment to be employed. Several conferences were dedicated to these topics [4,5]. This chapter provides information on the composition and properties of the solid residue from fluidized bed combustion of coal using either the AFBC or PFBC technologies. Also discussed will be the potential uses of these materials. The aim of this report is to generate interest in these materials whose volume is expected to increase steadily in the coming years.

As products of fluidized bed combustion exhibit hydraulic and/or pozzolanic properties, their use as a binder may result in a reduced consumption of Portland cement whose production is associated with a significant consumption of energy. This in course may also contribute to a reduction of the CO_2 emission. A cooperation of power plant operators, ecologists, material scientists, civil engineers and also legislators will be required to achieve an acceptance and a wide use of fluidized bed combustion ashes.

STARTING MATERIALS AND THE COMBUSTION PROCESS

Solid fuel

Bituminous or sub-bituminous coal, lignite and even different waste products from coal mining [19,28] may be employed as fuels in the combustion process. It is obvious that the quality of the fuel and in particular the composition of its inorganic fraction determines to a great degree the composition and quality of the produced ash. Bituminous coal is the fuel used most widely in the industry. Its ash content lies typically at around 8 to 15%. The sulfur content of different coals may also vary greatly, which requires in most instances a de-sulfurization, to keep the SO_2 content in the flue gas low enough. Occasionally fuels with a content of inorganic constituents of up to 50% are used which results in an increased volume of produced ash. Up to several hundred tons of ash may be produced per day in some units. On the other hand, a certain amount of inorganic constituents in the fuel is necessary to assure the formation of a fluidized bed.

The chemical composition of the produced ash is independent on the employed combustion technology whereas the phase composition may be affected considerably. The factor that affects greatly the phase composition of an ash is the temperature of combustion which in the case of fluidized bed technology lies at around 850 °C. Another factor that also determines both the chemical and phase composition of the final product is the quality and quantity of the sorbent intermixed with the fuel.

The coal in fluidized bed combustion must be ground to a fineness below 4 mm and the amount passing the sieve 0.5 mm is typically around 20% . Finely ground limestone is the sorbent most widely used for the sorption of sulfur oxide.

At the temperature existing in the fluidized bed, limestone decomposes and converts to highly porous calcium oxide (lime) which reacts with sulfur oxide produced in the oxidation of the fuel. It is assumed that sulfur oxide also reacts with ground limestone directly in a topochemical reaction. The sorbent is being interground with the fuel to an appropriate fineness. Sometimes water is added to the mix to produce a slurry of desired consistency.

Alternatively, dolomite may also be used as a sorbent of SO_2 instead of limestone. The selection of the sorbent is usually done depending on the local availability of the material. Magnesium carbonate present in dolomite together with calcium carbonate decomposes at a lower temperature, namely already at about 500 °C, while calcium carbonate in limestone decomposes at 700 - 800 °C. At temperatures existing in the fluidized bed, i.e. at 850 °C, magnesium oxide looses its reactivity and tends to react with sulfur oxide only incompletely. A significant fraction of it remains in the combustion residue in non-reacted form as crystalline periclase. Thus the content of free MgO in the product of combustion may often exceed 10% , if dolomite is used as a sorbent. Contrary to that, the periclase content of ashes produced by adding limestone to the coal rarely exceeds 2.5%. Periclase reacts in a very slow reaction with water yielding magnesium hydroxide $Mg(OH)_2$ (brucite). As this reaction is associated with an expansion which may cause cracking of building materials made from high periclase ashes, the potential of using such ashes is limited. One area of possible uses of high MgO ashes is in agricultural applications [29,30].

The combustion temperature of 850°C used in fluidized combustion has been selected as at this temperature the absorption of sulfur oxide by the added sorbent is most effective. Under these conditions a separate de-sulfurization of the residual flue gas becomes unnecessary. To attain a sufficient binding of sulfur, however, the sorbent has to be added in an stoichiometric excess, to increase the effectiveness of the sorption process. Obviously, the actual amount of limestone that has to be introduced increases with increasing sulfur content of the fuel. An addition of 4% of limestone is typical, if a coal with a sulfur content of around one percent is used.

The extent to which the added sorbent affects the oxidic composition of the produced de-sulfurization residue will depend on the ash content of the fuel and also on the amount of sulfur present in the fuel and thus on the amount of limestone that had to be added to attain a sufficient de-sulfurization. Typically the amount of CaO + MgO coming from the added sorbent represents about 8-15% of the total amount of the solid de-sulfurization residue, if a coal with an ash content of 15% is employed. If fuels are employed that contain a significant fraction of calcium carbonate among its constituents, the amount of added sorbent may be reduced accordingly. This may be the case especially in sub-bituminous coals which contain up to 10% of calcium carbonate.

The Combustion Process

Most widely used technology at present in fluidized bed combustion of solid fuels is the AFBC process (atmospheric fluidized bed combustion). Another technology, the pressurized bed process PFBC, that has been developed by the ABB company in Sweden [3] is being employed presently only at three different sites worldwide. In the recent approach the combustion takes place at a pressure of around 1.2 MPa. This makes it possible to reduce the height of the fluidized bed reactor significantly.

In both the AFBC and PFBC processes the combustion takes place in a circulating fluidized bed in which the fuel returns repeatedly into the reactor. Under these conditions the overall time in which the combustion takes place may reach up to several hours. In the course of combustion the primary formed sulfur dioxide SO_2 is converted to anhydrous calcium sulfate $CaSO_4$, anhydrite II. The oxidation of S^{+4} to S^{+6} is catalyzed by heavy metals present in the ash.

Due to the low burning temperature the reactors and cyclones employed in the AFBC and PFBC processes do not require any refractory linings. A cooling of the internal walls with water is sufficient. Nitrogen oxides NO_x are formed only from nitrogen bound within the organic constituents of the fuel. Thus, both the SO_2 and NO_x content in the flue gas, produced in these processes is very low, well below levels required by the existing specifications. Fine solid constituents present in the flue gas are being separated by appropriate gas filters and must be disposed to be utilized in a meaningful way.

Burning in a circulating fluidized bed results usually in a thorough oxidation of the organic constituents of the fuel, so that the content of residual carbon may be easily kept below 1. Only if the rate of combustion is intentionally reduced the carbon content in the residual ash may increase. The greatest part of residual carbon in the ash is present in an amorphous form and exhibits a large specific surface area. As a consequence of it the water requirement in the production of mortar or concrete may be increased if a high volume of fluidized bed ash with a significant carbon content is added to the mix.

The particle size distribution of the residual material produced in fluidized bed combustion varies in a wide range and fractions of different granulometry are precipitated and separated at different sites. The coarsest fraction is the bed ash which represents usually 30 to 40% of the total amount of the solids. About 60% of the solid material is separated in cyclones whereas the fine fraction precipitated in last electrostatic precipitators or hose filters rarely exceeds 5%. Figures 1-3 show the particle size distributions of different kind of ash fractions (PFBC, Finspong, Sweden). It is also noteworthy to mention that different fractions differ not only in their fineness, but also in chemical and mineralogical composition and thus may be suitable for different applications. Compositions of various ashes and fractions are given in Tables 1-3.

Different ash fractions exhibit also differences in their net and bulk densities and specific surface areas. From a comparison of the existing particle sizes and specific surface areas it is obvious that different ash fractions differ also greatly in their porosities. The exact values will depend on the quality of the employed fuel and the quantity and quality of the sorbent employed.

Chemical Reactions and Phase Transformations in the Course of Combustion

The main constituents of the inorganic part of bituminous and sub-bituminous coals are clay minerals (illite, kaolinite, montmorillonite, halloysite etc.) - up to 50% , quartz - up to 15% and then iron oxides (hematite, limonite, magnetite), iron sulfides (pyrite, markasite) and carbonates (calcite, dolomite, magnesite). The low content of calcium sulfate varies. Minor constituents include various

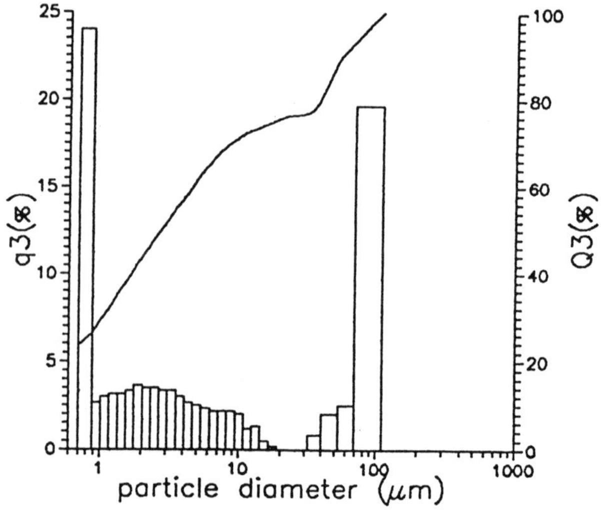

Figure 1. Particle size distribution (q3) and its integral curve (Q3) of PFBC ground bed ash: 25% < 0.97 μm, 60% < 5.00 μm, specific surface 7.96 m^2/g (BET), specific gravity 2718 kg/m^3.

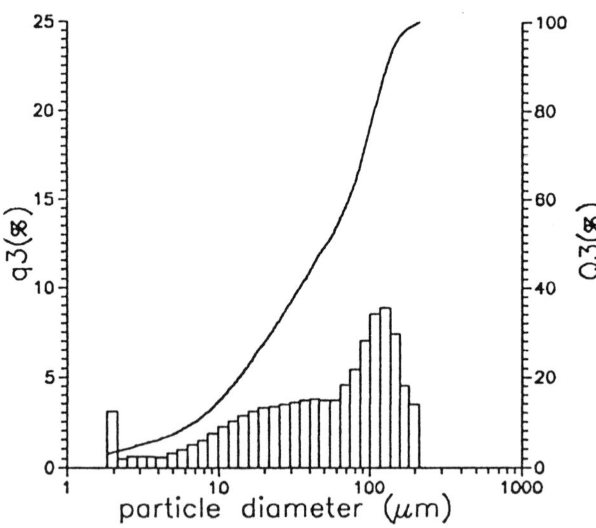

Figure 2. Particle size distribution (q3) and its integral curve (Q3) of PFBC cyclone fly ash: 25% < 15.1 μm, 60% < 53.6 μm, specific surface 4.72 m^2/g (BET), specific gravity 2692 kg/m^3.

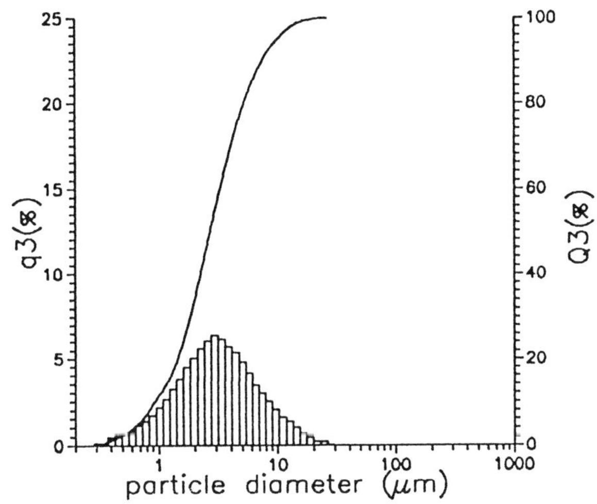

Figure 3. Particle size distribution (q3) and its integral curve (Q3) of PFBC filter fly ash: 25% < 1.46 μm, 60% < 3.46 μm, specific surface 6.39 m^2/g (BET), specific gravity 2580 kg/m^3.

Table 1. Chemical Analysis of Individual Fractions of PFBC Ground Bed Ash (Finspong, Sweden). *L.O.I. = loss on iginition

	Percent Fraction, mm			
	<0.040	0.040-0.090	0.090-0.125	>0.125
SiO_2	23.4	40.3	42.5	49.8
CaO	34.0	21.2	18.6	13.5
MgO	2.0	2.3	2.6	2.5
Al_2O_3	9.4	13.7	15.5	17.0
Fe_2O_3	3.0	4.3	3.8	3.8
TiO_2	0.7	0.8	0.9	0.9
P_2O_5	0.2	0.2	0.3	0.2
MnO	0.1	0.1	0.1	0.1
Na_2O	0.7	0.9	0.8	0.8
K_2O	0.9	1.6	1.7	2.1
C	4.4	2.9	2.4	1.2
S	3.7	1.5	1.4	1.8
L.O.I.* (1000°C)	18.8	12.5	10.9	7.7

Table 2. Chemical Analysis of Individual Fractions of PFBC Cyclone Fly Ash (Finspong, Sweden).

	Percent Fraction, mm			
	<0.040	0.040-0.090	0.090-0.125	0.125-0.200
SiO_2	36.8	37.0	33.7	31.7
CaO	17.3	21.6	26.6	29.5
MgO	3.7	3.5	3.3	2.9
Al_2O_3	18.3	14.3	11.6	10.5
Fe_2O_3	7.2	6.3	5.5	4.2
TiO_2	1.1	1.0	0.8	0.7
P_2O_5	0.3	0.3	0.2	0.1
MnO	0.2	0.2	0.2	0.1
Na_2O	1.0	0.9	0.6	0.6
K_2O	1.7	1.5	1.3	1.3
C	2.3	3.3	4.2	4.9
S	2.4	1.5	1.3	1.1
L.O.I. (1000°C)	7.9	11.2	14.9	17.5

Table 3. Chemical Analysis of Individual Fractions of AFBC Fly Ash (Trinec, Czech Republic).

	Percent Fraction, mm		
	<0.040	0.040-0.063	>0.063
SiO_2	44.8	48.4	56.1
CaO	7.9	8.9	5.4
MgO	3.4	4.3	3.2
Al_2O_3	22.5	21.0	22.6
Fe_2O_3	7.4	6.3	5.3
TiO_2	1.3	1.0	1.2
P_2O_5	0.7	0.5	0.4
MnO	0.2	0.2	0.2
Na_2O	0.7	0.7	0.7
K_2O	2.4	2.6	2.4
C	4.3	0.9	0.2
S	1.1	1.7	0.7
L.O.I. (1000°C)	7.5	1.5	0.8

phosphates, chlorides, humic acids and also free and chemically combined water.

The following chemical reactions take place in the course of coal burning in a fluidized bed in the presence of limestone: First, free water evaporates, followed by the escape of water bound within the clay minerals. The temperature of decomposition varies in different clay minerals: The interlayer water of halloysite escapes already at around 140°C, whereas its dehydroxylation starts at 450°C. Kaolinite looses its bound water at 500-650°C and converts to metakaolinite, which is an unstable highly reactive (x-ray amorphous) aluminum silicate with a large specific surface area. Montmorillonitic clays (bentonites) lose their water present in interlayer spaces at around 150°C; at this temperature the dehydration of sodium and potassium ions proceeds. At somewhat higher temperatures this is followed by an irreversible escape of water combined with cations of alkaline earths. Finally, at 600-700°C water bound as a constituent of the crystalline lattice is liberated. Illitic clays behave similar to montmorillonite, however, their decomposition proceeds more slowly and is often incomplete. The amorphous aluminosilicate phases and amorphous silicone dioxide, formed in the thermal decomposition of clay minerals, are responsible for the hydraulic and/or pozzolanic reactivity of the ashes seen after mixing them with water and other constituents.

Magnesium carbonate $MgCO_3$ decomposes at temperatures above 500°C and calcium carbonate $CaCO_3$ above 700°C. The rate of decomposition is accelerated if alkali ions are present. Pyrite decomposes at around 400-500°C and its decomposition is associated with an oxidation of the present sulfur and iron.

The calcium oxide produced in the decomposition of limestone exhibits a large surface and reacts readily with sulfur oxide formed in the combustion process in the temperature range 700-850°C. In the presence of excessive amounts of oxygen, anhydrite II (calcium sulfate, $CaSO_4$) rather than calcium sulfite $CaSO_3$ is formed as the final product of reaction. The excessive amount of the sorbent which was added to attain an effective de-sulfurization of the flue gases, remains in the ash in the form of free lime (in the AFBC process) or calcium carbonate (in the PFBC process). To maintain a high reactivity of the produced fluidized bed ash it is important not to exceed significantly the combustion temperature of 850°C. This low temperature assures a higher specific surface area and thus a greater reactivity of this type of ashes as compared to ordinary fly ashes formed at combustion temperatures reaching 1600°C, when a liquid melt is formed. The final conventional fly ash consists mostly of spherical particles that contain, besides of the glass phase, also crystalline constituents such as mullite, anorthite or spinel. The pozzolanic reactivity of this ash is mainly due to the glass phase, whereas the majority of the formed crystalline phases is nonreactive. If coals containing calcium carbonate are utilized, also here calcium oxide is formed in its decomposition and reacts with SO_2 and O_2, yielding $CaSO_4$. However, non-reacted CaO stays in the ash as dead burnt lime. This phase, as well as free MgO, are undesirable constituents of fly ashes as they may react slowly with water, yielding hydroxides. This reaction may be associated with an expansion and cracking of materials produced from such ashes.

THE CHEMICAL AND PHASE COMPOSITION OF AFBC AND PFBC ASHES

The solid residue formed in the AFBC and PFBC processes may be considered to be blend of the inorganic fraction of the employed fuel, the compound of sulfur produced in the desulfurization reaction and of the excessive sorbent. Their oxidic composition may be estimated from the composition of the fuel, of the interblended sorbent and their mutual ratio. The chemical composition is similar to that of fly ashes received by conventional high temperature combustion of the same coal, except higher CaO (or MgO) and SO_3 contents (see Table 4).

The chemical composition of the ash alone does not suffice for assessing the quality and thus the potential uses of the material. In addition to the overall amount of present CaO it is important to

Table 4. Chemical Composition (in oxides) of Bed Fly Ash [11,43]. (Compared to Ignited Samples).

	Bituminous Coal					Sub-bituminous Coal		
	Slag-tap boiler	Atmo-spheric fluidized bed FA	Pressurized fluidized PFBC solid residue Finspong			Slag-tap boiler	Atmospheric AFBC solid residue	
			Bed ash	Cyclone FA	Filter FA		Bed ash	FA
t, °C	>1300	<900	<900	<900	<900	>1300	<900	<900
SiO_2	40-50	33-48	35.4	37.09	38.48	20-70	7-17	80
Al_2O_3	23-35	13-20	12.88	15.63	23.39	1-15	4-6	0.5
Fe_2O_3	4-17	5-10	3.34	6.02	6.86	1.5-20	8-10	1
CaO	1-8	8-18	23.49	20.29	10.12	8-40	36-47	11
MgO	0.8-4.8	1-2	1.62	2.71	2.70	0.5-7.0	8-9	1
K_2O	1.5-5.5	2-4	1.57	1.64	2.12	0.1-1.5	0.3-0.7	0.1
Na_2O	0.1-3.5	0.4-0.7	0.60	0.89	1.82	0.1-2.0	0.1-0.2	0.2
SO_3	0.1-2.0	3-12	10.59	5.66	11.63	1.5-15	10-13	1.3
TiO_2	0.5-1.3	0.5-1	0.60	0.60	1.39	0.1-1	-	-

distinguish between the amount present in the form of free lime and those bound within the existing calcium sulfate, calcium carbonate or clay mineral decomposition products. It also does not suffice to know the loss on ignition (as determined at 975°C in line with EN 196-2) as this value includes the amount of residual carbon, combined water and CO_2 bound within the present calcium or magnesium carbonate. It may also be important to distinguish between the divalent Fe^{2+} and trivalent Fe^{3+} forms of iron as the oxidation of the former is associated with an undesired expansion [43].

A very important indicator of the quality of the ash is its residual carbon content which varies over a wide range (0.1-10%). Due to its large specific surface area, even small variations may significantly affect the amount of water required in the production of fresh mortar or concrete mixes containing the ash. This may affect the resultant strength as well. In addition, chemical admixtures may be adsorbed on the carbon surface, which may again affect the properties of the mix. Finally, the presence of carbon may affect the color of the ash. Increasing amounts of carbon cause a grey coloration of the material, similar to the presence of magnetite. A reddish color is due to the presence of iron (III) oxide. The XRD patterns show distinct lines of calcite, anhydrite and quartz. Small amounts of hematite, feldspar and illitic clay decomposition products also appear to be present. The PFBC material collected in the hose filter does not contain calcite. However, all other phases seem to be present in different proportions. Figures 4, 5 and 6 show x-ray patterns of PFBC ashes. As observed by optical microscopy, about 25% of calcite in the bed and cyclone ashes were found. The amount of quartz varied by approximately 15%. Around 18% of anhydrite was found in the bottom ash, but only 7% in the cyclone ash. The content of iron containing phases was ~ 7% and the amount of clay constituents varied between 17 and 26%. A microscopic analysis of the ash from the hose filter was not possible due to the fine particle size, however, the pale brown color suggests the presence of trivalent iron and the absence of significant amounts of residual carbon.

An x-ray diffraction analysis of a cyclone fly ash produced in the AFBC process revealed the presence of significant amounts of calcium oxide (Figure 7). High amounts of anhydrite and quartz and minor amounts of hematite and calcite were also present. The same ash, after reaction with water, contained significant amounts of ettringite, portlandite and gypsum (Figure 8).

Differential thermal analysis (DTA) and thermogravimetric analysis (TGA) are other methods well suitable for the study of the phase composition of the ashes. A comparison of a bed ash and cyclone ash from a PFBC process (Figures 9,10) revealed differences

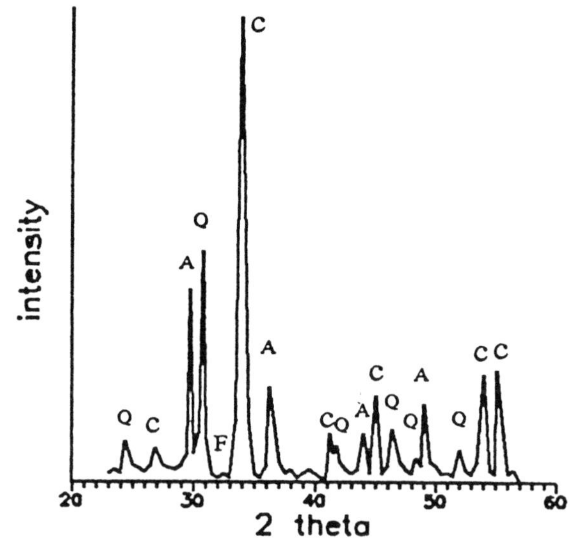

Figure 4. X-ray diffraction pattern of PFBC bed ash (Q-quartz; A-anhydrite II; C-calcite; F-feldspar).

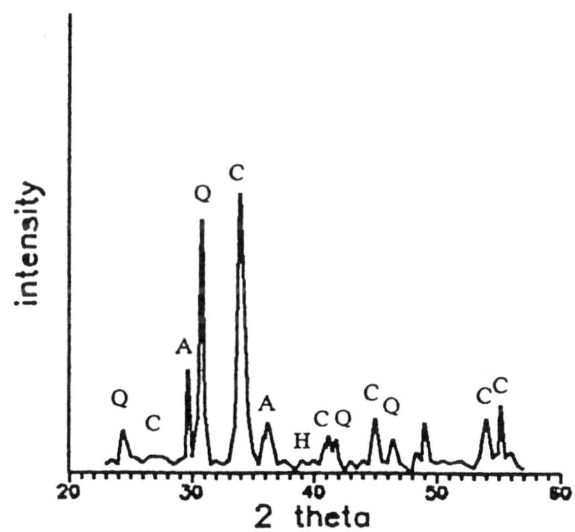

Figure 5. X-ray diffraction pattern of PFBC cyclone fly ash (Q-quartz; A-anhydrite II; C-calcite; H-hematite).

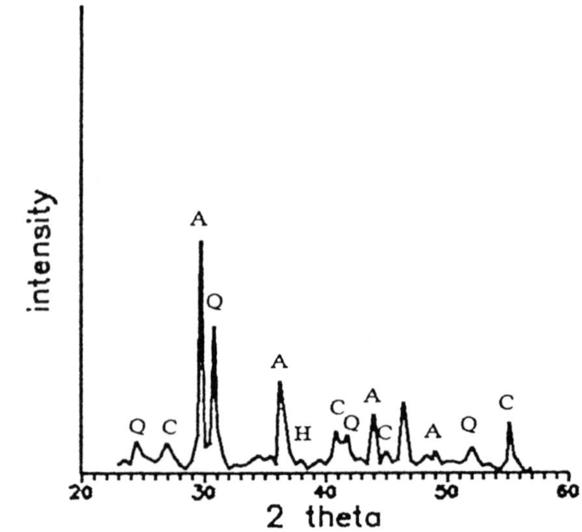

Figure 6. X-ray diffraction pattern of PFBC filter fly ash (Q-quartz; A-anhydrite II; C-calcite; H-hematite).

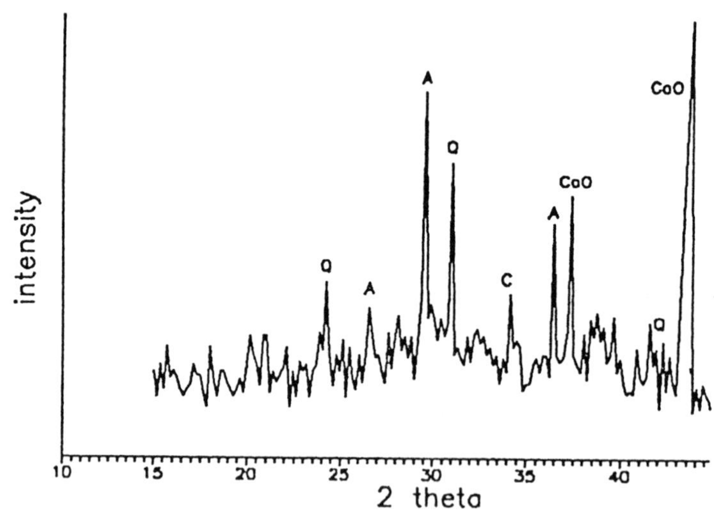

Figure 7. X-ray diffraction pattern of AFBC fly ash (Q-quartz; CaO-calcium oxide; A-anhydrite II; C-calcite).

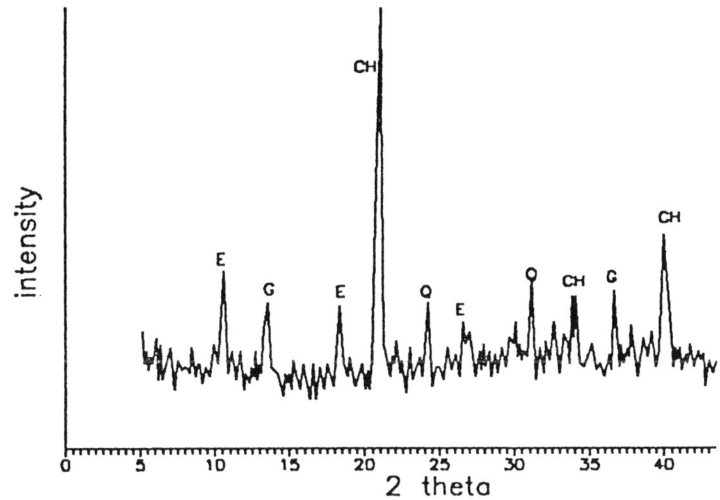

Figure 8. X-ray diffraction pattern of AFBC fly ash after hydration process (Q-quartz; E-ettringite; CH-portlandite; G-gypsum).

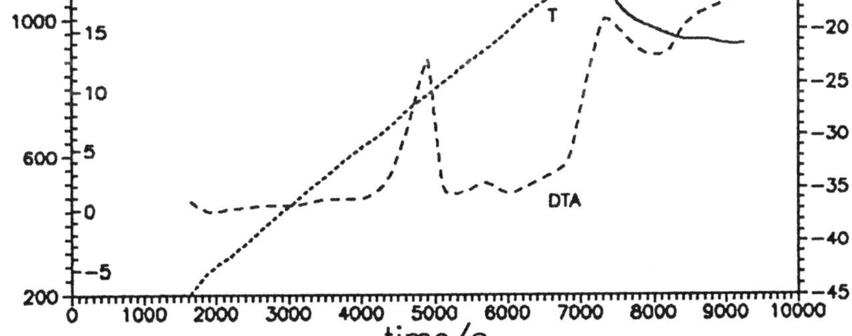

Figure 9. DTA-TG curves of PFBC bed ash.

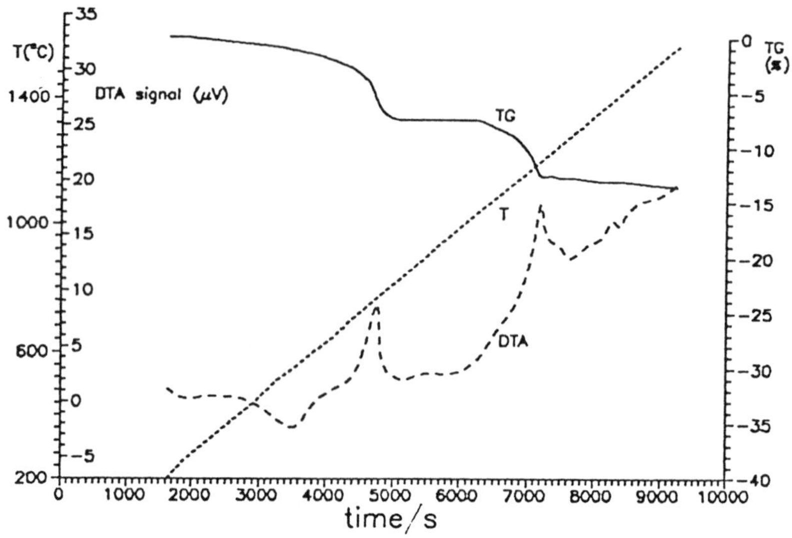

Figure 10. DTA-TG curves of PFBC cyclone fly ash.

in their calcium carbonate and anhydrite contents characterized by thermal effects at 700-800°C and 1100°C, respectively. A very moderate mass loss at around 600°C indicates the presence of small amounts of residual carbon and/or magnesium oxide. In the fine ash from the hose filter (Figure 11) the above mentioned phases were also found in smaller amounts.

The data generated by DTA and TGA analysis of fly ash produced in the AFBC process are different (Figure 12). An exothermic effect associated with a mass loss appears to be due to the oxidation of residual carbon. Above 450°C an endothermic decomposition of residual magnesite may be observed. The exothermic effect at 510°C associated with an increase of mass is due to the oxidation of divalent iron. Parallel to it and at even higher temperatures, the oxidation of residual carbon continues. Finally, above 700°C a decomposition of the residual calcite may take place.

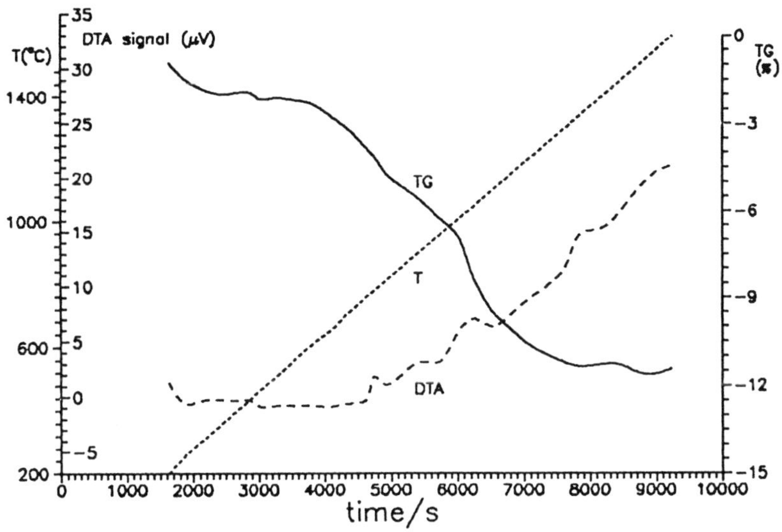

Figure 11. DTA-TG curves of PFBC filter fly ash.

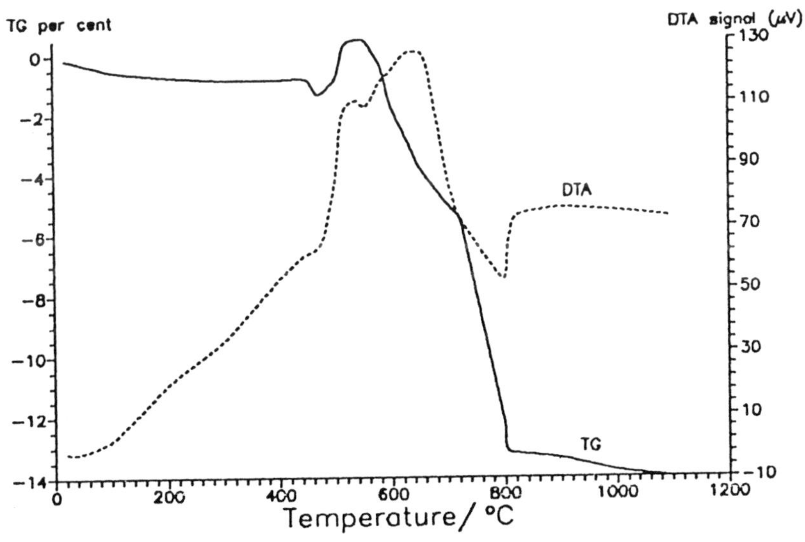

Figure 12. DTA-TG curves of AFBC fly ash.

Table 5. Combustion Technology and FA Phase Composition.

	High Temperature Combustion	Fluidized Bed Combustion Atmospheric	Pressurized
Temperature, °C	>1200	850	850
glass phase, %	<50	0	0
mullite, %	20	0	0
x-ray amorphous AS	0	++	++
clay, shale	0	+	+
dead burnt lime	+	0	0
free reactive lime	m	++	m
periclase, MgO	+	m+	m+
magnetite	+	+	0
anhydrite II	+	++	++
calcite	m	m	++
quartz	+	++	++
unburnt carbon	+	+	+

++ = main component; + = present; m = minimum amount

A thermal decomposition of anhydrite may be expected above about 1150°C. In samples stored in humid air a peak at around 500°C due to the decomposition of formed calcium hydroxide may also be observed. The combustion technology related to fly ash phase composition is given in Table 5.

THE REACTIVITY OF FLUIDIZED BED ASHES

A principal difference between AFBC and PFBC ashes consists in the absence of distinct amounts of free lime in the PFBC ash whereas the amount of this phase may exceed 15% in AFBC ashes. The amount of anhydrite is similar in both the materials. In presence of water the CaO and $CaSO_4$ of the AFBC ash undergo hydration and react with reactive forms of Al_2O_3 that have been formed in the thermal decomposition of clay minerals, yielding ettringite, $3CaO \cdot Al_2O_3 \cdot 3CaSO_4 \cdot 32H_2O$. The excess of free lime and calcium sulfate remains in the system in the form of portlandite and gypsum. This reaction is virtually absent in PFBC ashes due to the absence or only very limited presence of free lime in this type of material.

Another important reaction which may take place in both types of ashes is the formation of an amorphous CS(A)H gel from the present amorphous SiO_2 and aluminosilicate phases. Both the formation of ettringite and CS(A)H are due to the reactions associated with setting and hardening of fluidized bed ash - water mixes. The hardening process is distinctly faster in mixes of AFBC ashes as the formation of ettringite proceeds much faster than that of the CS(A)H gel. Another reaction that also contributes to setting and hardening is the hydration of anhydrite to gypsum, however, this process is of importance only in AFBC ashes with high calcium sulfate contents and in PFBC material.

The hydraulic properties of the ashes may be modified by adding to the system additional calcium hydroxide and/or gypsum, thus increasing the amount of ettringite that may be formed, which is dependent on the alumina content in the ash. An addition of calcium hydroxide may not be necessary in AFBC ashes with high free lime contents but is essential in PFBC ashes. The addition of calcium hydroxide may stimulate not only ettringite formation, but also the CS(A)H phase.

There exists also the possibility to combine AFBC and PFBC ashes with granulated blast furnace slag. In this case, the latent hydraulicity of the slag is activated by the free lime and calcium sulfate present in the ash. Finally, AFBC and PFBC ashes may be combined also with Portland cement, or even better just with ground Portland clinker. In such instances, ashes react both as a hydraulic/pozzolanic constituent and source of calcium sulfate.

Present heavy metal ions are immobilized by zeolitic AS(A)H gels, and some toxic anions (borates, arsenates) may be incorporated and bound within the structure of the formed ettringite. This opens the possibility for the use of fluidized bed ashes in environmental applications.

THE ANALYSIS OF ASHES AND TESTING OF ASH MIXES

Fluidized bed combustion residue are materials which started to be produced in larger volumes only in recent years. Specifications have not been developed for assessing their quality and there is little experience available as to their potential uses.

The content of the elements present in AFBC and PFBC ashes may be determined most conveniently by x-ray fluorescence analysis. To determine the amount of constituents that may be dissolved in water (especially heavy metals), atomic adsorption spectrometry appears most suitable. Other indicators important to assess the properties of the ash include the content of residual carbon, calcium carbonate, free calcium oxide and divalent iron. Also important may be the determination of pozzolanicity [18,23,27,31-36].

Loss on Ignition

The outcome of the determination of the loss on ignition may depend on the temperature employed. Virtually the whole amount of residual carbon is oxidized at temperatures below 700°C. At about 500 °C a decomposition of magnesium carbonate takes place. The decomposition of calcium carbonate starts at around 700°C, however, larger particles require somewhat higher temperatures. A decomposition of calcium sulfate in the ashes takes place only at temperatures well above 1100°C, just as a complete dehydroxylation of some clay minerals. The loss on ignition at 975 ± 25°C as suggested by many specifications is an arbitrary value and one has to keep in mind that it is the result of several processes taking place in the material.

Determination of Residual Carbon

Devices are available on the market designed for an automatized determination of free carbon, usually simultaneously with total sulfur. They are usually based on the thermal decomposition of the sample and determination of the liberated CO_2 (and SO_2) by infrared spectroscopy. Free carbon may also be determined as the loss on ignition of a sample in which calcium and magnesium carbonates had been decomposed with hydrochloric acid.

Determination of CO_2 of the Carbonates

The CO_2 bound within the carbonates may be determined by an acceptable precision by thermogravimetry as the mass loss between 700 and 900°C. Alternatively, it may be determined voluminometrically by measuring the volume of CO_2 liberated upon decomposition of the sample with an acid.

Determination of Divalent Iron

Significant amounts of divalent iron may be expected only in AFBC ashes where it is present in the form of magnetite. Its determination is important as its oxidation to iron (III) oxide may be accompanied by an expansion of mortar or concrete containing such ash. The procedure for determination of Fe^{2+} is based on the titration with dichromate after the sample has been decomposed with an acid in the absence of air.

Free Lime Determination

The determination of free lime is indicated in AFBC ashes whereas PFBC ashes contain only a few tenths of per cent of it, at best. Free lime including hydroxide may be determined conveniently by titration with an acid after extraction with a water solution of sucrose. An extraction into the ethyl acetate in butanol enables the determination of free calcium oxide (EN 451-2).

Determination of Sulfates

The amount of sulfates present is usually estimated from the sulfur content of the coal. Calcium sulfate may be determined also directly by quantitative x-ray diffraction analysis. The European standards EN 196-2 recommend gravimetric method. The SO_3 bound in calcium sulfate may be determined by TG analysis as the mass loss between 1100 and 1250°C.

Pozzolanic Activity

The pozzolanic activity [37-41] of the material is its ability to react with calcium hydroxide in the presence of water to form hydrates possessing cementitious properties. The products of hydration are calcium silicate (aluminate, alumino-silicate) hydrate phases that exhibit setting and hardening of the system. High-lime fly ashes have self-hardening properties. The pozzolanic reactivity depends not only on the chemical and phase composition of the material, but also on its degree of dispersion, especially on the specific surface area. Thus it is important to test the ash in the form as it will be used. A variety of methods has been developed for assessing the pozzolanicity of the material. The methods mostly used are based on the measuring of compressive strength of specimen, where the fly ash is activated by lime or by Portland cement. More favorable methods such as the ASTM C 311 or the European EN 450 describe the determination of strength activity index where the compressive strength of control OPC mortar is compared with the strength of a mixture where a part of OPC is replaced by fly ash, at ages of 7, 28 or 90 days. Some accelerated methods using elevated temperature may give misleading results. In some European countries, rapid thermochemical test of pozzolanic activity of FA by reaction of powdered sample with a mixture of hydrochloric and hydrofluoric acid is commonly used [37-39].

Determination of Volume Stability

The presence of some constituents in fluidized bed ashes may cause expansion (unsoundness) of the materials made from them, such as of mortar and concrete. To assess the expansivity due to the presence of free lime common in AFBC ashes, the Le Chatelier test (EN 196-3) appears most suitable. The expansivity due to both free calcium oxide and magnesium oxide may be determined by the ASTM autoclave test. An expansion may also be expected if the ash contains distinct amount of divalent iron.

An important property of concrete or mortar is the durability of hardened composite to the repeated cycles of freezing and thawing. It is reported that fly ash has no apparent effect on air voids in the hardened material, when a proper volume of air is entrained. This property of material containing fluidized bed ashes has not yet been sufficiently examined so far, the outdoor exposure tests are also needed to be performed. The air content in the fresh mix is dependent on the residual carbon in the ash used.

UTILIZATION OF ASHES WITHOUT ADDITIVES

The volume of ashes produced either in conventional combustion processes or in fluidized bed combustion is large and there is necessity to find suitable ways of their utilization. Presently the largest fraction of ashes produced is used for the filling of underground spaces left from coal mining. Other common utilizations include their use in the re-cultivation, sludge stabilization, base constructions of communications and their use in concrete/mortar mixes.

One important indicator that determines the possibility of using the materials in many applications is the quality and quantity of water soluble constituents present. The amount of undesirable constituents that may enter the liquid phase upon contact with water or water solutions must be low in those applications in which the danger of a contamination of underground waters exists. Significant amounts of anhydrite, sulfate and calcium ions present will enter the liquid phase and will cause a corresponding increase of its electrical conductivity. The attained Ca^{2+} and SO_4^{2-} concentrations in the leachate may exceed those permitted by specifications for drinking water. In addition, AFBC ashes also contain significant amounts of free lime which may increase the pH of the liquid phase up to a value of 11 to 12.5. However, both calcium sulfate and free lime may react with other ash constituents upon storage with water, yielding less soluble reaction products. Consequently, the amount of constituents entering the leachate including trace elements will decline significantly after several weeks. Sometimes, the concentrations of individual ions in the leachate of hardened fluidized ashes are lower than those in the leachate of some soils [7]. Ettringite and a CS(A)H gel are the main reaction products formed. The formation of these phases in fluidized bed ashes is significantly greater than in conventional fly ashes, that contain less SO_3, significant amounts of non-reactive mullite and slowly reacting glass phase. In the mixtures of PFBC ashes and water without free lime, the setting of pure pastes is brought about without ettringite formation, the 28 day compressive strengths reach up to 12 MPa [34]. The setting and hardening process of AFBC ashes is strongly influenced by their composition [31,32]. Due to the different content of free lime, anhydrite and divalent iron, some pastes of AFBC ashes with water exhibit slower setting, the 28d strengths differ and may reach up to 14 MPa [31]. Figure 13 shows the heat evolution of fluidized ashes and/or with other mineral constituents in contact with water measured in an adiabatic calorimeter [34,42], which approved to be very useful for the study of the courses of reactions in the setting and hardening mixtures of different compositions. If required, the resultant strengths may be reduced by adding of some less reactive materials (e.g. fly ash from dumping sites) to the system. This may be the case in some road construction applications.

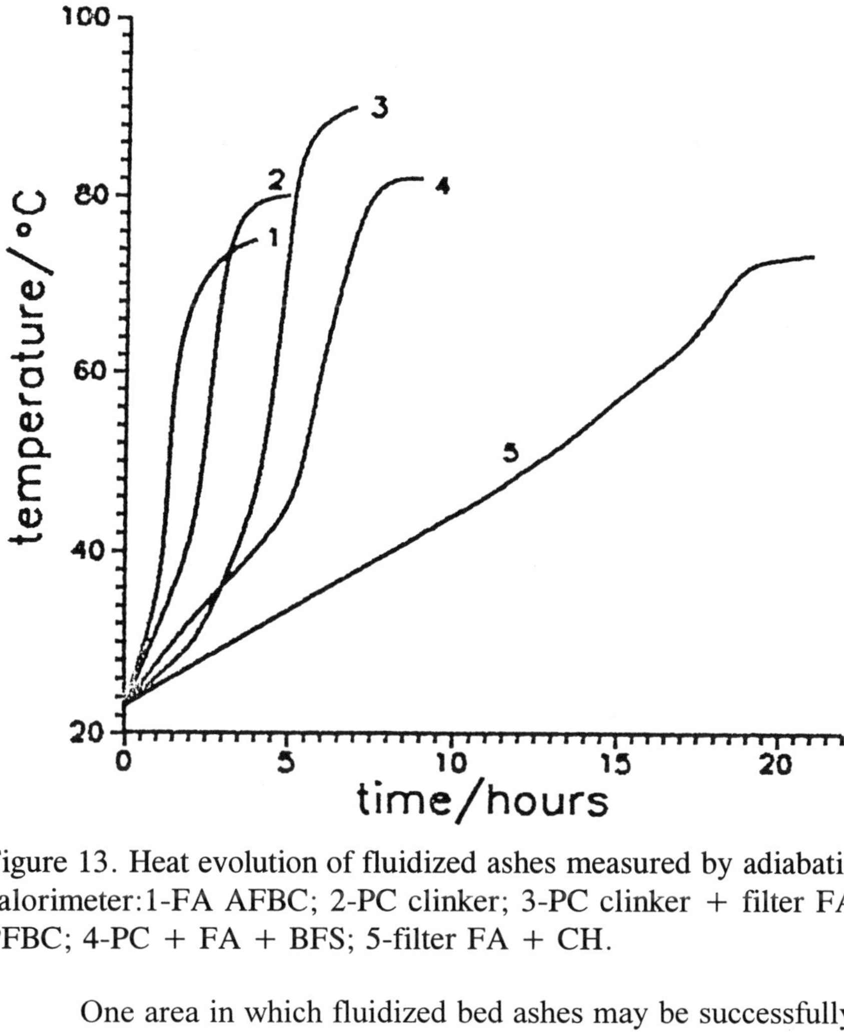

Figure 13. Heat evolution of fluidized ashes measured by adiabatic calorimeter: 1-FA AFBC; 2-PC clinker; 3-PC clinker + filter FA PFBC; 4-PC + FA + BFS; 5-filter FA + CH.

One area in which fluidized bed ashes may be successfully used is the improvement of soil quality [15,29,30]. For this kind of application, ashes produced with dolomite rather than with limestone should be preferentially used. One but not the sole beneficial effect of AFBC ashes consists in the neutralization of soils acidized by acid rains, brought about by the presence of free lime. Beneficial may also be the ion exchange characteristics of the zeolitic reaction products formed in the hydration of the ash. There are indications of a possible immobilization of some toxic trace elements present in the soil, such as, V, B, Cr, Se etc. by the ettringite [43], however, this subject has not been studied systematically so far.

The use of AFBC ashes for treatment (neutralization) of industrial waters e.g. in the chemical industry was tested. To avoid dusting during transportation and further applications, it may be meaningful to convert the ashes into a pelletized form, rather than to use them as a fine powder.

Some kinds of AFBC of sub-bituminous coal contain up to 60% of total CaO and their pretreatment before its use into the PC concretes is recommended to prevent swelling of hardening mixtures. For the quenching process, pressurized unit was constructed, which also assures the hydration of periclase [44].

PORTLAND FLY ASH CEMENTS AND COMPOSITE CEMENTS

Additions that may be combined with Portland clinker include, according to the draft of the European specifications ENV 197-1, the following materials: natural pozzolans of volcanic origin, artificial (industrial) pozzolanic materials (thermally treated clays), siliceous fly ashes of type V (type F according to ASTM) with a free lime content below 5%, lime-rich fly ashes of type W © with a higher free lime content and a residue on a sieve 40 μm between 10 and 30%. The SO_3 content of the ash has to be taken into consideration and the addition of calcium sulfate to the cement must be reduced accordingly. The possible additives include: calcined shale produced at about 800°C, limestone with a $CaCO_3$ content of at least 75%, microsilica, inorganic fillers with latent hydraulic or pozzolanic properties and finally natural or industrial calcium sulfate (dihydrate, hemihydrate or anhydrite).

The AFBC and PFBC ashes exhibit a composition that corresponds to a blend of several materials listed above. As to the possibility of using them as a constituent of a blended cement, the amount of ash that may be combined with Portland clinker is limited by the presence of SO_3. It may be considered to use AFBC and PFBC ashes for producing Portland fly ash cements (Type II/B-W) with an ash content between 21-35%. At the same time, it will be necessary to reduce or avoid completely a separate addition of calcium sulfate, in order not to exceed the permitted overall SO_3 content in the produced cement which is limited to 4.5%. In producing Portland fly ash cements the ash has to be interground with Portland clinker and eventually with reduced amounts of gypsum to obtain the required specific surface area. If the fineness of the ash is high enough, an interblending of the ash with pre-ground clinker may also be considered.

In our own experiments the effect of added fluidized bed ashes on cement strength was studied using a Portland clinker of a composition (%): CaO 66.8, SiO_2 21.1, Al_2O_3 4.96, Fe_2O_3 3.02, MgO 1.00, SO_3 0.59, K_2O 1.35, Na_2O 0.11, MnO 0.13, TiO_2 0.19, LOI 0.37. The clinker was combined with a bed ash ground to 340m^2/kg (Blaine) and a cyclone ash having a fineness of 330 m^2/kg (Blaine) or with the fine ash from the hose filter (d_{50} = 3.1 μm). Also employed was ground gypsum that contained 46.6% of SO_3. All the used ashes were collected from an industrial PFBC unit (Sweden). The experiments were performed on 40x40x160 mm test specimens made from a standard mortar. The obtained results are summarized in Table 6. It appears that even cements with high ash contents yielded fairly high strengths. The results obtained with the fine filter ash were most favorable, unfortunately, however, this ash represents only about three per cent of the total amount of ash produced. Additional characteristics of the hardened materials such as soundness, frost resistance, corrosion resistance and long term performance under different curing conditions are being still studied and the obtained results will be published at a later time.

Table 6. Replacement of Gypsum in Cement by PFBC Ashes. (Composition of cement in %, standard mortar tests, $\sum SO_3$: 3.3-3.6%)

	OPC	A	B	C	D	E	F
P-clinker	93	70	86	45	71	72	70
gr. bed ash	-	30	10	-	-	-	-
cyclone FA	-	-	-	55	25	20	-
filter FA	-	-	-	-	-	5	30
gypsum	7	-	4	-	4	3	-
Compressive Strengths, MPa							
1d	19.5	19.2	22.8	10.0	21.6	17.5	22.7
28d	56.7	54.2	64.0	42.6	59.5	61.0	62.3
90d	66.7	65.7	70.6	56.0	74.2	76.0	85..0

POSSIBLE USE OF AFBC AND PFBC SOLID RESIDUE AS CONSTITUENTS OF CONCRETE MIXES

Traditional high temperature fly ashes are being commonly used as concrete mix constituents already for decades [22]. Here, they act as both a fine filler and a pozzolanic constituent, thus improving the quality of the final product based just on pure Portland cement, in several respects.

Fluidized bed ashes have the potential to be employed in a similar way. Their advantage is a higher reactivity as compared to conventional fly ashes. Unfortunately, they do not meet all the requirements of the existing ASTM and EN specifications for ashes to be employed in concrete mixes. An amendment of these specifications would be indicated to enable to use also fluidized bed ashes, besides of conventional fly ashes, for this purpose.

The loss on ignition of the fly ash to be used in concrete is limited by EN 450 to 5%. This value was set, to limit the amount of residual carbon in the material. PFBC ashes may contain up to 20% $CaCO_3$. Calcium carbonate in the ash does not affect adversely the properties of a concrete mix, however, the loss on ignition is increased accordingly. Thus, that part of the loss on ignition value, that is due to the presence of calcium carbonate in the ash should be subtracted from the total LOI value to obtain reliable information on the amount of residual carbon in the material and thus on the suitability of the ash for concrete production. It must also be mentioned in this context that up to 35% of calcium carbonate may be interground into the cement according to ENV 197-1 (cement II/B-L).

The content of SO_3 in the ash as a constituent in OPC concrete mixes is limited to 3% by EN 450, just as by numerous national specifications. A more realistic value of 5% is permitted by ASTM C-618. Both values are lower than the amount of SO_3 found in most fluidized bed ashes. This handicap could be overcome by using blends of a fluidized bed ash and a low SO_3 fly ash, rather then fluidized bed ash alone, as a concrete additive. Such blending could be combined by a joint grinding whenever a grinding of the ash to be employed is needed to meet the requirement of ASTM in which the residue on the sieve 45 µm is limited to a maximum of 34%.

The amount of free lime in the ash to be used in concrete mixes is limited by EN 450 to 1 or 2.5%. Such limitation is well justified for ashes produced by traditional combustion technologies in which the constituents of the ash have been exposed to high temperatures. This may cause a volume expansion of the hardened concrete, due to a delayed hydration of CaO. Some fly ashes produced from sub-bituminous or lignitic coals may contain free lime exceeding these limits and then they cannot be used as concrete constituents. Contrary to that free lime in AFBC fluidized bed ashes can be considered usually harmless, due to their thermal history. Free lime formed at temperatures around 850°C as common in fluidized bed combustion is namely highly reactive and in fresh concrete mixes hydrates to $Ca(OH)_2$ almost quantitatively prior to setting, without causing an expansion in the later stages of hardening. At the same time, the calcium sulfate content must be taken into the consideration.

There exists the possibility to modify the fluidized AFBC ash by pre-treating it with a limited amount of water. This way free lime hydrates to calcium hydroxide, anhydrite to gypsum and both of them may react to yield ettringite. For the high-lime AFBC ashes, their pretreatment by pressurized quenching is recommended [44]. For storage and transportation purposes a pelletization of the ash may be meaningful.

To study the suitability of fluidized bed ashes as concrete additions, experiments were performed. Mortar compositions with Portland cement contents that varied between 100 and 300 kg/m^3 and PFBC ash additions that were varied between 1400-1600 kg/m^3 were prepared and studied on 40x40x160 mm prisms, which were stored in humid air. As expected, the mortars exhibited a moderate expansion due to higher SO_3 content. Nevertheless, the strength of the material was not affected adversely during the testing period (up to 540 days).

Some mortars were prepared with a very low content of Portland cement and high-volume of PFBC ashes. Therefore, all of the mortars did not meet the standard requirements on SO_3 content. The preliminary results presented in Table 7 appear to be promising. At present, long term durability of the composites are being studied.

It is also conceivable to use fluidized bed ashes in combination with inorganic binders other than Portland cement, such as lime, slag or gypsum plaster.

THE USE OF FLUIDIZED BED ASHES IN MULTI-COMPONENT PORTLAND CLINKER FREE CEMENTS

Portland cement is, and is expected to remain, the number one binder in the world. However, other binder systems with cementitious properties are being used in increasing amounts, for both economic and ecological reasons [21,45,46]. These include: fly ashes, slags produced in metallurgical processes, natural and industrial pozzolanic materials and other industrial waste products

Table 7. Mortars With PFBC Residue With Los Content of OPC. (Composition in kg/m^3).

	A	B	C
OPC 42.5	108	54	102
gr. bed ash (340m^2/kg)	544	-	-
bed ash, as rec'd	1020	1050	1030
cyclone ash (330m^2/kg)	-	530	515
water	268	319	328
Compressive Strengths, MPa			
1d	n*	1.49	1.78
7d	17.6	14.2	15.8
28d	29.0	19.3	20.1
90d	31.9	22.7	33.1

*n = not measured

that contain free lime or compounds of calcium, sodium or potassium. By-products produced in fluidized bed coal combustion and desulfurization processes are also included in this category.

Among materials with cementitious characteristics, granulated blast furnace slag is most common and is available in the largest volumes. To activate the cementing properties of the slag, it must be combined with a suitable activator [34,47-49]. Portland clinker is most commonly used, yielding Portland slag cements (ENV 197-1: type II/A-S and II/B-S) or blast furnace cement (type III). Other suitable activators include: free lime (alone or in combination with calcium sulfate), and AFBC solid residue (as free lime and calcium sulfate are both found in the ash). In a similar manner, some conventional ashes from fossil fuel combustion or some slags from other metallurgical processes can also be activated.

Table 8 shows the composition and properties of selected mixes made with a variety of by-products. Plasticizer were added to some mixes to improve the rheology. In others, high shear mixing was employed to reduce porosity and to increase strength. Good results were obtained with two-stage mixing as well. Here a good flowing mix, produced in the first stage, was mixed with a proper amounts of aggregate in the second-stage mixer. As a result of this procedure, compositions with very low water-to-cement ratios, yielding high strength concretes were produced.

From Table 8, it is apparent, that the resultant strength increased with increasing amounts of slag in the mix. A compressive strength of 42 MPa after 180 days was attained with just 196 kg/m^3 of slag using a two stage mixing procedure. The mix contained almost 50 % of non-reactive conventional concrete aggregate.

Table 9 summarizes results obtained on slag/conventional FA mixes activated with added PFBC fluidized ashes and lime. The specimens were either cured at ambient temperature or in an autoclave at high temperatures. The results available thus far indicate acceptable strength properties and suggest the possibility of using such mixes in the production of lime-slag-ash bricks.

AUTOCLAVED PRODUCTS

Autoclaved products are commonly manufactured by treating blends of quartz sand, serving as a source of silica, with calcium hydroxide and water in autoclave at temperatures between 100 and 200°C. It may be possible to replace the lime, either partially or fully, with Portland cement. Under these conditions, low CaO crystalline calcium silicate hydrate phases, especially tobermorite or xonotlite, which are responsible for the strength of final product, are formed. This way, either compact blocks (lime-silica bricks) are produced from a mechanically pre-compacted, moist raw materials blend, or

Table 8. Fly Ash/Slag Composites (OPC Free). (Composition in kg/m^3).

	A (paste)	B (mortar)	C (concrete)
PFBC cyclone FA	653	474	392
PFBC filter FA	16	12	10
FA conventional	490	355	294
ground BFS	163	118	196
aggregate 0-4mm	-	594	196
aggregate 4-8mm	-	-	734
superplasticizer	12	9	7
water	322	252	203
w/FA + BFS	0.25	0.27	0.24
Volume mass, kg/m^3	1656	1814	2080
<u>Strengths Flexural/Compressive, MPa</u>			
3d	0.53/1.0	0.47/0.9	0.99/3.3
7d	0.88/2.1	0.67/1.5	1.64/5.4
28d	1.98/7.6	1.52/4.7	3.37/15.4
90d	3.06/18.4	3.27/14.9	6.20/33.7
180d	3.96/22.4	3.63/19.6	6.97/42.6

an aerated lightweight product may be manufactured by autoclaving a raw material slurry in which a cellular structure had been produced prior to autoclaving by adding aluminum powder to the mix.

In principle, quartz sand commonly used as the silica source may be replaced by other SiO_2 containing materials and in fact, ordinary fly ash, produced in thermal power plants as a by-product is being widely employed for this purpose, especially in the production of cellular autoclaved products [17,18,26,50,51]. Laboratory tests

Table 9. Autoclaved Slag - Fly Ash - Lime Pastes. (Composition in kg/m^3; PFBC, Finspong, Sweden).

		A	B	C
cyclone FA		726	698	545
filter FA		439	425	320
ground gran. BFS		303	303	641
hydrated lime		45	91	48
Portland Cement (42.5)		-	-	48
water		393	405	385
w/FA+GBFS+CH		0.26	0.27	0.24
Soluble SO_3, %		6.27	5.92	4.57
Vol. mass, fresh		1906	1922	1987
Vol. mass, dry		1569	1622	1709
Compressive Strengths, MPa				
autoclaved		25.2	40.8	49.1
moist curing	2d	5.45	7.49	8.13
	28d	21.5	42.9	42.0
	60d	23.8	43.7	41.2

indicate the possibility of replacing part of the sand in the starting mix by SO_3 - bearing fluidized bed ash without effecting adversely the strength properties of the resultant material. In doing so, one has to bear in mind that ashes of this type may contain significant amounts of both combined and free lime which participates in the formation of crystalline calcium silicate hydrate phases responsible for the final strength of the material. The amount of added lime hydrate in the starting mix has to be reduced accordingly, to avoid an undesirable high CaO/SiO_2 ratio in the product, associated with a decline in strength. Fluidized bed ashes have a lower density than quartz and would require greater pressure in order to achieve the same degree of compaction. Thus, one has to expect a gradual decline of bulk density and along with this, a decline in strength with increasing ash additions, if constant compaction pressure is maintained in the production of compact raw mix blocks to be autoclaved. If a high pressurized fluidized bed ash with a very low free CaO content is employed, the material may be used in the production of autoclaved products without any pre-treatment except grinding to a fine particle size. If a high proportion of starting materials and proper autoclaving regime are employed, tobermorite with a rather high degree of (SiO_4)-(AlO_4) substitution in its crystalline lattice is formed as the main reaction product in the autoclaving process and the final material attains a fairly high strength. However, the attained strength will depend on the degree of compaction of the starting materials. Calcium sulfate is incorporated (during autoclaving) to a certain extent into the crystalline lattice of tobermorite. If an atmospheric pressure rather than a high pressure fluidized bed ash is used in the production of autoclaved products, precautions should be taken to avoid crack formation in the pre-compacted raw blocks during autoclaving. Such cracking may take place due to the hydration of calcium oxide, which is present in significant quantities in atmospheric fluidized bed ashes, and during autoclaving, reacts readily with water in a topochemical reaction associated with a significant increase in the volume of the solid phase.

$$CaO + H_2O \rightarrow Ca(OH)_2$$

To avoid crack formation in the course of autoclaving, it may be necessary to store a moistened ash at room temperature for several days to allow this reaction to be completed prior to autoclaving.

Under these conditions, the formed calcium hydroxide may further react as shown in the following chemical equation, with calcium sulfate and aluminum oxide (loosely bound in the ash), to form ettringite ($6CaO \cdot Al_2O_3 \cdot 3SO_3 \cdot 32H_2O$, AFt phase):

$$3\ Ca(OH)_2 + 3\ CaSO_4 + Al_2O_3 + 29\ H_2O \rightarrow 6CaO\cdot Al_2O_3\cdot 3SO_3\cdot 32H_2O$$

$$3\ CH + 3\ C\overline{S} + A + 29\ H \rightarrow C_6A\overline{S}_3H_{32}$$

The pre-treated ash may be then employed in the autoclaving process, however, both its density and compactibility become reduced due to ettringite formation in the pre-curing phase.

In the course of autoclaving, the tobermorite phase is formed as the main reaction product and the one primarily responsible for strength. At the same time, the ettringite phase formed in the course of pre-curing loses its thermodynamic stability and converts to monosulfate: $4CaO\cdot Al_2O_3\cdot SO_3.12H_2O$ (AFm phase)

$$6CaO\cdot Al_2O_3\cdot 3SO_3\cdot 32H_2O \rightarrow 4CaO\cdot Al_2O_3\cdot SO_3\cdot 12H_2O + 2CaSO_4\cdot 1/2H_2O + 19\ H_2O$$

$$C_6A\overline{S}_3H_{32} \rightarrow C_4A\overline{S}H_{12} + 2\ C\overline{S}H_{1/2} + 19\ H$$

and in the following step to hydrogarnet and anhydrite:

$$4CaO\cdot Al_2O_3\cdot SO_3\cdot 12H_2O \rightarrow 3CaO\cdot Al_2O_3\cdot 6H_2O + CaSO_4 + 6\ H_2O$$

$$C_4A\overline{S}H_{12} \rightarrow C_3AH_6 + C\overline{S} + 6\ H$$

As this reaction proceeds, a significant fraction of water originally bound within the crystalline lattice of ettringite is converted into free liquid water vapor by the autoclaving process. This results in further lowering the bulk density and increases the porosity of the final product. With a massive or complete replacement of quartz sand by an atmospheric fluidized bed ash, it may become possible by using a technology normally employed for dense lime-silica blocks to produce autoclaved products whose bulk density and strength is comparable with those of commonly produced autoclaved aerated concrete blocks. Figure 14 shows the bulk densities and compressive strengths for blends with increasing replacement of sand with a pressurized fluidized bed ash and lime (10% CH of the mineral constituents) at constant compaction pressure 5 MPa. Similar results may also be obtained with atmospheric fluidized ashes. Under these conditions, base ettringite is formed during room temperature storage which decomposes during autoclaving.

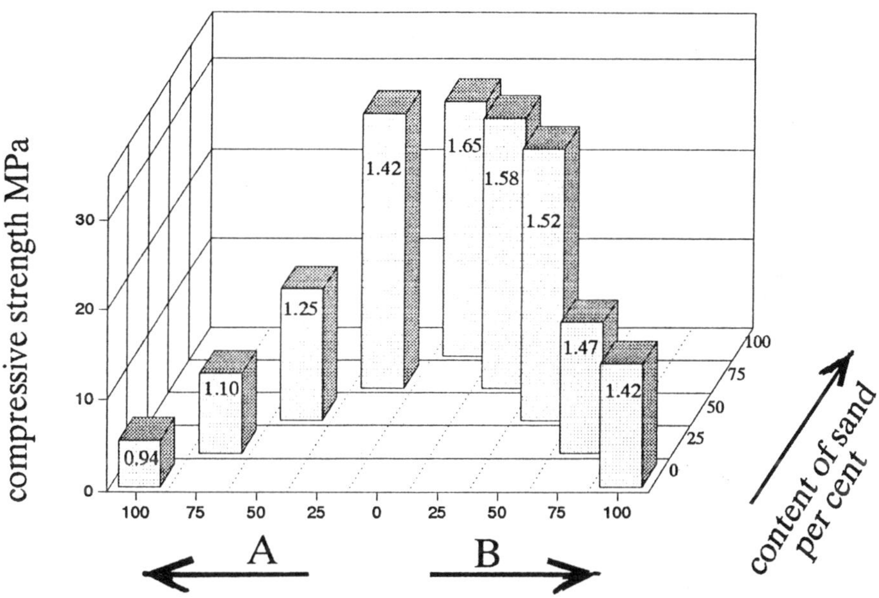

Figure 14. Dependences of compressive strengths of autoclaved samples on content of PFBC ashes and sand for non-prehydrated (A) and prehydrated (B) systems. The values on diagram represent the bulk densities of autoclaved samples.

The rate of ettringite formation at room temperature from PFBC ashes and calcium hydroxide was evaluated using x-ray diffraction. As shown in Figure 15, where two mass ratii of water to solid phase *(w/s)* were used, the dependence of conversion with time indicates different tendency of reaction. The formation of a diffusion resistance near sources of sulfate ions at *w/s* = 0.5 is indicated by lower conversion ($\alpha \sim 8$ %). Crystallization of ettringite is more apparent at higher content of water. At *w/s* = 2, the level of conversion was ~ 76% after 4 hours and 90% after 40 hours. Figure 16 demonstrates that the maximum conversion occurs at approximately 4 hours.

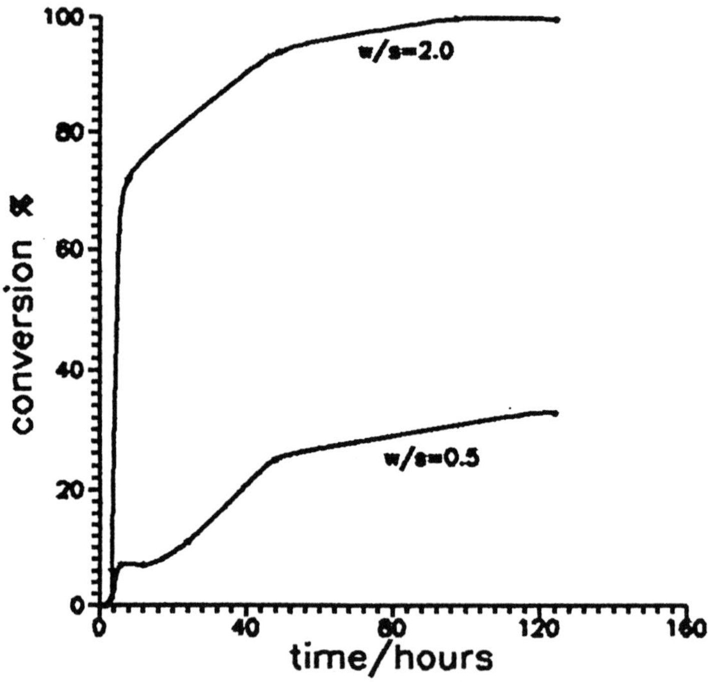

Figure 15. Dependences of conversion of ettringite formation vs. time for various water/solid phase ratio.

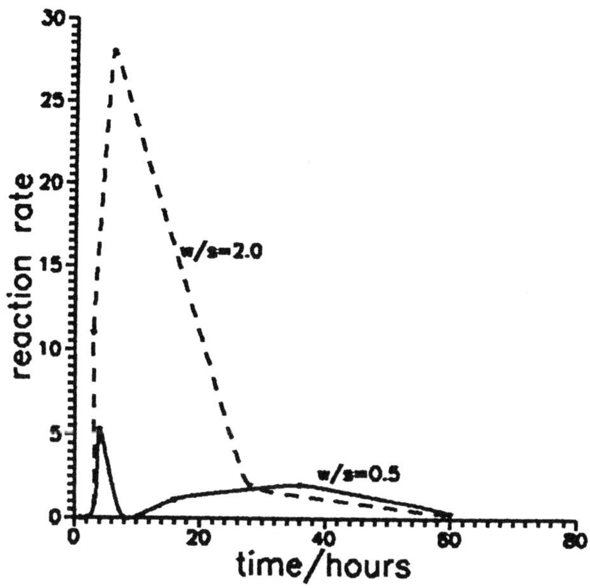

Figure 16. Dependences of reaction rate of ettringite formation vs. time for various water/solid phase ratio.

Water content is decisive for the rate and mechanism of hydration process. Both processes are controlled by slow nucleation of ettringite during the first 4 hours. Higher water content positively influences the crystallization process. In comparison with *w/s* = 2, kinetics of hydration at *w/s* = 0.5 is characterized by two periods of ettringite formation. Retardation of the process is influenced by deficiency of water. The reaction begins to slow down after 12 hours hydration (step 2).

The limiting factor appears to be the amount of alumina available in the fly ash, corrections for lime and/or gypsum content are often necessary [20,23,52].

The remaining uncertainty is the long-term durability of autoclaved products based on fluidized bed ashes. This is because there is a potential for a renewed delayed ettringite formation that may or may not be associated with crack formation and/or volume instability (soundness).

ARTIFICIAL CONCRETE AGGREGATE

In principle, artificial concrete aggregate may be produced from powdered waste materials by several methods:

1) pelletizing a blend of the waste material with an inorganic binder and curing at ambient temperatures,
2) thermal sintering of the pelletized waste powder, and
3) high pressure compaction of blends of the waste material and inorganic binder.

Of these three alternatives, methods 1 and 2 yield porous, lightweight products which may be considered only for lightweight concrete, whereas an artificial aggregate produced by the third method may exhibit densities comparable to those of natural rock.

In processing fluidized bed ashes containing significant amounts of $CaSO_4$, thermal sintering cannot be considered, as at the temperatures needed to attain an appropriate strength (1100-1250°C), a large fraction of the calcium sulfate would undergo thermal dissociation associated with a release of SO_2.

In producing an artificial aggregate from blends of ashes with inorganic binders, Portland cement is usually the binder of choice. However, to avoid sulfate expansion, the SO_3 content in the artificial aggregate should not exceed 5 %. Thus it may become necessary to blend the fluidized bed ash with another low SO_3 waste product such as fly ash or bottom ash and even to use ground Portland clinker instead of Portland cement [55].

For the production of lightweight aggregate, the amount of added Portland cement should be between 10-25% depending on the required final strength. To obtain acceptable strength properties, the pelletized material must be moist-cured for several weeks before use to attain a sufficient degree of hydration. Typical bulk densities of the products are between 1800 and 1400 kg/m^3.

For the production of a compact concrete aggregate an addition of 10-15% Portland cement to the starting mix appears sufficient in most cases. The water content of the mix should be around 10-15%. For compaction, a roller-type press appears most suitable. The compaction pressure should be between about 600-1100 MPa. To attain a sufficient hydration and thus an appropriate strength, the produced briquettes should be kept for a few weeks under water or sprinkled with water if cured in air, to attain a sufficient degree of hydration of the employed binder. In this method, an artificial concrete aggregate may be obtained with densities of 2000-2200 kg/m^3 and a compressive strength ≥ 100 MPa [18,55]. The produced briquettes may be crushed to produce a final product with a wide particle size distribution, suitable for concrete production.

Concrete mixes prepared with such artificial aggregate exhibited consistencies comparable with similar mixes made using natural rock (at equal w/c). The strengths of hardened concrete are usually higher (~20-50 %) due to a better bond between the aggregate particle and cement paste [18,55].

A conversion of fluidized bed ashes into artificial concrete aggregate may be justified in situations where there is a shortage of natural rock aggregates or where the necessity exists to use large amounts of ash to avoid the expenditures associated with disposal.

FLUIDIZED BED ASHES AS CONSTITUENTS OF PORTLAND CEMENT CLINKER RAW MEAL

Ash from coal combustion, especially fly ash, are widely used as constituents of raw meals for Portland cement clinker manufacture [54,56]. The advantage of such mix is that the CaO of the ash is present in decarbonized form. Moreover, the ash may contain variable amounts of residual carbon which may serve as an energy source.

Unlike other ashes, the use of fluidized bed ash in cement raw meals is limited by the amount of SO_3 in the material. The SO_3 in raw meal in amounts above about 1% is undesirable as it hinders or prevents the formation of tricalcium silicate, the main constituent of Portland cement clinker. A very important role is played by the concentration of alkalies and sulfates in other components of the raw mix. Moreover, the $CaSO_4$ present in fluidized bed ashes is thermally decomposed at temperatures that occur within a cement kiln. Finally, the amount of SO_3 from the ash in the raw meal is limited by the requirement that the overall SO_3 content in the final Portland cement cannot exceed 3-4 %.

However, the potential exists to use larger amounts of fluidized bed ashes in raw meals of binders other than Portland cement (where the allowable amount of SO_3 is higher). Cements of this kind do not meet the standard requirements for cements. This may be a serious problem in their marketing. Among cements of this kind, "sulfobelitic cements" containing the phases C_2S and $C_4A_3\overline{S}$ (and C_4AF) as the main constituents can be produced at temperatures about 300°C lower than normal Portland cement [56]. Another type of sulfoaluminate-based cement is "high SO_3 Portland cement" containing the phases C_3S, $C_4A_3\overline{S}$ and C_4AF. In some of the sulfoaluminate cements, calcium sulfosilicate $C_5S_2\overline{S}$ and $C\overline{S}$ is also present, but the contribution of $C_5S_2\overline{S}$ to cementitous properties was not evident. Ettringite and groups of AFt phases, generated by hydration of $C_4A_3\overline{S}$ and $C\overline{S}$ are responsible for the high early strengths of these cements. These ettringite-containing cements do not expand and their volume stability is similar to Portland cement. Nevertheless, durability problems may arise due to their relatively poor resistance to carbonation. Because calcium sulfoaluminate hydrate monosulfate AFm possesses higher resistance to carbonation, some investigations are focused on a cement with a higher content of brownmillerite, C_4AF, whose presence promotes the conversion of ettringite to monosulfate during hydration [56]. Using fluidized bed ashes, the production of these types of cements at lower temperatures can reduce energy, limestone consumption and emissions of CO_2.

LEGAL AND STANDARDIZATION ASPECTS, PERFORMANCE REQUIREMENTS

The development and implementation of new standards often lags behind technical achievements and the same can be said for developments in the area of industrial by-products and waste utilization. By-products and waste materials that could be utilized are often the result of new technologies introduced in a variety of industries. The composition of these waste products may differ from that of more commonly used materials. Difficulties may arise if the existing specifications do not cover these new materials.

Fluidized bed combustion residues have been recognized as an additive material for about twenty years. Their compositions may vary greatly, depending on the quality of the fuel, the composition and amount of added sorbent, and on the type of combustion technology used. Thus far, little has been published on their composition, properties and possible uses, especially PFBC ashes.

At present, there are no specifications which cover fluidized bed combustion ashes or other combustion residue which are formed in the de-sulfurization of the flue gas. The existing specifications for combustion residue refer almost exclusively to conventional fly ashes and generally are not applicable to fluidized bed combustion material.

Since it is the mineralogical composition, and not the chemical composition, which would govern the pozzolanic and cementitious behavior of a mineral admixture, classifications and specifications emphasizing the chemical composition are more of a hindrance than a help in promoting the use of mineral admixtures in cement and concrete industries [27]. New classifications, specifications and accelerated tests relating the desired performance criteria to the microstructure of the hydrated mixes of binders containing different mineral admixtures are needed [34,40,57-60]. Lack of international standards is also one of the obstacles preventing not only the large-volume use of supplementary cementing materials, but also commerce and export.

A critical examination of the chemical and physico-mechanical requirements showed that many tests in the current standards are unnecessary. In testing conventional fly ash for the purpose of quality assurance and uniformity of produced material, the loss on ignition, fineness and strength activity index appear to be sufficient. Other requirements should be performed at the option of the producer or user depending on the end-use of the material, especially if no-cement mixtures are prepared. For fluidized ashes, the following additional tests are recommended: unburnt carbon along with total carbon, free lime, divalent iron and SO_3 content. These specifications must take into account the anomalies of these by-products which will be produced in steadily increasing amounts. Better utilization of these waste materials will help to save dwindling natural resources.

CONCLUSIONS AND RECOMMENDATIONS FOR FUTURE RESEARCH

The increasing number of power plants using "clean" fluidized bed coal combustion technology focuses attention on the properties and use of the solid residue of this process, the composition of which differs substantially from the conventional fly ashes. Due to the low combustion temperature, the mineralogical composition of the material produced brings about its self-cementitious properties and more versatile utilization in comparison to high temperature fly ashes. The two types of fluidized combustion technologies employed (at atmospheric pressure and elevated pressure) bring about differences of the phase composition of the bed ash and fly ash. For example, free lime is found in AFBC material and calcium carbonate is found in PFBC ash. While there are a few publications dealing with AFBC material, the properties of PFBC ashes were presented only in [34,46] thus far. Therefore, much of the data in this publication can be considered as a preliminary, further research is needed.

At present, many producers and users of supplementary cementing materials are convinced that many regulations and standards are more prescriptive than performance based, hindering the use of some modifying mineral admixtures in concretes and the large-volume utilization of by-products including fly ash in general. New classifications, specifications and accelerated quality tests in international cooperation must be elaborated upon. Another problem is lack of information on the properties and use of many promising new materials. It appears that the information is not reaching the right people: in spite of papers published in technical journals and books, marketing plays the decisive role. The last link of the chain has to be permanently sustainable development.

REFERENCES

1. Manz, O.E. in, *Proceedings of the Tenth International Ash Use Symposium,* Vols. 1-3, American Coal Association, Orlando, FL, pp. 64.1 - 64.12 (1993).
2. *Proceedings of the Tenth Annual International Pittsburgh Coal Conference, "Coal - Energy and the Environment,"* The Univ. of Pittsburgh (1993).
3. Gustafsson, O., PFBC for Combined Heat and Power Generation, *Fernwärme International - FWI* 22 (5/6): 210-217 (1993).
4. *Proceedings of the 12th International Conference on Fluidized Bed Combustion,* San Diego, CA (1993).
5. *Proceedings of the 13th International Conference on Fluidized Bed Combustion, Continuing Evolution of FBC,* Kissimmee, FL (1995).
6. Schemenau, W., Druckwirbelschichtfeuerung-fortschrittliche Kohlenveströhmung mit Betriebserfahrung, *BKW* 45: 23-26 (1993).
7. vom Berg, W. and Puch, K.- H., Utilization of Residues from Fluidized Bed Combustion Plants. *VGB Kraftwerkstechnik* 73 (4): 330-334 (1993).
8. Berry, E.E. and Anthony, E.J., in Fly Ash and Coal Conversion By-Products:Characterization, Utilization and Disposal III,

Materials Research Society Symposium Proceedings, Vol. 86, pp. 353-364 (1986).

9. Fly Ash and Coal Conversion By-Products: Characterization, Utilization and Disposal. *Materials Research Society Symposium Proceedings*: I, Vol. 43 (1985); II, Vol.65 (1986); III, Vol. 86 (1987); IV, Vol. 113 (1988); V, Vol. 136 (1989); VI, Vol. 178 (1990).
10. Jahn, P. and Weiss, P., Grundlagenuntersuchung zur Verwertung von zirkulierenden Wirbelschichtaschen, *VGB Kraftwerkstechnik* 69: 924-929 (1989).
11. Kautz, K., Verwertung von Wirbelschichtfeuerungsanlagen. Rückstanden, *VDI-Berichte*, 601:319-332 (1986).
12. Lotze, J. and Wargalla, G., Kenndaten und Verwertungsmöglichkeiten von Aschen aus einer Feuerungsanlage mit zirkulierender Wirbelschicht. Teil I: Verfahrenstechnik und Charakteristik, *Z-K-G* 38 (5): 239-243 (1985);Teil II: Eignung und Anwendung in der Bauwirtschaft, dtto, 38 (7): 374-378 (1985).
13. Odler, I., in, *Proceedings of the Tenth International Ash Use Symposium,* Vols. 1-3, American Coal Association, Orlando, FL, pp. 63.1 - 63.6 (1993).
14. Odler, I. and Skalny,J. in, *Materials Science of Concrete*, Vol. III, American Ceramic Society, Westerville, OH, pp. 319-336 (1992).
15. *Proceedings: 11th International Symposium on Use and Management of Coal Combustion By-Products (CCBs)*, Vols. 1-3, American Coal Association, Orlando, FL (1995).
16. *Proceedings of the Tenth International Ash Use Symposium;* Vol. 1., High-Volume Uses (Concrete Applications);Vol. 2., Ash Use R-D and Clean Coal By-Products;Vol.3., Supplementary papers, American Coal Association, Orlando, FL (1993) .
17. Verwertungskonzept für Reststoffe aus Kohlenkraftwerken in der Bundesrepublik Deutschland, Teil 2, Rückstände aus der Verbrennung-Aschen, *VGB-TW* 702 2.1-2.36 (1992).

18. Zysk, K.-H., Charakterisierung von Aschen aus Kraftwerken mit zirkulierender Wirbelschicht und ihre Verwertungsmöglichkeiten, Dissertation, Technische Universität Clausthal (1990).
19. Environmental Aspects of Construction with Waste Materials, *Proceedings of the World Meeting WASCON 94*. Maastricht, June (1994).
20. Chandra, S. and Bernstsson, L., in, *Mineral Admixtures in Cement and Concrete* (Sarkar, S.L., ed.), ABI Books Ltd., New Delhi, India, pp. 494-559 (1994).
21. Hemmings, R.T., Berry, E.E. and Golden, D.M. in, *Proceedings: 11th International Symposium on Use and Management of Coal Combustion By-Products (CCBs)*, Vols. 1-3, American Coal Association, Orlando, FL, pp. 79-1 (1993).
22. Malhotra, V.M. and Ramezanianpour, A.A., Fly Ash in Concrete, pp. 307, *CANMET*, Ottawa (1994).
23. Malhotra, V.M. (ed.), Supplementary Cementing Materials for Concrete, pp. 438, *CANMET*, Ottawa (1987).
24. Materials and Process for Environmental Protection, *MRS Symposium Proceedings,* Vol.344, Pittsburgh (1994).
25. *Proceedings of the CANMET/ACI Conferences "Fly Ash, Silica Fumes, Slag and Natural Pozzolans in Concrete,"* Ed.V.M.Malhotra, 1st: Montebello, Canada (1983); 2nd: Madrid, Spain (1986); 3rd: Trondheim, Norway (1989); 4th: Istanbul, Turkey (1992); 5th: Milwaukee, USA (1995).
26. Wesche, K., (Ed.), Fly Ash in Concrete, Properties and Performance, *Report of Technical Committee 67-FAB Use of Fly Ash in Building,* E-FN Spon,. Chapman-Hall, London, pp. 255 (1991).
27. Mehta, P.K., in *Proceedings of the CANMET/ACI Conference"Fly Ash, Silica Fumes, Slag and Natural Pozzolans in Concrete,"* Ed.V.M.Malhotra, Montebello, Canada, pp. 1-46 (1983).

28. Montgomery, D.G., in, Fly Ash and Coal Conversion By-Products: Characterization, Utilization and Disposal, I. *Materials Research Society Symposium Proceedings,*Vol. 43, pp. 119-128 (1985).
29. Faber, W., Verwendung von Braunkohle als Kalkdünger, *Braunkohle* 44 (1/2) : 25-30 (1992).
30. Stout, W.L., in, *Proceedings: 11th International Symposium on Use and Management of Coal Combustion By-Products (CCBs)*, Vols. 1-3, American Coal Association, Orlando, FL pp. 4-11 (1995).
31. Odler,I. and Zysk, K.- H., "Hydraulic Properties of Fluidized Bed Combustion Ashes," in, ref (9), Vol. 178, pp. 189-196 (1989).
32. Berry, E. E. and Anthony, E. J., "Properties and Environmental Considerations Related to AFBC Solid Residues," in, ref (9), Vol. 86, pp. 49-58 (1986).
33. Berry, E.E., Jones, C.E., Rose, J.G. and Harness, J.L., "Performance Characteristics of Concrete Produced with Fluidized Bed Combustion Residue," in ref (9), Vol. 136, pp.23-36 (1989).
34. Brandštetr, J. and Šimek, Z., in, "Composition and Properties of Solid Residues of Fluidized Bed Coal Combustion, " in, ref (49), pp. 559-584 (1994).
35. Gribat, A-G., Mineralische Rückstände aus Wirbelschicht-feuerungsanlagen. Proceedings of the VGB-Conference Circulating Fluidized Bed Combustion, *VGB Power Station Technique*, Essen, (BRD) (1990).
36. Roy, D.M., Luke, K. and Diamond, S., "Characterization of Fly Ash and Its Reactions in Concrete," in ref (9), Vol. 43, pp. 3-20 (1985).
37. Brandštetr,J., Determination of the Pozzolanic Activity of Aluminosilicate Materials, *Czech.Pat.* 231 365 (1987).
38. Brandštetr, J., Huleja, J. and Kupec, J., Apparatus for Thermochemical Quantitative Analysis, *US Patent* 4, 379, 775 (1982).

39. Kautz, K., Oberheuser, G., Sajó, I. und Zobel, W., Zur Bestimmung der Aktivität von Steinkohlenflugaschen mit AKTIMET oder über Mörteldruckfestigkeiten, *VGB Kraftwerkstechnik* 65 (11): 1044-1051 (1985).
40. Manz, O.E., "Tests and Specifications Pertinent to Coal Ash Utilization," in ref (15), pp. 51-1 (1995).
41. Sybertz, F., "Comparison of Different Methods for Testing the Pozzolanic Activity of Fly Ashes," in, ref (25), Trondheim, pp. 477-97 (1989).
42. Brandštetr,J., Calorimeter. *Czech Pat.* 250 428 (1986).
43. Kumarathasan, P., McCarthy, G. J., Hassett, D.J. and Pflughoeft- Hassett, D.F., "Oxyanion Substituted Ettringites: Synthesis and Characterization and Their Potential Role in Immobilization of As, B, Cr, Se and V," in, ref (9), Vol. 178, pp. 83-104 (1989).
44. Blondin, J., Akman, M.S. and Anthony, E. J., "A New Hydration Process to Improve Quality of Silico and Sulfocalcic Ashes," in ref (25), Milwaukee, Supplementary Papers, pp.219-232 (1995).
45. Chandra, S. and Ohama, Y., *Polymers in Concrete*, pp. 204. CRC Press, Boca Raton, FL (1994).
46. Brandštetr, J. and Drottner, J., "Composites Based on Solid Residues of Fluidized Bed Coal Combustion and Other Products," in, ref (23), Milwaukee, Suplementary papers, pp. 386-412 (1995).
47. Davidovits, J., in, *Proceedings of the First International Conference on Alkaline Cements and Concretes,* Vol. 1, Sci. Res. Inst. of Binders, Kiew, Ukraine, pp. 131-162 (1994).
48. Krivenko, P.V., "Fly Ash - Alkali Cements and Concretes," in, ref (25), Istanbul, Supplementary papers, pp. 721-734 (1992).
49. *Proceedings of the First International Conference "Alkaline Cements and Concretes,"* Vols. 1-2, pp. 1128. Sci. Res. Inst. of Binders, Kiew, Ukraine (1994).

50. Moraitis, Ch., Einflüsse auf die Phasenbildung und die Festigkeit von hydrothermal gehärteten Kalksandprodukten unter besonderer Berücksichtigung der Auswirkung von Verunreinigungen in Form von Tonmineralien und Sulfat. Diplomarbeit, Technische Universität Clausthal (1995).
51. Sauman, Z., Bedingungen der Entstehung der Hauptmineralphasen in autoklavbehandelten Produkten, *Silikattechnik* 24 : 272 - 279 (1973).
52. Gabrišová, A., Havlica, J. and Sahu, S., *Cem. Concr. Res.* 21: 1023 - 1027 (1991).
53. Havlica, J. and Sahu, S., *Cem. Concr. Res.* 22 : 661-667 (1992).
54. Sahu, S., Havlica, J., Tomková, V. and Majling, J., *Thermochimica Acta* 175 : 45-52 (1991).
55. Neumann-Vanevere, S.V., Zysk, K.-H. and Odler,I., *Cem. Concr. Res.* 21: 262-268 (1991).
56. Belz, G., Beretka, J., Marraccoli, M., Santoro, L., Sherman, N. and Valenti, G. L., "Use of Fly Ash, Blast Furnace Slag and Chemical Gypsum for the Synthesis of Calcium Sulfoaluminate-Based Cements," in, ref (25), Milwaukee, pp. 513-530 (1995).
57. vom Berg, W., "New European Standards Concerning the Use of Fly Ash in Concrete," in, ref (16), Vol. 3, pp. 93-12
58. Bijen,J. and Sellevold, E.J., "European Standardization: Use of Additions," in, ref (25), Istanbul, Supplementary Papers, pp. 23-32 (1995).
59. Brendel, G. F., "Development of an ASTM Standard on Coal Ash Structural Fills," in, ref (16), Vol.1, pp. 17.1-17.5 (1993).
60. Majko, R.M., "Status of ASTM and Other National Standards for the Use of Fly Ash Pozzolans in Concrete, in, ref (9), Vol. 86, pp. 293-306 (1986).

2

PRODUCTION AND USE OF BY-PRODUCT GYPSUM IN THE CONSTRUCTION INDUSTRY

Wojciech Roszczynialski, Marek Gawlicki **and** *Wieslawa Nocuń -Wczelik*

INTRODUCTION

Two groups of gypsum by-products have been identified: chemical gypsum and desulphogypsum. The first is produced by chemical processing, the second originates from the flue gas desulphurization installations in power stations. The compositions of the by-products differ substantially, primarily with respect to impurity content. Chemical gypsums are contaminated with the raw materials used in particular technologies, as well as the products of chemical processes. These contaminants affect the properties of by-product gypsum and subsequently formed binders. The desulphogypsum produced by the wet process is highly pure calcium sulphate dihydrate. Some impurities do not influence the reactions in the $CaSO_4$-H_2O system. However, if the SO_2 sorbing agent is introduced to the flue gas stream without de-dusting or directly to the furnace, complex desulphurization products are formed. The products are mixtures of calcium sulphates with different water contents, calcium sulphide, along with significant amounts of fly ash or fly ash with bottom slag.

CHEMICAL GYPSUMS

Calcium sulphate dihydrate, $CaSO_4 \cdot 2H_2O$, is a basic component of chemical gypsums. Apart from the $CaSO_4 \cdot 2H_2O$, the hemihydrate gypsum, $CaSO_4 \cdot 0.5H_2O$, or anhydrite, can also be observed. The names of chemical gypsums are derived from the primary manufacturing products created during chemical processing [1]. The chemical reactions resulting in the different gypsum by-products are given below:

a) Phosphoric acid: Reaction between sulphuric acid and phosphorites. By-product: Phosphogypsum.

$$Ca_9(PO_4)_6 \cdot CaF_2 + 10H_2SO_4 + 20H_2O \rightarrow \\ 10CaSO_4 \cdot 2H_2O + 6H_3PO_4 + 2HF$$

b) Fluoric acid. By-product: Fluorogypsum:

$$CaF_2 + H_2SO_4 \rightarrow 2HF + CaSO_4 \\ CaSO_4 + 2H_2O \rightarrow CaSO_4 \cdot 2H_2O$$

c) Titanium white from ilmenite. By-product: Titangypsum:

$$FeOTiO_2 + 2H_2SO_4 \rightarrow TiOSO_4 + FeSO_4 + 2H_2O \\ TiOSO_4 + 2H_2O \rightarrow TiO(OH)_2 + H_2SO_4 \\ H_2SO_4 + CaCO_3 + 2H_2O \rightarrow CaSO_4 \cdot 2H_2O + H_2O + CO_2 \\ FeSO_4 + Ca(OH)_2 + 2H_2O \rightarrow CaSO_4 \cdot 2H_2O + Fe(OH)_2$$

d) Boric acid. By-product: Borogypsum

$$Ca_2B_6O_{11} \cdot 5H_2O + 2H_2SO_4 + 6H_2O \rightarrow \\ 2(CaSO_4 \cdot 2H_2O) + 6H_3BO_3$$

e) Organic acids (e.g. citric acid). By-product: Citrogypsum:

$$(C_3H_4OH\ (COO)_3)_2Ca_3 + 3H_2SO_4 + 6H_2O \rightarrow$$
$$2C_3H_4OH\ (COOH)_3 + 3CaSO_4{\cdot}2H_2O$$

f) Phenol. By-product: Phenologypsum:

$$(C_6H_5O)_2Ca + H_2SO_4 + 2H_2O \rightarrow$$
$$2C_6H_5OH + CaSO_4{\cdot}2H_2O$$

g) Neutralization of sulphuric acid containing liquid wastes and galvanic sludges by use of lime milk or limestone:

$$H_2SO_4 + Ca(OH)_2 + 2H_2O \rightarrow CaSO_4{\cdot}2H_2O$$
$$H_2SO_4 + CaCO_3 + 2H_2O \rightarrow CaSO_4{\cdot}2H_2O + H_2O + CO_2$$

h) Production of soda using the Solvay method. Neutralization of calcium chloride. By-product: Sodagypsum:

$$CaCl_2 + Na_2SO_4 + 2H_2O \rightarrow CaSO_4{\cdot}2H_2O + 2NaCl$$

I) Recovery of NaCl from sea water. By-product: Saltgypsum:

$$MgSO_4 + CaCl_2 + 2H_2O \rightarrow MgCl_2 + CaSO_4{\cdot}2H_2O$$

Phosphogypsums

Phosphogypsum is the most common of the by-product chemigypsums. It is produced in the fertilizers plants as a result of the reaction between sulphuric acid and natural phosphorites. The chemical compositions of different phosphogypsum materials produced using a variety of methods are given in Table 1. Calcium sulphate dihydrate forms as a main component of the by-product material and the impurity content is relatively small. Phosphogypsums originating from phosphorites can be contaminated with radioactive elements and organic substances (about 0.3%) [2]. Phosphogypsums from apatite concentrates (deposits at Kola Peninsula in Russia) show a significant amount of strontium and rare earth elements [3]. In practice, every 1 ton P_2O_5 combined in the form of phosphoric acid corresponds to 4 - 6 tons phosphogypsum [4]. Therefore, with an annual production of P_2O_5 of approximately 20 million tons, about 100 million tons phosphogypsum are produced [5]. The chemically bound water content in calcium sulphate from the phosphogypsum depends on the method of phosphoric acid production, temperature and the sulphate and phosphate ion concentration in the solution at later

Table 1. Chemical Composition of Phosphogypsum from Various Sources.

Component	Content (%)			
	*PG1	PG2	PG3	PG4
H_2O_{cryst}	19.5	19.5	20.0	18.0
SO_2	43.2	45.2	44.0	43.6
CaO	32.2	32.1	31.0	32.0
MgO	0.01	0.10	--	0.40
Al_2O_3+Fe_2O_3	0.27	0.36	0.14	1.82
SiO_2 ins. in HCl	1.51	0.72	2.40	1.64
Na_2O	0.47	0.23	0.18	0.36
P_2O_5 total	1.01	1.65	0.78	1.03
F total	1.76	0.29	0.57	0.76
Organic Matter	0.08	--	0.24	0.26

**PG1 - Phosphogypsum from Morocco; dihydrate process [13]*
PG2 - Phosphogypsum from Kola apatites; dumped [3]
PG3 - Phosphogypsum from Florida; dihydrate process [2]
PG4 - Phosphogypsum from India; dihydrate process [7]

stages of the process. The storage of phosphogypsum in open stockpile leads to combination with water and subsequent transformation of $CaSO_4 \cdot 0.5H_2O$ and anhydrite to $CaSO_4 \cdot 2H_2O$.

The most common means of producing phosphoric acid is the dihydrate method. The finely divided, crystalline $CaSO_4 \cdot 2H_2O$ is highly contaminated. Because of the high water content, it is difficult to find an effective method for re-processing this by-product. The less troublesome by-product Phosphogypsums are those originating from the hemihydrate method or combined methods (hemihydrate-dihydrate process or dihydrate-hemihydrate process). With the above mentioned processes, the gypsum crystals are better developed, moisture content is reduced and the level of contamination due to impurities is reduced as well [6].

Impurities in Phosphogypsums. The impurities occurring in phosphogypsum can be classified as follows:

—External substances from the phosphogypsum raw materials which do not react with the sulphuric acid. These substances are accumulated in phosphogypsum and stored together on the stockpiles;

—Compounds produced as a result of ore processing - products of reaction between sulphuric acid and the components of phosphorite, the residue of sulphuric acid, product of corrosion in installation.

The type and amount of impurities in the first group depends primarily on the phosphate ore origin and method of ore processing. The second group impurities in turn, depend on the phosphoric acid production method, as well as on the temperature and intensity of washing operation.

From the point of view of phosphogypsum application in the building materials industry, the most convenient is the classification of impurities for the two groups. In the first group, are the water insoluble species which do not take part in the reaction between binding the material and water. The second group consists of the soluble components. These are chemically active during the setting and hardening process of materials produced from phosphogypsum. Among all the impurities present in phosphogypsums, the most important are the phosphorus- and fluor-containing compounds along with the organic matter. These materials strongly affect the setting of gypsum products and cement manufactured with phosphogypsum. These impurities are determined as P_2O_5, F and "organic matter," respectively. The phosphate residue consists of the phosphoric acid, $Ca_3(PO_4)_2$, $Ca(H_2PO_4)_2$ and $CaHPO_4{\cdot}2H_2O$.The fluor-containing compounds are NaF, Na_2SiF_6, Na_3AlF_6, Na_3FeF_6 and CaF_2 [7]. The organic matter in phosphogypsum is composed of aliphatic compounds of carbonic acids, amines and ketons with an average of 10-30 carbon atoms [7]. Along with the impurities mentioned above, phosphogypsums contain minor amounts of quartz, chalcedone and feldspar, as well as aluminum, iron and alkaline sulphates. Traces of chlorides are also found.

The impurities are classified as follows by Singh [8]:

- water soluble,
- insoluble,
- present inside the $CaSO_4{\cdot}2H_2O$ clusters, and
- solid solutions in $CaSO_4{\cdot}2H_2O$.

The most commonly occurring soluble impurities are: H_3PO_4, H_2SO_4, $Ca(H_2PO_4)_2{\cdot}H_2O$, NaF and Na_3AlF_6. These compounds are easily adsorbed on the surfaces of calcium sulphate grains. The insoluble impurities consist of unreacted phosphorite residue, mainly the fluoroapatites. The foreign substances present in the clusters, irrespective of their solubility, cannot be removed easily by simple washing operations. The leaching out of soluble substances trapped in gypsum clusters appears to be possible after the transformation of dihydrate gypsum to hemihydrate or anhydrite and subsequent intensive washing. The contaminations present in the $CaSO_4{\cdot}2H_2O$ solid solution, such as $CaHPO_4{\cdot}2H_2O$, sodium orthophosphate and other anions substituting for the sulphate anions ($[AlF_5]^{2-}$ and $[FPO_3]^{2-}$) can be removed in the same way [2,9]. The addition of lime milk during the calcination of phosphogypsum brings about the neutralization of residual phosphoric acid. At the same time, the acid phosphates transform to the insoluble $Ca_3(PO_4)_2$ and the fluorides, fluorosilicates and fluoroaluminates form CaF_2. The impurities occurring in phosphogypsum from different sources are shown in Table 2.

Table 2. Impurities Occurring in Phosphogypsum.

Component	Content (%)		
	*PG1	PG2	PG3
Water-soluble P_2O_5	0.36	0.24	0.25
Co-crystalline P_2O_5	0.50	0.20	0.51
Water-insoluble P_2O_5	0.06	0.03	0.26
Total P_2O_5	0.92	0.47	1.02
Water-soluble F	1.03	0.41	0.24
Water-insoluble F	0.47	0.45	0.55
Total F	1.50	0.86	0.79
Organic Matter	0.13	0.44	0.38

**PG1 - Phosphogypsum from India; dihydrate process [7]*
PG2 - Phosphogypsum from India; dihydrate-hemihydrate process [7]
PG3 - Phosphogypsum from Turkey; dihydrate process [4]

In many phosphogypsums produced from the phosphate rocks a relatively high concentration of radioactive elements, such as U^{238}, U^{234}, Ra^{226}, Pb^{210}, Po^{210} can be detected. As a consequence, the Rd^{222} emission is further observed [10,11]. The contents of radioactive elements vary in a wide range depending on the origin of phosphate rock. Therefore the phosphogypsum based materials should be carefully examined from the point of view of their radioactivity.

The heavy metals contents in phosphogypsums are usually very low [12]. However, some phosphate rocks, especially those occurring at Kola Peninsula show a significant strontium and rare elements level, up to 3% [2] so as these elements are recovered from the concentrates.

The percentage of impurities in phosphogypsum is strongly affected by the size and microstructure of $CaSO_4 \cdot 2H_2O$ crystals and, as a consequence by the specific surface exposed to the adsorption from the post-reaction solutions. This in turn depends on the deposit used. For example, the specific surface of phosphogypsum produced by dihydrate method in one fertilizer plant in Poland is 4500 - 6000

cm^2/g when the rocks from Kola Peninsula are processed, 4000 cm^2/g at the phosphorites from Togo processing, 2500 cm^2/g - at the phosphorites from Morocco and finally, about 2000 cm^2/g in the case of phosphorites from Florida [2].

The water content in the phosphorites of largest surface area exceeds 40%, while the water content in the materials of the lowest specific surface is about 20%. It should be also mentioned that a large part of impurities occurs in the particle size fraction above 160 μm and below 25 μm. The particle size fractions larger than 160 μm are enriched with fluorine, silicon and sodium oxide compounds, while the fractions smaller than 25 μm are contaminated with organic substances and phosphate [13].

Requirements for Phosphogypsums. Most of phosphogypsums, especially those produced by dihydrate method, cannot be used in the construction industry in a raw form. The undesirable contaminations are, most of all, those containing the phosphate and fluoride ions, both adsorbed and incorporated to the structure, which are transferred to the solution. The total phosphate compounds in phosphogypsums, calculated as P_2O_5, attains incidentally 5%, including 3% of "soluble" P_2O_5. In the requirements from the users the other impurities, such as chlorides and alkalis are also limited. Obviously, the pH value and the radioactive elements contents must be limited too. In the case of phosphogypsum used in the production of gypsum binders the organic impurities are restricted. The organic compounds affect the whiteness and are undesirable in the production of gypsum panels. The following requirements for the phosphogypsum used in the processing for gypsum binder must be fulfilled in Poland [14]:

$CaSO_4 \cdot 2H_2O$ content ≥	92 %
fraction > 0,1 mm <	1 %
Blaine specific surface	3000-6000 [cm^2/g]
P_2O_5 content:	
Total	< 0.20 %

Soluble	< 0.10 %
F^- content	< 0.05 %
Alkalis as Na_2O content	< 0.03 %
$Al_2O_3 + Fe_2O_3$ content	< 0.15 %
Radioactivity factors:	
f_1	$\leq$ 1
f_2	$\leq$ 185

The f_1 and f_2 factors are used in Poland for all types of construction materials as a standardized evaluation of radioactivity based upon the measurements of gamma radiation emitted by the isotopes K^{40}, Ra^{226} and Th^{232}. The value of f_1 factor is determined as follows [15]:

$$f_1 = 0.00027\ S_K + 0.0027\ S_{Ra} + 0.0043\ S_{Th} \leq 1$$

where S_K, S_{Ra}, S_{Th} - are the intensities of K^{40}, Ra^{226} and Th^{232} radiation in Bq/kg.

The f_2 factor is determined by Ra^{226} radiation and consequently serves as a measure of radon Rn^{222} emission.

Therefore : $f_2 = S_{Ra} \leq 185$

where $S_{Ra} = Ra^{226}$ radiation intensity in Bq/kg.

The requirements for the phosphogypsum used as a cement set controlling agent can be rigorous especially when the cements must show rapid setting. In this case the application of phosphogypsum as a set controlling agent results in prolonged, regular setting.

Removal of Impurities. The treatment and further reuse of phosphogypsum by-products results from the economic, local conditions and ecological reasons. Because of the relatively low value of material thus recovered, the operation aimed with the phosphogypsum purification must be carried out at little cost. In practice it is done in hydrocyclones by use of water at temperature 60-90°C. Thus the soluble phosphorus and fluor compounds are removed [16]. The organic impurities , alkalis and chlorides are also leached out. At final stage of leaching, the lime milk is added to neutralize the acids present in phosphogypsum and the insoluble phosphates and calcium fluoride precipitate from the solution. The potassium hydroxide or carbonate can also be used as neutralizing agents [17]. Very good results can be achieved by washing with diluted ammonium hydroxide or ammonium sulphate. Among the phosphates and fluorides $Ca(H_2PO_4)_2 \cdot H_2O$ and NaF are the most easily soluble in ammonium solutions. The soluble ammonium salts thus produced can be easily removed from the reagents. The fluorosilicates, fluoroaluminates and fluoroferrites react also with NH_4OH giving the soluble ammonium salts [18]:

$$Na_2SiF_6 + 2NH_4OH \rightarrow (NH_4)_2SiF_6 + 2NaOH$$
$$Na_3AlF_6 + 3NH_4OH \rightarrow (NH_4)_3AlF_6 + 3NaOH$$
$$Na_3FeF_6 + 3NH_4OH \rightarrow (NH_4)_3FeF_6 + 3NaOH$$

Purification can be carried out simultaneously with $CaSO_4 \cdot 2H_2O$ dehydration in liquid medium at ambient pressure or in hydrothermal conditions. The impurities adsorbed on the surfaces of calcium sulphate crystals and those trapped inside the crystals pass to the liquid phase. On $CaSO_4 \cdot 2H_2O$ dehydration the decomposition of solid solutions also occurs. In the presence of lime milk, the neutralization and precipitation of CaF_2, $Ca_3(PO_4)_2$ and hydroxy-apatite takes place during the washing operation. The formation of hydroxyapatite follows the equation [5]:

$$10Ca^{2+} + 6(PO_4)^{3+} + 2OH \rightarrow Ca_{10}(PO_4)_6(OH)_2$$

The effectiveness of calcium phosphate precipitation is the highest at pH=10-11. At basic environment, the conversion of fluorosilicate compounds takes place with the precipitation of CaF_2 and SiO_2. The dehydration of phosphogypsum in acid environment

(H_2SO_4 solution, pH=3-4) leads also to the leaching out of phosphate ions and $Ca_3(PO_4)_2$, as well as CaF_2 precipitation [16]. The phosphates, fluorides and organic substance can also be reduced by treating the phosphogypsum with H_2SO_4 and SiO_2 [19].

The effective purification can be achieved in some cases by separation of fine and coarse particle size fractions. This can be done using different methods, such as screening, flotation or separation in hydrocyclone [13].

Utilization of Phosphogypsums in Building Materials Production. When the purification of phosphogypsums is completed, the waste materials are transformed into standard calcium sulphate, which can be used for most applications, unlike natural gypsum [20,21]. However, because of the high humidity and very often unsatisfying results of the "deep neutralization" process, accompanied by high costs of the waste material treatment, the quality of recovered material, ecological requirements and costs must be carefully compromised. It is evident that the phosphoric acid technology that supplies the phosphogypsum by-product for reuse would be an optimum solution but, it is possible only for some combined methods [6]. However, taking into account the annual output of phosphogypsum, as well as the requirements for the phosphogypsum materials, it seems that this by-product can be most of all consumed by cement industry as a set controlling agent in the Portland and blended cements production. Another important way of phosphogypsum utilization is the production of gypsum plasters and different types gypsum elements. Some amount of phosphogypsum can be taken into account as raw materials in the associated technologies of cement (lime) and sulphuric acid production, as a natural gypsum replacement in cellular concrete manufacture and several other technologies listed above. The radioactivity of phosphogypsum must be carefully determined while the possibility of application in the construction materials is considered [10,22].

Use of Phosphogypsum in Cement Technology. The role of gypsum in cement hydration has been widely reported and discussed. Briefly, setting of cement, as well as the other properties, including mechanical strength are strongly influenced by the presence of sulphate ions in the cement-water system. Obviously, the natural gypsum replacement by the phosphogypsum by-product brings about significant changes so pronounced as both materials differ one from another. The potentially detrimental effects are connected with the high humidity, the presence of impurities affecting the cement hydration and, finally, the grain size composition differing from those for natural gypsum. The impurities become active on cement hydration and subsequently influence the setting and strength development [23,24]. The excess water content in phosphogypsum may cause some difficulties in transportation, storage and dosage. The lower grindability of phosphogypsum should be taken into account in cement production. Both the features - high humidity and fineness create some technological problems but are of limited significance as it relates to cement properties.

The use of phosphogypsum in cement industry is determined first of all by the chemical composition. The hydration process is strongly influenced by the presence of phosphate, fluorosilicate and fluoride particularly those present in soluble form and adsorbed on the surfaces of $CaSO_4 \cdot 2H_2O$ crystals. The impurities incorporated in the structure are of less importance. As it is commonly accepted the presence of $[H_2PO_4]^-$, $[HPO_4]^{2-}$, $[PO_4]^{3-}$, $[SiF_6]^{2-}$ and F^- ions in the highly alkaline solution (pH=12-13) and supersaturated with calcium ions leads to the formation of insoluble calcium phosphates, fluorides and fluorosilicates covering the hydrating cement grains. The cement hydration process is therefore retarded by a protective coating mechanism [1,25]. Some authors are of the opinion that the precipitation products form the bridges reinforcing the hydrating cement structure [24]. The influence of monocalcium phosphate, dicalcium phosphate and sodium fluorosilicate which can be considered as model, soluble impurities [26,27] cause the prolonged setting and compressive strength decrease with increasing admixture content. It should be mentioned that the strength development decrease is less pronounced than the setting time increase and occurs in different way for particular cements [1,28-30]. However, the effect of phosphogypsum set controlling agent on the compressive strength is found only within the first 7 days (Table 3).

Table 3. Properties of Cement Pastes Prepared With Different Set Controlling Agents [29].

	Set Controlling Agent		
Property	Phospho-Gypsum	FGD Gypsum	Natural Gypsum
Initial setting [h^{min}]	2^{30}	1^{44}	1^{54}
Final setting [h^{min}]	3^{40}	2^{33}	2^{54}
Compressive Strength (MPa)			
1 day	8.0	8.4	8.6
3 days	18.4	19.8	18.8
7 days	20.1	22.5	20.1
28 days	37.0	38.9	37.0
60 days	38.1	39.2	38.1

Since the delayed hydration of the phosphogypsum containing cements results from the formation of impermeable layers composed of products forming between the calcium hydroxide and impurities from phosphogypsum, one should expect that all the operations reducing the durability of these layers would result in the hydration acceleration. However, it has not been confirmed. Similarly, the retarding effect at low temperature on the phosphogypsum containing cement has also not been confirmed. One should underline that the substitution of natural gypsum by the phosphogypsum in Portland and metallurgical cements subsequently used in concrete technology brings about only a small changes in consistency, almost independent of the type and percentage of several impurities.

Production of Sulphate Binders. The $CaSO_4{\cdot}2H_2O$ dehydration is a basic process in the production of gypsum binders from phosphogypsum. Depending on the parameters of the process, the following types of gypsum materials are formed: α-$CaSO_4 \cdot 0.5H_2O$, β-$CaSO_4 \cdot 0.5H_2O$, anhydrite II or estrich- gypsum. The problems arose with the phosphogypsum utilization are as follows:

- high moisture content and too high fineness,
- presence of impurities on the level above the limits.

The phosphates, fluorosilicates and fluorides bring about the setting time increase and lower the strength development. Chlorides and alkalis form efflorescence on the elements. The organic compounds lower the whiteness. The residual acids cause the corrosion of installation. The dehydration of relatively pure phosphogypsums, produced by some new technologies or purified occurs in the same way as natural gypsum.

The materials showing the high moisture content and fineness must be dried and briquetted before heating in the grate furnace. Thus the energy consumption per 1 ton of plaster powder is much higher (40-45 kWh) than for natural gypsum (about 20 kWh) [22]. Gypsum materials are considered to be rapid setting and the presence of retarders is undesirable. Therefore, all the impurities, such as phosphates, fluorosilicates and fluorides should be removed and it is "conditio sine qua non" of phosphogypsum utilization in gypsum materials technologies where the gypsum components occur as almost 100% component. The optimum solution is to carry out purification and dehydration simultaneously.

Production of α-$CaSO_4 \cdot 0.5H_2O$. The α-$CaSO_4 \cdot 0.5H_2O$ can be produced by phosphogypsum dehydration in liquid medium in hydrothermal conditions or at the atmospheric pressure in the solutions of salts. The soluble impurities and most of insoluble ones are removed during the phosphogypsum processing [2,6,31]. There are several solutions realized on industrial scale. The Giulini method [6] in which the preliminary washed phosphogypsum is subjected to the autoclaving at 105-140°C is one of them. The phosphogypsum is slurried in water. The suspension of phosphogypsum is then subjected to flotation (first step of purification) and subsequently pumped to a buffer tank. From the buffer tank, the material is transported to the scrubbing tower and, after introducing of nucleating agent, passes to the autoclave for $CaSO_4{\cdot}2H_2O$ dehydration. The suspension containing α-$CaSO_4{\cdot}0.5H_2O$ formed in autoclave is centrifuged at 90-100°C to separate the excess liquid phase. The hemihydrate is washed with hot water and dried at 90°C. The α-$CaSO_4{\cdot}0.5H_2O$ production following the Giulini method can be carried out without autoclaving, in concentrated solutions (e.g. 1% KCl+ 23% $CaCl_2$) near the boiling point. This technology is however inconvenient because of huge amount of waste liquid, resulting from the intensive washing process. The high pressure modification of Giulini method is most often applied and the output of gypsum producing plants attains several hundred tons per 24 hours. The Russian WNIISTROM process, similar to the Giulini one is carried out in autoclave in the presence of surface active agents. The α-$CaSO_4{\cdot}0.5H_2O$ separation is done on vacuum filters or by centrifuging. The phosphogypsum slurry is introduced to the decantation and subsequently the liquid is neutralized and the concentrated phosphogypsum slurry is pumped to the autoclave together with 0.1-0.3% of surface active admixture. The slurry passes from autoclave to the heat exchanger in which the reduction of pressure occurs. The excess water is removed on vacuum filter. The hemihydrate thus obtained is dried and ground. The compressive strength of gypsum paste is approximately 20 MPa and the flexural strength - 4 MPa respectively [2].

The high quality α-$CaSO_4{\cdot}0.5H_2O$ can be produced by phosphogypsum dehydration in sulphuric acid or calcium hydroxide solutions [6].

Production of β-$CaSO_4{\cdot}0.5H_2O$. There are several technologies available for the production of β-$CaSO_4{\cdot}0.5H_2O$ phosphogypsums. Two groups can be distinguished:

—one-step calcium sulphate dihydrate dehydration,

—two-step calcium sulphate dihydrate dehydration with an intermediate rehydration of hemihydrate.

The first stage dehydration and subsequent rehydration leads to purification of material, the second stage to the formation of β-$CaSO_4 \cdot 0.5H_2O$ product.

The Eipeltauer method [2] is a one-step process in which particular operations are: washing with water, filtration, centrifuging, neutralization of acid impurities by KOH or K_2CO_3, drying, redehydration of purified phosphogypsum. In the presence of potassium ions the insoluble syngenite $K_2SO_4 \cdot CaSO_4 \cdot H_2O$ is formed thus inhibiting the potassium ions leaching and the formation of efflorescence on the surface of final gypsum elements.

The Rhône-Poulenc technology also includes the intensive washing to remove the impurities [6]. This washing operation is done on the vibrating sieves, in hydrocyclones or by flotation. The slurry from phosphogypsum of low impurity content, is pumped through the vibration sieve. The water consumption is 1.6-3 tons per 1 ton of phosphogypsum and allows to remove 80% of soluble phosphates and fluorides. At higher impurity content in phosphogypsum the washing operation is carried out in hydrocyclone giving the 10 times lowering of impurity level. The water consumption is high: 6-10 tons per 1 ton of phosphogypsum. The flotation method is a low water consuming one. At 2-3.3 tons of water per 1 ton of phosphogypsum, the approximately 90% of impurity removal is achieved. The separation of purified phosphogypsum is carried out on vacuum filter or by centrifuging before drying and dehydration to β-$CaSO_4 \cdot 0.5H_2O$.

Among the two-step dehydration methods the solution proposed by Prayon [6,16] should be mentioned, in which the first phosphogypsum dehydration occurs in H_2SO_4 solution at 110-120°C. The α-$CaSO_4 \cdot 0.5H_2O$ thus obtained converts to $CaSO_4 \cdot 0.5H_2O$ and is subjected to redehydration at ambient pressure.

The technology worked out by Knauf [16] is another example. The first phosphogypsum cleaning dehydration is carried out in standard conditions by dry method. The β-hemihydrate rehydration occurs in the presence of $Ca(OH)_2$. The calcium sulphate dihydrate thus produced is redehydrated to β-hemihydrate and it can be used as a component of plasters.

Anhydrite Production. The anhydrite ($CaSO_4$ II) production from phosphogypsum can be done by wet or dry method. In dry processes the cleaning of phosphogypsum can be carried out as a first operation. In the case of relatively pure phosphogypsum ($F^- < 0.2\%$, $P_2O_5 < 0.8$ %) the purification may be neglected. After drying, the material is subjected to the calcination at the temperature 300-500°C [32], 600-700°C [33,34] or even at 900°C [35]. This process is carried out in the rotary or grate kiln. On burning, the recrystallization of anhydrite together with the transformation of phosphates to the inert β-$Ca_2P_2O_7$ [34] , as well as the decomposition of fluorosilicates and other fluor compounds takes place. These fluor compounds are removed on washing thus creating some environmental hazard. The anhydrite is ground with 1-3 wt.% of activator improving the subsequent hydration. Potassium sulphate, calcium hydroxide, Portland cement, ferric sulphate, aluminum potassium sulphate and other substances can be used as activators. The properties of anhydrite binders thus produced are influenced by the type of phosphogypsum raw material and can be modified by preliminary treatment (briquetting), burning temperature, use of several activators and grinding of final product (Table 4) [35].

Table 4. Mechanical Properties of Anhydrite Binders.

	Type of Binder		
Property	A1	A2	A3
Initial Setting (min)	80	135	90
Final Setting (min)	60	100	140
Flexural Strength (MPa)			
3 days	1.6	1.3	3.0*
14 days	7.9	6.8	--
28 days	8.1	7.6	4.2
Compressive Strength (MPa)			
3 days	5.3	4.8	13.5*
14 days	27.7	24.3	--
28 days	42.8	39.1	17.7

* *7 days strength*
A1 - anhydrite from wet process; 1.5% K_2SO_4 *[33]*
A2 - anhydrite from wet process; 2% K_2SO_4 *[33]*
A3 - phosphogypsum burned at 300 °C (55% anhydrite + 45% hemihydrate) activated with 1% CaO + 2% OPC [35]

A significant moisture resistance increase in the case of anhydrite binder from dry process can be done by the addition of granulated blast furnace slag together with some calcium hydroxide and alkaline sulphates [36]. The wet process consists in the heating of phosphogypsum suspension in the alkaline or acid solutions, liquid phase separation, drying and burning. The $[HPO_4]^{2-}$ ions are removed from the calcium sulphate dihydrate structure with increasing temperature of preliminary treatment. The presence of sulphuric acid on dehydration brings about the removal of other ions present in phosphogypsum. In the wet processing of phosphogypsum Kola the recovery of rare earth elements is possible [33,34]. The phosphogypsum is firstly stirred at 40°C in 12% H_2SO_4 solution, at the liquid/solid ratio equal 2:1 and after filtration, treated again with H_2SO_4 at 50°C. After 2 hours anhydrite is separated from the solution, neutralized and calcined at 800°C. The final product is ground with 1.0-1.5% K_2SO_4. The material thus obtained exhibits relatively short setting time and the 28 days compressive strength up to 42 MPa. The flexural strength is 8 MPa [33]. It should be mentioned that the technology of phosphoric acid and anhydrite production from phosphorites has also been worked out but it has no practical significance because of the high energy consumption [34].

The anhydrite binders are used in the production of plasters, wall panels, materials for self leveling floors, to fill excavations in mines and in sealing rock masses [10,37].

Other Possibilities for Phosphogypsum Utilization Among the possibilities for using phosphogypsum, special attention should be paid to the associated technology of cement and sulphuric acid. In this technology, limestone is substituted for phosphogypsum.

Calcium oxide from gypsum is bound with other oxide components (SiO_2, Al_2O_3, Fe_2O_3), in Portland cement clinker phases at 1400°C. Some amount of coal is introduced to the raw mixture as a reducing component. The formation of SO_2 occurs according to the following reaction:

$$2CaSO_4 + C \rightarrow 2SO_2 + CaO + CO_2$$

The dosage of coal to the raw mixture must be controlled very carefully in compliance with the other chemical processes which may potentially occur in the kiln, e.g. the reduction of CO_2 to CO or sulphates to sulphide [38]. The content of coal in the raw mixture is on the level 5.1-5.2% by weight of calcium sulphate. The excess of coal in the kiln brings about the S^{2-}formation. In turn, at low coal content, the excess sulphate formation is observed and the consequences in both cases have a negative influence on cement properties. At the excess sulphate phase, the set controlling process is seriously disturbed. In the presence of sulphide ions the volume changes and the corrosion of reinforcement are expected. The high moisture content in phosphogypsum is the reason of its less applicability in this technology. Furthermore, the phosphate and fluoride impurities lead to the decrease in the clinker quality . The P_2O_5 content in the clinker is limited to 2-2.5 wt.%. The cement pastes from this cement reveal prolonged setting and slow rate of strength development within the first 28 days. At later ages their compressive strengths exhibit normal values.

The efforts to produce quicklime and sulphuric acid on industrial scale failed because of the high energy consumption and other difficulties.

Some amount of phosphogypsum can be introduced to the batch for expansive cement clinker, to form the calcium sulphoaluminate.

The other examples of phosphogypsum utilization are as follows:

- cellular concrete production [2],
- production of blended binders (with fly ash, lime, cement, granulated blast furnace slag) [36,39-43],
- production of elements reinforced with glass fibre, organic fibers [44-46], different gypsum composites [22].

Utilization of Other Chemical Gypsums

Calcium sulphate dihydrate or anhydrite is the main constituent (>90%) of all chemical gypsums; the differences between them result from the presence of the other components, which are not inactive on hydration of the gypsum materials. The most important chemical gypsums are described above. All the remarks concerning the possibilities of phosphogypsum utilization, after removal of impurities, can be related also to the other chemical gypsums. The purified chemical gypsums can substitute the natural gypsum [47]. The economical and ecological reasons are the decisive factors, indicating the possibility of waste chemical gypsum utilization. Three types of chemical gypsums are discussed: fluorogypsum, titangypsum and borogypsum. Their chemical compositions are shown in Table 5.

Fluorogypsums. Fluorogypsums are produced during fluoric acid manufacturing. The anhydrite transforming subsequently to dihydrate is the basic constituent, usually exceeding 95%. The impurities occurring in fluorogypsum are: residual fluoric acid, unreacted sulphuric acid and the other inorganic salts, active on hydration. These impurities can be removed by washing. The fluorogypsums are used with gypsum or anhydrite as set controlling admixtures to cement. As it is reported [1], the effect of fluoride impurities from fluorogypsum on hydration is relatively smaller than that of impurities from phosphogypsum. This may result from the different behavior of hydrofluoric acid in an alkaline medium as compared to fluorosilicate from phosphogypsum. The setting process is retarded to some extent, but strength development is not greatly

Table 5. Chemical Composition of Chemical Gypsum.

Component	Content (%) Fluorogypsum [13]	Titangypsum [48]	Borogypsum [49]
SO_3	56.4	52.3	44.2
CaO	40.8	36.4	28.8
MgO	0.13	0.40	1.70
Al_2O_3/Fe_2O_3	0.27	0.21	2.15
SiO_2	0.69	9.50	8.98
Na_2O	0.03	0.20	--
F	0.92	--	--
B_2O_3	--	--	11.26

affected [45]. Energy consumption for cement grinding in the presence of fluorogypsum has been also reported [1].

Titangypsums. These waste materials are produced during basalt weathered rock processing to obtain titanium compounds. The impurities accompanying the gypsum component are: SiO_2, TiO_2 and Fe_2O_3. Titangypsum used as a set controlling agent in cement has no influence on the properties of cement as compared to natural gypsum [1]. The mixed 60% titangypsum containing binders show fairly good mechanical properties [48].

Borogypsums. Borogypsums consist of calcium sulphate dihydrate and boron containing compounds of the content exceeding in some cases 10 wt.% [49]. Therefore they can be processed to obtain B_2O_3. A significant amount of lithium (0.4 wt.%) is also detected and this element can be also recovered [50]. The boron containing by-products are the hazardous wastes. When added to cement, they cause prolonged setting but slightly improved hardening as compared to reference materials [49].

Other Chemical Gypsums. The data dealing with the other chemical gypsum utilization are not readily available. This results from the local meaning of particular materials and their limited output. The preparation of blends with natural gypsum and subsequent use in cement industry for cement set controlling admixture seems to be the suitable method for chemical gypsum utilization. It should be emphasized that some impurities can provoke false setting (alkalis from sodagypsum) [1] or retard the hydration of cement (citrogypsum). The heavy metal level in chemical gypsums from neutralization of galvanic and other industrial sludges must also be carefully monitored.

FLUE GAS DESULPHURIZATION PRODUCTS

The flue gas desulphurization products were initially produced in Japan in the 1960s as a result of purification of gases from power plants. The goal of the purification was the reduction of SO_2 emissions to the atmosphere. In the 1970s, environmental protection equipment was installed in European Community countries, the United States, Canada, South Africa, China, Taiwan and India. The flue gas desulphurization with gypsum as the end product is currently being introduced in the east and central Europe. The data shown in Figure 1 and Table 6 show that efforts to decrease air pollution via emission purification have been successful.

Desulphogypsums

In the desulphurization process, SO_2 from flue gas is bound in the form of a solid desulphurization product. From an ecological point of view, this product should be fully reused. The possibility of the flue gas desulphurization (FGD) products application depends on their composition and properties .

The optimum FGD product is produced in wet-lime method installations, in which the by-product, so-called desulphogypsum or FGD-gypsum consists almost entirely (>95%) of the calcium sulphate dihydrate. In this method, the de-dusted gases are put through the

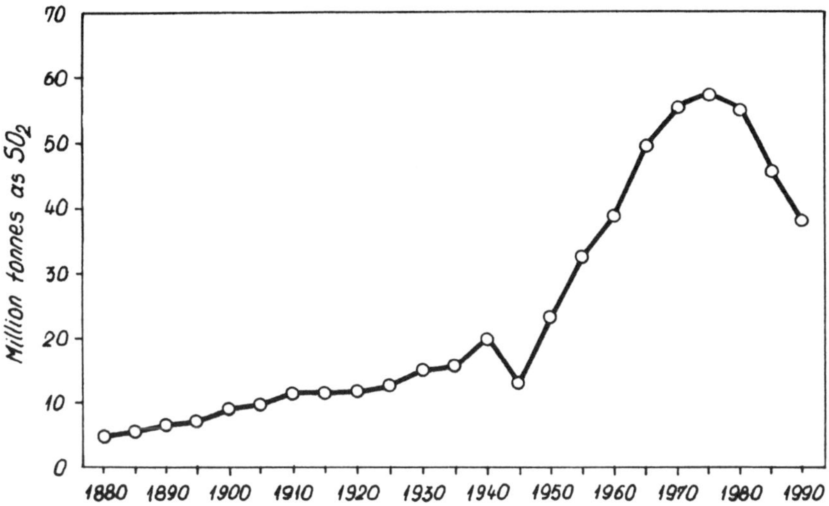

Figure 1. Emission of sulphur dioxide in Europe [51].

Table 6. Emissions of Sulphur in Europe [52].

Country	Emission of Sulphur (1000 tons S/year) 1980	1992
Albania	25	25
Austria	199	42
Belgium	414	224
Bulgaria	1025	830
Czechoslovakia	1551	1110
Denmark	226	122
Finland	292	97
France	1669	685
Germany (E)	2150	--
Germany	1597	2870
Greece	200	250
Hungary	816	505
Iceland	3	3
Ireland	111	95
Italy	1900	1090
Luxembourg	12	8
Netherlands	233	101

Table 6. Continued,

Norway	70	24
Poland	2050	1498
Portugal	133	102
Romania	900	900
Spain	1660	1158
Sweden	260	85
Switzerland	63	31
Turkey	138	171
United Kingdom	2449	1887
Byelorussia	370	298
Ukraine	1925	1269
Moldavia	75	53
Russia	3581	2106
Estonia	81	57
Latvia	56	39
Lithuania	143	100
Slovenia	110	125
Croatia	183	208
Bosnia/Hercegovina	64	72
Yugoslavia	295	334
Remaining area	256	256
Int. Trade Baltic Sea	36	36
Int. Trade North Sea	87	87
Int. Trade rem. Atlantic	158	158
Int. Trade Med.	6	8
Int. Trade Black Sea	--	--
Biogenic Sea	362	362

scrubber in which $CaCO_3$ or CaO is sprayed out. The sorption of SO_2 occurs via the following reactions:

$$CaCO_3 + 2H_2O + SO_2 + 0.5O_2 \rightarrow CaSO_4{\cdot}2H_2O + CO_2$$

or

$$CaO + 2H_2O + SO_2 + 0.5O_2 \rightarrow CaSO_4{\cdot}2H_2O$$

The stages of this process can be written as follows:

dissolution of sorbent: $CaCO_3 \rightarrow Ca^{2+} + CO_3^{2-}$

$CaO + H_2O \rightarrow Ca(OH)_2$
$Ca(OH)_2 \rightarrow Ca^{2+} + 2OH^-$

sorption of SO_2: $SO_2 + H_2O \rightarrow H^+ + HSO_3^-$

$HSO_3^- \rightarrow H^+ + SO_3^{2-}$

oxidation: $SO_3^{2-} + 0.5O_2 \rightarrow SO_4^{2-}$

precipitation and crystallization:

$$Ca^{2+} + SO_4^{2-} + 2H_2O \rightarrow CaSO_4{\cdot}2H_2O$$

The gypsum/water suspension containing 25 wt.% of solid phase is dewatered to a moisture content of less than 10 wt.% The wet lime desulphurization process is shown in Figure 2. This type of process has been implemented for the Bełchatów power plant in Poland, the largest lignite-based power plant in Europe [53]. The chemical compositions of desulphogypsums originating from the coal and lignite are given in Table 7 and their physical properties are shown in Table 8, together with the data for high quality natural gypsum. The desulphogypsums are chemically pure, similar to the natural gypsums of high quality. The grain size composition of FGD gypsums differs significantly from that of the natural, ground gypsum. In the FGD gypsum the 20-60 μm fraction dominates (see also Figure 3).

The microstructure of desulphogypsums observed under SEM reveals the presence of well-developed 15-100 μm crystals while the ground natural gypsum grains appear as pieces of larger crystals, with fractured surfaces (Figure 4).

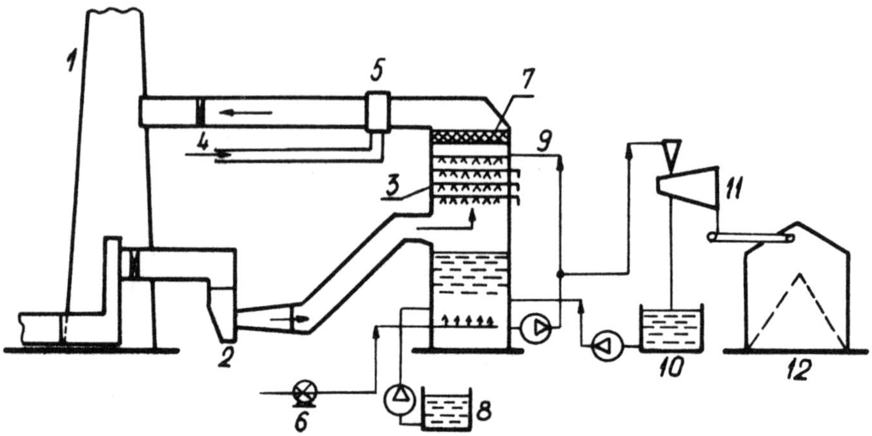

Figure 2. Schematic of wet lime process desulphurization [53]. 1) chimney; 2) fun; 3) absorber; 4) flue gas; 5) preheater; 6) oxidizing; 7) water drop eliminator; 8) sorbent; 9) recirculation of sorbent; 10) solution from dewatering of FGD gypsum; 11) dewatering of FGD gypsum; 12) FGD gypsum storage.

Table 7. Chemical Composition of FGD Gypsums.

Component	Chemical composition of gypsum material [%]		
	FGD gypsum from coal	FGD gypsum from lignite	Natural gypsum
L.O.I. (400°C)	20.66	20.52	19.50
SiO_2	0.44	0.88	2.87
Al_2O_3	0.25	0.30	1.05
Fe_2O_3	0.15	0.20	0.48
CaO	31.08	31.80	31.85
MgO	0.04	0.02	0.15
SO_3	45.42	45.24	43.43
K_2O	0.04	0.03	0.05
Na_2O	0.06	0.07	0.08
TiO_2	0.01	0.01	0.02
F^-	0.03	0.04	0.01
Cl^-	0.03	0.05	0.01
C	0.15	0.12	-
	Traces [ppm]		
Sr	147	235	2340
Cd	0.5	0.4	0.3
Pb	5.8	3.3	2.9
Cu	7.0	9.0	6.5
Zn	19.4	12.5	9.8
Hg	2.6	3.1	< 0.1
As	4.6	4.3	3.3
Co	0.5	< 0.1	0.5
Ni	2.5	3.5	5.2
Se	6.9	2.4	2.5
Sb	< 0.1	0.2	0.4
Cr	6.9	8.1	12.1
Mn	68.4	60.5	47.5
V	19.7	23.1	11.8
Sn	< 0.1	< 0.1	< 0.1
Be	< 0.1	< 0.1	< 0.1
pH	6.4	6.6	6.5

Table 8. Physical Properties of FGD Gypsum.

	Type of gypsum		
	FGD gypsum from coal	FGD gypsum from lignite	Natural gypsum
Fraction			
> 60 μm	6.4	17.2	12.5
60 - 20 μm	68.3	67.2	17.8
20 - 10 μm	17.5	10.1	20.4
10 - 5 μm	6.3	4.5	27.1
< 5 μm	1.5	1.0	22.2
Specific density [g/cm^3]	2.33	2.31	2.37
Moisture [%]	7.5	9.3	0.7
Whiteness [%]	78	55	81

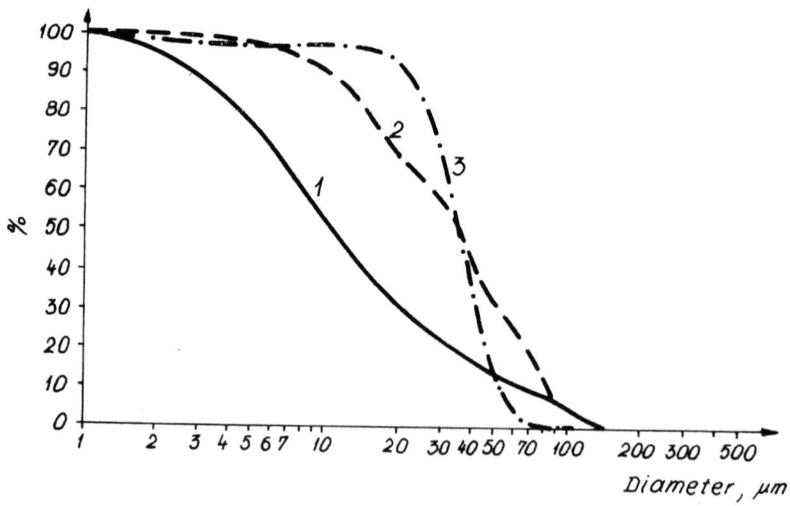

Figure 3. Grain size distribution curves; 1) ground natural gypsum; 2) FGD gypsum; 3) phosphogypsum.

(a)

(b)

Figure 4. Microstructure of desulphogypsum (a,b).

(c)

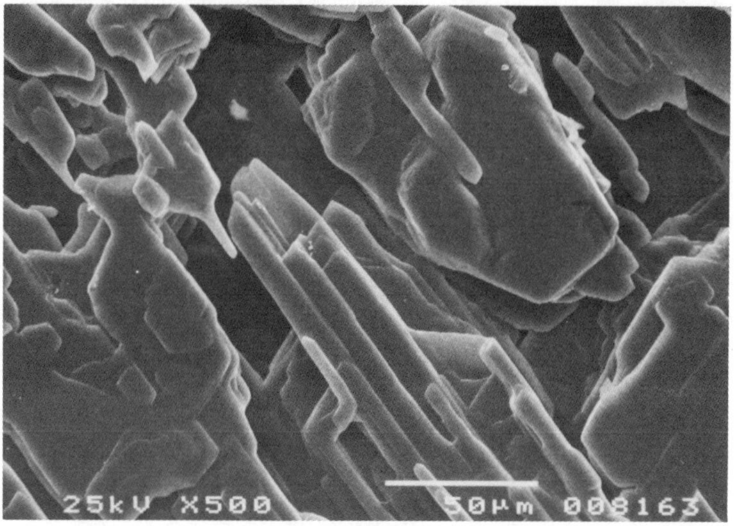

(d)

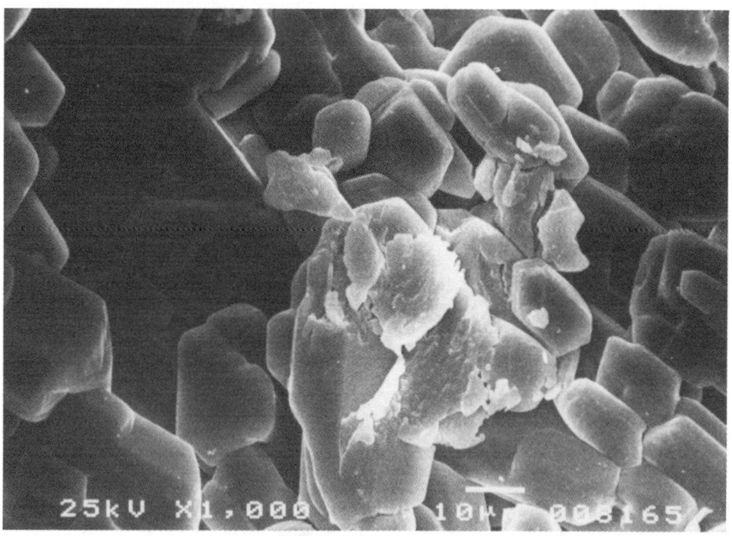

Figure 4. Continued, Natural gypsum (c,d).

The x-ray diffraction (XRD) studies show that desulphogypsum materials consist of the pure $CaSO_4 \cdot 2H_2O$ (the weak peaks attributed to $CaCO_3$ or SiO_2 can also be distinguished). The XRD patterns of desulphogypsum and natural gypsum are shown in Figure 5.

The radioactivity factors, determined for desulpho-gypsums and other materials following the standard polishing procedure (described earlier) are shown in Figure 6. It can be seen that the values obtained for the desulphogypsum materials are similar to those obtained for the natural gypsums and are significantly lower than those found for other building materials.

The chemical and physical characteristics of desulphogypsums derived from the wet lime method presented above were the basis for the FGD gypsum definition proposed by EUROGYPSUM (Working Community of the European Gypsum Industry) and have been accepted by ECOBA (European Association for the Utilization of Residual Materials from Coal-Fired Power Stations) and EURELECTRIC (The European Association of Electricity Power Suppliers) [55]:

"*Gypsum from the flue gas desulphurization plants (FGD gypsum, desulphogypsum) is moist, finely divided and crystalline calcium sulphate dihydrate ($CaSO_4 \cdot 2H_2O$) with a high level of purity. It is obtained deliberately during wet flue gas desulphurization using washing processes with lime(stone) after oxidation in air and subsequent processing of the gypsum.*"

This FGD gypsum meets all the EUROGYPSUM requirements given in Table 9 and cannot be considered as a waste but as a secondary raw material or by-product. Therefore it is not included in the European Waste Catalogue (EWC) nor in the OECD (Organization for Economic Cooperation and Development) Waste List [55-59]. The wet lime desulphurization method is most frequently applied in the countries devoting large financial expenditures to environmental protection [60] as shown in Figure 7, illustrating the scope of desulphurization operation through the consumption of sorbents in Japan, USA and Germany.

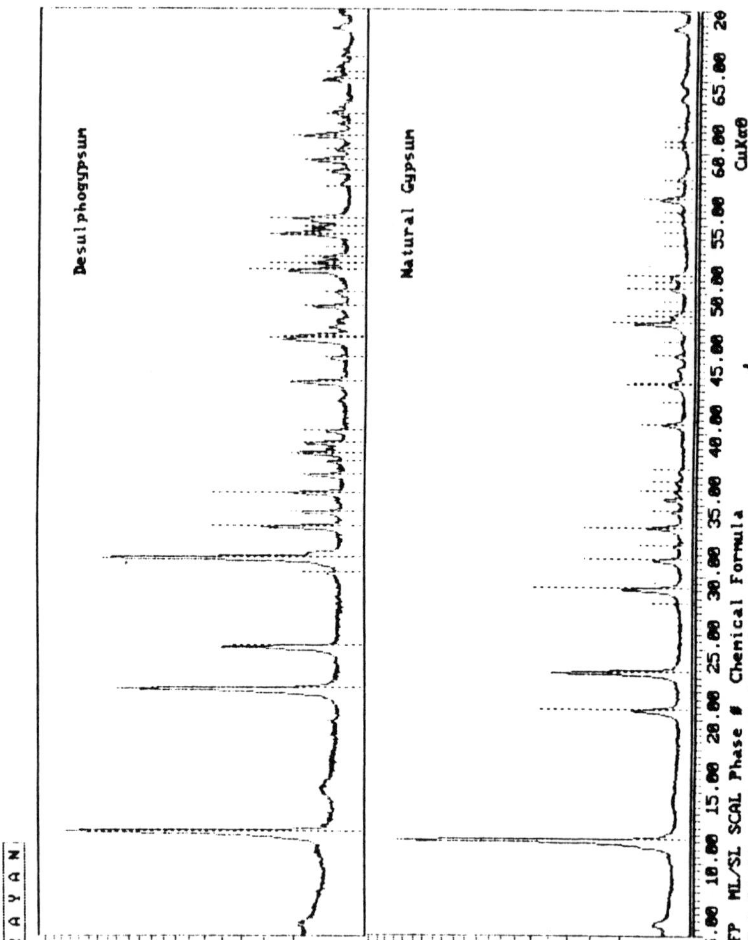

Figure 5. X-ray diffraction patterns of FGC gypsum (1) and natural gypsum (2).

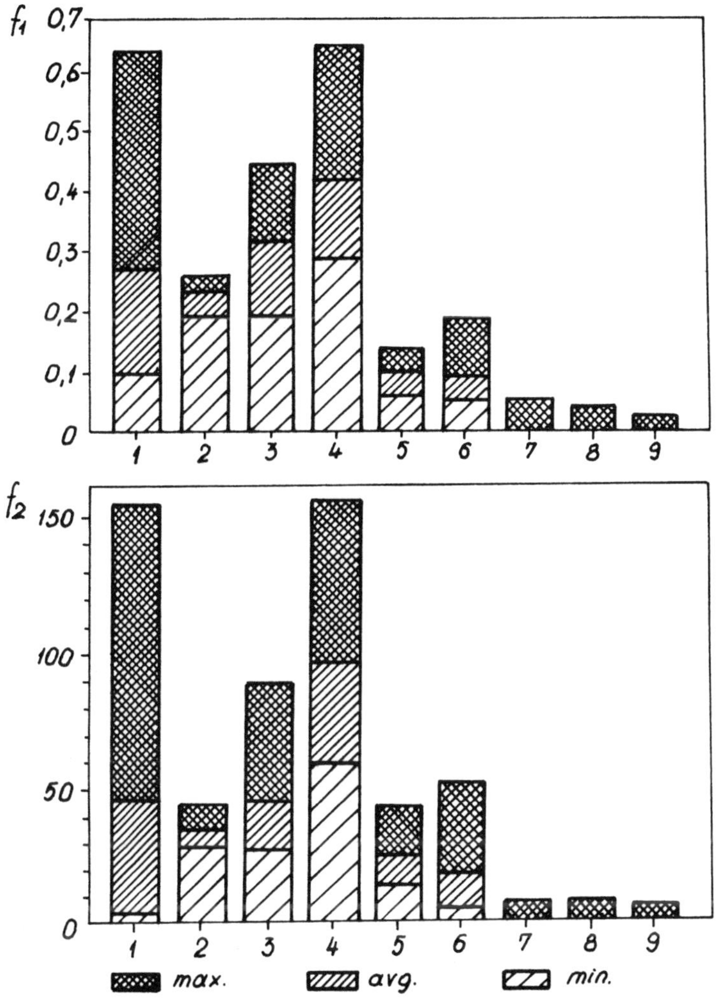

Figure 6. Radioactivity factors for gypsum materials and some building materials [54]. 1) cement, total; 2) cement 35, no admixtures; 3) cement 35 with admixtures; 4) metallurgical cement; 5) quick lime; 6) gypsum binder; 7) natural gypsum; 8,9) desulphogypsums.

Table 9. Quality Standards for FGD Gypsum [55-58].

Property	Requirement
Free moisture	< 10%
Gypsum content	> 95%
pH	5 - 8
Whiteness	> 80%
Odor	neutral
32 μm sieve residue	> 60%
Minor components	< 5%
Water soluble MgO	< 0.10%
Water soluble Na_2O	< 0.06%
Water soluble K_2O	< 0.06%
Water soluble Cl^-	< 100 ppm
$CaSO_3 \cdot 1/2H_2O$	< 0.50%
Organic components	< 0.10%
Al_2O_3	< 0.30%
Fe_2O_3	< 0.15%
SiO_2 - HCl insoluble	< 2.5%
$CaCO_3+MgCO_3$	< 1.5%
NH_3+NO_3	0

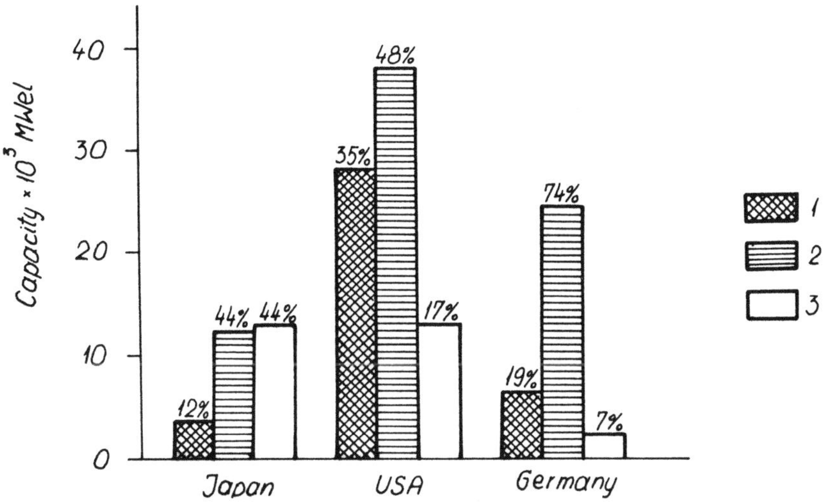

Figure 7. Capacities of FGD in different countries [61]. 1) SO_2 - scrubbing with CaO; 2) scrubbing with $CaCO_3$; 3) other FGD products.

Fluidized Bed Desulphurization Products

The fluidized bed combustion (FBC) installations are a proposed solution in which solid fuels with high sulphur content and low calorific value can be applied. The combustion and desulphurization operations are carried out simulatneously inside the furnace. Therefore, the expensive auxillary desulphurization equipment is not necessary.

High purity ground limestone, dolomite [62,63], and calcium hydroxide are used as sorbents [64]. The sorbent is supplied together with the pulverized coal to the fluidized bed, as shown Figure 8. In the presence of $CaCO_3$ used as a sorbent SO_2 is combined according to the following reaction:

$$CaCO_3 \rightarrow CaO + CO_2$$
$$CaO + SO_2 + 0.5O_2 \rightarrow CaSO_4$$

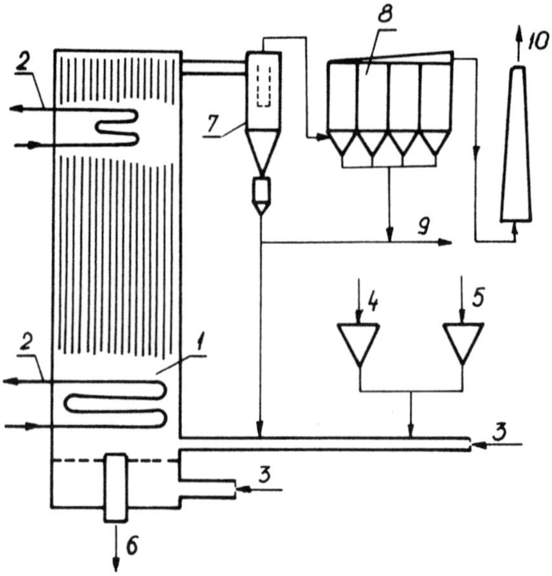

Figure 8. Schematic of FBC installation [65]. 1) fluidized bed; 2) steam generator; 3) air; 4) coal; 5) limestone; 6) bed ash; 7) cyclone; 8) electrofilters; 9) cyclone ash; 10) purified gas.

This reaction occurs on the surface of lime sorbent grains exposed to SO_2, and after the $CaSO_4$ coating formation, the diffusion of SO_2 and O_2 inside the grains is strongly retarded. Therefore, the properties of sorbent (specific surface area and porosity) are of special importance.

Two types of solid by-products are formed in the FBC installation:

- bed ash (BA) composed of 0.3-6 mm particles, constituting 30-60 wt.%,

- cyclone ash (CA), composed of 1-300 μm particles, constituting 40-70 wt.%.

Both by-products are a mixture of sulphate and sulphite phases, residual sorbent and fly ash. Their properties are influenced by the following parameters:

- type of solid fuel,
- type and percentage of unburned substance,
- sulphur content in the fuel,
- type of sorbent,
- characteristics of the combustion process.

The fluidized bed installation must be exploited in such a way so as to enable the formation of anhydrite. In practice, the bed ash and cyclone ash are of varying chemical compositions as illustrated by the data presented in Table 10. The typical crystalline phases present in both by-products are: anhydrite (desulphurization product), $CaCO_3$ (unreacted sorbent) and β-quartz (constituent of fly ash).

Products from Other Desulphurization Installations

Desulphurization by-products from lesser operations are of somewhat less importance as they are not as frequently applied and the materials obtained are of limited practical applicability. Among these, the by-products from the semi-dry lime desulphurization should be mentioned. This method consists of spraying a lime-containing solution or suspension in the absorption chamber to the stream of gases. The most frequently used solutions attributed to the semi-dry method are the "dry scrubbing" and "drypac" [69-71]. The chemical and phase composition of by-products from semi-dry lime desulphurization are presented in Tables 11 and 12. These data indicate that the by-products are composed of sulphate material, fly ash and residual sorbent.

Similar components are present in the by-products originating from the dry lime method and those from the bi-alkaline method. In the former, the limestone, dolomite, quick lime and hydrated lime of high specific surface area are used as sorbents [72,73]. In the latter -

Table 10. Chemical Composition of Dedusting and Desulphurization Products from Fluidized Bed Combustion [65-68].

By-product		Component [weight %]							
		L.O.I.	SiO_2	Fe_2O_3	Al_2O_3	CaO_t	CaO_f	MgO	SO_3
Hard coal installation - [65]	BA	12.0	40.6	2.8	4.9	29.2	9.7	0.1	10.3
	CA	24.5	13.5	11.2	6.2	29.9	8.8	0.5	14.1
Lignite installation - [65]	BA	5.1	19.9	4.3	5.8	44.2	15.3	0.6	17.7
	CA	24.2	12.0	12.0	6.0	30.7	9.9	0.4	14.1
Bituminous shale installation - [66]	BA	12.0	6.1	1.2	1.0	53.8			25.5
	CA	16.8	12.3	9.5	4.0	41.4			15.2
Different coal installation - [67]	BA + CA	0.8-18	17-50	5-15	15-25	6-29	0.5-9	1-3	0.1-19
Coal instal. - [68]	BA + CA	12-18	33-48	5-10	13-22	8-18		1-2	3-12
Lignite instal. - [68]	BA + CA	14-27	10-25	2-10	2-6	11-40		2-9	10-12

Table 11. Chemical Composition of Flue Gas Desulphurization Products [72,76,77].

Type of FGD method	Component [weight %]							
	L.O.I.	SiO_2	Fe_2O_3	Al_2O_3	CaO	MgO	SO_3	SO_2
Semi-dry lime method "dry scrubbing"								
- desulphurization of dedusted flue gas	27.8	7.4	1.3	5.0	36.8	1.2	4.0	16.1
- desulphurization of flue gas with fly ash	4.4	25.4	4.2	14.2	43.4	1.0	1.3	5.9
Dry lime method								
- desulphurization of flue gas with fly ash	1.2	37.7	12.4	17.6	22.6	2.1	0.9	5.0
Bi-alkaline method								
- desulphurization of dedusted flue gas	26.0	5.9	1.1	2.4	41.5	2.1	3.8	16.7

Table 12. Phase Composition of Flue Gas Desulphurization Products [72,76,77].

Type of FGD method	Component [weight %]							
	$CaSO_3 \cdot 1/2H_2O$	$CaSO_4 \cdot 2H_2O$	$CaSO_3$	$CaSO_4$	$Ca(OH)_2$	$CaCO_3$	$CaCl_2$	FA
Semi-dry lime method "dry scrubbing"								
- desulphurization of dedusted flue gas	49	13	-	-	8	3	-	24
- desulphurization of flue gas with fly ash	13	3	-	-	-	7	-	75
Dry lime method								
- desulphurization of dedusted flue gas	-	-	61	12	12	6	4	4
Bi-alkaline method								
- desulphurization of dedusted flue gas	40	8	-	10	15	15	-	10

SO_2 is sorbed by Na_2CO_3 or NaOH solution [74,75]. The chemical and phase compositions of by-products are given also in Tables 11 and 12.

Utilization of the Flue Gas Desulphurization By-Products - General Remarks

A variety of by-product gypsum materials are generated in different installations. The wet-lime desulphurization method provides a material of high purity, meeting all the standard quality requirements. This desulphogypsum, being practically "pure" $CaSO_4 \cdot 2H_2O$ can be easily processed to standard gypsum binders and materials [78,79]. Excavation of natural gypsum can be reduced and this fact is also important from the environmental point of view. However, other desulphurization by-products are very troublesome in most cases, because of their composition and properties. Some of them cannot be refined or enriched. The utilization of mixtures composed of fly ash, different sulphate and sulphite phases and sorbent is possible only on a limited scale. Therefore, special attention must be paid to the utilization of desulphogypsum processed to other standard gypsum materials. However, the total annual output of FGD gypsum by the wet-lime method continues to increase (up to several million tons) as shown in Figure 9 and Table 13, in which the data dealing with FGD output in Germany and FGD gypsum production predicted for 1996 for several countries are shown. The utilization of FGD gypsum in the gypsum plants built near power stations has become an insufficient solution and the new exploition strategies for FGD gypsum are required [55].

Studies on desulphogypsum utilization in other areas such as mining and road engineering should result in the solutions dealing with the FGD gypsum disposal on a larger scale.

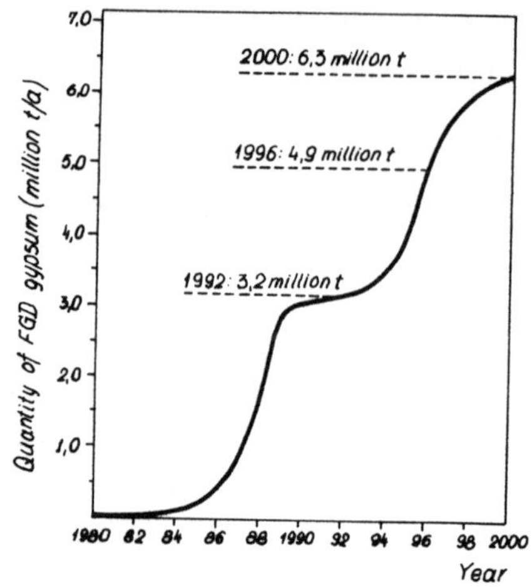

Figure 9. Quantities of FGD gypsum from coal-fired and lignite-fired power station in Germany 1980-2000 [55].

Table 13. Predicted Quantities of FGD Gypsum from Coal-Fired and Lignite-Fired Power Stations in 1996 [55].

Country	Quantity of FGD gypsum [t]
Germany	4 920 000
Denmark	340 000
Finland	190 000
Great Britain	1 200 000
Italy	40 000
Netherlands	360 000
Austria	100 000
Poland	770 000
Russia	60 000
Czechoslovakia	40 000
Ukraine	85 000
Total for Europe	8 225 000
USA and Canada	3 000 000
Japan and Taiwan	2 300 000

Based on the available data, the general outlook indicating the possibilities of FGD gypsums and other desulphurization by-products application was plotted in Figure 10. Apart from the commonly known traditional technologies, future solutions are also highlighted.

The wet desulphogypsum, with a moisture content in the range 6-10 wt.% must be refined in such a way that it can be substituted for natural gypsum. Following the studies on drying procedures, primary attention was paid to the briquetting operation as a most promising agglomeration method. Agglomeration allows for transportation and storage of material which would be otherwise impossible. The following requirements were determined for FGD gypsum briquetting operations:

- processing without binders or other additives,
- high efficiency at high output,
- possibility to produce briquettes of various size and density.

These requirements are fulfilled by use of roll presses [80,87] or edge-runners [82,83], which are schematically shown in Figures 11 and 12. In Table 14 the properties of FGD gypsum briquettes are given. The moist FGD gypsum compacted by roll-press must be dried before pressing. In turn, the material pressed in edge-runners is dried after briquetting.

The technology of FGD product granulation to produce lightweight aggregates for structural concrete and masonry units has been developed in the United States [84]. This aggregate meets the ASTM C-90 and ASTM C-332 standards.

As shown in Figure 10, the desulphogypsums produced from installations working at lignite-fired power stations reveal a limited applicability as raw materials in binders and gypsum elements manufacturing. In spite of the relatively high purity, exceeding 95%, the calcium sulphate dihydrates originating from lignite are not sufficiently white. The dark color (degree of whiteness between 40 and 70%) is caused by impurities incorporated to the gypsum

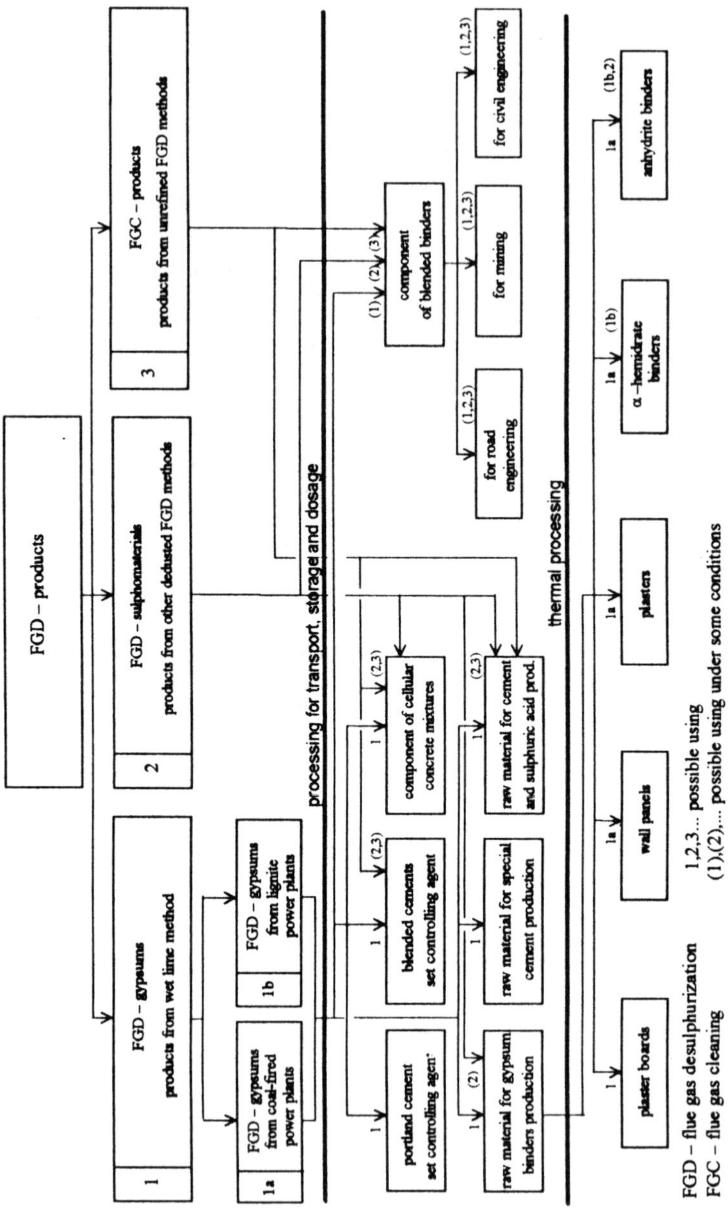

Figure 10. Flowchart showing utilization of gas desulphurization products.

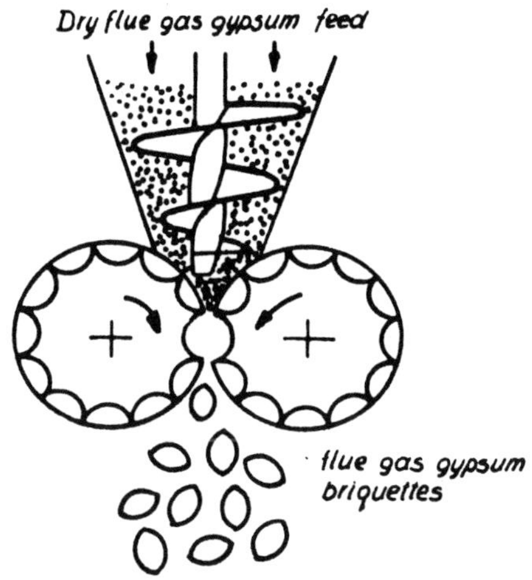

Figure 11. Diagram of compaction of FGD gypsum with roller press [82].

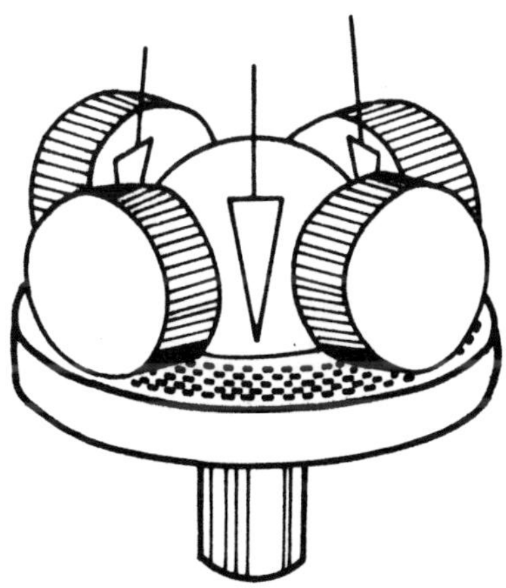

Figure 12. Diagram of compaction of FGD gypsum with edge-runner extruding press [82].

Table 14. Physical Characteristics of Briquettes from Different Briquetting Installations [80,83].

	Installation	
	Roller press	Edge-runner press
Material	dried FGD gypsum	FGD Gypsum with moisture content 8-10% (max. 15%)
Energy consumption	ca. 10 KW/t	ca. 25 KW/t
	Properties of briquettes	
Shape	sticks	cylindrical
Dimensions:		
diameter	21 × 14 mm	10 - 20 mm
length	15 - 60 mm	10 - 30 mm
Specific density	> 2.1 g/cm³	1.9 - 2.0 g/cm³
Bulk density	1.0 - 1.2 t/m³	1.05 - 1.3 t/m³
Sieve residue		
4 mm		< 3 wt. %
1 mm	< 5 wt. %	
Point compressive strength		
in wet state	> 500 N	> 40 N
in dry state	> 500 N	> 200 N

structure, primarily by iron compounds. This is because the desulphogypsum from lignite power stations is usually washed for longer times than the by-product from hard coal and it is not only subjected to attack by impurities but also recrystallization. Thus, the formation of large gypsum-contaminated crystals takes place. The microstructures of both desulphogypsums are shown in Figure 13. In Figure 14, the FGD gypsum crystals from the lignite combustion power plant with a coloring inclusion is presented. In Figure 15, the EDX spectra for this inclusion and for surrounding calcium sulphate dihydrate are plotted.

(1)

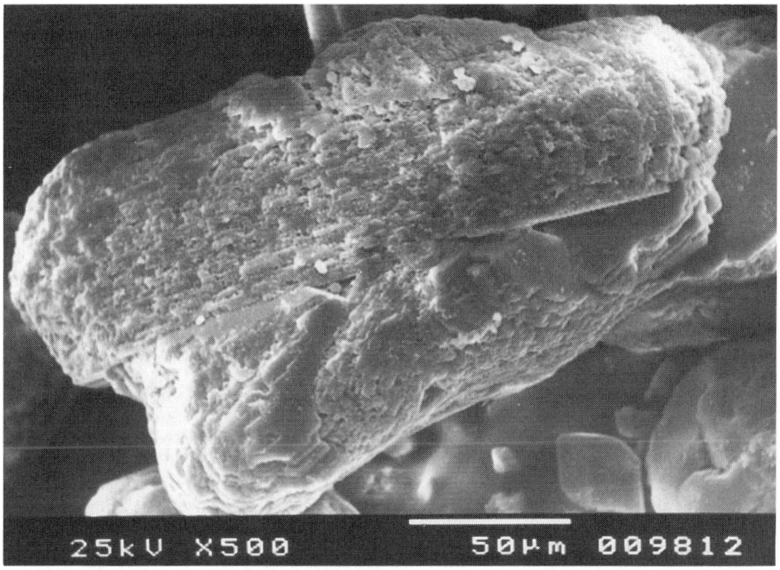

(2)

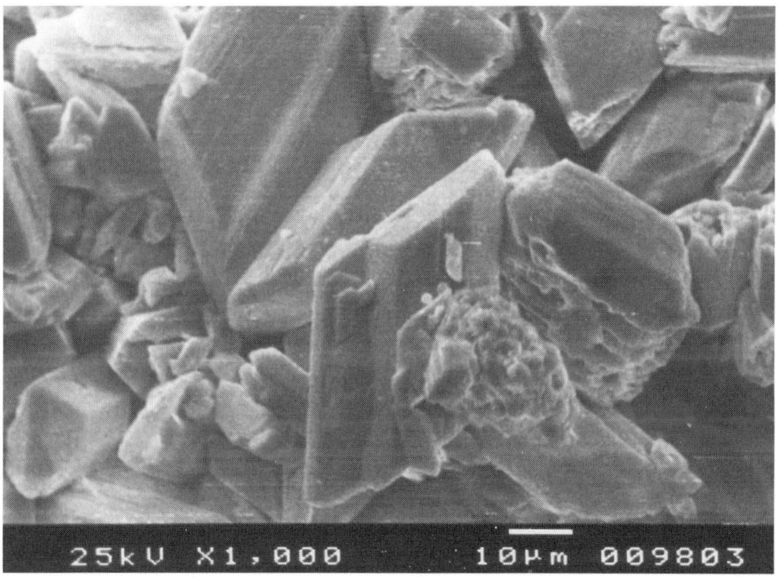

Figure 13. SEM micrograph of FGD gypsum. 1) FGD gypsum from lignite-fired power plant; 2) FGD gypsum from coal-fired power plant.

Figure 14. SEM micrograph of FGD gypsum from lignite-fired power plant Belchatów with a coloring inclusion (χ).

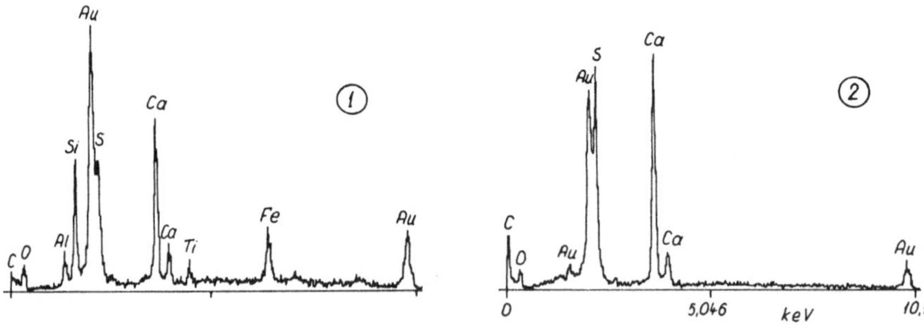

Figure 15. EDX plots for FGD gypsum Belchatów; 1) composition of inclusions (Ca, S, Si, Fe, Ti); 2) composition of gypsum crystal.

The gypsum of low whiteness can only be utilized on a limited scale due to customer preferences and requirements from building materials producers. Hence, this gypsum is not used in plaster, wall panels or the other special elevation materials manufacture but it can be converted into α-hemihydrate and used in mining, for plasterboard and self-levelling floors.

The overflow cleaning method has been used to reduce the coloring admixtures [55]. A flow chart of this method is shown in Figure 16. The purification of gypsum is carried out in hydrocyclone; the impurities do not go back to the scrubber but are separated as a sludge in a thickener of high efficiency. The impurity level is reduced by approximately 70% . The results can be seen in Table 15. Overflow cleaning installations could improve the usage of FGD gypsums on a large scale.

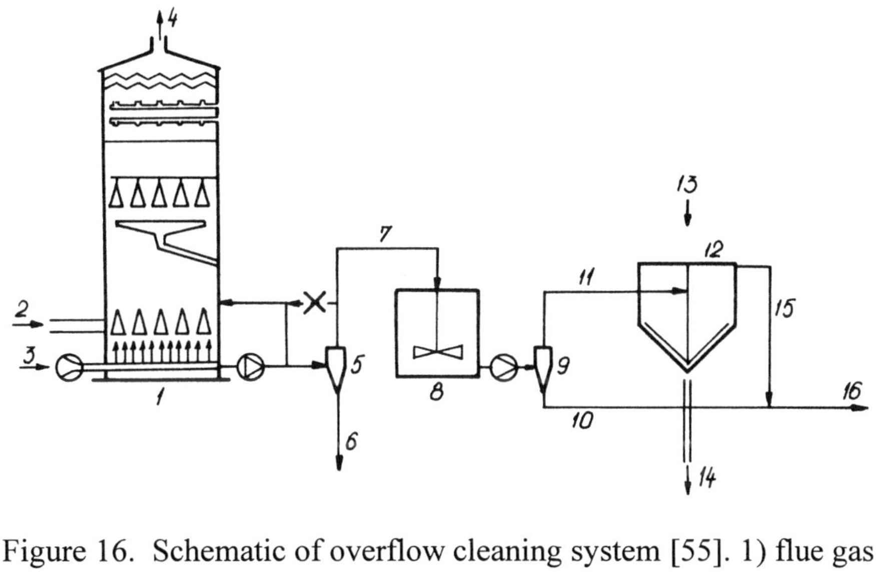

Figure 16. Schematic of overflow cleaning system [55]. 1) flue gas scrubber; 2) flue gas; 3) air; 4) clean gas; 5) hydrocyclone; 6) FGD gypsum to filter; 7) overflow from flue gas scrubber; 8) container; 9) hydrocyclone; 10) underflow; 11) overflow; 12) thickner; 13) flocculent; 14) sludge for moistening ash; 15) clear flow; 16) return to gas scrubber.

Table 15. Improvement in Quality of Lignite FGD Gypsum by Overflow Cleaning [55].

Quality parameter	With overflow cleaning [%]	Without overflow cleaning [%]
Free moisture	9.5	10
Calcium sulphate-dihydrate	97.3	96.3
Inert substances	0.44	1.45
Iron III oxide	0.04	0.18
Aluminum oxide	0.02	0.12
Carbon	0.04	0.29
Colour (standard colour value)	65	33

Methods for FGD by-products utilization in the production of binders and other building materials are discussed in subsequent sections. First, the technologies dealing with the FGD raw by-products, in which FGD materials are not processed by thermal treatment are presented. The FGD products may be only dried, briquetted or granulated for transportation, storage and dosage, if required. Gypsum binders and building materials produced from burned FGD products will also be described.

Cementitious and Building Materials Produced With Flue Gas Desulphurization By-Products and Without Thermal Treatment

FGD Gypsums as Cement Set Controlling Admixtures. The FGD products can be used as a set controlling admixture to cement and is an important means of recycling materials. The high purity FGD gypsums successfully substitute for the natural gypsum material in cement and its applicability has been fully confirmed in studies on cement paste hydration and other critical properties [85-88]. In Table 16, the results of strength tests and setting time determination are given for cements prepared using natural gypsum, natural anhydrite and desulphogypsum at 3% SO_3. The differences between the particular cements, based on the same clinker, are negligible. However, the heat evolution curves practically interfere (see Figure 17). One can conclude that the FGD gypsum is neither a better or worse set controlling agent than the natural gypsum or anhydrite. The amount of sulphate component should be carefully calculated in relation to the tricalcium aluminate content in cement clinker. This is important from the point of view of hydration kinetics, strength development and durability of hardened material [89,90]. The grinding of cement should be carried out in such a way as to avoid dehydration of gypsum false setting [91]. In most cases, the desulphogypsum is supplied to cement plants as briquettes, however it is possible to supply wet FGD gypsum as well [92].

The other desulphurization by-products, particularly those containing fly ash may eventually be used in blended cement production as an admixture to pozzolanic or metallurgical cements. However, additional research will be required prior to implementation with special attention given to the chemical stability of the composition.

FGD products as a component of blended binders. The cementitious mixtures composed of different materials, including the waste and by-product gypsum have been the subject of numerous projects. The binders for particular purposes can be produced by proper selection of components and mix proportions.

The following types of blends have been developed in the programs dealing with the utilization of harmful wastes: sulphate-fly ash-cement, sulphate-slag-cement, sulphate-slag materials. The wastes from the flue gas desulphurization process can be

Table 16. The Properties of Cements with Different Set Controlling Agents (3% SO_3)

	OPC clinker + FGD gypsum	OPC clinker + natural gypsum	OPC clinker + natural anhydrite
Initial setting [h^{min}]	1^{34}	1^{54}	1^{17}
Final setting [h^{min}]	2^{33}	2^{54}	2^{11}
Flexural strength [MPa] at age:			
- 1 day	2.5	2.6	2.8
- 3 days	4.9	5.1	4.9
- 7 days	6.0	6.5	5.9
- 28 days	8.8	8,6	8.4
- 60 days	8.9	8.6	8.5
Compressive strength [MPa] at age:			
- 1 day	8.4	8.6	9.6
- 3 days	19.8	18.8	19.3
- 7 days	22.5	20.6	24.0
- 28 days	38.9	37.4	37.7
- 60 days	39.2	38.6	39.0

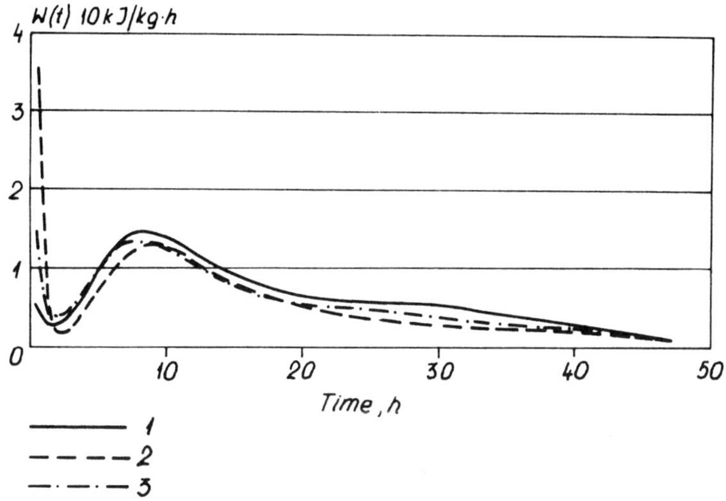

Figure 17. Microcalorimetric curves for cement with: 1) FGD gypsum; 2) natural gypsum; 3) anhydrite (3 wt% SO_3).

successfully used as a sulphate component. This possibility is of special importance in the production of sulphate -fly ash-cement binders. First of all, the wastes produced without dust separation can be used. Also in this case, all types of by-product gypsum, i.e. those from the wet, dry or semi-dry method can be applied. It has been shown [93] that the sulphate-fly ash-cement materials produced at approximately equal weight proportions of particular components reveal the best mechanical and physical properties. The characteristics of blended binders produced with FGD gypsum and natural ground anhydrite used as reference are presented in Table 17 and Figure 18. The data for ordinary Portland cement used as a component of blends are also reported. The properties of FGD gypsum were given in a previous section. The blended FGD gypsum containing material presented here shows rather slow strength development. This strength development has been observed during a 2-year storage in water. The strength values are slightly lower than those for the samples stored in air at 100% RH. The FGD gypsum containing materials reveal lower

Table 17. The Properties of Sulphate - Fly Ash - Cement Binders.

Property		Composition of binder		
		35% FGD gypsum 30% Fly ash 30% OPC	35% natural anhydrite 30% Fly ash 30% OPC	100% OPC (reference)
Flexural strength [MPa] - Water curing, 20°C at age:				
3	days	0.6	1.0	3.7
7	days	0.9	1.5	5.5
28	days	1.5	3.3	9.6
104	days	2.5	5.0	9.6
700	days	3.6	5.8	9.8
Compressive strength [MPa] - Water curing, 20°C at age:				
3	days	2.3	3.6	13.0
7	days	2.7	4.9	21.2
28	days	5.7	12.7	44.9
104	days	10.4	20.7	48.1
700	days	22.3	41.6	50.2
Water content [%]		23	22	27
Initial setting [h^{min}]		4^{50}	2^{15}	3^{00}
Final setting [h^{min}]		7^{20}	3^{50}	4^{40}

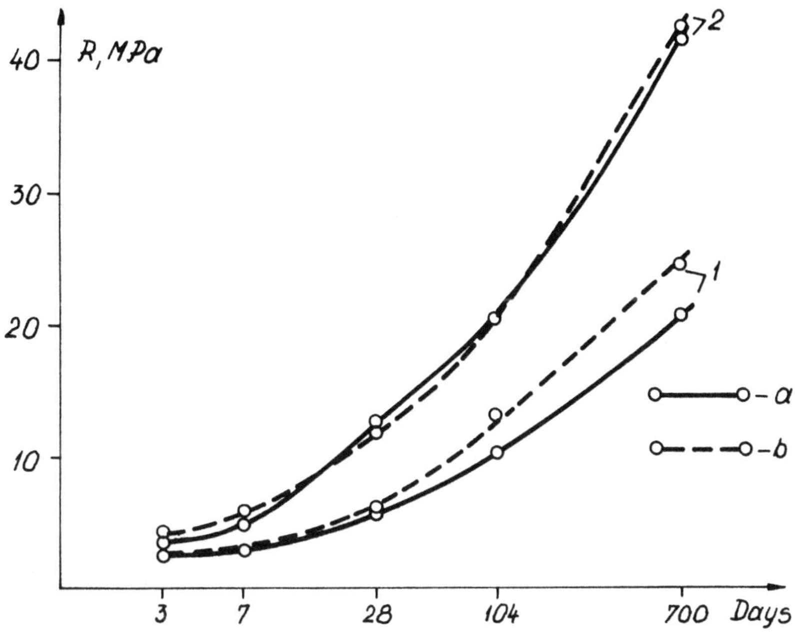

Figure 18. Compressive strength of sulphate-fly ash cement mortars. 1) 35% FGD gypsum- 35% fly ash-35% OPC; 2) 35% anhydrite-30% fly ash-35% OPC; a) cured in water; b) cured in air at 100% RH.

strength than the reference with the same anhydrite content. These differences result from the mechanisms and kinetics of the hydration process. For FGD gypsum, hydration and hardening of the mixture proceeds as follows: calcium silicates from Portland cement react with water to form a calcium silicate hydrate phase, which has an affect on the mechanical properties. Tricalcium aluminate from Portland cement clinker ($3CaO{\cdot}Al_2O_3$) reacts with the sulphate component ($CaSO_4{\cdot}2H_2O$) to form ettringite ($3CaO{\cdot}Al_2O_3{\cdot}3CaSO_4{\cdot}30\text{-}32H_2O$). In the presence of increased sulphate content in the blended binder, continuous formation of ettringite takes place from the early stages of hydration until the excess $CaSO_4 \cdot 2H_2O$ reacts with $3CaO \cdot Al_2O_3$ in the hardened paste. Generally, this would result in a collapse of the hardened structure because of the substantial volume increase accompanying ettringite formation, but the presence of fly ash acts as a reaction modifier. The active, pozzolanic constituents of fly ash (SiO_2 active and Al_2O_3 active) reduce the calcium hydroxide concentration in the liquid phase, as a consequence of the pozzolanic reaction. Hence, the solubility of calcium aluminate increases and the large crystals of ettringite do not form on the aluminate grains. First, small ettringite particles precipitate from the pore solution into the pores, thus tightening and reinforcing the paste microstructure. This classical theory proposed by Lafume [94] has been confirmed in numerous large crystals studies[95]. The results of SEM examinations on 2-year old materials indicate the formation of finely dispersed ettringite (see Figure 19). It should be noted that the strong structure of sulphate-fly ash-cement materials is formed as a result of cement-fly ash pozzolanic reaction products (calcium silicate hydrates and hydrated calcium aluminosilicates). In the presence of natural anhydrite, the hydration of this component in the alkaline environment occurs, producing gypsum, irrespective of the hydration and hardening of the other components. The additional hydration product formed improves the bonds in the hardened material and brings about a significant strength increase when compared to the FGD gypsum containing material. Therefore, one can conclude that the anhydrite containing FGD material should be more efficient as a sulphate component in blended, sulphate-fly ash-cement materials.

These sulphate-fly ash-cement binders, like the sulphate-slag-cement or sulphate-slag materials can be successfully used in road making as a stabilizing agent and in road foundation structures. The utilization of blended binders in road building has been the subject of several patents and reports [96-101]. Anhydrite has been used until recently as the sulphate component in sulphate-fly ash-cement binders [96-99]. The results obtained for the FGD gypsum containing binders

(1)

(2)

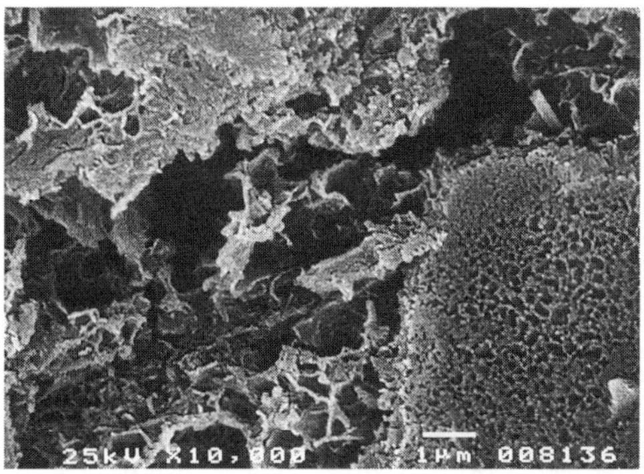

(3)

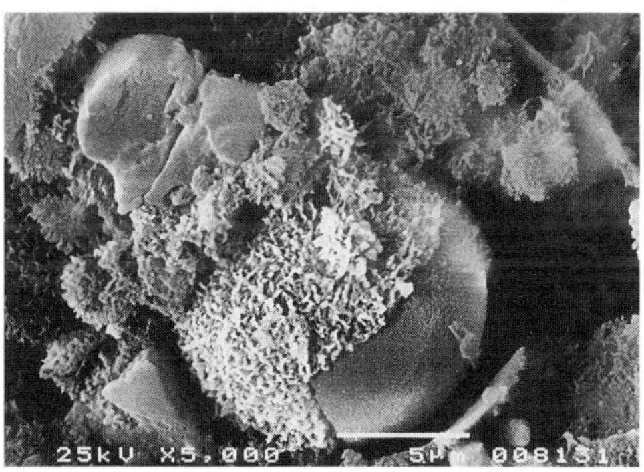

Figure 19. SEM micrographs of slag fly ash paste (cured 2 yrs). 1) ettringite; 2) CSH; 3) fly ash grain with hydration products.

have been reviewed while investigating materials for soil stabilization [102]. Experiments consisted of preparation of soil-binder samples, compressive strength measurements and freeze-thaw resistance tests. The sand-gravel mix was taken as "soil." The binder was composed of 35% FGD wet by-product, 30% fly ash, 35% Portland cement. The results are shown in Figures 20 and 21. As a reference, the soil-Portland cement mixture was used. The strength of mixtures containing blended binders as cement replacement was reduced within the first 28 days of curing when compared with the reference. However, at later aging times, significant strength increases were observed.

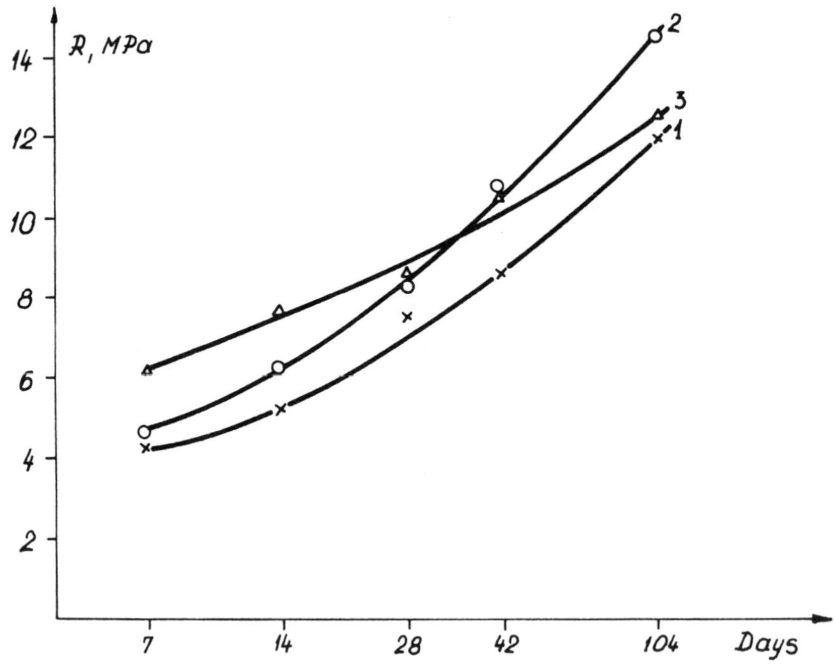

Figure 20. Compressive strength of soil-binder blends cured at 100% RH, 20°C. Proportions between soil and binder: 1) 100:12; 2) 100:14.3; 3) 100:5 (cement).

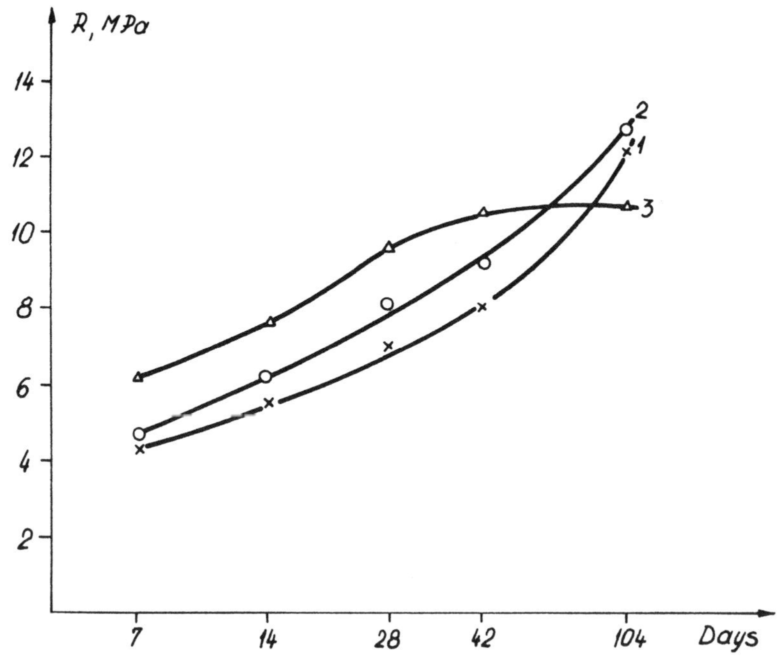

Figure 21. Compressive strength of soil-binder blends cured 14 days at 100% RH: subsequently stored in water, 20 °C. The same blends as those shown in Figure 20.

The sulphate-slag binder with the sulphate component produced in the FGD alkaline process can also be used in soil stabilization. Due to the alkaline residual components present in this material, the blast furnace slag is well activated and cement additions were not necessary. The soil-binder mixtures with the sand-gravel component acting as a substitute for the soil and sulphate-slag, blend as a cementitious agent (30% by-product gypsum from the alkaline FGD process and 70% granulated blast furnace slag) reveal a low strength increase within the first two weeks and a substantial strength increase at later times (see Figure 22).

The sulphate-slag binder presented above can be successfully used in the autoclaved cellular concrete manufacture [77]. The best properties of concrete elements have been achieved for the following proportions for 1m^3 of concrete:

312 kg	-	sulphate slag binder
308 kg	-	fly ash
310 kg	-	water
0.38 kg	-	aluminium powder as a pore forming agent
0.62 kg	-	superplasticizer

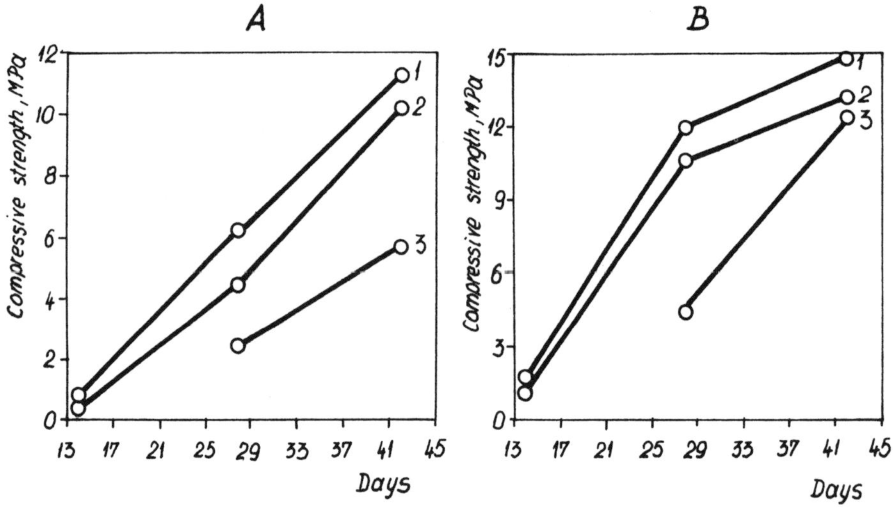

Figure 22. Compressive strength and freeze-thaw resistance of soil-binder blends. A) soil: sulphate slag binder = 100:10; B) soil: sulphate slag binder = 100:12; 1) stored in air (100%RH, 20°C); 2) stored in air (100%RH, 20°C) for 13 days, then in water; 3) stored in air (100%RH, 20°C) and subsequently 14 cycles of freeze/thaw.

The FGD products containing blended cement materials can be widely used in mining. Good results have been obtained using sulphate-containing wastes from desulphurization of unrefined flue gas, so-called flue gas cleaning (FGC) products [103]. The mixtures for mining works, depending on their use, contain 25-50wt.% Portland cement, 10-15% α-$CaSO_4 \cdot 0.5H_2O$ [67] positive results were achieved with 10-15% addition of phosphogypsum briquettes as a filler [104]. The FGC product-based blended materials mixed with finely grained fillers, such as fly ash, ground stone, post-flotation wastes, were transferred to underground excavations to form load-bearing structures, excavation fillers and reconsolidation works. The various applications for these materials in underground mining are schematically presented in Figure 23. In the case of a FGD installation working in conjunction with a lignite-fired power plant, it is possible to produce a mixture of fly ash and FGD gypsum with a relatively high free-CaO content. This mixture reveals hydraulic properties and can be successfully used in sealing and filling of excavations within the lignite mine supplying the fuel [105].

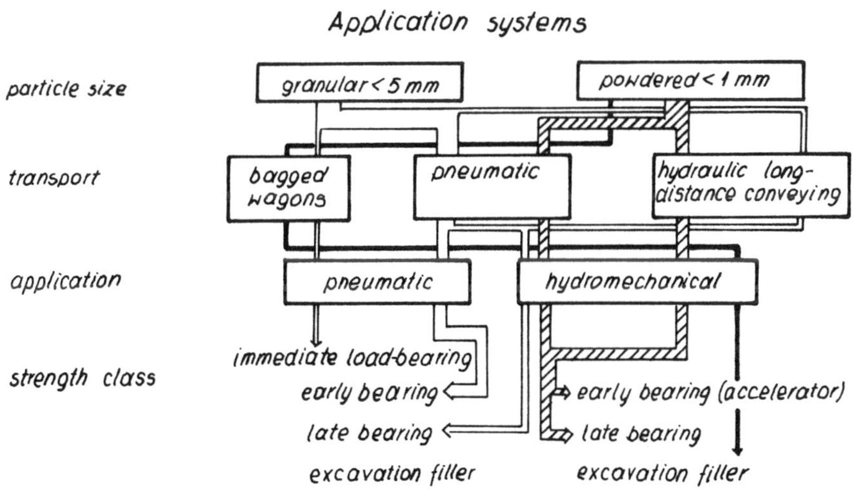

Figure 23. Use of construction materials in underground mining [67].

Desulphogypsums can substitute for natural gypsum in the slag-sulphate-cement materials used for injections in structures exposed to sulphate corrosion. The calcium sulphate dihydrate acts as a slag activator to form the ettringite as a result of reaction with Al_2O_3 from slag. The most efficient injection pastes were produced from 80-85% granulated blast furnace slag of the Blaine specific surface 5000-6000 cm^2/g, 10-15% FGD gypsum, 5-10% Portland cement [106].

Binders and Building Materials from Dehydrated FGD Gypsum

The many different ways flue gas desulphurization products could be utilized presented in the previous section indicate the possibility to reuse both chemically "pure" desulphogypsums as well as FGC-products being incorporated with mixtures of fly ash, sulphate components and residual sorbents. In the proposed uses, the purity of desulphogypsums was not critically important. It is commonly known that fly ash does not occur in desulphogypsums but is a constituent of FGC-products. Therefore, the production of gypsum binders and materials is the most effective method of desulphogypsum utilization. Their applicability in gypsum technology results in high purity and consequently, stability of chemical and physical properties attributed to the one-phase material. The desulphogypsums usually show better quality than the natural raw gypsum and additionally show a finely divided form and do not require any crushing or grinding. On the other hand, processing of desulphogypsums to binders and building materials requires serious investments and marketing of final products. Consequently, the choice of desulphogypsum processing technology should be carried out with the demand for the final product in mind.

There is a mutual relationship between the thermal treatment and properties of gypsum products. The phase transformations occurring on heating are given in Figure 24. In practice, the phase transitions in the $CaSO_4$-H_2O system are strongly influenced by several parameters and conditions in which the process is carried out.

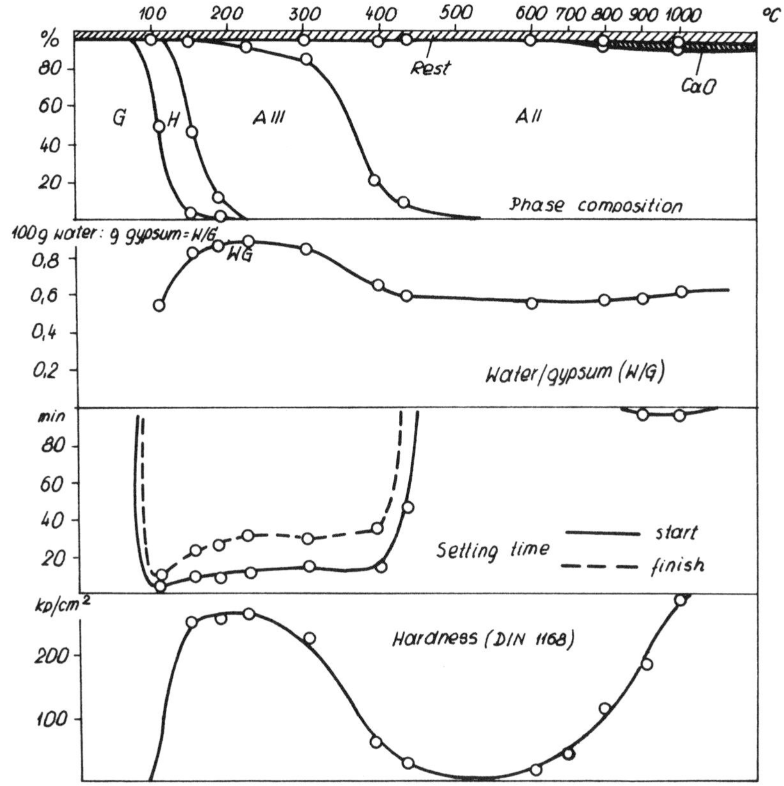

Figure 24. Phase composition and properties of gypsum binder as a function of burning temperature [107].

Among them, the type and quality of particular gypsums are important. There are some controversial data and opinions stating that the desulphogypsum dehydration proceeds slowly as compared to the reaction occurring in natural gypsum and that the dehydration degree is lower [108]. However, other results (see Figure 25) do not confirm this opinion [109]. The degree of calcium sulphate dihydrate dehydration isa function of grain size as it has been reported for several gypsums from different sources [110]. This is illustrated in Figure 26. The energy consumption increases during the FGD

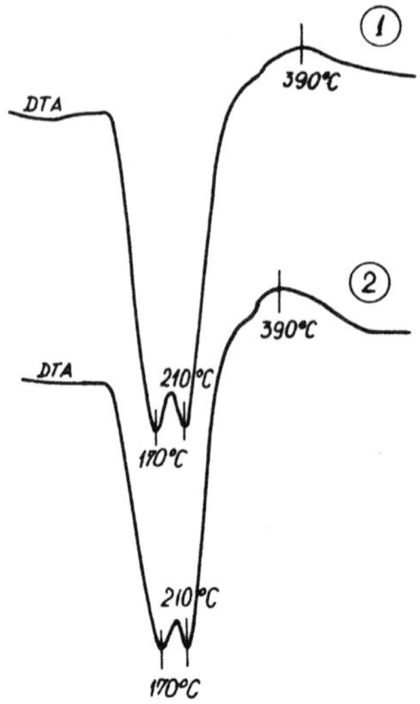

Figure 25. DTA curves of FGD gypsum (1) and natural gypsum (2). Heating rate 10°C/min.

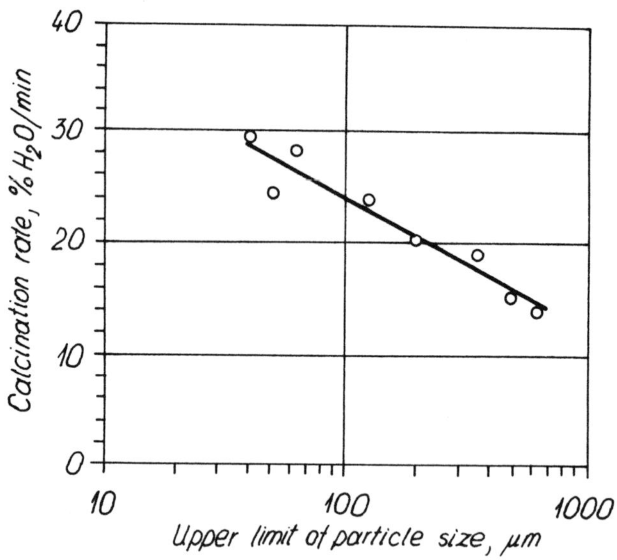

Figure 26. Calcination rate for different particle sizes of gypsum (calcination temperature = 127°C).

gypsum thermal processing results from a relatively high moisture content (6-10%). The simplified heat balance for dihydrate drying and calcination is shown in Figure 27. The total heat consumption per 1 ton of desulphogypsum is 1130 MJ while the heat consumption per 1 ton of raw natural gypsum is 800 MJ [111] and these values have been confirmed in several published works, though the differences, resulting from the variety of methods are acceptable [112-114].

There are several processing technologies leading to gypsum binder manufacture. In the most frequently used process, the following methods are used: calcination kettle, rotary kilns, travelling grate kilns, autoclaves.

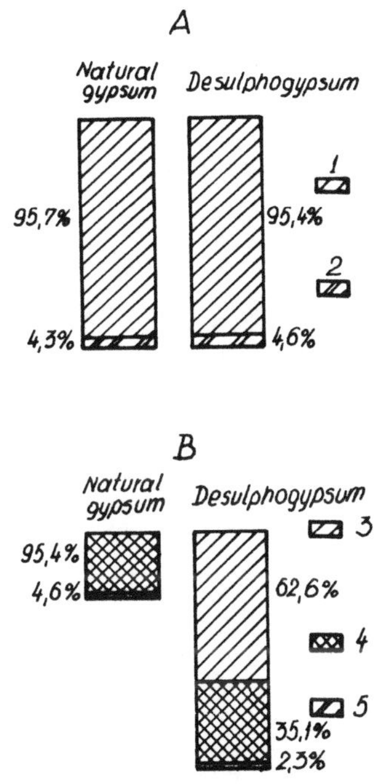

Figure 27. Simplified heat balance during calcination and drying of gypsum [111]. A) heat consumption on dehydration to β-hemihydrate; 1) heat consumed on calcination; 2) heat from air heated in the cooler. B) heat consumption on drying; 3) heat from hot technological gases; 4) heat of flue gas from furnace; 5) secondary air heat.

The β-hemihydrate with good uniformity and reactivity is produced in calcinators [115]. In Figure 28, a SEM micrograph of β-hemihydrate from desulphogypsum is shown. This material is used primarily in the production of plasterboards, the most common gypsum material used worldwide. The hemihydrate of lower whiteness, from the FGD gypsum produced in a lignite-fired power plant, can also be used for this purpose. The hemihydrate from calcinators can be further directed to wall panels and other gypsum elements producing plants.

In rotary kilns, the burning of β-hemihydrate is accomplished. Rotary kilns are used at higher temperatures for the burning of hemihydrate and anhydrite II. The latter is often used as gypsum plaster.

Figure 28. SEM micrograph of β-hemihydrate produced from FGD gypsum.

The traditional anhydrite binder burning in a traveling grate kiln must be preceded by a briquetting operation. The desulphogypsum in the raw powdered form can be used in anhydrite production following the Dutch POLCAL technology [55]. This energy-saving process is carried out in a system of pipelines and cyclones. Because of a burning temperature of 800 °C, the product is relatively pure anhydrite II. The blend of anhydrite II and β-hemihydrate can be successfully used as a plaster of good quality. Anhydrite II with activators can also be used as a material for self-levelling floors as well as in some excavation filling works in mining. For this latter purpose, α-hemihydrate produced from desulphogypsum in an energy saving, continuous autoclaving can also be produced. The comparative tests results for α- and β-hemihydrates from desulphogypsums are given in Table 18.

Table 18. Comparison of Test Results for β- and α-Hemihydrate Gypsums.

Property	β-hemihydrate	α-hemihydrate
Water demand [%]	0.62	0.39
Initial setting [min]	8	14
Final setting [min]	15	27
Flexural strength [MPa]		
- after 2 hours	3.4	8.8
- after drying to constant weight	6.6	13.8
Compressive strength [MPa]		
- after 2 hours	7.0	19.7
- after drying to constant weight	16.1	41.5

Effect of Hydration Kinetics and Microstructure on the Properties of Hydrated Gypsum Materials

The hydration process of gypsum binders produced from FGD gypsum occurs in a similar manner to the hydration of materials produced from natural gypsums and follows the commonly known reaction:

$$CaSO_4{\cdot}0.5H_2O + 1.5H_2O \rightarrow CaSO_4{\cdot}2H_2O$$

The product is formed by nucleation followed by precipitation and recrystallization of calcium sulphate dihydrate, resulting from the $CaSO_4{\cdot}2H_2O$ lower solubility in relation to $CaSO_4{\cdot}0.5H_2O$. The solubility curves for α-, β-hemihydrate, anhydrite II and dihydrate are shown in Figure 29. The reaction kinetics curve is of a sigmoidal nature (Figure 30). The rate of hydration is relatively high and the reaction is almost complete within 40-60 minutes [119-121]. The most important factors affecting the hemihydrate gypsum hydration kinetics are as follows:

- grain size composition,
- water/gypsum ratio,
- temperature,
- admixtures to hydrating suspension and admixtures present as calcium sulphate solid solution.

The process can be significantly retarded by proper admixture [122]. The effect of fineness, W/G ratio and temperature can be investigated through observations of the microstructure. Comparative studies reveal that the well developed β-hemihydrate crystals produced from FGD gypsum transform to hydrated paste of higher porosity than the paste produced from finely grained, ground β-

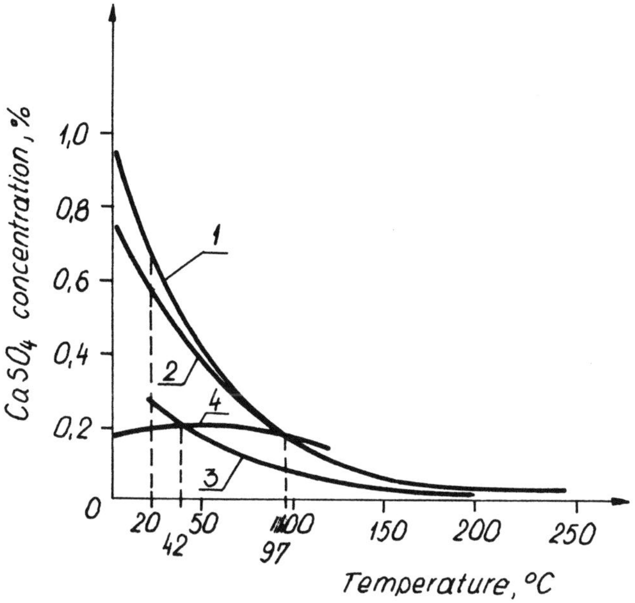

Figure 29. Solubility in water [117,118]. 1) β-hemihydrate; 2) α-hemihydrate; 3) anhydrite II; 4) dihydrite.

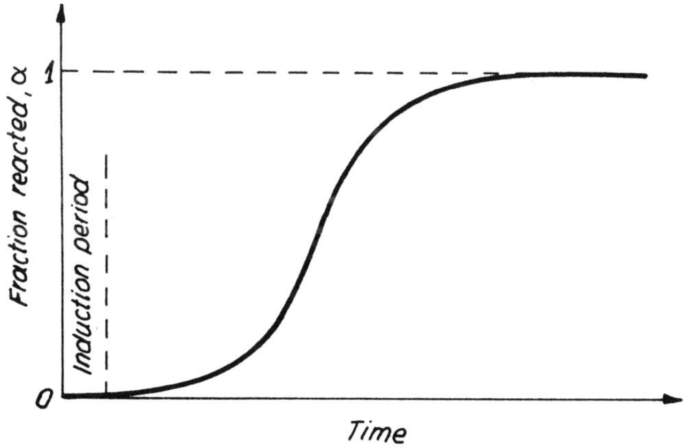

Figure 30. Kinetics of hemihydrate hydration [118].

hemihydrate from natural gypsum (see Figures 31, 32). The relation between the water/gypsum ratio and porosity is given in Figure 33. The increased porosity with increased W/G ratio leads to strength lowering as shown in Figure 34. The W/G ratio also affects other properties of hydrated gypsum materials, such as plasticity modulus, density, thermal conductivity and gas permeability [125]. The relations for both materials produced from FGD and natural gypsums are of the same character. The change of porosity as a function of setting temperature is plotted in Figure 35.

Recently, special attention has been paid to the setting and hardening modifying admixtures [126] widely used in both natural gypsum and FGD gypsum originating materials. Set retarders are added in most cases to the gypsum plasters and binders. Plasterboards are produced with set accelerators (e.g. calcium sulphate dihydrate). Some admixtures improve the properties of the final products (plasticizers, foaming agents, hydrophobic and aerating admixtures). The following mechanisms are responsible for the action of admixtures:

- adsorption on the hemihydrate grains,
- change of equilibrium in the $CaSO_4$-H_2O system,
- nucleation.

The admixtures usually influence the microstructure and strength of hydrated material (see Figure 36). The effect of total porosity on the strength of hydrated gypsum material is more pronounced than the pore size distribution.

In practice, the admixtures acting simultaneously as set retarders and plasticizers are used. In Figures 37 and 38 the effect of methyl cellulose and starch on the workability of fresh pastes and strength of hardened pastes are shown [127]. Citric acid and the alkaline citrates are also commonly used as gypsum set retarders [128,129]. Their effectiveness is influenced by pH as shown in Figure 39. In turn, the pH value is affected by the impurities present

(1)

(2)

Figure 31. SEM microstructure of hydrated hemihydrates: 1) FGD gypsum; 2) natural gypsum.

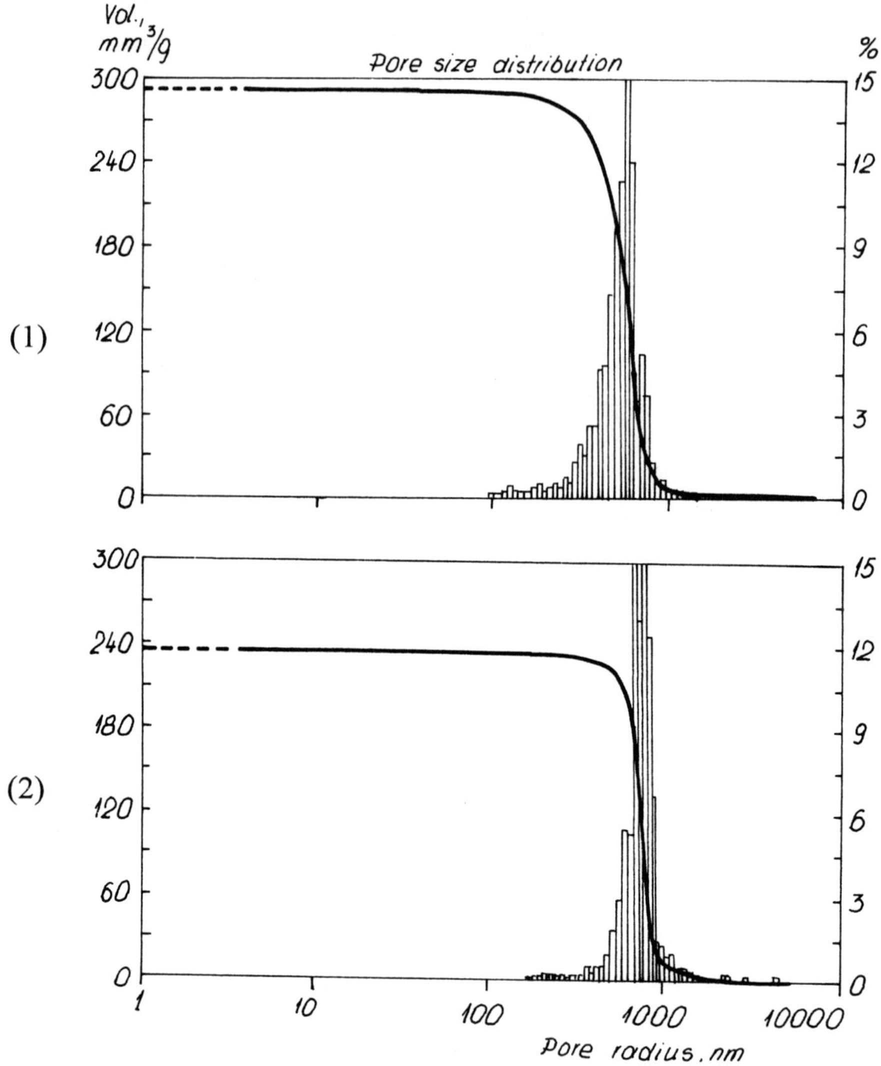

Figure 32. Pore size distribution curves for hydrated hemihydrates: 1) FGD gypsum; 2) natural gypsum.

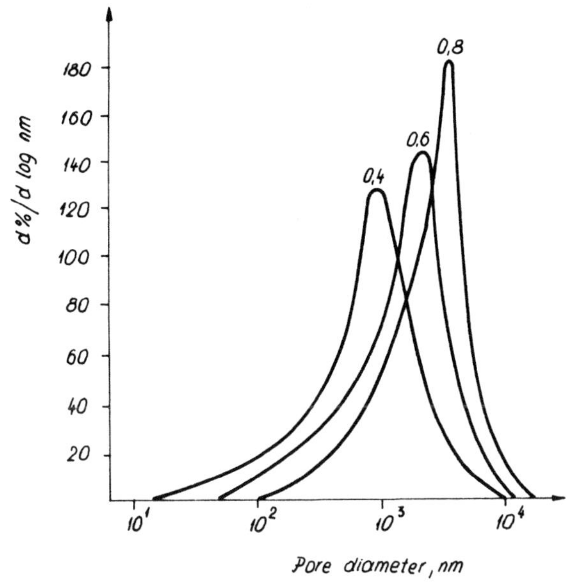

Figure 33. Pore size distribution as a function of W/G ratio [123].

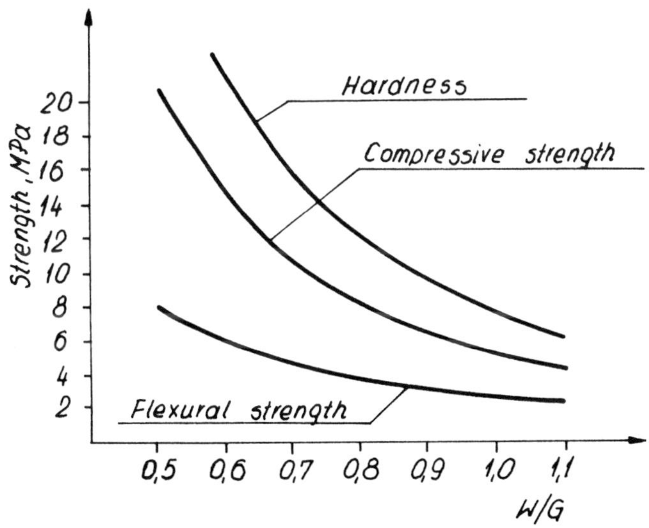

Figure 34. Mechanical properties of hardened plaster products (dry) as a function of water/solid ratio [124].

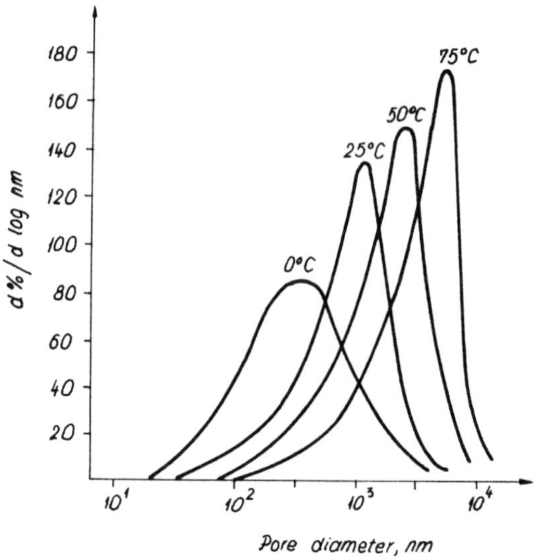

Figure 35. Influence of setting temperature on pore size distribution; initial product - β-hemihydrate [123].

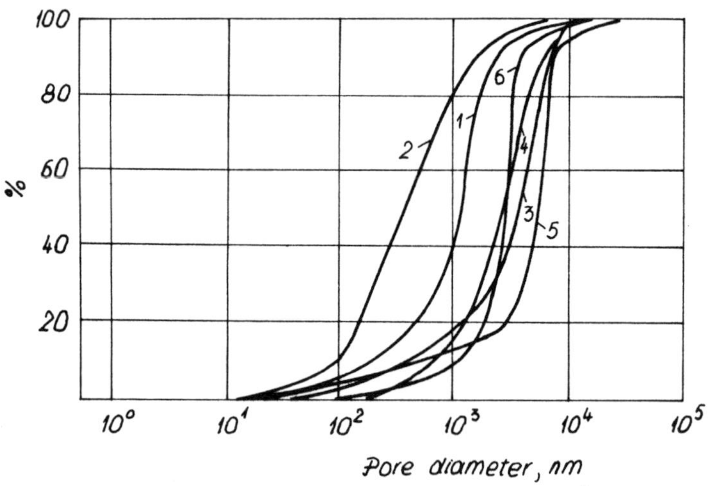

Figure 36. Effect of various admixtures on the pore size distribution in hardened hemihydrate (W/G = 0.45, T= 25°C) [126]. 1) no admixture; 2) $CaSO_4{\bullet}2H_2O$; 3) K_2SO_4 (1%); 4) citric acid (1%); 5) pepton; 6) sodium lignosulphonate (0.5%).

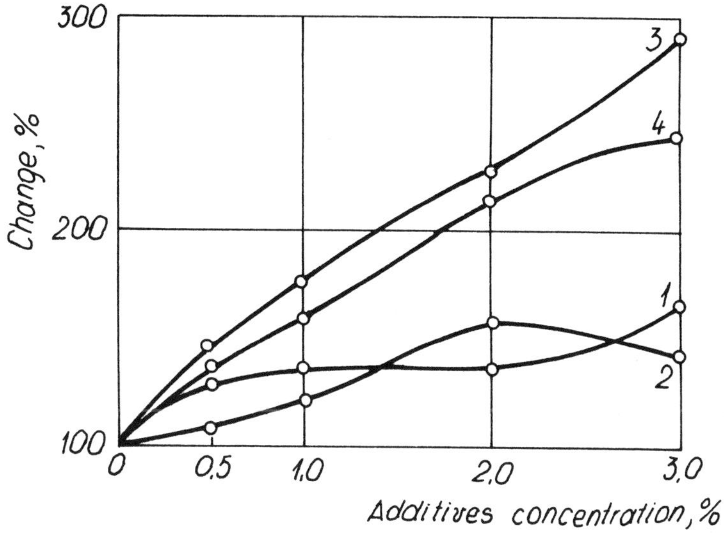

Figure 37. Change in workability properties of fresh gypsum plaster with additives [127]. 1) methyl cellulose, initial set; 2) methyl cellulose, spreadability; 3) starch, initial set; 4) starch, spreadability.

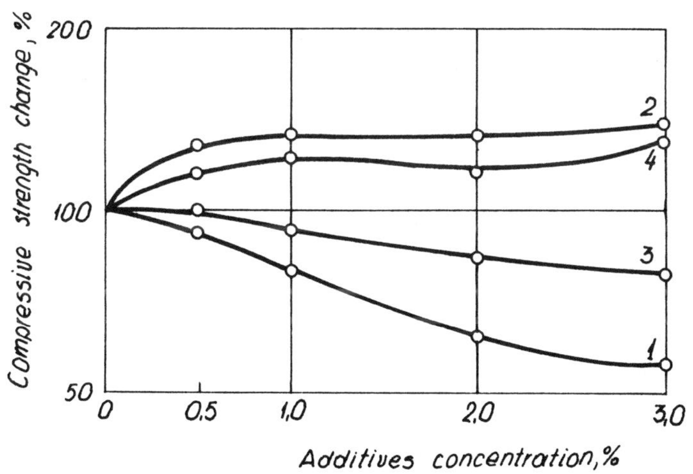

Figure 38. Compressive strength of gypsum prisms with additives. 1) methyl cellulose, 2 hrs; 2) methyl cellulose, air-dry prisms; 3) starch, 2 hrs; 4) starch, air-dry prisms [127].

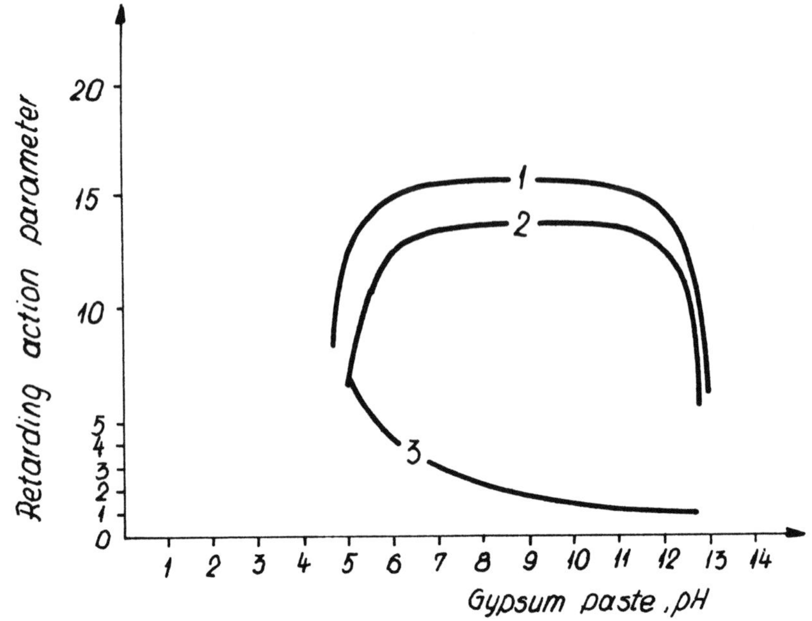

Figure 39. Relation between retarding action and pH value of gypsum paste [130]; 1) citric acid (0.2%); 2) sodium citrate (0.2%); 3) calcium phosphate (0.2%).

gypsum material. However, most of by-product gypsum materials, like natural gypsum, show a neutral pH value close to 7.

In the case of desulphogypsum, the production of multiphase binders (with better mechanical properties) is possible by adding clayey minerals, as with some natural gypsums, containing over 10 wt.% of clayey component [131]. The dehydration process can also be controlled by temperature in such a way that the formation of the proper amount of anhydrite III admixture is possible. This admixture is detectable by calorimetric measurements, as can be seen in Figure 40. The anhydrite admixture impacts positively on the properties of binder.

The anhydrite material produced from desulphogypsum differs from that of natural anhydrite or natural gypsum, but because of different hydration kinetics resulting from the solubility, it must be activated by CaO (estrich-gypsum) or by use of chemical activators. This anhydrite activated binder exhibits very good mechanical properties, improved moisture resistance and reduced shrinkage.

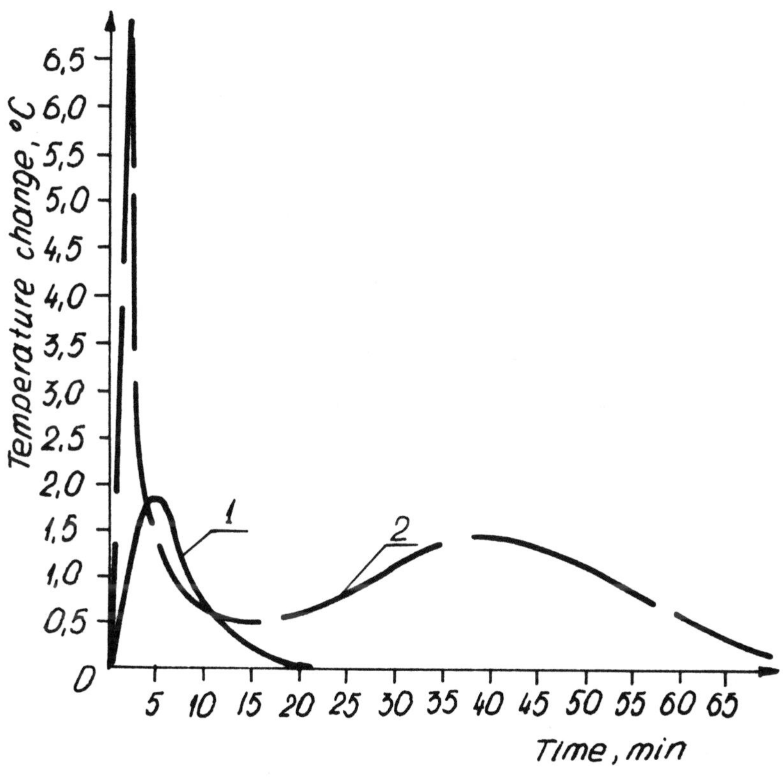

Figure 40. Heat evolution on β-hemihydrate rehydration. 1) FGD gypsum calcined within 1.5 min at 150°C; 2) FGD gypsum calcined within 10 min at 150°C.

REFERENCES

1. Murakami, K., By-product Gypsum from Various Chemical Industries, as a Retarder for the Setting of Cement, *Proceedings of the Fifth International Symposium on the Chemistry of Cement,* 4:457-503 Tokyo (1968).
2. Osiecka, E., *Phosphogypsums - Binders and Building Materials* (Arkady, ed.) pp. 45-66, Warszawa (1980).
3. Jarosiński, A., Mazanek, Cz., Utilization of Phosphogypsum in Poland, *Szklo i Ceramika,* 44(2):13-18 (1993).
4. Yilmaz, V.T., Isidak, Ő., Influence of Some Set Accelerating Admixtures on the Hydration of Portland Cement Containing Phosphogypsum, *Advances in Cement Research,* 5(20):147-150 (1993).
5. Jarosiński, A., Mazanek, Cz., Phosphogypsum - Polish and World Problem, *Szklo i Ceramika,* 43(5):19-22 (1992).
6. Jarosiński, A., Mazanek, Cz., Solution of Phosphogypsum Problem-a World Review, *Szklo i Ceramika,* 44(1):9-13 (1993).
7. Singh, M., A Chemical Process for Purifying Phosphogypsum, *Indian J. Environ. Hlth.* 25(4):300-306 (1983).
8. Singh, M., Processing of Phosphogypsum for the Manufacture of Gypsum Plaster, *Research and Industry,* 27(6):167-169 (1982).
9. Eipeltauer, E., The Applicability of Phosphoric Acid - Gypsum Sludge, *Tonindustrie Zeitung* 97(1):4-8 (1973).
10. Lopez, M.A., Roger, K.S., The Environmental and Geotechnical Aspects of Phosphogypsum Utilization and Disposal, *Mediterranean Conference on Environmental Geotechnology.* Cesme 25-27 May 1992, (M.A. Usmen and Y.B. Acar, eds.) pp.437-443, A.A.Balkema/Rotterdam/ Brookfield (1992).
11. Rihanek, St., Radioactivity of Phosphate Plaster and Phosphate Gypsum, *Tonindustrie Zeitung* 95(9): 264-270 (1971).
12. Taha, R., Roger, K.S., Phosphogypsum Literature Review IRM Report 1-90-4 LSU, Baton Rouge.

13. Lutz, R., Preparation of Phosphoric Acid Waste Gypsum for Further Processing to Make Building Materials, *Zement Kalk Gips* 47(12):690-696 (1994).
14. Instytut Techniki Budowlanej w Warszawie, *The Qualitative Requirements for By-product Gypsums as Raw Material in Building Materials Production*, Warszawa (1991).
15. Instytut Techniki Budowlanej w Warszawie, *Assesment of Natural Radioactivity of Raw Materials and Building Materials*, Warszawa (1980).
16. Ostrowski, Cz., Utilization of Phosphogypsums in Gypsum Plaster Manufacturing, *Cement Wapno Gip,* 37/51(2): 37-40 (1984).
17. Őlmez, H.,Yilmaz, V.T., Infrared Study the Refinement of Phosphogypsum for Cement, *Cement and Concrete Research,* 18:449-454 (1988).
18. Singh, M., Garg, M., Rehsi, S.S., Purifying Phosphogypsum for Cement Manufacture, *Construction and Building Materials,* 7(1):3-7 (1993).
19. Singh, M., Rehsi, S., Taneja, C., Rendering Phosphogypsum Suitable for Plaster Manufacture, *Indian Journal of Technology* 22;28-32 (1984).
20. Wirsching, F., Chemical Gypsum as Basis for High Quality Building Gypsum, *Tonindustrie Zeitung* 95(1) 14-1 (1971).
21. Singh, M., Rai, M., Autoclaved Gypsum Plaster from Selenite and By-Product Phosphogypsum, *J. Chem. Tech. Biotechnol.* 43:1:12 (1988).
22. Gutt, W., Nixon, P.J., Use of Waste Materials in Construction Industry, *Materials and Structures*, 12(70):255-306(1979).
23. Singh, M., Influence of Phosphogypsum Impurities on Two Properties of Portland Cement, *The Indian Concrete Journal* (7):186-190 (1987).
24. Tabikh, A.A., Miller, F.M., The Nature of Phosphogypsum Impurities and Their Influence on Cement Hydration,*Cement and Concrete Research*, 1:663-678 (1971).

25. Ostrowski, Cz., The Ways of Phosphogypsum Utilization, *Cement Wapno Gips* 40/54(4-5):83-88 (1987).
26. Kobayashi, K., Influence of Impurities of Phosphogypsum on the Development of Strength and the Setting Time of Portland Cement, *Tsemento Gijutsu Nenpo,* 18:79-85 (1963).
27. Mori, H., Sudoh, G., Influence of Phosphates on the Hardening of Portland Cement and Blast Furnace Slag Cement, *Tsemento Gijutsu Nenpo,* 14:67-81 (1960).
28. Gawlicki, M., Nocuń-Wczelik, W., Roszczynialski, W., Use of Chemical Gypsums as a Portland Cement Set Controlling Agent, *Prace Instytutu Mineralnych Materialów Budowlanych,* (16):45-52 (1994).
29. Gawlicki, M., Nocun´-Wczelik, W., Roszczynialski, W., Utilization of Chemical Gypsums in Cement Industry, Ceramics 44, *Polish Ceramic Bulletin,* (8):331-335 (1994).
30. Őlmez, H.,Erdem, E., The Effects of Phosphogypsum on the Setting and Mechanical Properties of Portland Cement and Trass Cement, *Cement and Concrete Research,* 19:377-384 (1989).
31. Haerter, M., Experience in Preparing Chemical Gypsum in Germany, *Tonindustrie Zeitung* 95(1):9-13 (1971).
32. Wirshing, F., The Phases of the System $CaSO_4$ -$CaSO_4{\cdot}2H_2O$, *Zement Kalk Gips* 55/19(10):487-492 (1966).
33. Jarosiński, A., Properties of Anhydrite Cement Obtained from Apatite Phosphogypsum, *Cement and Concrete Research,* 24:99-108 (1994).
34. Jarosiński, A., The Methods of Phosphoanhydrite Manufacture and Use, *Szklo i Ceramika,* 44(4):20-23 (1993).
35. Ostrowski, Cz., Production of Anhydrite II from the Briquetted Phosphogypsum, *Cement Wapno Gips* 44/58(3):66-70 (1991).
36. Singh, M., Garg, M., Reshi, S.S., Durability of Phosphogypsum Based Water-resistant Anhydrite Binder, *Cement and Concrete Research,* 20:271-276 (1990).

37. Kumbhojkar, A.S., Utilization of Phosphogypsum as Construction Material, *Mediterranean Conference on Environmental Geotechnology.* Cesme 25-27 May 1992, (M.A. Usmen and Y.B. Acar, eds.) pp. 563-572, A.A.Balkema/Rotterdam/Brookfield (1992).
38. Peukert, St., Utilization of Phosphogypsum in Associated Manufacturing of Cement Clinker and Sulphuric Acid, *Cement Wapno Gips* 41/55(9):185-189 (1988).
39. Valentini, G.L., Cioffi, R., Influence of Chemical and Physical Properties of Italian Fly Ashes on Reactivity Towards Lime,Phosphogypsum and Water, *Cement and Concrete Research,* 18:91-102 (1988).
40. Valentini, G., Santoro, L., Hydration of Granulated Blast Furnace Slag in the Presence of Phosphogypsum, *Thermochimica Acta,* 78:101-112 (1984).
41. Santoro, L., Aletta, I,. Hydration of Mixtures Containing Fly Ash, Lime and Phosphogypsum, *Thermochimica Acta,* 98:71-80 (1986).
42. Erden, E., Ölmez, H., The Mechanical Properties of Supersulphated Cement Containing Phosphogypsum, *Cement and Concrete Research,* 23:115-121 (1993).
43. Singh, M., Garg, M., Activation of Gypsum, Anhydrite-Slag Mixtures, *Cement and Concrete Research,* 25:332-338 (1995).
44. Singh, M., Garg, M., Gypsum Binders and Fibre-reinforced Gypsum Products, *The Indian Concrete Journal* (8):387-392 (1989).
45. Singh, M., Garg, M., Gypsum-based Fibre Reinforced Composites an Alternative to Timber, *Construction and Building Materials* 8(3):155-159 (1994).
46. Singh, M., Garg, M., Investigation of a Durable Gypsum Binder for Building Materials, *The Indian Concrete Journals* 6(1):52-56 (1992).

47. Takemoto, K., Ito, I., Suzuki, S., Influence of Chemical ByProduct Gypsum on the Quality of Portland Cement,*Tsemento Gijutsu Nenpo* 11:55-61 (1957) .
48. Derdacka-Grzymek, A., Grzymek, J., Gawlicki, M., Konik, Z., Stok, A., Use of By-product Raw Materials in the Associated Manufacture of Alumina, Iron, Titanium and Binders, *Ceramika, Prace PAN - Oddz. Kraków*(34):67-96 (1984).
49. Erdogan, Y., Genc, H., Demirbas, A., Utilization of Borogypsum for Cement. *Cement and Concrete Research,* 22:841-844 (1992).
50. Erdogan, Y.,Demirbas, A., Genc, H., Partly-Refined Chemical By-product Gypsum as Cement Additives. *Cement and Concrete Research,* 24:601-604 (1994).
51. Mylona, S., Trends of Sulphur Dioxide Emissions, Air Concentrations and Depositions of Sulphur in Europe Since 1980, *EMEP/MSC-W Report 2/93,* Oslo (1993).
52. Elvingson, P., EMEP Report *Acids News,* (5):16-17 (1993).
53. Kucowski, J., Laudyn, D., Przekwas, M., *Power Industry and Environmental Science,* (Wydawnictwo Naukowo-Techniczne, ed.) pp. 73 - 108, Warszawa (1993).
54. Mikoś, J., Wiśniewski, A., Use of Gypsum By-Products in Construction Industry, *Proc. of Conference on Gypsum* Materials, 1-10, Katowice (1994)
55. Hamm, H., Coping with the FGD Gypsum Problem - A Task for the European Gypsum Industry, *Zement Kalk Gips,* 87/51(8):443-451 (1994).
56. Flue Gas Gypsum Quality, *Zement Kalk Gips,* 82/46(5): 223-228 (1989).
57. Jurkowitsch, H., Hüller, R., New Plasterboard Plants Using Natural Gypsum and Different Kinds of Synthetic Gypsum, *Zement Kalk Gips,* 83/47(12):583 -588 (1990).
58. Wirsching, R., FGD-gypsum - Production, Properties and Use, *Baustoffindustrie,* (1):11-13 (1990).

59. Wirsching, R., Hüller, R., Olejnik, R., Gypsum from Flue Gas Desulphurization Plants - Definitions and Legislation in the European Communities in the OECDand in Germany, *Zement Kalk Gips*, (2):65-69 (1994).
60. Hamm, H., Dealing with the Desulphogypsum Problem in the Federal Republic of Germany from the Technical, Economic and Market-Orientated Points of View, *Zement Kalk Gips*, 84/48(12): 593-604 (1991).
61. Mittelbach, G., Desulphurization of Flue Gases on the Basis of Lime or Limestone, *Commission of Flue European Communities Joint Research Center*, Ispra, (1990).
62. Bulewicz, E.,M., Juryś, Cz., Kandefer, S., Fluidized Bed Combustion - Possibility to Limit Sulphur Emission, *Air Pollution Control* (2):43-47 (1987).
63. Warych, J., Desulphurization of Fluidized bed Combustion Gases, *Proc. of Conf. Problemy ochrony środowiska przy wytwarzaniu energii elektrycznej i cieplnej*, (165-173), Bielsko Biala (1988).
64. Yoo, H.J., Ganley, P.,J., Steinberg, M., Calcium Silicate Cements for Desulphurization of Combustion Gases, *Semiannual Progress Report*, US Dept. of Energy, Brookhaven National Lab. (1981).
65. Brylicki, W., Physical and Chemical Characteristics of Fluidized Bed Combustion Gases Desulphurization Products, *Proc of Conf. Sorbenty do odsiarczania spalin*,(1-12), Tarnów Opolski (1995).
66. Majling, J., Roy, D.,M., The Potential of Fly Ash for Cement Manufacture, *Amer. Ceram. Soc. Bull.*, (10):77-80 (1993).
67. Zysk, K.,H., Schroer, D., Possible Applications for Flue Gas Desulphurization Product in Coal Mining, *Zement Kalk Gips*, 86/50(7):192-198 (1993).
68. Cermak, J., Use of Waste Materials from Power Plants in Building Materials Production, *Stavivo*, (7-8): 275-277 (1990).

69. Mosch, H., The Suitability of Anhydrite from Flue GasSulphur Dioxide as a Sulphate Admixture, *Zement Kalk Gips,* 79/43(1):33-35 (1986).
70. Grzeszczyk, S., The Binders Produced from Fly Ash and FGD Products from the Semi-dry Installation,*Cement Wapno Gips,* 48/62(1):8-12 (1995).
71. Durych, A., Laszuk, A., Wiechowski, A., Dry-scrubbing - The New Method of Flue Gas Desulphurization, *Ochrona Powietrza,* (3):45-50 (1985).
72. Ludwig, H., Hetschel, M., Fitjer, H., Residual Products from Flue Gas Desulphurization by Spray - Dryer Method - Technical and Economic Aspects of their Disposal and Recovery for Utilization, *Proceedings of the II European Conference on Environmental Technology,* pp: 165-181, Amsterdam (1987).
73. Pietroń, J., Olejarz, J., The Evaluation of the FGD Gypsum Utilization from the Economic, Environmental and Technical Point of View, Report, Inst. Miner. Mat. Budowl., Kraków, (1992).
74. Brodziak, A., Bi-alkaline Flue Gas Desulphurization Methods - Technical Aspects, *Ochrona Powietrza,* (6):156-161 (1978).
75. Kaplan, N., Introduction to Double - Alkali Flue Gas Desulphurization Technology, *Symposium on Flue Gas Desulphurization,* Washington, (1976).
76. Brylicki, W., Gawlicki, M., Roszczynialski, W., Malolepszy, J., Nocuń-Wczelik, W., Theoretical Basis for the Flue Gas Desulphurization Products Management, Report, Akademia Górniczo-Hutnicza (1991).
77. Brylicki, W., Lagosz, A., Use of FGD Products in Cellular Concrete Production, *Proc. of Conf. Zagospodarowanie pyłów i popiołów z energetyki i cieplownictwa,* (31-39) Sopot, (1995).

78. Pietroń, J., Olejarz, J., Wdowiak, N., Cwakliński, A., Tomasik, D., Synthetic Gypsum from the First in Poland Flue Gas Desulphurization Installation in Belchatów Power Plant, *Cement Wapno Gips*, 48/62 (3): 82-89 (1995).
79. Wirsching, F., Production and Use of Industrial Calcium Sulphates, the Properties of the Gypsum Building Materials Produced from Them, *Zement Kalk Gips,* 84/48(5):248-252 (1991).
80. Stahl, H., Jurkowitsch, H., Briquetting of Flue Gas Desulphurization Gypsum, *4th Internat. Symp. on Agglomeration,* Toronto (1985).
81. Zisselmar, R., Use of Roller Press in Processing of FGD-Gypsum, *Zement Kalk Gips,* 78/42(5):243-249 (1985).
82. Weiler, H., Hamm, H., Hüller, R., Processing of Moist Finely Divided Flue Gas Gypsums, *Zement Kalk Gips,* 76/40(11): 609-915 (1983).
83. Günter, F., Conditioning of Wet Flue Gas Gypsum for Re-Utilization by Pelleting, *Proceedings of The 1991 SO_2 Control Symposium* pp.33-43, Washington (1991).
84. Schmith, Ch., L., Commercial Aggregate Production from FBD Waste, *Proceedings of The 1991 SO_2 Control Symposium pp.*489-504, Washington (1991).
85. Šauman, Z., Šiške, V., Use of FGD-gypsum as a Set Controlling Agent, *Stavivo* (1):1-5 (1990), *Stavivo* (5):179-184 (1990).
86. Odler, I., Bloss, W., Utilization of Flue Gas Desulphurization Gypsum in Cement Industry, *Tonindustrie Zeitung,* 103(12): 747-751 (1979).
87. Jungmann, J., Utilization of Flue Gas Desulphurization Product in Clinker Production, *Stavivo,* (11):452-454 (1988).
88. Gawlicki, M., Nocuń-Wczelik, W., FGD-gypsum as a Set Controlling Agent, *Prace Komisji Nauk Ceramicznych, Ceramika* (41):192-194, Kraków (1992).
89. Sersale, S., Cioffi, R., Relationship between Gypsum Content, Porosity and Strength in Cement, *Cement and Concrete Research,* 21(1):120-126 (1991).

90. Gustaw, K., Roszczynialski, W., Libura, J., Optimum Gypsum in Portland Cement as a Function of Calcium Aluminate Content, *Baustoffindustrie*, (3):84-85 (1985).
91. Wolter, H., Influence of Calcium Sulphate Form and Mixing Time on Stiffening and Setting of Cement, *Zement Kalk Gips*, 80/46 (7): 372-375 (1989).
92. Krogbeumker, H., Storage and Metering of Moist Desulphogypsum at the Phoenix Cement Works, *Zement KalkGips*, 83/49(3): 105-109 (1992).
93. Roszczynialski, W., Grzybowska, W., Gawlicki, M., Sulphate-Fly Ash-Cement Binders for Road Making, *PolishCeramic Bulletin, Ceramics* (46): 375-380 (1994).
94. Lea, F.M., *The Chemistry of Cement and Concrete,* (Academic Press ed.) pp. 411-413, London (1956).
95. Teoreanu, I., Dumitrescu, C., Mechanisms of Controlled Expansion in Cement Hardening, *Cement and Concrete Research,* 12(2):141-155 (1982).
96. Asanagi, E., Okabayashi, I., Terasima, O., Mitsu, K., Adzuma, T., Stabilization of Peat Soil, *Jap. Pat.,* No. 59-75957, Tokyo (1984).
97. Simada, T., Charato, S., Medzima, K., Composition for Soil Solidifying, *Jap. Pat.*, No. 57-102985, Tokyo (1982).
98. Hyde, A.,F.,L., Some Preliminary Test Data on Cement -PFA-Anhydrite Mixes, *Mag. Concr.*, (36):174-18 (1984).
99. Roszczynialski, W., Gawlicki, M., Grzybowska, W., Fonferko, L., Anhydrite-Fly Ash-Cement Binders in Road Making, *Proc. of Conf. Zagospodarowanie pyłów I popiolów z energetyki i cieplownictwa,* 31-44, Sopot (1993).
100. Iwatami, K., Kagaja, K., Kamada, G., Binding Material with Anhydrite Component, *Jpn. Pat.*, No. 56-96757, Tokyo (1981).
101. Toda, N., Iwata, T., Sediment Solidifying Compositions, *Jpn. Pat.*, No. 60-137496, Tokyo (1985).

102. Roszczynialski, W., Grzybowska, W., Gawlicki, M., Stabilization of Soil Using the Fly Ash and FGD-Gypsum Containing Binders, *Proc. of Conf. Zagospodarowanie pyłów I popiolów z energetyki i cieplownictwa,* 40 - 46, Sopot (1995).
103. Demmick, J., Weißflog, E., German Experience of FGD By - product Disposal and Utilization, *The 1991 SO_2 Control Symposium,* 397 - 419, Washington (1991).
104. Thien, P., Wirsching, F., Utilization of the Flue Gas Desulphur-ization By-products, *Glückauf,* (19/20):899-905 (1991).
105. Bloss, W., Problems of Flue Gas Gypsum Disposal as Exemplified by Rhenish Lignite District, *Zement Kalk Gips,* 77/41(10):540-546, (1984).
106. Trautmann, J., Knöfel, D., Development of Mortar Consisting of Slag, Gypsum and Portland Cement for Injection into Multiple-leaf Masonry, Part. I: Optimizing the Binder, *Zement Kalk Gips,* 86/50(4):E115-E120 (1993).
107. Schwiete, H.,E., Knauf, A.,H., Gips. The Production and Properties of Gypsum Materials - an Overview and New Data, Merziger Druckerei und Verlags GmbH (1971).
108. Scholze, H., Hurbanic, M., Ruf, H., Comparative Considerations on the Behaviour of Natural Gypsum and Flue Gas Gypsum, *Zement Kalk Gips,* 78/42:431-436 (1985).
109. Stenström, S., Influence of Temperature, Particle Size and Humidity on the Gypsum Dehydration Process Investigations in a Cahn Thermobalance, *Zement Kalk Gips,* 85/49(9):478-482 (1992).
110. Hilbert, H., T., Lempfer, K., Heine, W., Use of Gypsum Binder from FGD Installations in the Production of Plasterboards, Report, Fraunhofer-Institut für Holzforschung, Braunschweig (1988).
111. Jurkowitsch, H., Hüller, R., New Plasterboard Plants Using Natural Gypsum and Different Kinds of Synthetic Gypsum, *Zement Kalk Gips,* 83/47(12):583-588 (1990).

112. Roddewig, D.L., Operating Experience with the Calcination of Natural Gypsum and Desulphogypsum in a Grinding and Calcining Plant, *Zement Kalk Gips,* 84/48(8):416-420 (1991).
113. Gawlicki, M., Possibility of Natural Gypsum Replacement by FGD Products, *Proc. of Conf. Sorbenty do odsiarczania spalin*, 30-38, Tarnów Opolski (1995).
114. Brosig, A., Wolff, G., Commissioning and Operating Experience at a Gypsum Calcining and Gypsum Chipboard Plant, Part I - Calcining Plant, *Zement Kalk Gips* 81/45(7): 158-159 (1988).
115. Roszczynialski, W., About the Further Development of Gypsum Thermal Treatment Operation in Gypsum Plant "Stawiany", Report, Akademia Górniczo-Hutnicza, Kraków, Poland (1993).
116. Kappe, J., High Quality Anhydrite the from Flue Gas Desulphurization Gypsum, *Environment and Technology* (1):3-7 (1991).
117. Akerman, K., *Gypsum and Anhydrite*, (Pan´stwowe Wydawnictwo Naukowe, ed.), pp. 35-54, Warszawa (1964).
118. Hand, R.J., The Kinetics of Hydration of Calcium Sulphate Hemihydrate. A Critical Comparison of the Models in the Literature, *Cement and Concrete Research,* 24(5):885-895 (1994).
119. Kopp, U., Photometric Method for Characterizing the Progress of Setting in α- and β-hemihydrate Gypsum Plasters, *Zement Kalk Gips,* 87/51(9):239-241 (1994).
120. Gerard, H., Measurements of Temperature and Setting Time Behaviour for Evaluating Gypsum Hydration Characteristics, *Zement Kalk Gips,* 84/48(8):426-429 (1991).
121. Thole, V., Determination of the End of Hydration of Gypsum Plasters at Room Temperature Using Temperature - Time Curves, *Zement Kalk Gips,* 87/51(4):E99-E103 (1994).
122. Fischer, H., B., Henning, O., Investigating the Relationship between the Stiffening and the Electrical Conductivity of Gypsum Plaster Mixtures with Additives, *Zement Kalk Gips,* 87/51(8):452-455 (1994).

123. Rößler, M., Odler, J., Relationship between Pore Structure and Strength of Set Gypsum Pastes. Part I: Influence of Water/Gypsum Ratio and Temperature, *Zement Kalk Gips,* 82/46(2):96-100 (1989).
124. Sattler, H., *Recycling in the Construction Industry*, EF - Verlag für Energie und Umwelttechnik, Berlin, (1987).
125. Teschner, A., Changes in the Physical Properties of Hydrated β-hemihydrate on Drastic Lowering of the Water/Plaster Ratio, *Zement Kalk Gips,* 84/48(8): 430-434 (1991).
126. Odler, I., Rößler, M., Relationship between Pore Structure and Strength of Set Gypsum Pastes. Part II: Influence of Chemical Admixtures, *Zement Kalk Gips,* 82/46(8): 419-424 (1989).
127. Siewert, K., Henning, O., Chemical Additives with Water-Retaining Properties for Gypsum Binders, *Zement Kalk Gips,* 81/45(4):173-174 (1988).
128. Roszczynialski, W., Technology of Hemigypsum for Ceramics, report, Akademia Górniczo-Hutnicza, Kraków (1985).
129. Henning, O., Brockner, O., The Optimum Retarding Action of Citric Acid on the Hydration of Gypsum, *Zement Kalk Gips,* 83/47(7):357-360 (1990).
130. Mallon, T., Retarding Action of Gypsum Plaster Retarders of Various Chemical Composition in Relation to the pH Value of the Plaster, *Zement Kalk Gips,* 81/45(6):309-311 (1988).
131. Murat, M., Attari, A., Modification of Some Physical Properties of Gypsum Plaster by Addition of Clay Minerals, *Cement and Concrete Research,* 21(2/3):378-387 (1991).

3

FLY ASH IN CONCRETE

Aimin Xu

INTRODUCTION

At the 6th international symposium on the chemistry of cement held in Tokyo, 1966, Kokubu [1] gave an extensive review of the studies on fly ash used for cement-concrete up to that time, including ash characteristics, testing pozzolanic reactivity, and influences of fly ash, physical and chemical, on properties of concrete. At the 7th international congress on the chemistry of cement held in Paris, 1980, Takemoto and Uchikawa [2] further explored the mechanisms of the reactivity of pozzolanic materials in various cementitious systems. Berry and Malhotra [3] made a comprehensive review of the principal advances in research, development and practical application of fly ash in concrete from 1976 to 1984.

Since the 1980s, with the increasing concern about durability of concrete structures, more and more researches have been directed to study the performance of concrete with an incorporation of fly ash as a supplementary cementing material. Using more fly ash in concrete (high volume fly ash concrete) have also been extensively studied in some major fly ash producing countries. Many studies have been devoted to this material and its effect on the various systems that make use of it. Characterization of fly ash, including methods of assessment, experimental data and literature surveys, in accordance with their utilization as a pozzolanic material or for ecological concerns, are compiled in a proceeding edited by McCarthy et al. [4].

The objective of this chapter is to provide a cohesive literature survey on studies published in recent decades, accounting for the mechanisms and the contribution or behavior of fly ash in concrete, rather than an exhaustive report summary of all research work. Since fly ashes have been increasingly used in nearly all types of concrete to enhance durability and/or to reduce structure cost, any attempt to include all relevant studies would result in a work too lengthy for this publication.

In elaboration on the effect of fly ash, some earlier work, mainly characterization of fly ash and hypotheses on hydration mechanisms for cement systems containing fly ash, are also referred to. The fact that fly ash is a by-product whose quality, in terms of demand for high quality concrete, varies over a rather wide range which in a number of cases makes laboratory research results contradictory. Thus, the order of materials is arranged such that a reasonable explanation may be drawn.

FLY ASH CHARACTERIZATION

Fly ash, or pulverized fuel ash, is a residue derived from the combustion of pulverized coal in furnaces of thermal power plant. The characteristics of fly ash varies according to the combustion operation system as well as the coal composition.

Various suspension-firing systems, e.g. vertical firing and horizontal firing, etc., have been widely used, which afford a high steam-generation capacity and quick response to load changes. The combustion temperature is high (approximately 1200°C) and the ash, in finely divided form (usually less than 100 μm in size) is carried along in the air stream, collected by electrical or mechanical precipitators (dry process) while it is quickly cooled. In some power stations, the old wet collection process is still in use (see Figure 1). Fly ash collected by a dry process is usually rather homogenous in particle size; whereas that collected by a wet process is more segregated due to the fact that the sedimentation speed is lower for smaller or lighter particles, and it contains large quantities water. The coarser portion of the coal ash (~ 15-20% by mass [5]) is heavy and falls to the bottom of the furnace and thus is called bottom ash. Residue of combustion consists of about 85% fly ash and 15% bottom ash [5].

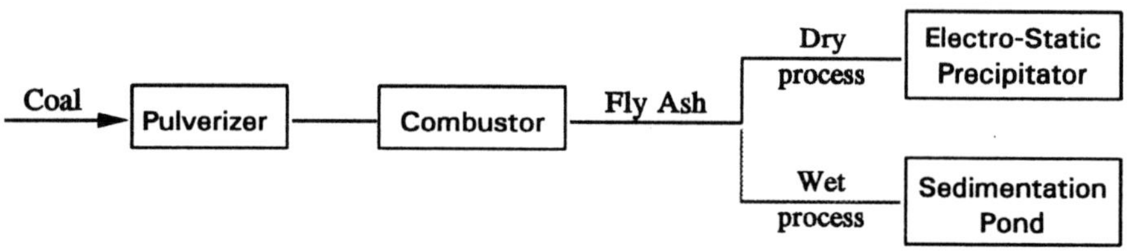

Figure 1. Schematic diagram of coal burning and ash collection processes.

With the ignition of burnable matter, under high temperature, the inorganic compounds in coal melt and the liquid assumes a spherical shape forming fly ash glass embedded with various minerals. The collision between particles results in some larger particles or particles made up of several smaller ones bonded together. It is often seen under microscopic examination, that some particles have enclosed pockets of air, or appear to be hollow spheres. Various shapes of fly ashes are shown in Figure 2.

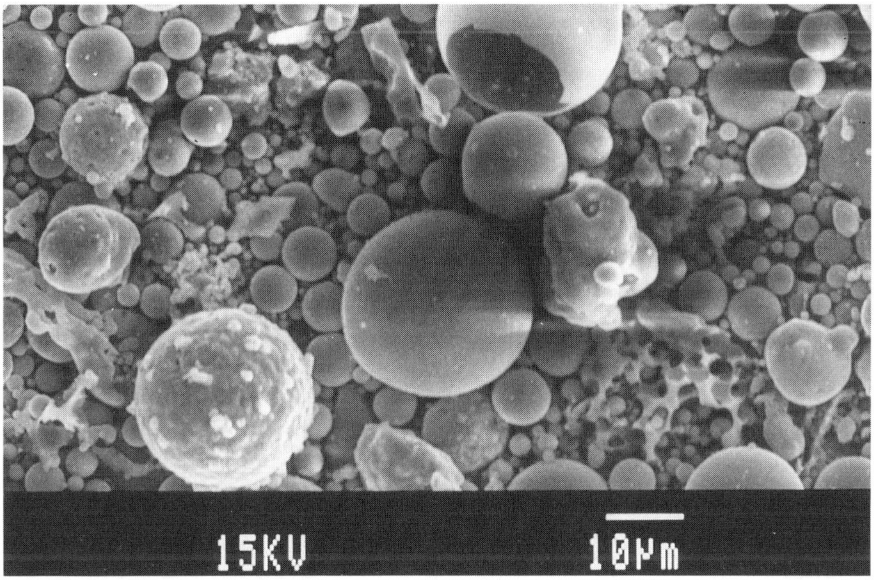

Figure 2. SEM micrograph of fly ash particles.

Partly due to the rapid cooling from high temperatures, the fly ash glass phase, or even some mineral phases, are thermodynamically metastable and possess hydrolysis potential. The experience of nearly a half century's utilization of fly ash has confirmed that fly ash is reactive in the presence of lime and water, or, possesses pozzolanic reactivity. This can be largely attributed to the glass in fly ash. The glass content can be determined by measuring the x-ray diffraction (XRD) background intensity in the range 15-35 $°2\theta$ (for Cu Kα) over which a diffuse pattern is formed (Figure 3), as described by Stevenson and McCarthy [6] and Hemmings and Berry [7]. Glass content in fly ash varying from 50% to 90% has been reported [8].

The combustion system directly influences ash composition. A form of coal burning referred to as *fluidized-bed combustion* has been in use since the 1970s. This process has a high heat-transfer rate so that it uses lower temperatures (approximately 1000°C). Consequently, the formation of nitrogen oxides is lower and the volatilization of potassium, sodium and sulfur found in the coal are also lowered. While the low temperature process produces less air pollution, it also results in a different fly ash composition when compared to ashes produced under higher processing temperatures. In recent years, in order to reduce air pollution, limestone powder is introduced into the combustion system to absorb the sulfur released from burning high sulfur coals, which forms calcium sulfide - sulfate and lime [5,9].

The chemical composition of an ash depends on the coal composition, while its mineralogical composition may vary considerably from the original matter and is also a function of the fineness of the fuel and the combustion operation. In response to the demands for power, the ignition efficiency changes and peak temperature varies. This results in a varying amount of unburned carbon in the ash. Furthermore, it has been reported that the residence time of fly ash in furnace causes variations in the alkali content and the vitrification degree of the fly ash [10].

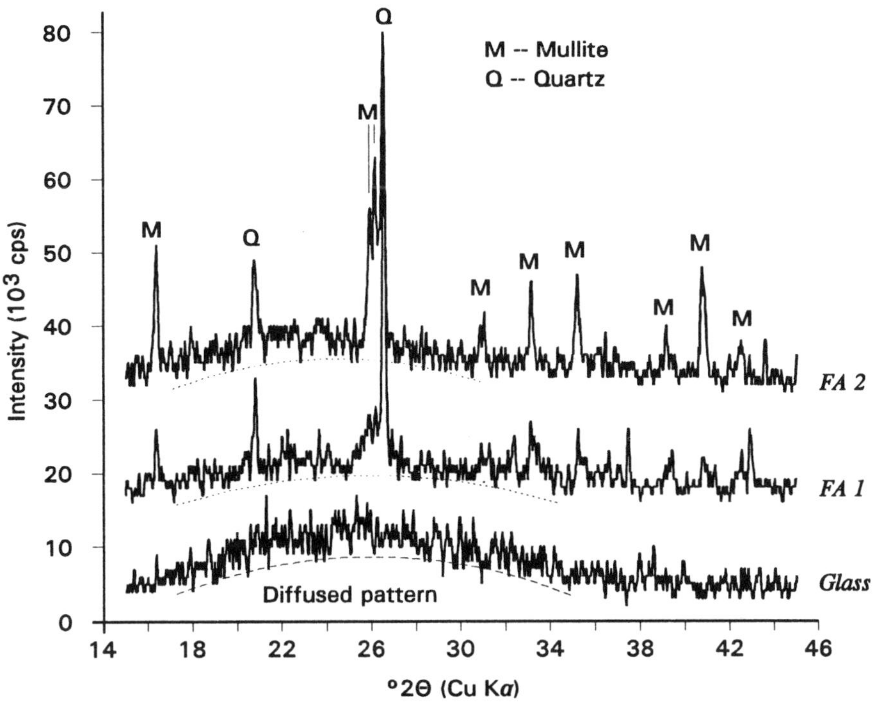

Figure 3. X-ray diffraction patterns of a glass plate and low lime fly ashes. FA1 is produced in Sweden; FA2 in the US.

Chemical and Mineralogical Compositions and Structure of Fly Ash

Coals used in thermal power stations are mainly anthracite (hard coal, with carbon content higher than 80%, burning with little smoke), bituminous coal (composed of carbon and volatile matter in similar proportions, burning with a great deal of smoke), sub-bituminous coal (similar in composition to the bituminous coal, with more moisture), and lignite coal (brown coal, similar to sub-bituminous coal in solid matter composition, usually with a high moisture content). The composition of coal and that of the ash vary considerably. The first two types of coals usually produce low-calcium ashes, whereas the latter two lead to high-calcium ashes.

Coals are associated with the many minerals in soil, e.g. clay (illite, Kaolinite, etc.), quartz, iron, calcium carbonates, feldspar and sulfur containing compounds [11,12]. The minerals are nonvolatile in the combustion temperature range (except that $CaCO_3$ releases CO_2), but undergo some oxidation and mineralization processes and remain in the ash. It may be noted that the lime produced in ashes comes from limestone or dolomite, the latter also results in MgO. Joshi and Lohtia [5] showed a fly ash of lignite origin containing as high as 32% CaO and 11% MgO.

For the use of fly ashes in concrete, the ASTM C 618 specification classifies an ash as Class F when the sum of SiO_2, Al_2O_3, and Fe_2O_3 is not less than 70%, or Class C if the sum is not less than 50% (Table 1). The difference in oxide content is due to the high lime content in the latter ash. Specifically, it states that:

Class C pozzolan - fly ash normally produced from lignite or subbituminous coal, in addition to having pozzolanic properties, also has some cementitious properties. Some Class C fly ashes may contain lime higher than 10%.

Table 1. Chemical Requirements of ASTM C618-91 Specification.

	Mineral Admixture Class		
	N	F	C
Sum of SiO_2, Al_2O_3, and Fe_2O_3, min %	70.0	70.0	50.0
SO_3, max., %	4.0	5.0	5.0
Moisture content, max., %	3.0	3.0	3.0
LOI*, max., %	10.0	6.0**	6.0

* LOI = loss on ignition
* *The use of Class F pozzolan containing up to 12% loss on ignition may be approved by the user if either acceptable performance records or laboratory test results are made available.

Although not specified, the lime (CaO) content 10% mentioned here has been used as a criterion for distinguishing a "high- lime fly ash" from a "low-lime fly ash."

The difference in CaO, as identified by Diamond [11], associates with a wide range of chemical composition for fly ashes produced in each country. This was also shown by Rehsi [8] who summarized the chemical compositions of the fly ashes of some major ash producing countries.

The chemical compositions of low- and high-lime fly ashes produced in North America are shown in Figure 4. It appears that when compared with the low-lime fly ashes, the high-lime fly ashes have a similar alumina content but a much lower iron-oxide content. The silica content in high-lime fly ashes varies, but is approximately the same as that of the low-lime ashes. This brings about an important feature: the ratio of calcium to alumina or the ratio of calcium to silica in high-lime ashes is much higher, and calcium aluminates and/or calcium silicates can form during the heating-cooling process. In addition, the presence of a high amount of calcium alters the structure of fly ash glass, as reviewed by Xu and Sarkar [13]. Other possible fly ash glass compositions for the $CaO-Al_2O_3-SiO_2$ system are illustrated in a ternary phase diagram by Hemmings and Berry [7].

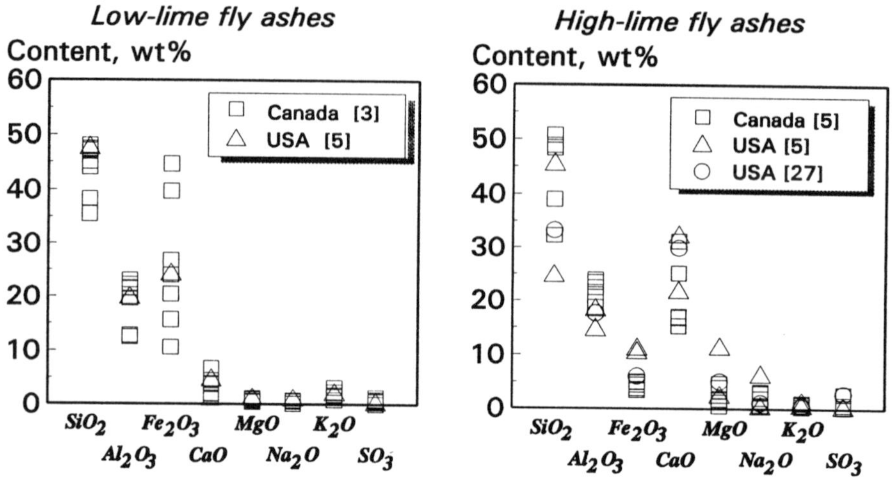

Figure 4. Chemical compositions of several fly ashes.

The minerals in Class F fly ashes are typically quartz or cristobalite, mullite, hematite, magnetite, lime, calcium sulfates and alkali sulfates. In high- lime fly ashes, calcium silicates or calcium aluminates are formed. Stevenson and McCarthy [6] showed the presence of dicalcium-silicate or alike and other calcium bearing phases. Most of the minerals are embedded in the glass, whereas alkali sulfates deposit on the surface of fly ash particles. Iron in fly ash is usually concentrated in individual particles (either there is little or very rich iron in a particle). Unburned carbon is present in fly ash in various amounts and the carbon grains often contain many small spherical holes which appear to be previously accommodated by glass particles.

Diamond stated that quartz, in most cases, is a residue that pre-existed in the coal, whereas other minerals are formed during the heating-cooling process [11]. The actual minerals depend on the mole-ratios between chemicals in the coal, and the size of the coal grains because very small grains may contain only one substance while larger ones may contain several. In addition, the combustion temperature also affects the mineralogical composition of the ash. It is well known in cement technology that calcium aluminates, calcium silicates and calcium aluminosilicates begin to form at relatively low temperatures (<1000°C), whereas mullite is formed at higher temperatures.

The chemical composition of fly ash has been known to vary between particle and particle due to phase separation and the small particle sizes [11]; and vary from batch to batch due to the coal resources and firing operation as stated previously. However, there are many instances demonstrating that the composition of the fly ash produced from the same power plant can remain constant for many years [14]. And when one looks at a broader view, the variation is even smaller. Dhir et al. [15] showed the oxide compositions in Class F fly ashes produced from several major fly ash producing countries are generally similar. For use in concrete, consistency in overall composition is more important than small scale variations.

As ultimate users of fly ash, concrete engineers focus their attention on the bulk physical properties of the ash: the specific gravity and particle size, and the water requirements of the concrete as influenced by the utilization of fly ash.

Particle Size and Density of Fly Ash

Fly ash, as it comes from the combustion of pulverized coal (e.g. in an Australian power plant, the coal powder is required to be more than 70% smaller than 70 μm [10]), is a powder with particle sizes similar to that of cement. Generally, more than 70% of an ash can pass through a 45 μm sieve, and a fraction of particles are smaller than 3 μm, as shown by Joshi and Lohtia [5]. Unburned coal particles are usually larger than 100 μm. The Blaine Fineness of typical Class F fly ash is in the range of 300 - 400 m^2/kg.

During the combustion-cooling process, smaller fly ash particles may cool more rapidly and remain more glassy [11] and thus, have a higher reactivity. Because of the glassy nature of smaller fly ash particles, the surface is the most reactive part [5]. Rehsi [8] shows the lime reactivity of fly ashes is higher for the percent of ash with particle sizes < 45 μm, which appears to be the only parameter (compared with silica content, silica-alumina content, and the Blaine specific surface) that has significant relation to the reactivity. According to ASTM C 618, fly ashes used in concrete should possess no more than 34% of particles retained by the 45 μm sieve. Butler and Mearing [16], in reviewing the use of fly ash fractions for different purposes, consider that particles in the range of 10 to 50 μm mainly act as void fillers in concrete, whereas the particles smaller than 10 μm are more reasonably classified as "pozzolanic reactive". It should be noted that this effect is largely physical, because for a unique mass, the smaller the particles, the higher the specific surface which provides a larger interface available for reaction. As an evidence of this point, Carles-Gibergues and Aitcin [14] showed that there is no significant differences in chemical or mineralogical compositions for a Class F fly ash separated by an air classifier.

Hemmings and Berry [7] reported chemical and mineralogical compositions of a fly ash (CaO content 10%) separated by solutions of specific gravity from 0.79 to 2.85 (determined by a sink-float method). Although the density of each portion of the ash is influenced by the presence of voids in ash particles, thus each portion contains different sizes of particles observed by SEM, it does show that in the portions of middle and small density, the silica content is nearly the same, whereas the high density portions contain more Fe, Ca and Mg, but less alumina.

The average density of fly ash is about 2.3 g/cm^3 and increases with the content of iron-oxides and decreases with unburned carbon [13]. Densities lower than 2.0 g/cm^3 or higher than 2.5 g/cm^3 are rare. Since the densities of most of the minerals and glass in fly ash are higher than 2.6 g/cm^3, the density of the ash is comparatively low, indicating the presence of a significant amount of closed voids (more than 10% by volume[1]). Hollow particles of fly ash can be easily observed via microscopic examination [17,18].

EFFECT OF FLY ASH ON FRESH AND HARDENED CONCRETE PROPERTIES

Being a finely divided powder consisting of silica-alumina glass of various forms, fly ash in concrete functions as a filler between cement grains and aggregate, and as an effective binder providing a cementitious property.

In fact, when fly ash is a major constituent of concrete, it physically alters the properties of the system. There is also some evidence that fly ash influences the formation of hydration products at very early stages, e.g. by providing nucleation centers for the cement hydration products to precipitate which eventually will form a film or "double-layer" [11]. This film grows dense with time and persists for a long period, but the layer thickness quickly reaches a steady-state value and does not continue to grow past values observed at earlier stages (Figure 5)

Luke and Glasser [19] showed, in their study of ions concentration change, that such an influence for pastes with a water to binder ratio 0.60, cured at 18°C, started as early as 6 hours after mixing. They observed a persistent high concentration of Ca^{2+} and SO_4^{2-} in solution during the induction period, and postulated that the hydration products in the fly ash-cement paste are composed of non-crystalline phases of higher solubility than the crystalline phases such as ettringite in plain cement paste. The slow pozzolanic reaction of fly ash is due to the difficulty in dissolving the glass phase, which requires an increase in alkalinity of the pore solution. The reactivity of mineral admixtures attributed to the reaction of glass was presented by Bijen and Peterson [20] who examined the micro-

---

[1]Void (%) = 100 (1 - D_{App} /D_{Abs}), one obtains: Voids = 100 (1 - 2.3/2.6) = 11.5%

Figure 5. Cement paste with a low-lime fly ash ((30%) at 7 days. A double layer about 1 μm thick is evident on the fly ash particle.

structure and properties of concrete. Fraay et al. [21] demonstrated that the solubility of silica and alumina from a fly ash remained very low until the pH of the solution increased to 13.37. To reach this pH level, a concrete must be hydrated for about one week. Fly ash with the dissociation of the glass phase will release alkalis, which subsequently will add to the solution's alkalinity and the reaction of fly ash will increase. This also suggests that the pore solution ionic strength changes due to the presence and reaction of fly ash, thus the electric resistivity of concrete, which is a function of concentration of electrolytes and porosity, can provide information on the reactivity of fly ash. Such an evaluation has recently been used by Tashiro et al. [22].

Fresh Concrete

Incorporation of fly ash in concrete, either by replacing a part of cement or by replacing a part of fine aggregate [3], changes the concrete proportion. More important, it alters the water to cement ratio (w/c). As stated by Xu [17], the water to binder ratio is related to w/c by

$$\frac{w}{c+F} = \frac{w}{c}\left(1 - \frac{F}{c+F}\right) \tag{1}$$

This means that when the water to binder ratio is the same, the concrete with more fly ash has a higher water to cement ratio. Thus, a 50% replacement leads to a doubled w/c ratio.

The hydration of cement minerals can be explained, at least in part, by the through-solution reaction mechanism. At the initial stage, ions go into solution until a certain supersaturation degree for the hydration products is reached, as was shown by studies of ionic concentration [19] or the solution conductivity [23,24]. At a higher water to cement ratio, the time needed for getting sufficient ions into solution will be longer, and the setting time is longer. This is one of the reasons for the delayed setting of concrete containing Class F fly ashes. Setting delay due to increase in fly ash - cement replacement, and the water to binder ratio was reported by Hwang and Shen [25].

Although fly ash particles provide nucleation points for the formation of hydration products [17] and enhance the hydration process, the overall reduction in cement content in the system leads to a lower evolution of heat. The early hydration proceeds slowly compared to the plain cement system with the same water content, because the evolution of heat leads to a higher temperature which in turn accelerates the hydration. At a higher w/c, there is less heat source material (reacting cement) but more heat-sink (water, fly ash, etc.). Thus, it is often observed that the final setting of fly ash-cement concrete is more retarded than the initial setting [26]. In contrast, high-lime fly ashes, which contain lime and generate heat while it is in contact with water, often exhibits less of a retarding effect [3].

The effect of fly ash on early hydration is also attributed to the presence of various compounds in the ash. Roszczynialski [28] observed a delayed heat evolution due to the presence of fly ash and speculated that unburned coal particles, which are hydrophobic, disturbed the wetting of cement grains and consequently hindered the hydration.

There are reasons to believe that water soluble chemicals in fly ash, typically sulfates and alkalies, also influence properties of fresh concrete [14]. Soroushian and Alhozaimy [29] showed a high-lime fly ash reduced water requirement of concrete from 0.53 to 0.47 at a fly ash to binder ratio of 0.30. This ash contains 5.1% SO_3 , whereas two low-lime low-sulfate fly ashes showed little or no negative effect on workability. Ravina and Mehta [26] also demonstrated that initial setting of concrete was delayed by two high-lime fly ashes which contained a higher amount of sulfate.

The effect is reportedly different for different types of cement, and for different strength grades of concrete as it also depends on the w/c. It was shown that in high strength concretes whose w/c is very low, there was less sulfate going into solution due to a lack of water [30].

One important feature, especially for the Class F fly ash, is that the water required to obtain the desired consistency is somewhat less than its plain cement counterpart. Dhir et al. [15], using fly ash produced in the United Kingdom, demonstrated the fly ashes' effect on water requirements, and concluded that most fly ashes result in a lower water requirement (Figure 6). The researchers proposed that the lower water requirement was due to the dispersion of cement grains by fly ash particles, which was enhanced by the hydrophilic feature of fly ash glassy particles. They considered that this model was more reasonable than the "ball-bearing" model to demonstrate the effect of fly ash on concrete workability proposed by previous researchers (e.g. summarized by Kokubu [1]).

It may be pointed out that the water reducing effect of fly ash decreases with the ash fineness (Figure 6). Because the fly ash particles are larger than the cement particles, have a comparatively smooth surface, and have a smaller absorption capacity, less water is required to wet the fly ash. However, more water is needed to wet the same mass of finer ash. The variation of water requirement occurring within each particle size range can be attributed to differences in morphology and surface area of ash particles. The water requirement has been shown to be lower with an increase in fly ash content. Leshchinsky and Velichko [31] showed the water requirement for concretes with 25, 50 and 75% fly ash were 90, 85 and 77.5% respectively.

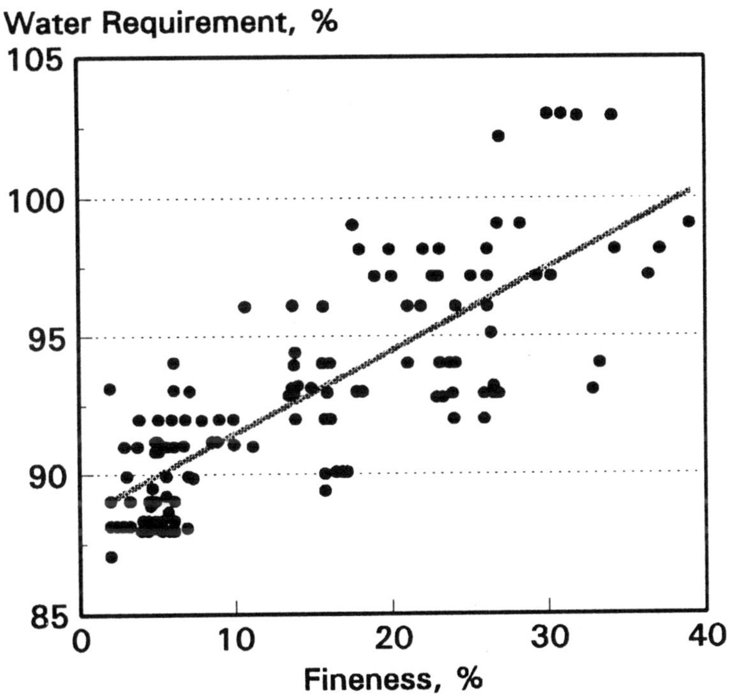

Figure 6. Relation between water requirements and fly ash particle size. Adapted from Dhir [15].

Another important aspect is the surface charge of the grains which is higher on angular surfaces than on the spherical surfaces. This has been observed in the stiffening mud used in oil-well drilling, which often has negatively charged faces and positively charged edges leading to an edge-to-surface aggregation [32]. However, this effect has not been extensively studied in concrete. The work of Nägele and Schneider [33] on the zeta-potential of suspensions of cementitious materials may shed some light in this regard, for example, that fly ash has a more negative zeta-potential than cement in the presence of sulfate ions.

Fly ash has a higher specific volume, the replacement of cement with fly ash by an equal mass increases the paste volume. This offers concrete, especially the lean concrete, an improved workability, as shown by Ravina and Mehta [26]. They also showed that the bleeding of concrete with fly ash is higher, which decreases with an increase in ash fineness. Since bleeding is due to the sedimentation of the solid part in concrete, it depends on the hydration of the cementitious materials. Fly ash retards the reaction, and results in a longer times for building up a rigid skeleton formed by the hydration products and other solid particles.

Hardened Concrete

The reactivity of fly ash develops slowly, primarily due to the large particle size and the stability of the glass particles. By studying the depletion of calcium hydroxide (CH) process in paste containing two types of fly ashes, Marsh and Day [34] observed that CH began to decrease in about 14 days. It has been shown that the 28-day strength of concrete with fly ash is lower [35], whereas the long-term strength of concrete with fly ash is higher [3,36,37] as illustrated in Figure 7. Though this is not always found, for example, the long-term strength of concrete with fly ash is approximately the same as that of the control [38].

Marsh and Day [34] showed that after one year, the CH content per mass of cement was about 22%, 14% and 4% for the pastes with 0, 30% and 50% fly ashes, respectively. In spite of the distinct difference in composition, the two fly ashes' consumption of CH was approximately the same. However, the development of non-evaporable water content and strength were shown to be much faster for the ash with high-lime content. They concluded that the high lime ash was partly cementitious whereas the low lime ash only pozzolanic. They showed the hydration products formed in fly ash added pastes at early age contained higher amount of non-evaporable water than that in plain OPC paste.

Calcium hydroxide crystals form preferentially on fly ash particles because of the latter's "nucleation center effect," which leads to the dispersion of CH with distribution of fly ash. This modifies the matrix of the concrete because CH would have preferentially formed on the aggregate surfaces otherwise, which may weaken the concrete. Even in fly ash concrete, in some zones, the CH concentration can be rather high, so that a phenomenon of "encapsulation" of fly ash particles by CH occurs (Figure 8).

The strength of concrete cured for up to 28 days is generally lower due to the replacement of cement with fly ash, as reported by Babu and Rao [39]. In such a concrete, under-reacted fly ash particles in the paste-aggregate interfacial zone weakens the structure, especially for concrete containing a high content of fly ash as shown by Xu [17], and also for high strength concrete containing 20% fly ash (w/c=0.32) as shown by Sarkar and Baalbaki [40]. The chemical composition of fly ash may provide some differences in this aspect, e.g. the dissolution of alumina and sulfate containing compounds from the ash may lead to formation of ettringite at early ages and increase the strength gain compared to the ash that has less soluble sulfates [14]. Grinding fly ash to increase the early strength of fly ash concrete has been shown to be effective even for ashes with very high carbon content. Grinding has been used since the 1980s in China; and work in this area has continued [41].

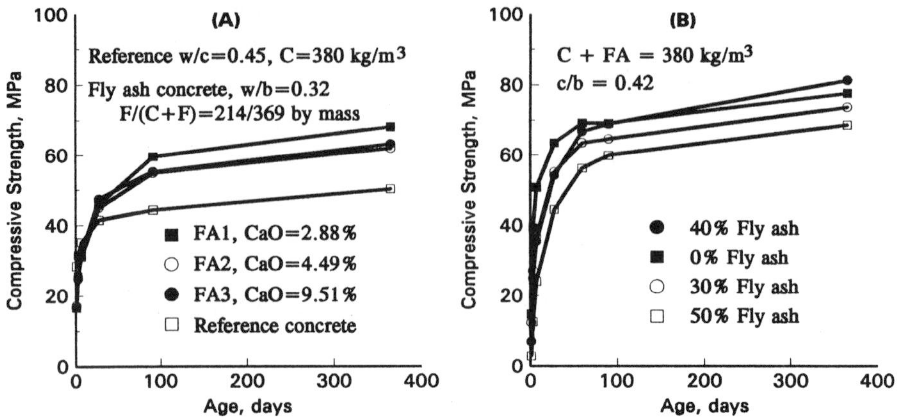

Figure 7. Strength development of concretes made with Type III cement. (A) Concrete with different fly ash compositions [36]; (B) Concretes with different contents of low-lime fly ash [37].

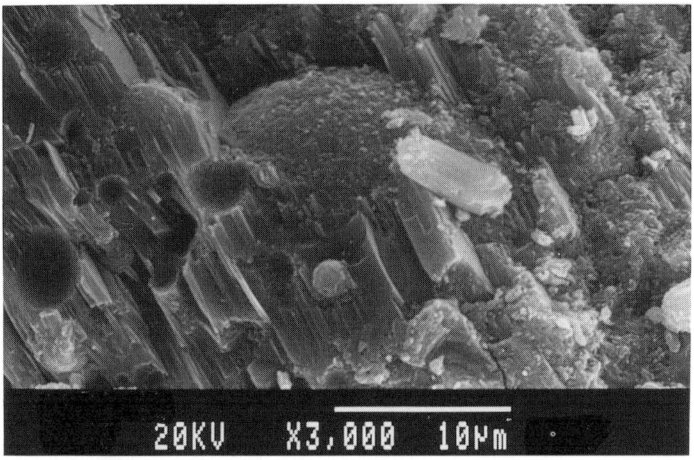

Figure 8. Encapsulation of fly ash particles by CH. Mortar containing 30% fly ash after 180 days. The fly ash particle in the center of the photo had already reacted.

The reaction of fly ash particles observed microscopically shows many variations, i.e. from a completely consumed skeleton, partially reacted spheres with dense surface coatings, to nearly intact particles with smooth surfaces [13] [42]. One can, with some patience, count the apparently reacted particles to assess the reaction of fly ash. Dhir et al. [18], demonstrated that the reacted particles increase from about 10% after 7 days to greater than 60% after 24 months. Figure 9 is a micrograph of a 90-day old paste containing fly ash particles reacted to various degrees. The high strength of the paste caused some fly ash particles to break into halves upon splitting the specimen (preparation of the sample for SEM study), and some particles to be "pulled-out" from the reacted shell. The latter was also shown by Grutzeck et al. in an earlier study [27].

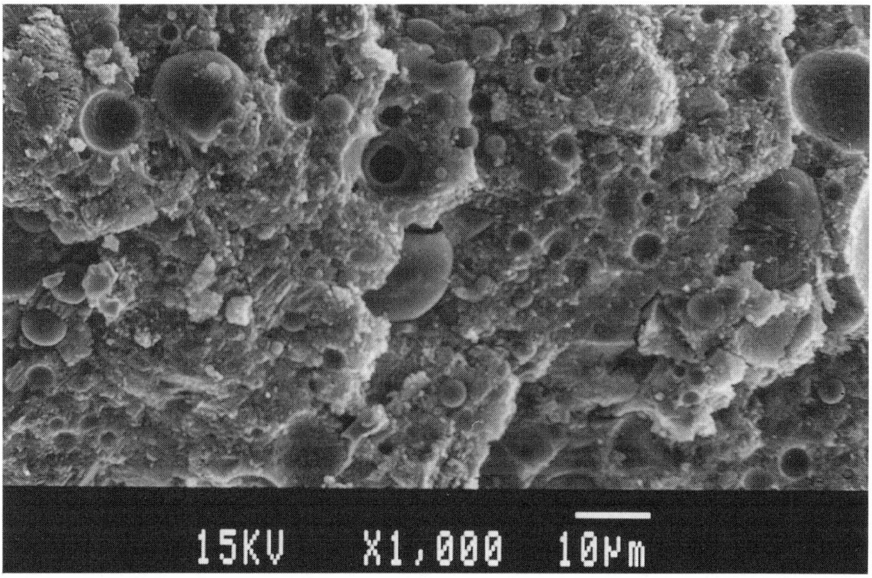

Figure 9. Cement paste containing 30% fly ash (Class F) cured in water for 90 days.

In order to estimate the hydration degree, Xu [17] proposed that non-evaporable water in concrete can be considered simply as a sum of the water held by cement and fly ash hydrates:

$$W_n = \alpha_C W^{\infty}_{nC} C + \alpha_F W^{\infty}_{nF} F \qquad (2)$$

where α is the degree of reaction, W_n^{∞} denotes the maximum non-evaporable water at complete reaction per unit mass, and subscripts C and F denote cement and fly ash respectively. Thus, the non-evaporable water per cement can be expressed by dividing the two sides of the equation by the cement mass c, and rearrange the coefficient:

$$W_n / c = a + k \cdot F / c \qquad (3)$$

where for a Portland cement, $a = 0.23\alpha_C$, and the ratio k/a is a measure of the effect of fly ash. The contribution of fly ash (evaluated by non-evaporable water content per cement mass, excluding that from calcium hydroxide) increases with curing age from 25% at 7 days to 43% at 90 days, which is higher for the lower *w/b* concretes [17]. Using a similar concept in evaluating the contribution of fly ash to strength leads to a 10% increase in strength at 7 days, and a 20% increase after 90 days. This indicates that the hydrates produced from the fly ash reaction have a higher water content, possibly due to the production of calcium aluminate hydrates which contain more interlayer water than C-S-H.

The overall behavior of concrete with fly ash has been reported to be satisfactory. Dhir et al. [43], determined the strength development and deformation behavior of steam cured concretes composed of rapid hardening cement and up to 45% Class F fly ash. The study concluded that the concrete containing fly ash performed as well as or even better than the control. Class C fly ash has exhibited a more pronounced effect, even with comparatively short curing periods, as demonstrated by the concrete permeability shown by Naik et al. [44].

As the pozzolanic reaction is a chemical reaction, its rate will increase with temperature [13] and thus, the dissociation of fly ash glass is enhanced [21]. Leshchinsky and Velichko [31] tested 20 to 40 MPa strength-grade concretes incorporated with 30 to 40% fly ash as binder, which were cured in 85-90°C for 8 hours followed by moist curing at 20°C. Both the post-heating and 28-day strength of the fly ash containing concretes were shown to be higher than that of fly ash-free concretes, however the strength growth during room temperature curing up till 28 days was lower for the former. The strength of fly ash concretes after 2.5 years was about 150% of that at 28-day.

Xu and Sarkar [42] studied the microstructure development of cement paste containing up to 60% fly ash (Class F), and the effect of gypsum on the reaction of fly ash. The results showed that in the presence of gypsum, fly ash particles' dissociation was enhanced and more hydration products formed. The major formation of ettringite (the anticipated phase) in the pastes with additional gypsum started at rather late ages, i.e., after 28 days. This corresponds to an increase in strength that can be attributed to the addition of gypsum for the parallel mortars tested. The calcium hydroxide in the pastes with a high volume of fly ash, as shown by DTA studies, not only decreases with age, but their lattice stability is also lower.

DURABILITY

Some environmental conditions, natural or those due to human activity, have been proved to cause the deterioration of concrete. It is difficult to generalize the mechanisms in a few words, however, it can be said that the durability of concrete is often associated with the porosity and transport (gas, ion, water) properties of concrete. This section will begin by reviewing these properties (some Cl^- ion diffusion studies are included in the section for reinforcing steel corrosion), whereas other specific properties are dealt with in separate subsections.

Using the helium pycnometric method and mercury porosimetry, Xu and Huang [45] presented that pore structure development of a fly ash blended cement (30% fly ash) and the relation between strength and porosity. Differences were observed between the fly ash cement and the plain OPC counterpart. Despite a higher total porosity, the paste with fly ash showed a lower porosity for pores larger than 1000 Å after 90 days curing when its strength was more augmented. It was postulated that strength of fly ash cement paste at early ages is negatively affected by the presence of gap between fly ash particles and the surrounding matrix, which is gradually filled with pozzolanic reaction products.

Kobayashi and Shuttoh [46] determined the diffusion coefficient of O_2 in cement pastes and concretes (moisture saturation degree 60%) and showed that the values for plain cement concretes were in range of 2.4 to 9.8×10^{-8} m^2/s where the low values were for the 28-day water cured samples and the high values for the 50% RH (20°C) cured samples. The corresponding values for concretes with 20% fly ash as binder were 1.8 to 12.3×10^{-8} m^2/s. This illustrates the importance of adequately curing the fly ash concretes and further shows that concrete with fly ash can be made better than ordinary concrete.

Yu and Page [47] determined oxygen diffusion in water through cement pastes, as well as that of Cl^-. Fly ash (20%) added pastes showed lower transport coefficients than the plain cement pastes with the same water to binder ratios. Interestingly, the oxygen diffusivity appears to be at least twice as high as that of Cl^-, and the difference increases to nearly 9-fold for low Cl^- diffusivity samples, in spite of the fact that the ionic diffusivity of oxygen and Cl^- in water are nearly the same. They postulated that this may be because of the formation of a double layer on the cement pastes, which in ideal case, may give rise to a thickness of 13 Å in a 1 M NaCl solution (the condition used in their test). However, more work must be done to verify this. A study by Chatterji and Kawamura [48] determined the surface charge of silica particles and C-S-H, and proposed a model to explain the effect of the double layer on ion transport as well as hydration..

Dhir and co-worker's investigation [49] on the effect of fly ash on Cl^- diffusion coefficient clearly shows a decrease in the diffusivity with an increase in fly ash glass content (which is associated with Al_2O_3 content), fineness, and fly ash content. It was confirmed that inadequate curing will cause a higher Cl^- diffusivity for the high volume fly ash concrete. However, the diffusivity determined was sufficiently low ($D < 2 \times 10^{-12}$ m^2/s), and concrete with a 75 mm cover may offer at least 50 years protection to reinforcing steel. It is important to note that the diffusion coefficient determined by the diffusion-cell method, as used by these authors, is lower than that in the real diffusion process by at least a porosity factor. Thus one should be cautious in using these data to predict the concrete "service life."

Freeze-Thaw Resistance

Nasser and Ghosh [50] demonstrated a generally good freeze-thaw resistance of high strength concrete containing lignite fly ash and silica fume (replacing less than 50% cement), and showed the microstructure of the concrete was modified by the presence of fly ash. A higher fly ash content resulted in a lower freeze-thaw durability. Based on microstructure study showing that ettringite formed in the pores after freeze-thaw cycles, it was postulated that the deterioration was a result of an increase in the amount of ettringite dissolved into the pore solution (because of the reduction or depletion of calcium hydroxide in the high volume fly ash concrete) and recrystallized in air-voids and caused cracks.

The replacement of cement with Class F fly ash often leads to a low resistance of concrete to freeze-thaw cycles. Machand et al. [51] reported a poor resistance to deicer scaling for a low w/b roller-compacted concrete containing fly ash. This is closely related to the low degree of pozzolanic reaction, as it is with strength, especially for the concrete surface which is subjected to drying at early ages.

Because high volume fly ash concrete is subject to deterioration from freeze-thaw cycles, many studies have been conducted to determine the cause. It has been shown that with adequate curing, Class F fly ash incorporated in concrete, even at a high volume, can perform satisfactorily. This is confirmed by the results shown by Sirivivatnanon et al. [52] who tested the effect of Australian fly ashes on durability as well as mechanical properties.

Bisaillon et al. [53], and Bilodeau and Malhotra [54] investigated the durability of high volume FA massive concrete which included: drying shrinkage, strength, Young's modulus of the concrete, and the effect of fly ash replacement on concrete temperature. They showed that the fly ash had a beneficial effect and attributed to a reduction in accumulated heat in young massive concrete and consequently, less crack formation with an increase in volume of fly ash. However, the results also strongly suggested that adequate moist curing of these high volume fly ash concretes played a key role in the excellent performance.

Selecting fine fly ash particles to enhance the concrete durability is another method that has been attempted by many researchers. For example, Ishii et al. [55] showed that fine particles of fly ash classified by an air-classifier performed better than raw ash. However, this can also be attributed to the fact that fly ash so treated is richer in glass content (the hollow-spherical glass particles which are low in density make up a large portion of the fines).

Sulfate Resistance

In a sulfate containing environment, concrete may deteriorate. Seepage water, ground water and alkali soil which contains a high sulfate salt content, commonly $MgSO_4$, Na_2SO_4 and $CaSO_4$, can lead to disintegration of concrete. For Portland cement, a high C_3A content has been recognized as the single reason that results in the lower resistance of concrete to sulfate attack. However, high alumina cement concrete used at relatively low temperatures has been proven to be immune from sulfate attack for extended periods of time [56].

Soroushian and Alhozaimy [29] investigated the effect of two low-lime fly ashes and a high-lime fly ash (18.6% CaO, 5.1% SO_3 and 8.5% alkalies) on the sulfate resistance of concretes. In term of reducing expansion, the replacements of 20% by the low-lime fly ash and 30% cement by the high-lime fly ash were shown to be effective; whereas at a 10% replacement level, the high-lime fly ash increased expansion. The finer low-lime fly ash showed a better effect. A correlation between expansion and "Cl^- permeability" (by ASSHTO T277 method) has been demonstrated. It should be noted that this method demonstrates the electrical conductivity of the specimen, which is a function of pore solution chemistry (mainly OH^- and Ca^{2+} concentration) as well as the pore structure. Thus, this result indicates the significance of OH^- and Ca^{2+} in the mechanism.

This may be compared to the findings of Nasser and Ghosh [50] who showed a high strength concrete containing 10% silica fume and up to 50% lignite fly ash performed well in sulfate attack test, or even considerably suppressed the attack. Note that the pozzolanic reaction of silica fume reduces OH^- in pore solution, thus offsets the possible negative effect brought about by the high-lime fly ash. The benefit of using silica fume together with fly ash has also been demonstrated by Marzouk and Jiang [57] who examined the strength and rupture energy, etc. of such concretes after 700 cycles freeze-thaw tests.

A negative effect of fly ash has also been reported. In a study conducted by Paillère [53], the resistance of mortar specimens to artificial seawater showed degradation for specimens where low C_3A cement (up to 30%) was replaced with fly ash, or only marginally improved for the mortars made with a high C_3A cement. This can be attributed to the short moist curing period, or the 50% RH conditioning after the water-curing, as the latter is a drying rather than a curing condition for concrete containing fly ash.

The type, or rather the chemical composition, of fly ash is important in the performance of concrete in the sulfate environment.

Based on the results of testing 57 concrete mixtures after 450 days of sulfate exposure (in 10% Na_2SO_4 maintained at pH < 10.0), Freeman and Carrasquillo [58] reported an improved performance of concretes with low-lime fly ashes, and more deterioration for the concretes containing high-lime fly ashes. They concluded that, the reaction between sulfate and the high calcium aluminosilicate glass in the high-lime fly ashes, and formation of gypsum by the reaction between calcium hydroxide and the sulfate ions are responsible for these phenomena. To some extent, there is a correlation between the sulfate resistivity and the content of alumina phases and free-lime in the concrete (see Figure 10) [59].

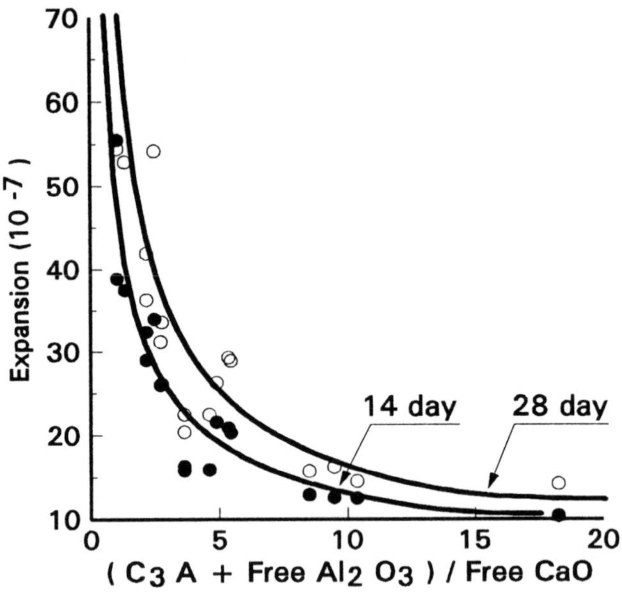

Figure 10. The relationship between sulfate expansion and C_3A, Al_2O_3 and CaO. Adapted from [59].

Alkali Aggregate Reaction

Alkali aggregate reaction has been reported to be remarkably reduced by the addition of Class F fly ashes [60]. This is generally true, despite that a portion of the alkalis in fly ash are readily soluble in water, and others present in the glass matrix will gradually be released into pore solution.

In fact, only part of alkalies in fly ash, even under leach test conditions as specified by ASTM C311, are released into solution. Data presented by Lee et al. [61] showed the leaching rate was high for ashes stored at 55°C than for those stored at 38°C. However, the amount of alkali released after six months testing was not significantly different, and in some cases, the samples cured at 55°C
showed a lower available alkali content at later ages than the 38°C counterparts. Obviously, the pozzolanic reaction that occurred during the process influences the ability of fly ash to retain alkalis.

The findings of Glasser et al. [62] who studied concrete pore solution, showed that fly ash lowered the alkali concentration in the solution at early ages; while the content of alkalis released from fly ash after long times reached steady- state conditions within 90 days. The researchers concluded that fly ash decreases the C/S ratio of the C-S-H phase and thereby improves its uptake potential for alkali, noting that the alkali partition into C-S-H is most effective for Cs, and decreases for K and Na.

Duchesne and Bérubé [63] showed that low-lime fly ashes, with hydration time consumed alkali in the pore solution and when the alkali concentration was lower than 0.65 N, the AAR expansion of concrete was reduced. It was believed that the pozzolanic reaction products entrapped alkalies. One high-lime fly ash studied, caused an increase in alkali concentration in pore solution, but did not lead to AAR expansion of a concrete made with lime stone as aggregate.

Qian et al. [64], used a low-lime fly ash and slag to partially replace cement to study their effect in reducing AAR. The expansion was reduced with an increase in the binder's initial acidity (fly ash or slag content). It was concluded that this was due to a decrease in reactivity between the reactive aggregate and alkali brought by the fly ash and slag. Hobbs [65], in a study of concrete containing cristobalite and stored in a moist environment at 20°C, showed that the onset of expansion was generally delayed in concretes with fly ash, but low (6%) replacement by the fly ashes led to greater AAR expansion. Higher replacements (25 and 40%) reduced expansion, depending on the alkali content of fly ash. Berra et al. [66], however, found no correlation between the performance of fly ashes and their available alkali content.

Steel Reinforcement Corrosion

There are two main focuses for the study of steel corrosion in concrete: passivation of steel in a certain alkaline condition, and the acceleration effect of chloride ions in pore solution.

By studying the pressed pore solution of pastes (w/b=0.50, age 14 months), Byfors [67] showed that the presence of 15% fly ash caused a pH decrease from 13.4 to 13.1. This reduction, which corresponds to a 50% reduction in OH^- concentration in the solution, is significant in terms of the initiation of corrosion. On the other hand, fly ash resulted in a significant reduction in Cl^- diffusivity (diffusion-cell method, tested in a 1.0 M NaCl solution at room temperature). Specimens containing 40% fly ash with w/b=0.40, 0.50 and 0.60 showed diffusivities of 0.79, 1.3 and 17 $\times$ 10^{-13} m^2/s, whereas their OPC counterparts had values 18, 68 and 187 $\times$ 10^{-13} m^2/s respectively.

The diffusion coefficient determined by different laboratories is not always in agreement. Nevertheless, it has been proven that the addition of fly ash reduces Cl^- diffusivity. The diffusivity of Cl^- in pastes determined by Kumar et al. [68] using a 0.05 M CsCl solution at 38°C was 129 $\times$ 10^{-13} m^2/s for OPC paste with w/c=0.35 and 13.5 $\times$ 10^{-13} m^2/s for its 30% fly ash counterpart. The explanation offered for the decrease in the ion diffusivity was that the fly ash gave rise to a fine pore system with a more torturous path.

In real concrete structures whichoften have inadequate early curing, the slow hydration rate of fly ash blended cementing system tend to cause shrinkage-induced cracks. This may completely offset the beneficial effect of fly ash in reducing Cl^- diffusion, as was reported by Mangat and Gurusamy [69].

Kawamura et al. [70], determined the pore solutions of mortars mixed with $CaCl_2$ or NaCl, reported that a 30% substitution of fly ash did not have effect on Cl^- concentration for the mortar with less than 2% Cl^- added in the form of $CaCl_2$; whereas fly ash considerably reduced Cl^- concentration in pore solution for the mortars mixed with NaCl. One can deduce from this that fly ash particles have a higher capacity in adsorbing free Cl^-.

Beaudoin et al. [71], showed that low C/S ratio C-S-H bound less Cl^-, which was in agreement with the study of silica fume added to concrete. This does not seem to be the case for concrete containing fly ash which also produces low calcium C-S-H [17]. This can be explained by a reaction mechanism between the chloride and alumina phases which are rich in most of the Class F fly ash blended concrete. In this line, it is noteworthy to mention that Glasser et al. [62] briefly stated the forms of reaction products of chloride and cement components as:

Reactants	Nature of Phase Formed
$Ca(OH)_2$, $CaSO_4$ and dilute $CaCl_2$ with aluminates	C_4A_{13} - Type: $C_3A \cdot 0.5Ca(OH,Cl)_2 \cdot 0.5CaSO_4 \cdot 12H_2O$ - $C_3A \cdot CaSO_4 \cdot 12H_2O$ solid solution
$Ca(OH)_2$, $CaSO_4$ and concentrated $CaCl_2$ with aluminates	α. β - $C_3A \cdot CaCl_2 \cdot 10H_2O$ and $C_3A \cdot 3CaCl_2 \cdot 30H_2O$ (below -10°C)

Hussain and Rasheeduzzafar [72] tested the effect of fly ash blending (30%) on the corrosion resistance of reinforced concretes made with Type I cement (C_3A = 14.00%) and Type V cement (C_3A = 2.43%). For specimens mixed with Cl^-, FA reduced the free chloride concentration in pore solution by 13-34% and 32-44% for Type I and V cement respectively being more effective at low Cl^- level while it reduced the OH^- concentration by 40% to 20%. They also showed that although the total porosity of the blended cement was slightly higher, the water permeability (at 180-day) and Cl^- diffusivity (90-day) were reduced by 5 and 6 times respectively. The time of corrosion initiation in the blended cement concretes was about twice as long as in their plain cement counterparts.

Likewise, numerous studies have shown that the incorporation of fly ash significantly reduced Cl^- diffusion and that this delayed Cl^- induced reinforcing steel corrosion. Kouloumbi et al. [73], reported the "anticorrosion" effect of fly ash in reinforced concrete. Montemor et al. [74], based on the electric (alternating current) impedance measurement results, demonstrated that the resistance of steel (concretes exposed to a 3% NaCl solution) to corrosion was improved with the increase in fly ash content in concrete and speculated that the concrete matrix was densified by fly ash. Polder et al. [75], presented the corrosion of steel decreased due to the presence of fly ash. Zhang and Wheat [76] studied the corrosion rate by polarization resistance, concluded that corrosion protection offered by fly ash could be further improved when superplasticizers were used.

Dhir et al. [77], demonstrated that the corrosion initiation was greatly increased because of fly ash, observed that the total quantity of Cl^- causing corrosion was not significantly different for the concrete with or without fly ash. This indicates that the reaction between fly ash and Cl^- is either largely a physical adsorption, or the reduction in pH value of pore solution due to the pozzolanic reaction which causes a lower threshold.

Carbonation

Due to the consumption of calcium hydroxide by pozzolanic reactions, fly ash containing concrete is likely to be more susceptible to carbonation. Studies in this area have been directed to examine the extent to which fly ash may affect the resistance to carbonation, and how to improve this property.

Concretes made with blended cements (10% and 25% fly ash) studied by Madge [78] were carbonated somewhat faster than plain cement concrete, even when the blended cements had higher fineness and the concretes made with it (the blended cement) had higher 28-day compressive strength. The relative rate of carbonation for the concretes exposed to atmospheric CO_2 in 50% RH compared with the protected (from rain) outdoor samples, and those outdoor samples without shelter were 100%, 88% and 66% respectively. This is attributed to the presence of different amounts of water in the concretes, which blocks the CO_2 diffusion path. Xu [17] revealed that even with identical RH conditions, the moisture content in concrete will vary significantly between concretes. This effect is a function of the moisture sorption capacity of concrete and the carbonated fly ash containing concrete is much more vulnerable than its plain concrete counterpart.

Data shown by Meland [79] on carbonation rate of concretes carbonated in 60% RH and 90% RH revealed a larger difference for paste with a water to binder ratio (w/b) = 0.5, but much less for that with w/b = 0.7. It was shown that the carbonation of cement-fly ash pastes had a relatively higher content of vaterite, whereas the plain cement paste contained more calcite.

The direct reduction in pH value of pore solution due to the presence of fly ash has been shown to be of only marginal significance. However, the rate of carbonation was higher for concrete containing more fly ash, as reported by Al-Amoudi et al. [80]. Apparently, this was further enhanced by the Cl^- contamination of concrete. Their data indicate that when the chloride content in the concrete with 40% fly ash is higher than 0.68% by mass of cement, the concrete is more easily carbonated, e.g., the depth of neutralization was 10.0 mm in 11 months, whereas the same type of concrete with less than 0.34% chloride per cement had a depth of about 7.0 mm. A generally high carbonation rate for fly ash containing concretes was explained by the higher permeability (from drying), but the effect of chloride was not clear. An alternative explanation to these data is, that carbonation results in a coarser pore structure and lower Cl^- binding capacity, and consequently the concrete allows a faster Cl^- diffusion.

In addition to the known effect on steel passivation, carbonation leads to a lower specific surface in hydrated cementitious materials. As demonstrated by Xu [17], a significant change in moisture sorption behavior was observed, especially for concrete containing fly ash. This also decreases the capacity of concrete in binding Cl^-, which depends primarily on the surface area of the cement paste, as is shown in the last section. The experimental evidence of the negative effect of carbonation on chloride binding has been recently shown in several studies, Kayyali and Haque [81].

OTHERS

The ability of fly ash to react with various chemicals has been utilized to solve some engineering problems such as improving materials for prefabricated building components.

Limestone for making cement often contains impurities. As a result, the cement may not comply with established standards for cement composition. Using fly ash and other pozzolanic materials in concretes made with such cements initially reduces ("dilution effect") the amount of harmful compounds in the system. This effect has been studied in order to improve the concrete performance. Rehsi and Garg [82] reported the use of a low-lime fly ash in a concrete made with a high magnesia cement (MgO 10% in clinker). The concrete with fly ash showed an expansion of 0.263% after 20 years curing in water at 27°C, which is remarkably less than the control concrete whose expansion was 1.833% after only 14 years. Apparently, there is less $Mg(OH)_2$ formed in fly ash concrete as revealed by DTA thermogram.

Lian and Chen [83] studied the expansion and strength of cement with extra gypsum (added as gypsum or anhydrite). Incorporation of a fly ash (low-lime, 0% SO_3) reduced the expansion and improved concrete strength. Other pozzolans were also used and showed a similar effect but to different extents, which indicated that the effect of pozzolan was not only due to the "dilution effect" of C_3A in the system.

A study conducted on the use of fly ash in lightweight aggregate (expanded clay) concrete subjected to heat treatment, Leshchinsky and Velichko [31], showed that use of 20% fly ash resulted in strengths twice that of normal values. This allows an increase in air-entraining while achieving the desired strength. This, and a partial replacement of quartz sand by fly ash leads to a reduction in concrete unit weight by 100 to 150 kg/m^3, and a slightly lower heat conduction in dry state (λ=0.25-0.27 W/m·°C). It was observed that the concrete with fly ash had a better air-holding ability (for transportation and handling). Likewise, Atzeni et al. [84] reported, for gas-concretes fabrication, concrete with 40% fly ash steam-cured at 70°C, and that with 60% fly ash autoclaved at 180°C showed a significantly higher strength than their plain cement counterparts.

REFERENCES

1. Kokubu, M., "Fly Ash and Fly Ash Cement", *Proc. 5th Int. Symp. Chem. Cem.*, Tokyo, Vol. IV, 75-113 (1968).
2. Takemoto, K., and Uchikawa, H., "Hydration of Pozzolanic Cement," *Proc. 7th Int. Cong. Chem. Cem.*, Paris, Vol. I, IV2/1-29 (1980).
3. Berry, E.E., and Malhotra, V.M., *Fly Ash in Concrete*,CANMET SP85-3 (1986).
4. *Fly Ash and Coal Conversion By-Products: Characterization, Utilization and Disposal II, MRS Symp. Proc.* Vol. 65, (Gregory J. McCarthy, Fredrik P. Glasser and Della M. Roy, eds) Boston, MA (1985).
5. Joshi, R.C., and Lohtia, R.P., "Types and Properties of Fly Ash," MACC, 118-157 (1993).
6. Stevenson, R.J. and MacCArthy, G.J., "Minerology of Fixed Bed Gasification Ash Derived from North Dakota Lignite," Ref. 4, 77-90.

7. Hemmings. R.T. and Berry, E.E., "Speciation in Size and Density Fractionated Fly Ash," Ref. 4, 91-104.
8. Rehsi, S.S., "Portland Fly Ash Cement", *MACC*, 158-173 (1992).
9. Culmo, R.F. and Fyans, R.L., "Thermogravity as a Tool for Determining Combustion Efficiency and Calcium Utilization of a Fluidized Bed Combustion Furnace," *Compositional Analysis by Thermo-gravimetry*, edited by CharlesM. Earnest, *ASTM STP 997*, 245-153 (1988).
10. Roper, H., Baweja, D. and Kirkby, G.A., "Production and Charecterization of Fly Ash from New South Wales Coals," Ref. 4, 115-124.
11. Diamond, S., "The Characterization of Flyashes," *Proc. MRS Symp. N.*, 12-23 (1991).
12. Finkelman, R.B., "Characterization of the Inorganic Constituents in Coal, Ref 4, 71-76.
13. Xu, A. and Sarkar, S.L., "Hydration and Properties of Fly Ash Concrete," *MACC*, 174-225 (1992).
14. Carles-Gibergues, A. and Aitcin, P.-C., "Analysis of the Cementitious and Pozzolanic Properties of a Silico-Aluminous (Class F) Fly Ash," Ref 4, 181-192.
15. Dhir, R.K., Hubbard, F.H., Munday, J.G.L. and Jones, M.R., "Characteristics of Low-Lime Fly Ashcs Significance to Their Use in Concrete," *ACI SP-91*, 693-721.
16. Butler, W.B. and Mearing, M.A., "Fly Ash Beneficiation and Utilization in Theory and in Practice," Ref 4, 11-17.
17. Xu, A., *Structure of Hardened Cement-Fly Ash Systems and their Related Properties*, P-92:7, Division of Building Materials, Chalmers University of Technology, Göteborg, (1992).
18. Dhir, R.K., Munday, J.G.L., Jones, M.R. and Duerden, S.L., "Contribution of PFA to Concrete Workability and Strength Development," *Cem. Conc. Res.* 18(2), 277-289 (1988).
19. Luke, K. and Glasser, F.P., "Chemical Change Occurring the Early Hydration of PFA-OPC Mixtures", Ref 4, 173-180.

20. Bijen, J. and Pietersen, H., "Mineral Admixtures : Reactions, Micro-Structure and Macro-Properties," *Adva. Cem.Conc.* 292-328.
21. Fraay, A.L.A., Bijen, J.M. and de Haan, Y.M., "The Reaction of Fly Ash in Concrete : A Critical Examination," *Cem. Conc. Res.* 19(2), 235-246 (1989).
22. Tashiro, C., Ikeda, K. and Inoue, Y., "Evaluation of Pozzolanic Activity by the Electric Resistance Measurement Method," *Cem. Conc. Res.* 24(6), 1133-1139 (1994).
23. Damidot, D. and Nonat, A., "A Method for Determining the Advancement of the Hydration of C_3S in Diluted Suspensions by Means of Simultaneous Conductimetric and Calorimetric Measurements," *Proc. 9th Int. Cong. Chem. Cem.*, New Delhi, Vol. IV, 227-233 (1992).
24. Vernet, C. and Noworyta, G., "Reactivity Test for Fine Silicas and Pozzolanic Mineral Additives," *Proc. 9th Int. Cong. Chem. Cem.*, New Delhi, Vol. III, 79-85 (1992).
25. Hwang, C.-L. and Shen, D.-H., "The Effect of Blast-Furnace Slag and Fly Ash on the Hydration of Portland Cement," *Cem. Conc. Res.* 21(4), 410-425 (1991).
26. Ravina, D. and Mehta, P.K., "Properties of Fresh Concrete Containing Large Amounts of Fly Ash," *Cem. Conc. Res.* 16(2), 227-238 (1986).
27. Grutzeck, M.W., Roy, D. M. and Scheetz, B. E., "Hydration Mechanisms of High-Lime Fly Ash in Portland Cement Composites," *Proc. MRS Symp. N.*, 92-101 (1981).
28. Roszczynialski, W., "The Influence of Some Fly Ashes and Blast-Furnace Slags on Blended Cement Hydration and Properties," *Proc. 9th Int. Cong. Chem. Cem.*, New Delhi, Vol. IV, 698-704 (1992).
29. Soroushian, P. and Alhozaimy, A., "Correlation Between Fly Ash Effects on Permeability and Sulfate Resistance of Concrete," *Proc. 9th Int. Cong. Chem. Cem.*, New Delhi, Vol. V, 196-202 (1992).

30. Tagnit-Hamou, A., Baalbaki, M. and Aitcin, P. C., "Calcium Sulphate Optimization in Low Water/Cement Ratio Concretes for Rheological Purposes," *Proc. 9th Int. Cong. Chem. Cem.*, New Delhi, Vol. V, 21-25 (1992).
31. Leshchinsky, M. Y. and Velichko, V. A., "The Use of Fly Ash in Concretes Subject to Heat," *Cem. Conc. Res.* 21 (2/3), 205-218 (1991).
32. Shaw, D.J., *Introduction to Colloid and Surface Chemistry*, 3rd Ed., Butterworths (1986).
33. Nägele, E. and Schneider, U., "The Zeta-Potential of Cement. Part V. Effect of Surfactants," *Cem. Conc. Res.* 18(2), 257-264 (1988).
34. Marsh, B.K. and Day, R.L., "Pozzolanic and Cementitious Reactions of Fly Ash in Blended Cement Pastes," *Cem. Conc. Res.* 18(2), 301-310 (1988).
35. Paillère, A.M., Platret, G., Roussel, P. and Gawsewitch, J., "Influence of Mechanical Strength and Curing Methods on Sea Water Durability of Mortars Containing Fly Ash and Slag," *ACI SP-132*, 179-204.
36. Bilodeau, A. and Malhotra, V. M., "Properties of High-Volume Fly Ash Concrete Made with High Early-Strength ASTM III Cement," *ACI SP-153*, 1-23.
37. Galeota, D., Giammatteo, M. M. and Marino, R., "Structural Concrete Incorporating High Volume of Fly Ash," *ACI SP-153*, 25-42.
38. Nepper-Christensen, P., Kristensen, B. W. and Rasmussen, T. H., "Long-Term Durability of Special High Strength Concretes," *ACI SP-145*, 173-190.
39. Babu, K. G. and Rao, G. S. N., "Early Strength Behaviour of Fly Ash Concretes," *Cem. Conc. Res.* 24(2), 277-284 (1994).
40. Sarkar, S. L. and Baalbaki, M., "The Influence of the Type of Cement on the Properties and Microstructure of High-Performance Concrete," *Proc. 9th Int. Cong. Chem. Cem.*, New Delhi, Vol. V, 89-94 (1992).

41. Monzó, J., Payá, J. and Peris-Mora, E., "A Preliminary Study of Fly Ash Granulometric Influence on Mortar Strength," *Cem. Conc. Res.* 24 (4), 791-796 (1994).
42. Xu., A and Sarkar, S.L., "Microstructural Study of Gypsum Activated Fly Ash Hydration in Cement Paste," *Cem. Conc. Res.* 21(6), 1137-1147 (1991).
43. Dhir, R.K., Munday, J. G. L. and Ho, N. Y., "PFA in Structural Precast Concrete: Engineering Properties," *Cem. Conc. Res.* 18(6), 852-862 (1988).
44. Naik, T. R., Singh. S. S. and Hossain, M. M., "Permeability of Concrete Containing Large Amounts of Fly Ash," *Cem. Conc.Res.* 24(5), 913-922 (1994).
45. Xu, L. P. and Huang, S. Y., "Relationship Between Strength and Porosity of Fly Ash Cement Pastes," *Proc. 9th Int.Cong. Chem. Cem.*, New Delhi, Vol. IV, 337-349 (1992).
46. Kobayashi, K. and Shuttoh, K., "Oxygen Diffusivity of Various Cementitious Materials," *Cem. Conc. Res.* 21(2), 273-284 (1991).
47. Yu, S.W. and Page, C.L., "Diffusion in Cementitious Materials: 1. Comparative Study of Chloride and Oxygen Diffusion in Hardened Cenemt Pastes," *Cem. Conc. Res.* 21(4), 581-588 (1991).
48. Chatterji, S. and Kawamura, M., "Electrical Double Layer, Ion Transport and Reactions in Hardened Cement Paste," *Cem. Conc. Res.* 22(5), 774-782 (1992).
49. Dhir, R. K., Jones, M. R. and Seneviratne, A. M. G., "Diffusion of Chloride into Concrete. Influence of PFA Quality," *Cem. Conc. Res.* 21(6), 1092-1102 (1991).
50. Nasser, K. W. and Ghosh, S., "Durability Properties of High Strength Concrete Containing Silica Fume and Lignite Fly Ash," *ACI SP-145*, 191-214.
51. Marchand, J., Pigeon, M., Boisvert, J., Isabelle, H. L. and Houdusse, O., "Deicer Salt Scaling Resistance of Roller-Compacted Concrete Pavements Containing Fly Ash and Silica Fume," *ACI SP-132*, 151-178.

52. Sirivivatnanon, V., Cao, H. T. and Nelson, P., "Mechanical and Durability Properties of High Volume Fly Ash Concrete," *ACI SP-145*, 967-984.
53. Bisaillon, A., Rivest, M. and Malhotra, V. M., "Performance of High-Volume Fly Ash Concrete in Large Experimental Monoliths," *ACI Mat. J.* 91(2), 178-187 (1994).
54. Bilodeau, A. and Malhotra, V. M., "High-Performance Concrete Incoporating Large Volume of ASTM Class F Fly Ash," *Adva. Cem. Conc.,* 508-523.
55. Ishii, M., Ukita, K. and Kohno, K., "Durability of Concrete Using Air-Classified Fly Ash," *ACI SP-145*, 985-1002.
56. Lea, F. M., *The Chemistry of Cement and Concrete*, 3rd Ed., Edward Arnold Ltd. (1970).
57. Marzouk, H. and Jiang, D., "Effects of Freezing and Thawing on the Tension Properties of High-Strength Concrete," *ACI Mat. J.* 91(6), 577-586 (1994).
58. Freeman, R. B. and Carrasquillo, R. L., "Effects of Intergrinding Fly Ash on the Sulfate Resistance of Fly Ash Concrete," *ACI SP-132*, 281-297.
59. Erdo_an, T. Y., Tokyay, M. and Ramyar, K., "Investigation on the Sulfate Resistance of High-Lime Fly Ash Incorporating PC-fa Mortars," *ACI SP-132*, 271-280.
60. Thomas, M., "Fly Ash and Alkali-Aggregate Reaction," *Adva. Cem. Conc.* 362-376.
61. Lee, S., Schlorholtz, S. and Demirel, T., "Available Alkalis in Fly Ash," Ref 4,125-130.
62. Glasser, F.P., Luke, K. and Angus, M.J., "Modification of Cement Pore Fluid Compositions by Pozzolanic Additives," *Cem. Conc. Res.* 18(2), 165-178 (1988).
63. Duchesne, J. and Bérubé, M.A., "The Effectiveness of Supplementary Cementing Materials in Suppressing Expansion due to ASR: Another Look at the Reaction Mechanisms Part 2: Pore Solution Chemistry," *Cem. Conc. Res.* 24(2), 221-230 (1994).

64. Qian, C., Guo, H. and Tang, M., "Mechanism of Mineral Admixture Suppressing Alkali-Silica Reaction: Part I, Corrosion Degree of Reactive Aggregate in Blended Cement Pastes and Its Correlations with Expansion Value and Electric Resistance Change," *Cem. Conc. Res.* 24(6), 1111-1120 (1994).
65. Hobbs, D: W., "The Effectiveness of PFA in Reducing the Risk of Cracking Due to ASR in Concretes Containing Cristobalite," *Mag. Conc. Res.* 46(168), 167-175, (1994).
66. Berra, M., Mangialardi, T. and Paolini, A. E., "Application of the NaOH Bath Test Method for Assessing the Effectiveness of Mineral Admixtures Against Reaction of Alkali with Artificial Siliceous Aggregate," *Cem. Conc.Comp.* 16(3), 207-218 (1994).
67. Byfors, K., "Influence of Silica Fume and Fly Ash on Chloride Diffusion and pH Values in Cement Paste," *Cem. Conc. Res.* 17(1), 115-130 (1987).
68. Kumar, A., Komarneni, S. and Roy, D. M., "Diffusion of Cs^+ and Cl^- Through Sealing Materials," *Cem. Conc. Res.* 17 (1), 135-160 (1987).
69. Mangat, P. S. and Gurusamy, K., "Chloride Diffusion in Steel Fibre Reinforced Concrete Containing PFA," *Cem.Conc. Res.* 17(4), 640-650 (1987).
70. Kawamura, M., Kayyali, O. A and Haque, M. N., "Effect of Fly Ash on Pore Solution Composition in Calcium and Sodium Chloride Bearing Mortars," *Cem. Conc. Res.* 18 (5), 763-773 (1988).
71. Beaudoin, J. J., Ramachandran, V. S. and Feldman, R. F., "Interaction of Chloride and C-S-H," *Cem. Conc. Res.* 20(6), 875-883 (1990).
72. Hussain. S. E. and Rasheeduzzafar, M., "Corrosion Resistance Performance of Fly Ash Blended Cement Concrete," *ACI Mat. J.* 91(3), 264-272 (1994).

73. Kouloumbi, N., Batis, G. and Malami, C., "The Anticorrosive Effect of Fly Ash, Slag and a Greek Pozzolan in Reinforced Concrete," *Cem. Conc. Comp.* 16(4), 253-260 (1994).
74. Montemor, M.F., Salta, M. M., Simões, A. M. P. and Ferreira, M.G.S., "Influence of Fly Ash Addition on Reinforced Concrete Corrosion. Comparison Between Laboratory and Field Tests," *Steel in Conc.* 2, 751-761.
75. Polder, R. B., Bamforth, P. B., Basheer, M., Chapman-Andrews, J., Cigna, R., Jafar, M.I., Mazzoni, A., Nolan, E. and Wojtas, H., "Reinforcement Corrosion and Concrete Resistivity," *Steel in Conc.* 1, 571-581.
76. Zhang. H. and Wheat, H. G., "Electrochemical Testing of Steel-Reinforced Concrete," *Steel in Conc.* 2, 721-730.
77. Dhir, R. K. Jones, M. R. and McCarthy, M. J., "PFA Concrete: Chloride-Induced Reinforcement Corrosion," *Mag. Conc. Res.* 46(169), 269-277 (1994).
78. Maage, M., "Carbonation in Concrete Made of Blended Cement," Ref 4, 193-198.
79. Meland, I., "Carbonation Effects in Hardened Fly Ash Cements," Ref 4, 199-206.
80. Al-Amoudi, O.S.B., Rasheeduzzafar. M. and Maslehuddin, M., "Carbonation and Corrosion of Rebars in Salt Contaminated OPC/PFA Concretes," *Cem. Conc. Res.*21 (1), 38-50 (1991).
81. Kayyali, O. A. and Haque, M. N., "Effect of Carbonation on the Chloride Concentration in Pore Solution of Mortars with and without Fly Ash," *Cem. Conc. Res.* 18(4), 636-648 (1988).
82. Rehsi, S.S. and Garg, S.K., "20-Year Study on Hydration of High Magnesia Cement with Fly Ash at Ambient Temperature," *Proc. 9th Int. Cong. Chem. Cem.*, New Delhi, Vol. IV, 684-690 (1992).
83. Lian, H. and Chen, E. "Inhibition of Blending Materials on Effect of Excess SO_3 Content in Cement," *Proc. 9th Int. Cong. Chem. Cem.*, New Delhi, Vol. IV, 691-687 (1992).

84. Atzeni, C., Massida, L. and Sanna, U., "Properties of Gas Concretes Containng High Proportion of PFA," *Cem.Conc. Res.* 21 (4), 455-461 (1991).

LIST OF ABBREVIATIONS FOR SOME RESOURCES:

ACI SP-91 *Fly Ash, Silica Fume, Slag, and Natural Pozzolans in Concrete,* Proceedings of CANMET/ACI 2nd International Conference, Madrid, 1986, *ACI SP-91*. Editor V.M. Malhotra. Two volumes, 1609 pages.

ACI SP-132 *Fly Ash, Silica Fume, Slag, and Natural Pozzolans in Concrete*, Proceedings of CANMET/ACI 4th International Conference, Istanbul, 1992, *ACI SP-132*. Editor V.M. Malhotra. Two volumes, 1681 pages.

ACI SP-145 *Durability of Concrete*, Proceedings of CANMET/ACI 3rd International Conference, Nice, 1994, *ACI SP-145*. Editor V.M. Malhotra. 1224 pages.

ACI SP-153 *Fly Ash, Silica Fume, Slag, and Natural Pozzolans in Concrete,* Proceedings of CANMET/ACI 5th International Conference, Milwaukee, 1995, *ACI SP-153*. Editor V.M. Malhotra. Two volumes, 1207 pages.

MACC *Mineral Admixtures in Cement and Concrete*, Vol.4, edited by Shondeep L. Sarkar, ABI Books Pvt. Ltd. New Delhi, 1992.

Adva. Cem. Conc. *Advances in Cement and Concrete*, *Proceedings of an Engineering Foundation Conference*, Durham, NH, 1994, Edited by Michael W. Grutzeck and Shondeep L. Sarkar, ASCE, New York.

MRS Symp. Proc. Materials Research Society Symposia Proceedings, MRS, Pittsburgh. *Effects of Flyash Incorporation in Cement and Concrete*, edited by Sidney Diamond, *Symposium N.*, Boston, 1981.

4

THE USE OF RICE HUSK ASH IN CONCRETE

Chao Lung Hwang and Satish Chandra

INTRODUCTION

Rice husks are by-products of rice paddy milling industries. For rice growing countries, rice husks have attracted more attention due to environmental pollution and an increasing interest in conservation of energy and resources [1,2]. There are several reasons that draw the attention of concrete researchers and are discussed below.

Proper Disposal

About 20% of a dried rice paddy is made up of the rice husks. The current world production of rice paddy is around 500 million tons and hence 100 million tons of rice husks are produced, as shown in Table 1 [1,3]. The rice husk has a large dry volume due to its low bulk density (90~150 kg/m^3), and possesses rouge and abrasive surfaces that are highly resistant to natural degradation. Disposal has become a challenging problem. It is recognized that only the cement and concrete industries can consume such large quantities of solid pozzolanic wastes.

Supplementary Binder

For developing countries where rice production is abundant, the use of rice husk ash (RHA) to partially substitute for cement is attractive because of its high reactivity. As the production rate of rice husk ash is about 20% of the dried rice husk, the amount of RHA generated yearly is about 20 million tons worldwide. Also, properly treated ashes have been shown to be active within cement paste. Hence, the use of rice husk ash in concrete is important. In this chapter, the characteristics, quality, hydration mechanism and influence of rice husk ashes on the quality of concrete are discussed.

CLASSIFICATION OF RICE HUSK ASH

The chemical composition of rice husk is similar to that of many common organic fibers and contains: a) cellulose ($C_5H_{10}O_5$), a polymer of glucose, bonded with B-1.4, b) lignin ($C_7H_{10}O_3$), a polymer of phenol, c) hemicellulose, a polymer of xylose bonded with B-1.4 whose composition is like xylem ($C_5H_8O_4$), and d) SiO_2, the primary component of ash [3]. The holocellulose (cellulose combined with hemicellulose) content in rice husk is about 54% [4], but the composition of ash and lignin differ slightly depending on the species, as shown in Table 2. The critical composition of rice husks from different species also varies slightly (Table 3) [4-7].

After burning, most evaporable components are slowly lost and the silicates are left. The characteristics of the ash are dependent on the components, temperature and time of burning [1]. In order to obtain an ash with high pozzolanic activity, the silica should be held in a non-crystalline state and in a highly microporous structure [1,3]. Hence, the burning process should be controlled to remove the cellulose and lignin portion while preserving the original cellular structure of rice husk. Traditional open-field burning can create air pollution that is suspected to cause lung and eye diseases within the human population, as well as damage to plant life [1,2]. Mehta's

Table 1. World Production Rate for Rice Paddy and Rice Husk (Million Metric Tons).

COUNTRY	RICE PADDY	RICE HUSK
Bangladesh	27	5.4
Brazil	9	1.8
Burma	13	2.6
China	180	36.0
India	110	22.0
Indonesia	45	9.0
Japan	13	2.6
Korea	9	1.8
Philippines	9	1.8
Taiwan	14	2.8
Thailand	20	4.0
US	7	1.4
Vietnam	18	3.6
Others	26	5.2
Total	500	100

Rice husk ash is approximately 20% of the rice husk, i.e., the total world production of RHA will be 20 million metric tons.

Table 2. Chemical Composition of Rice Husks.

Rice Husk	Extractives			Chemical Composition(%)			ref.
	Alcohol-benzene	1% NaOH	Hot Water	Holo-cellulose	Ash	Lignin	
Japonica	1.8	32.3	5.4	53.9	13.6	24.8	[3]
Indica	2.1	30.6	5.1	54.3	11.7	25.8	[3]
Anhydrous Rice Husk	-	-	8~15	40~50	15~20	25~30	[1]

Table 3. Ultimate Analysis of Rice Husk, Hwang and Wu [3].

Chemical Composition (Wt%)							
C	H	O	N	S	Cl	Ash	Ref.
38.3	5.7	39.8	0.5	0.0	0.0	15.5	[5]
39.4	5.5	36.1	0.5	0.2	0.2	18.2	[6]
39.5	5.5	37.7	0.8	0.0	0.0	16.5	[7]

fluidized bed furnace was designed in 1974 to produce energy and highly pozzolanic ash from incineration of rice husks [8]. Modern electric furnace has been used in order to maintain proper control of the burning process and to produce better quality ash [1].

The Effect of Burning Temperature

Reactions at different burning temperatures are summarized below, and the chemical compositions of RHA produced are shown in Table 4 [3].

Table 4. Chemical Compositions of RHA (Taiwan) under Different Burning Temperatures, Hwang and Wu [3].

		Temperature(°C)				
		<300	400	600	700	1000
Element (%)	Si	81.90	80.43	81.25	86.71	92.73
	K	9.58	11.86	11.80	7.56	2.57
	Ca	4.08	3.19	2.75	2.62	1.97
	Na	0.96	0.92	1.33	1.21	0.91
	Mg	1.25	1.20	0.88	0.57	0.66
	S	1.81	1.32	1.30	1.34	0.16
	Ti	0.00	0.00	0.00	0.00	0.45
	Fe	0.43	1.81	0.68	0.00	0.68
Oxide (%)	SiO_2	88.01	88.05	88.67	92.15	95.48
	MgO	1.17	1.13	0.84	0.51	0.59
	SO_3	1.12	0.83	0.81	0.79	0.09
	CaO	2.56	2.02	1.73	1.60	1.16
	K_2O	5.26	6.48	6.41	3.94	1.28
	Na_2O	0.79	0.76	1.09	0.99	0.73
	Fe_2O_3	0.29	0.74	0.46	0.00	0.43

(a) At 400°C, due to transglycosylation, polysaccharides begin to depolymerize, producing levoglucosan, monosaccharide derivatives, and oligosaccharides.

(b) Dehydration of the sugar units occurs above 400°C producing 3-deoxyglucosenone, levoglucosenone, funicular and furan derivatives.

(c) At 700°C, the sugar unit decomposes, producing some cabby compounds such as acetaldehyde, glyoxal and acrolein.

(d) At temperatures above 700°C, these unsaturated products react together and through free radical reaction, form a highly reactive carbonic residue [5, 8-11].

Differential thermal analysis (DTA), thermogravimetric analysis (TGA), and TMA showed the first peak at 95°C was caused by dehydration. Between 150°C and 250° C, a low intensity endothermic reaction was followed by one extreme at about 300°C, which may have been due to oxidation reactions. The amount of char formed was about 60% of the initial weight (Figure 1).

X-ray, chemical, and EDAX analyses revealed that the higher the burning temperature, the greater the Si content in the ash. The K, S, Ca, Mg as well as several other compounds were revealed to be volatile, as shown in Table 4 and Figure 2 [3]. The SEM micrographs in Figure 3a and 3b [3] illustrate how the surface fiber was burnt to ashes first, while the inner fiber had not been fully decomposed. Ground husk decomposed more completely at high temperatures (Figure 3c). It is clear that the heating rate and time affect the quality of ash produced. The amount of ash formed at around 600 °C was greater than that at 400°C and it was in a dispersed state (Figures 3d to 3 I). At 700°C, the char ash changed from a gray to a pink color. Furthermore, some spherical debris were observed (Figure 3j). The type of crystals formed at 700°C and 800°C were similar, however, some conversion to a coral-shaped configuration occurred (Figure 3 k). possibly due to different temperature effects. Between 900°C and 1000°C, the formation of coral-shaped crystals increased (Figure 3m to 3o). Although the crystals retained their coral shape at 1100°C, they were finer and melted considerably, as can be seen in Figure 3p or 3r.

A phase diagram of the CaO-SiO_2 system can be used to determine the type of phases formed. The types of silica formed between 900°C and 1000°C may be tridymite, wollastonite, β-Ca_3SiO_3 and β-quartz [12]. At 700°C to 800°C, α-quartz and wollastonite phases coexist. Because the clinkering temperature of rice husk is under 1500°C, polymorphism of silica such as quartz, tridymite, and cristobalite with open structured SiO_2 in RHA is produced which can promote pozzolanic reaction. However, this reaction may vary with the solubility of SiO_2, depending on both the soluble layer and the density of the silica [13]. According to Table 5 [3], the mean

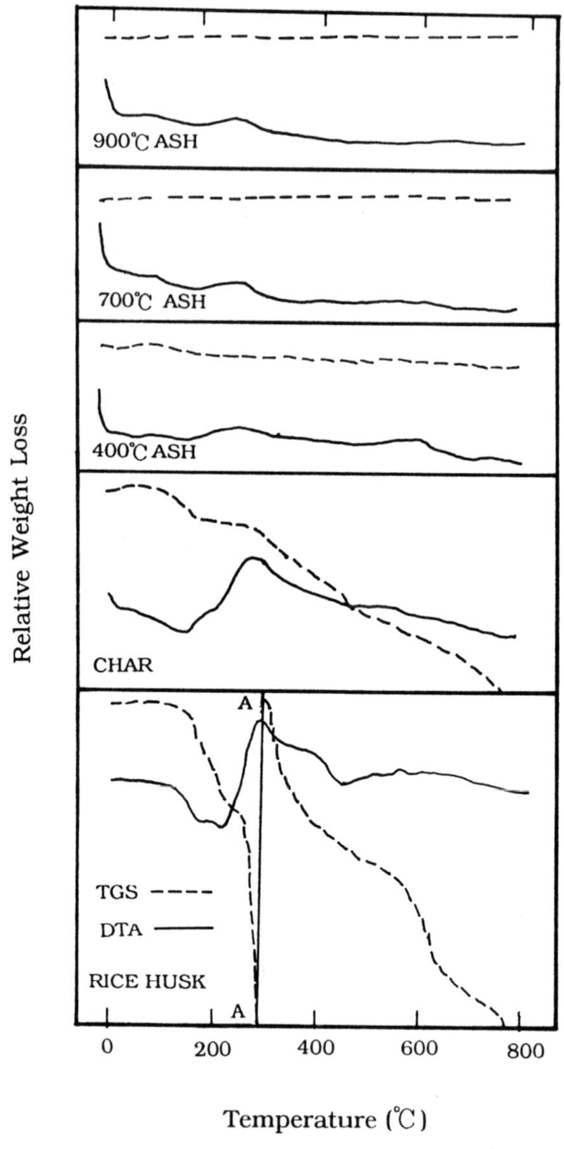

Figure 1. The DTA and TGA spectra for the rice husk under different burning temperatures. Adapted from Hwang and Wu [3].

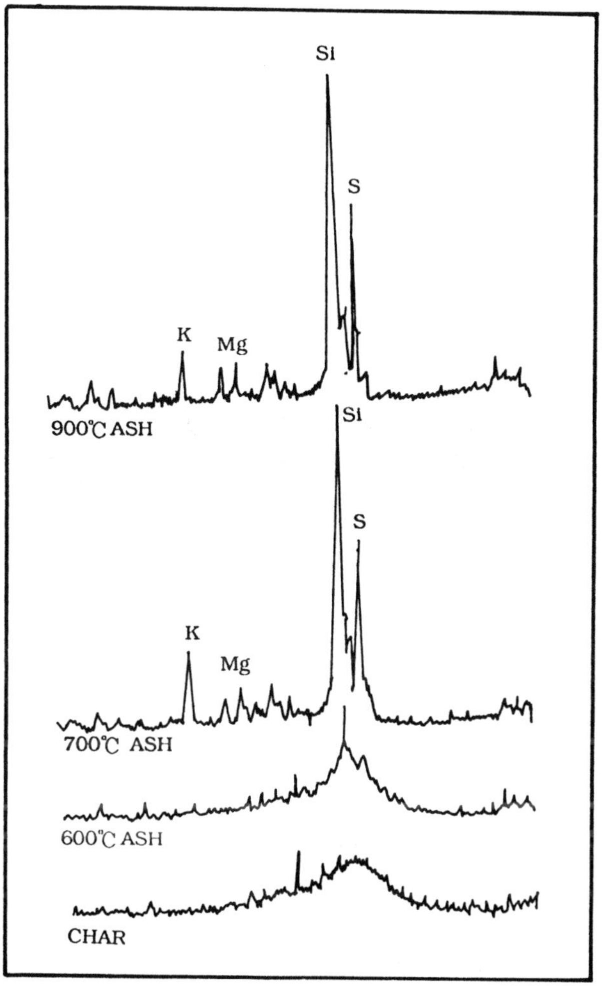

Figure 2. The x-ray diffraction pattern of RHA under different burning conditions. Adapted from Hwang and Wu [3].

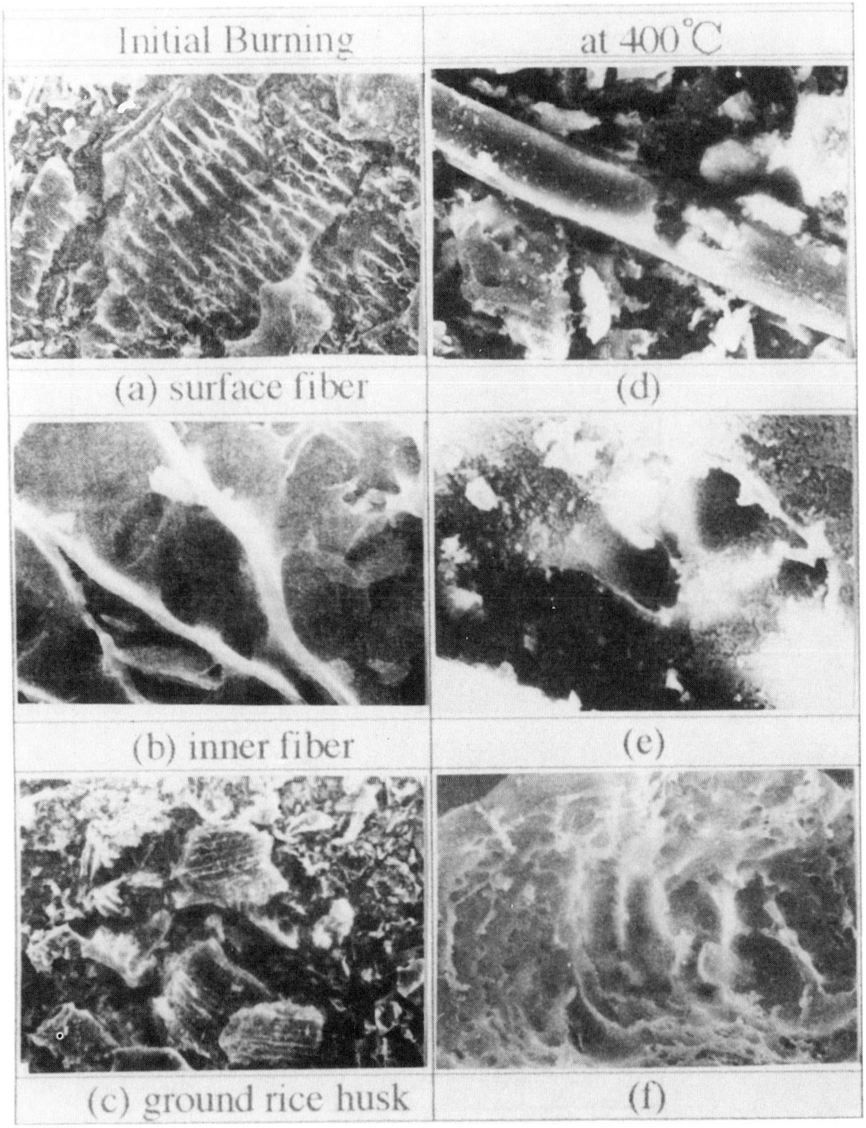

Figure 3a-f. The SEM micrographs of RHA burned at different temperatures, after Hwang and Wu [3].

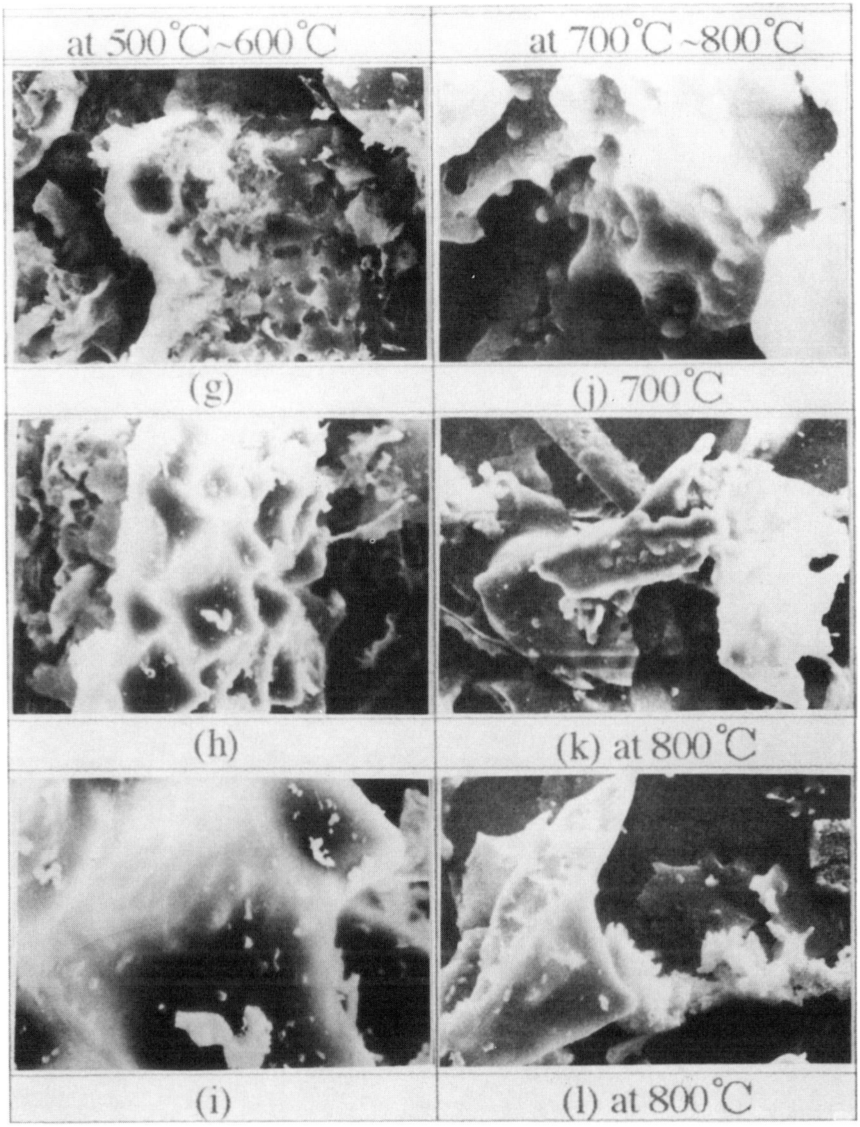

Figure 3g-l. The SEM micrographs of RHA burned at different temperatures, after Hwang and Wu [3].

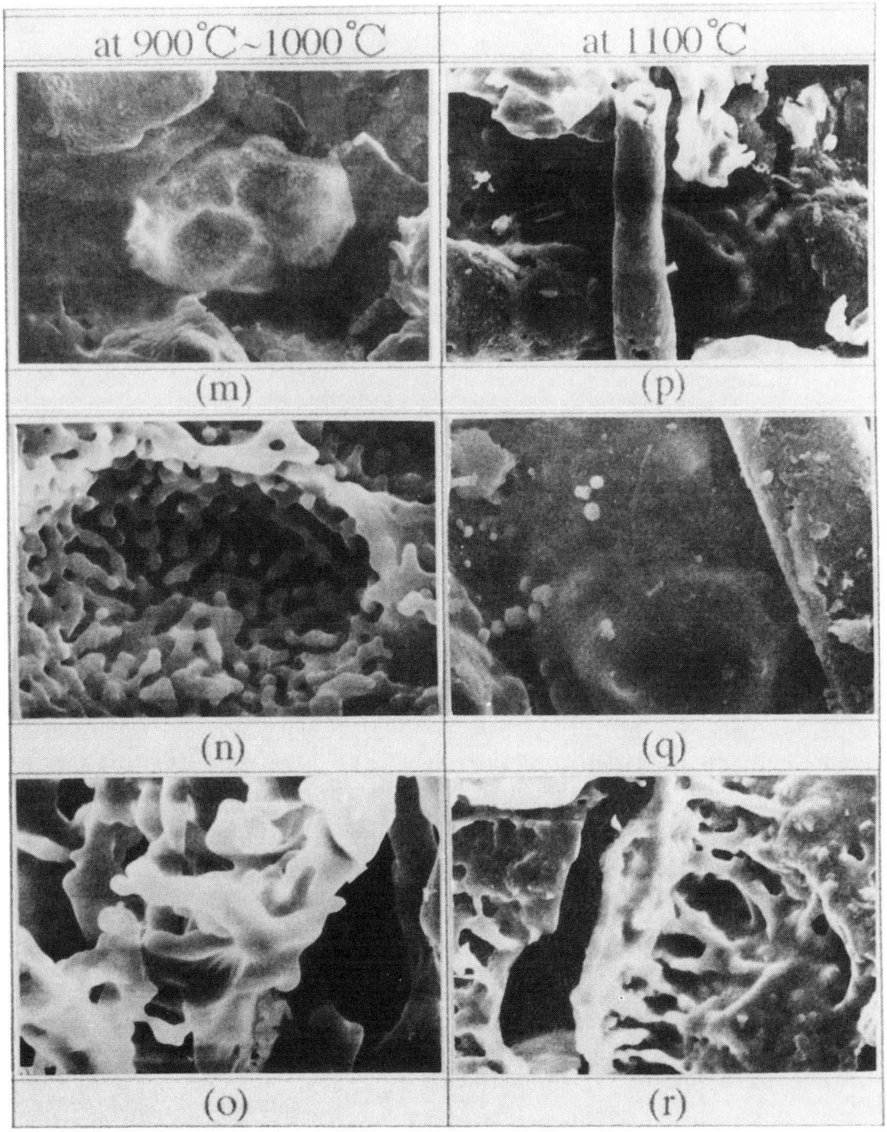

Figure 3m-r. The SEM micrographs of RHA burned at different temperatures, after Hwang and Wu [3].

diameter of pores of RHA between 600 and 700°C is highest and therefore, the pozzolanic reaction of ash formed at this temperature should also be greatest. The lower the burning temperature, the lower the energy consumption and thus, temperatures of 600 to 700°C for rice husk ash formation do not involve excessive energy.

Table 5. Pore Analysis of RHA Under Different Burning Temperatures, after Hwang and Wu [3].

Burning Temperature(°C)	Mercury Penetration Volume (cm^3/gm)	Mean Pore (Å)	Median Pore (Å)
Char	0.3567	5257	4585
400	0.3379	5954	6847
600	0.2912	6069	6847
700	0.0795	5867	5718
800	0.0643	5446	5250
900	0.0393	5150	4483
1000	0.0323	4972	3867
1100	0.0171	3900	9803

The Effect of Burning Time & Furnace Environment

Above 800°C, an increase in the burning temperature, time, and environment tend to cause a sintering effect (coalescing of fine particles) as was mentioned earlier, and is indicated by a dramatic reduction in the specific surface (Table 6) [14]. Combustion environment also plays an important role. It should be noted that a change in the rate of oxidation from moderately oxidizing conditions (CO_2 environment) to highly oxidizing conditions (oxygen environment) was responsible for the steep drop in the microporosity (Table 5) and surface area (Table 6).

The diffusion process for obtaining a reactive cellular rice husk is shown in Figure 4 and is based on the data presented in

Table 6. Effect of Burning Conditions on the Crystal Structure and Surface Area of Rice Husk Ash. Adapted from Ankra [14].

Burning Temperature	Hold Time	Environment	Properties of ASH	
			Crystalline	Surface Area, m^2/g
500~600°C	1 min	moderately oxidizing	non-crystalline	122
	30 min			97
	2 hrs			76
700~800°C	15min~1hr			100
	>1 hr	highly oxidizing	partially crystalline	6-10
>800°C	>1 hr		crystalline	<5

Tables 6 and 7. It shows that optimum incineration condition is important to obtain reactive rice husk ash with microporous and cellular structure. But, it is not suggested to burn rice husk above 800°C longer than one hour to obtain suitable pozzolanic reactivity. Also, it is believed that rapid cooling may increase the reactivity of RHA [15].

ANALYSIS OF THE QUALITY OF RHA

The quality of RHA actually depends on the method of ash incineration and the degree of grinding. It also depends upon the preservation of cellular structure and the extent of amorphous material within the structure.

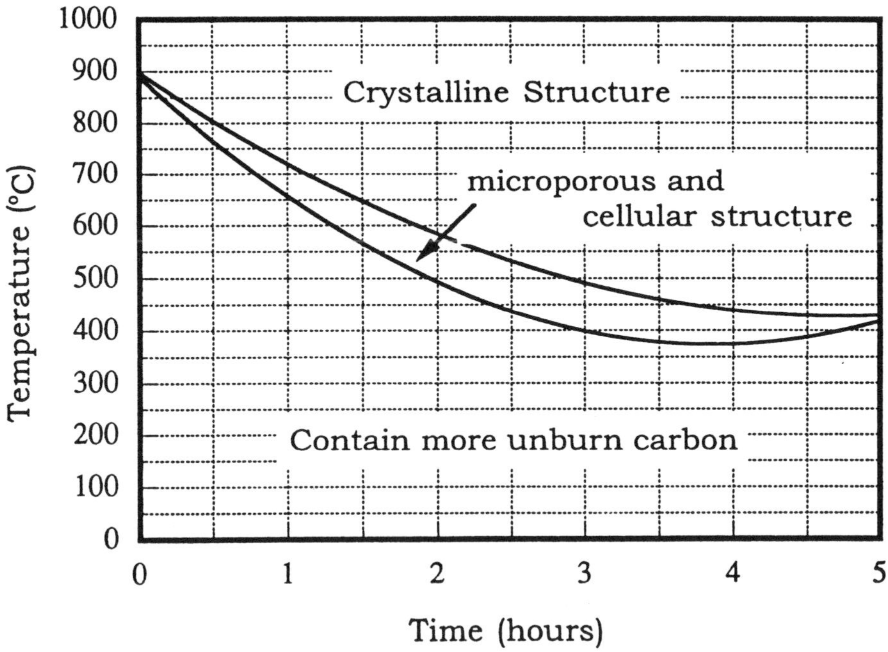

Figure 4. The optimum incineration condition curve for obtaining reactive cellular RHA.

Different Sources

The production of rice husk in the whole world is about 100 million metric tons a majority of it coming from countries in Southeast Asia. Depending on growing conditions and the variety of rice species, the quantity of straw also varies, but as indicated in Table 1, the chemical composition of different sources is quite similar. Typical components of RHA, as shown in Table 2 are [16]:

Cellulose	40~50%
Lignin	25~30%
Ash	15~20%
Moisture	8~15%

After burning at a suitable temperature and time (~ 600-700°C, 2 hours), in an industrial furnace, the rice husk is composed of 90-95% SiO_2, 1-3% K_2O and < 5% unburnt carbon, as shown in Table 7 [1,3, 17-21]. The quality of ash from different sources after proper burning varies only slightly. For use, the RHA should be amorphous and highly porous. The pozzolanic reactivity is also dependent on the surface area of RHA. Rice husk ash with a surface area of 50-100m^2/gm is now available. The summary of the research conducted from 1976 to 1991 is shown in Table 7. The pozzolanic activities for the RHA when mixed with < 70% (by weight) cement is greater than 100%.

Different Processes

Burning process affects the quality of rice husk ash produced. Research studies have shown the influence of processing on a variety of properties of the RHA, as shown in Table 7 [1-3, 7-12].

Table 7. The Incineration Conditions, Chemical Compositions, Characteristics and Strengths of Rice Husk Ash. Adapted from Hwang and Wu [3] and Mehta [2].

Author (year)	Burning Method	Incineration condition		Chemical Composition			Crystalline	Specific Characteristic	Specific Surface area (m^2/g)	Finer passing 75µm	Pozzolanic Index	Replacement % of Cement	Mortar specimen		
		temperature (℃)	Time (hr)	SiO_2	K_2O	unburn carbon							w/c	Strength (Mpa)	
														28d	90d
-	open-field	uncontrolled	-	88	5		cristobalite, tridymite	black in color, low reactivity							
Mehta (1976-1979)	fluidized-bed	700-800℃	-	80-95	1-2	3-18	amorphous (highly cellular)		50-60						
Cook (1976-1981)	control burning	450℃	4	93	2.3	3				32		10 20	0.49 0.53	41 34	46 37
Kanpur (1974-1976)		600-700℃	2				crystalline, amorphous		60			20	0.42-0.47	44-60	-
Yamamoto (1982)		400℃, cooled quickly	4						-		100-140	20-70		28-40	
Yousif (1984)		500℃	2			3.3			0.5-2.1		110				
James and Rao (1986)	automatically burning	700℃	1-12			3			100-150 at low temperature						
Boatend and Skeete (1990)		600-700℃	7-9			3-7					100		0.6		
Hwang and Wu (1989)		700℃	4	92	4		amorphous	mean pore 590Å, bulk density 0.14-0.29 g/cc, true density 2.22 g/cc				5-30	0.35-0.50	17-57	31-68
Mehta (1991)				91	2.1	5			24	25					
Typical	industrialized burning			90-95	1-3		amorphous and highly cellular form	bulk density 0.2 g/cc	50-100						

a. Open-Field Burning

Open-field burning of rice husk not only produces poor quality of ash, but is banned in many countries due to pollution problems. Uncontrolled burning results in a structure of highly crystalline form that is of low reactivity.

b. Fluidized-Bed Furnace Burning

Fluidized-bed furnace is designed to control burning of rice husk [17]. In the process, the heat from the combustion of rice husk was utilized to produce steam or electricity. A close control of the time-temperature parameter in the burning operation is maintained. A highly pozzolanic ash is produced. Highly pozzolanic RHA is created by maintaining husk combustion temperatures between 500 and 700°C for a relatively long period to remove most of the carbon [1], or at temperatures around 700-800°C for less than one minute. The chemical analysis of the ash samples produced by a fluidized-bed furnace showed 80-95% SiO_2, 1-2% K_2O, and 3-18% unburned carbon. The ash was highly cellular, with 50-60 m^2/g surface area measured by nitrogen adsorption. The fibrous texture of the silica in ash results presumably from its deposition between cellulose fibrils in the microvoids.

c. Industrial Furnace

A modern industrial furnace has been recently used due to environmental and economic reasons [1] (Table 7). Depending upon the efficiency of combustion, the silica content of RHA may be in the range of 90-95% with residual carbon as the main remaining ingredient. In addition to residual carbon, alkalis ranging from 1 to 3% form the other impurity. By controlled combustion in the industrial furnace, it is simple to produce RHA with silica in an amorphous and highly cellular form, with 50-100 m^2/g surface area. This type of rice husk ash is highly pozzolanic.

The Effect of Burning Time and Temperature on the Surface Area and Its Reactivity

The RHA must be burned for the correct time and temperature to achieve the requisite pozzolanic activity, as indicated in Table 7 and illustrated in Figure 4. Table 6 clearly indicates that not only the burning temperature [14], the burning time is equally important in remvoving carbon while keeping the silica in an amorphous and highly cellular form. The surface area of RHA burned at 500-600°C for 1 minute is as high as 122 m^2/g without causing crystallization, as shown in Table 6. Longer burning times will cause collapse of the cellular form and also coalescence of the fine pores [3], which consequently causes a reduction in surface area [1]. At higher temperatures with longer burning times, a crystalline structure is formed with a sharp reduction in surface area. This lowers the pozzolanic activity. Figure 4 indicates the ideal time/temperature path to obtain optimum quality rice husk ash with a microporous and cellular structure which is highly reactive.

HYDRATION MECHANISMS OF PASTE WITH RHA

Understanding the hydration of paste with RHA is important for using pozzolanic materials and controlling the properties of the paste [3]. The heat evolution curve, microstructure development, and ultrasonic velocity are important for studying hydration mechanisms in paste made with RHA.

Paste/RHA Heat Evolution Curve

The study of the rate of hydration can be used to predict strength development. However, if the interior temperature is too high, cracks may develop in the cement paste [3]. Figure 5 reveals that the heat evolution curve of cement paste with RHA is similar in shape to that without RHA. It also shows the effect of water to

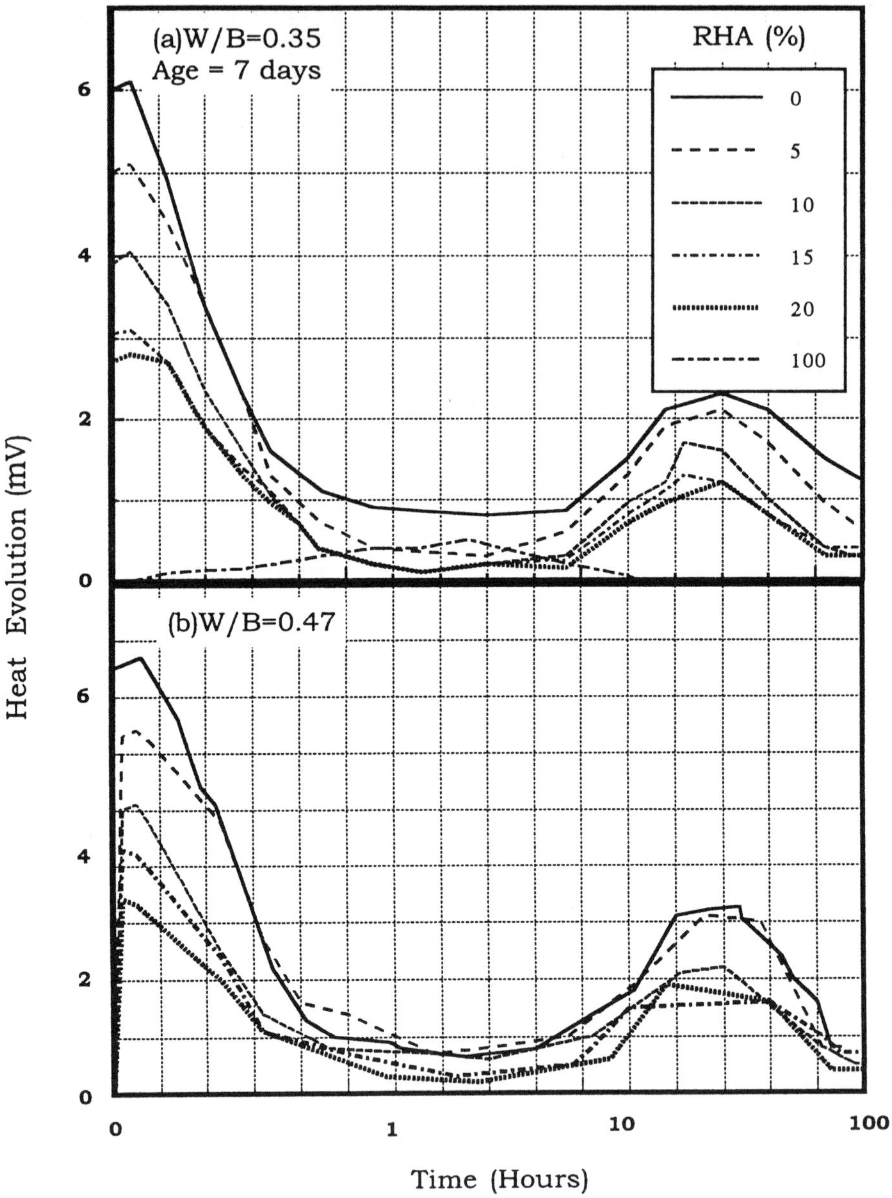

Figure 5. Heat evolution of cement paste with RHA. Adapted from Hwang and Wu [3].

cement ratio and the amount of replacement by RHA on heat evolution. The first peak is higher than the second peak. The larger the amount of RHA added, the lower the amount of heat evolved. At higher RHA contents, K^+ and SiO_2 react with Ca^{2+} to lower both the first and second peaks. High water to cement ratio may also lower the heat of hydration due to the diluting effect of water. At the second stage of hydration (dormant period), the concentration of Ca^{2+} decreases, thereby increasing the saturation time of ion and thus delaying the second peak. However, the pH value increases primarily due to the potassium content in RHA dissolved in water, as shown in Table 8. This compensates for the alkali concentration consumed by Ca^{2+} to nucleate earlier and to accelerate the second peak. The setting time of ordinary cement paste occurs before the second peak of the calorimetric curve; such a mechanism can be responsible for short setting time of cement paste containing RHA (Figure 6).

Table 8. The pH Values of RHA Dissolved in Water. Adapted from Hwang and Wu [3].

RHA/W	W/RHA	pH Value		
		1 day	3 days	7 days
0.05	20	9.91	9.59	9.55
0.1	10	9.98	9.82	9.75
0.15	6.67	10.00	9.91	9.80
0.2	5	10.12	9.91	9.86
0.25	4	10.14	9.98	9.91
0.3	3.33	10.21	9.98	9.91

* Ash burned at 700°C

Microstructure of Hydrated Cement Paste with RHA

Cement hydrates to C-S-H gel, CH, ettringite, and monosulfoaluminate. The type of capillary pores and unhydrated core varies with age and depends on cement composition and curing conditions. However, different additives may alter the crystalline

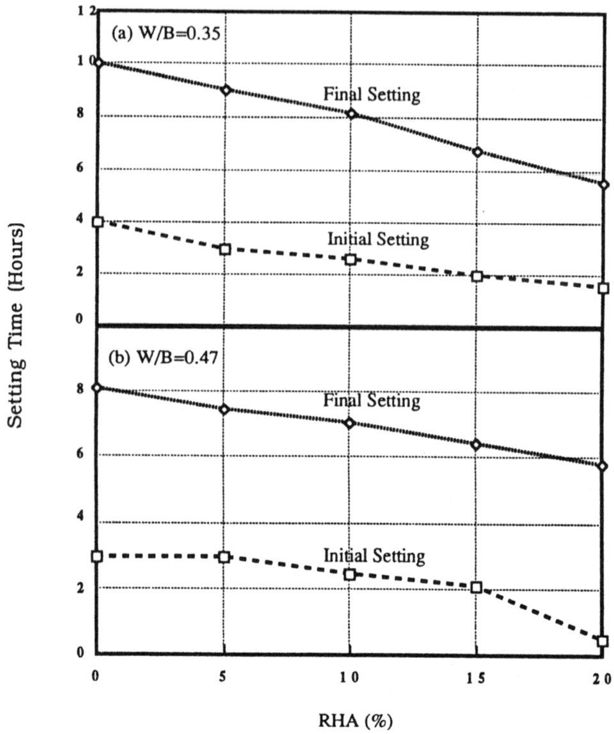

Figure 6. Setting time of cement paste with RHA. Adapted from Hwang and Wu [3].

nature of products, thereby resulting in the generation of more hydrates. From optical microscopy observations, a magnification exceeding 400X or an excessively high concentration of cement paste (low water to cement ratio), produces an unclear image. Only the development of CH crystals can be observed. A high water to cement ratio of approximately 0.6 is therefore used to obtain clear images. The symbol "→ " in Figure 7 indicates that a CH crystal appears between unhydrated cores and that C-S-H gel appears after 5 hours [3]. The crystal at 6.5 hours is 4 times the size of that after 5 hours; after 7.5 hours it is 8 times greater and after 12.5 hours, 26 times greater. The free water generates a significant amount of C-S-H gel, and consequently, RHA and unhydrated cores become large due to

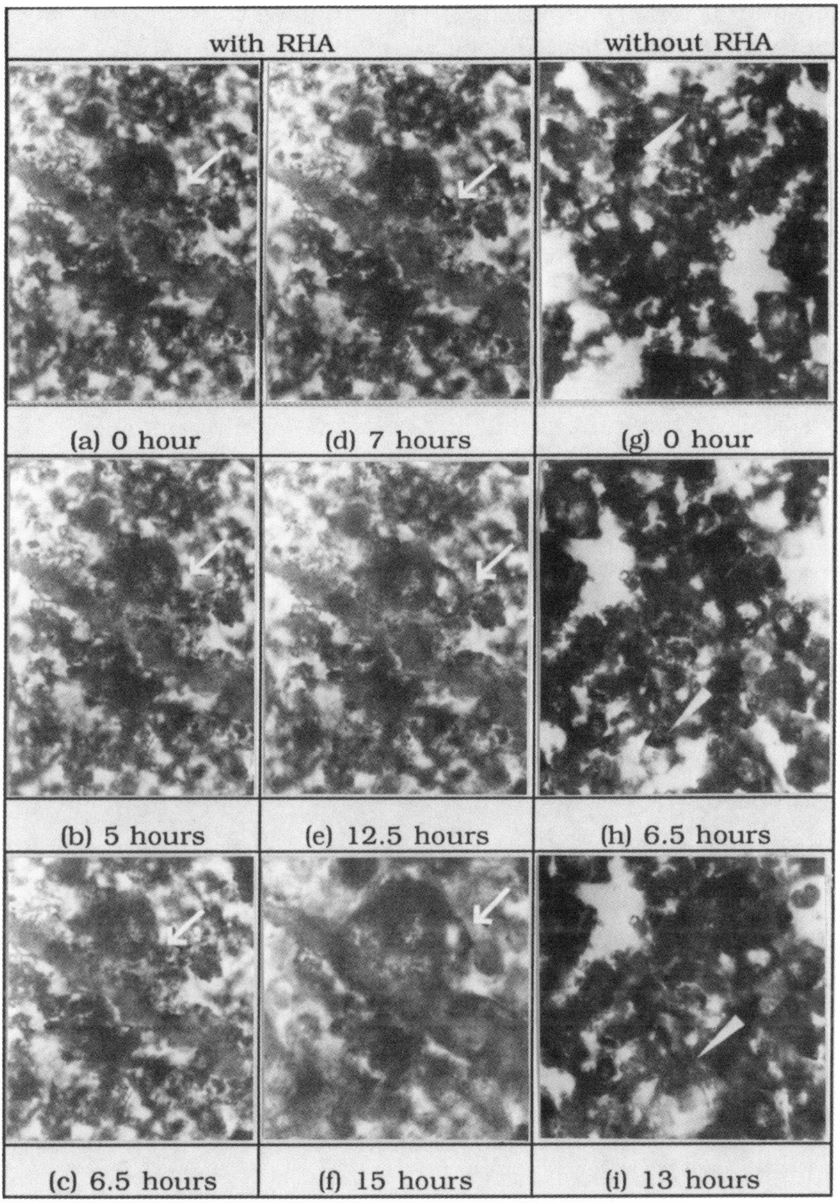

Figure 7. Optical micrographs of cement paste with RHA. Adapted from Hwang and Wu [3].

the build-up of hydrates. This may be related to the dense sponge matrix, as shown in the SEM micrograph of cement paste containing RHA. However, such a phenomenon does not occur in cement paste without RHA (Figure 8g-8h). The symbol "→" in Figure 8a to8f indicates the occurrence of a hexagonal crystal after 7 hours.

A cement paste formed with a water to cement ratio of 0.4 and containing 15% RHA was investigated using scanning electron microscopy and revealed a significant difference between pastes with and without RHA. After one day, the C-S-H gel was in the form of a dense sponge matrix (Figure 8a, b, c) [3]. Most of the CH may have reacted with RHA. Some hexagonal plates (about 0.1 μm) grow on the surface of the RHA similar to that of monosulfoaluminate (Figure 8d). After 3 days, the dense fibers bond with the matrix within large pores (Figure 8e). The fiber surface is no longer smooth and some of it is filled with C-S-H gel (Figure 8f). After 7 days, the crystal (1.5 μm) looks like a bundle of fibers, constrained in the middle, but flower-like at the ends (Figure 8g). After 28 days, the matrix has become more dense and the undecomposed CH crystal clogs the matrix (Figure 8h). After 60 days, the matrix bonds together and the presence of CH becomes unclear (Figure 8i). A cement paste with water to cement ratio of 0.47 and containing 10% RHA was also investigated. The porosity of this system was found to be higher than that of cement paste with a water to cement ratio of 0.35 which may result in lower compressive strength. The pore structure of the cement paste containing 10% and 15% RHA replacement are nearly identical (Figure 9) [3]. Therefore, theoretically, the compressive strengths of these pastes should not differ much from each other.

Hydration Mechanisms of Paste with RHA

Figure 10 reveals the hydration behavior which includes the calorimetric curve, the ultrasonic pulse velocity curve, the penetrative resistance curve, and the setting time of paste with 20% RHA (22). According to the hydration mechanism of ordinary portland cement without RHA (23), a transition zone exists in the ultrasonic pulse

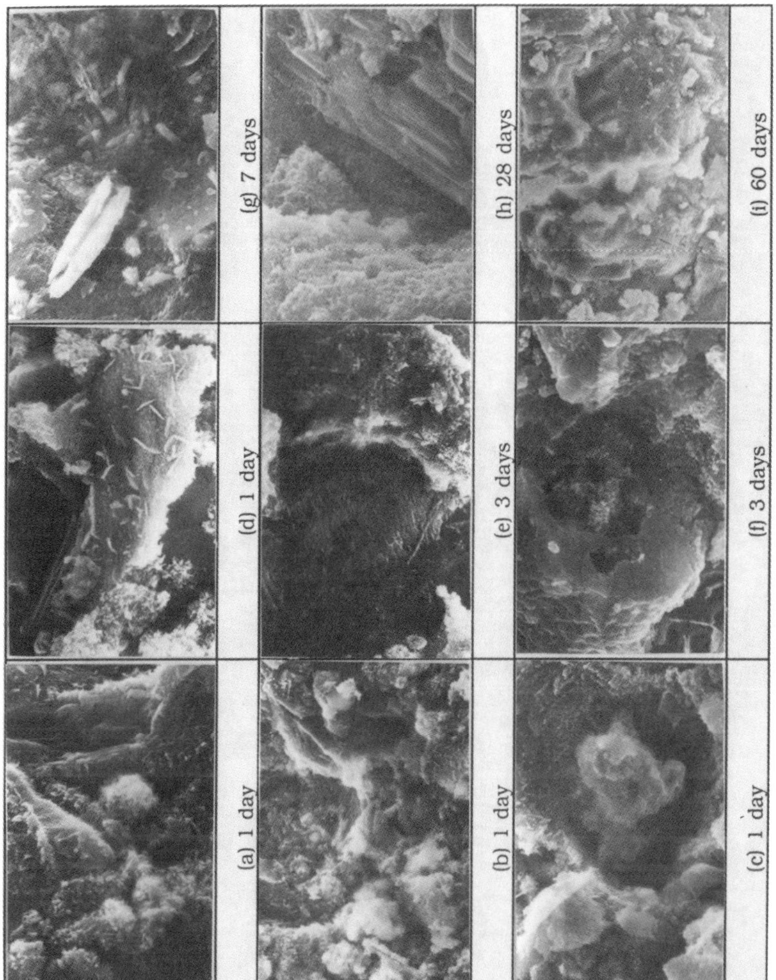

Figure 8. SEM micrographs of hydrated cement paste with RHA. Adapted from Hwang and Wu [3].

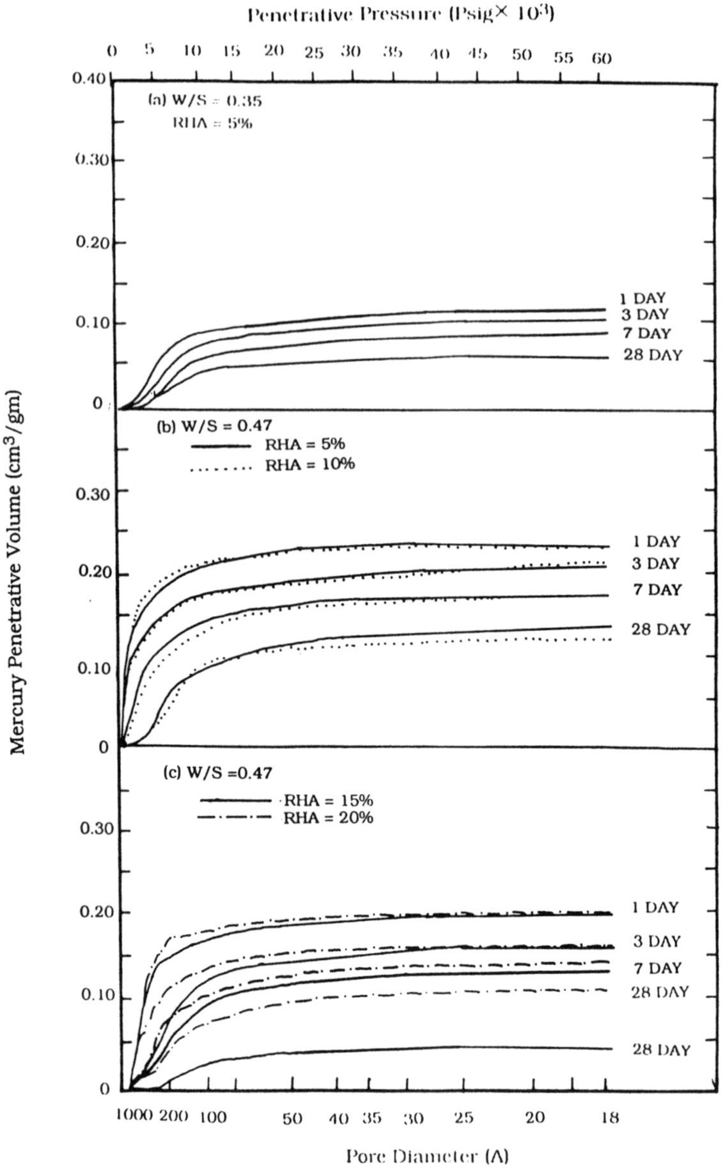

Figure 9. Pore analysis of cement paste. Adapted from Hwang and Wu [3].

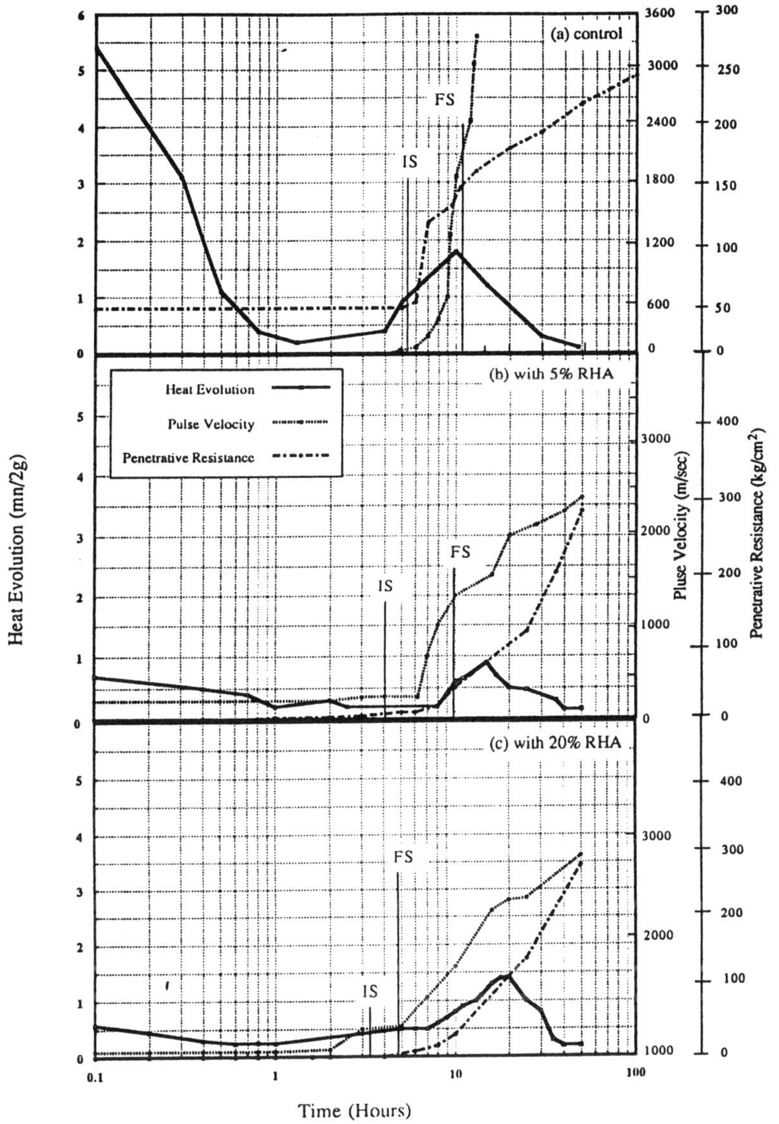

Figure 10. The relation of calorimetric curve, ultrasonic pulse curve, penetrative resistance strength curve and setting time (w/c=0.47), Hwang, Peng and Lin [22].

velocity curve which corresponds to the period between the dormant and deceleration periods in the calorimetric curve. The transition zone occurs roughly at the points of inflection in both the calorimetric and the penetrative resistance curves, with the setting time measured with a Vicat needle. Moreover, the time of appearance of the second peak in the calorimetric curve approximately coincides with the final setting time. Such a transition zone will be delayed with the increase in water to solid ratio. The addition of RHA will cloud the transition zone [22], but the hydration mechanism is similar to that of ordinary Portland cement. The final setting of paste with 5% and 20% RHA added begins at the second peak as shown in Figure 10b and 10c [22,23]. This implies that the addition of RHA will strengthen the matrix.

The hydration mechanism of paste with RHA can be hypothesized, may appear as shown schematically in Figure 11 [22], and may be described as follows:

The penetration resistance coincides with the growth of CH (Calcium hydroxide) up to 8 hours, and this is similar to the behavior of ordinary portland cement paste. The early resistance may be primarily due to the formation of CH crystal. From the pulse velocity curve as shown in Figure 10, the pulse change coincide with the early detection of penetration resistance. The formation of CH at the surface of rice husk ash may be due to the adsorption by cellular structure of RHA. In such case the bleeding water will be significantly reduced. The adsorbed water enhances the pozzolanic reaction inside the inner cellular spaces and gain significant strength. The ultra-sonic velocity gain rapidly after the formation of CH and the contact of RHA solid through the silica skeleton. This behavior is different from that of pure Portland cement paste. After 40 hours, the pozzolanic reaction further binds Si in RHA with CH to form C-S-H gel and solid structures. This means that RHA fills the finer pores and reduces the permeability, which may be beneficial to the durability.

EARLY CHARACTERISTICS OF CONCRETE WITH RHA

The early characteristics of concrete with RHA depends on the water to cement ratio, the amount of paste used, the amount of RHA added, other admixtures used and mixture proportion.

The Workability of Fresh Concrete with RHA

At a given water to cement ratio, small addition (less than 2 to 3 by weight of cement) of RHA may be helpful for improving the stability and workability of concrete by reducing the tendency towards bleeding and segregation [24]. This is mainly due to the large surface area of rice husk ash which is in the range of 50 to 60m^2/g [1]. Large additions would produce dry or unworkable mixtures unless water-reducing admixtures or superplastizers are used, as shown in Figure 12 [3]. Due to the adsorptive character of cellular rice husk ash particles, concrete containing RHA require more water for a given consistency. At high water-cement ratio, the workability tends to improve, as shown in Figure 12 [3]. The addition of sand will significantly reduce the flow table spread.

For a given consistency, the reduction of water requirement can lead to an overall improvement in many engineering properties. Granulometric characteristics of the coarse aggregate, fine aggregate, and cement particles influence the volume of voids and water requirement of a concrete mixture. The addition of fine particles of a mineral admixture, typically in the order of 1 to 20 μm in size, would supplement the cement grains in further reducing the volume of voids in the concrete mixture. Consequently, it will require less water to produce a concrete of a given consistency [25].

The workability of fresh concrete with RHA can be improved by densifying the mixture [25, 26]. The process uses cement and rice husk ash with water to fill the pores and voids within well-compacted aggregates. The density of concrete made by this process is higher

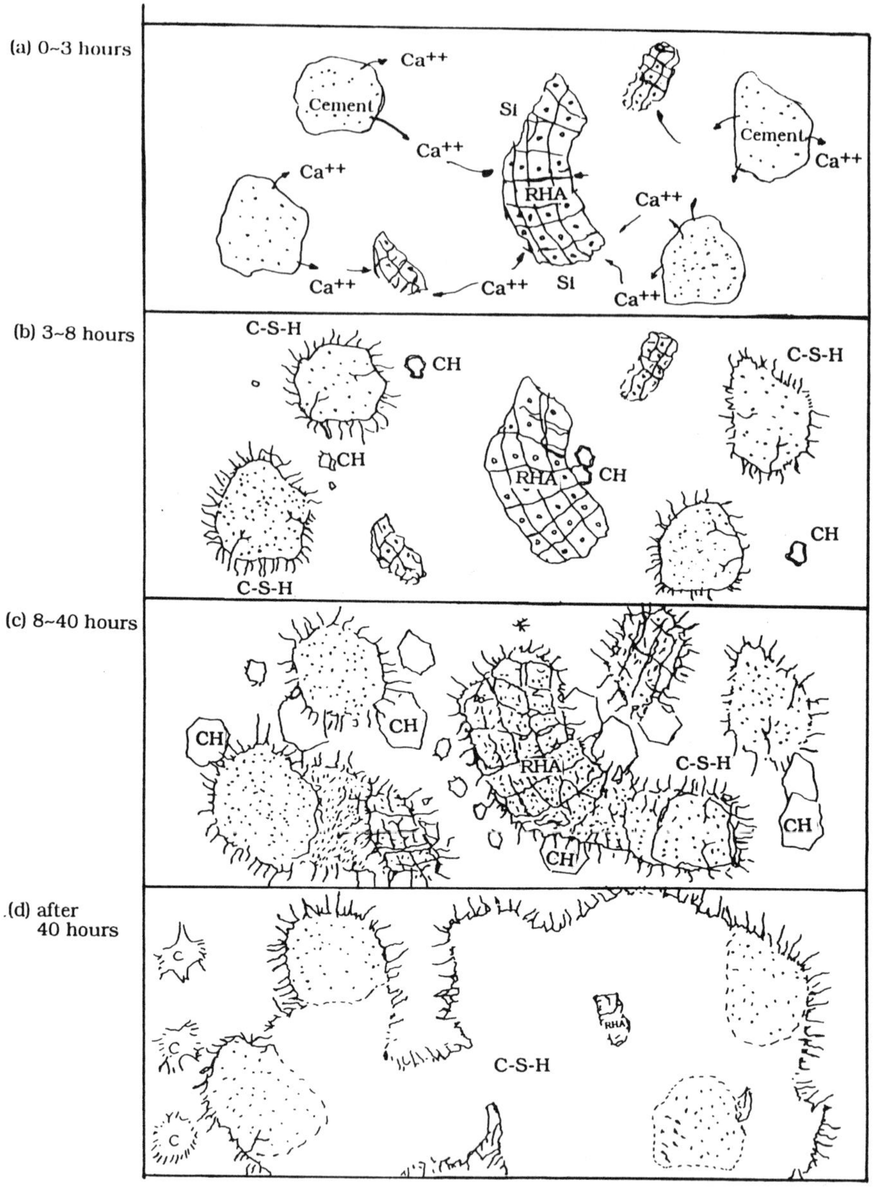

Figure 11. Schematic drawing of the hydration of cement paste with RHA. Adapted from Hwang, Peng and Lin [22].

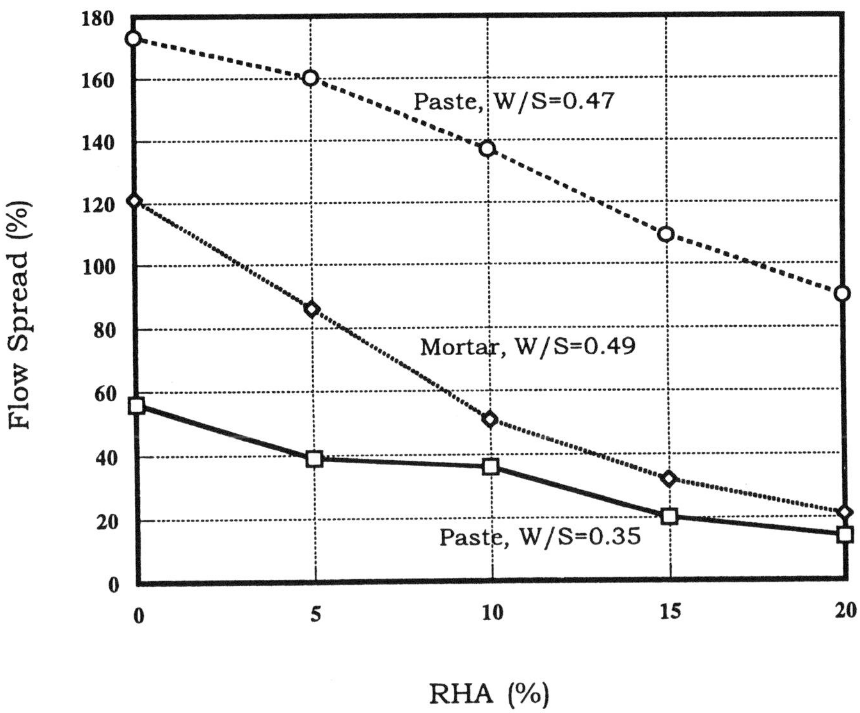

Figure 12. Flow table spread of cement paste and mortar with RHA. Adapted from, Hwang and Wu [3].

than that of conventional mixtures with more cement. Slump can be controlled to 250 ± 20mm range with excellent rheological properties and with small reduction in slump after 45 minutes [25].

The Setting Time of Concrete with RHA

Unlike other pozzolanic materials, rice husk ash tends to shorten the setting time, as shown in Figure 6. This may be due to the water adsorption ability of the cellular form of rice husk ash and hence, the surrounding water-to-cement ratio is reduced. It is further substantiated by the early detection of the ultrasonic pulse velocity, as shown in Figure 10, reflects that the rigid silica cellular skeleton also plays an important role in setting time. Higher water-to-cement ratio tends to increase the setting time because there is less contact between the open matrix and the silica cellular structure causes a reducuction in early strength development.

The Compressive Strength and Impermeability of Concrete with RHA

In normal concrete, the transition zone is generally less dense than the bulk paste and contains a large amount of plate-like crystals of calcium hydroxide, with the c-axis perpendicular to the aggregate surface [27]. This is suspected to induce microcracks due to the tensile stresses induced by thermal and humidity change. The structure of the transition zone is the weakest phase in concrete and has a strong influence on the properties of the concretes.

The addition of pozzolanic materials can affect both strength and permeability by strengthening the aggregate-cement paste interface and by blocking the large voids in the hydrated cement paste through pozzolanic reaction. This phenomenon is shown in Figure 13. It is known that the pozzolanic reaction modifies the pore-structure. Products formed due to the pozzolanic reactions occupy the empty space in the pore-structure which thus becomes densified. The porosity of cement pastc is reduced, and subsequently, the pores are refined. Mehta [1] has shown significant reduction in the porosity of cement paste with RHA additions and refinement in the pore-structure. Pozzolanic reaction is a slow process and proceeds with time. It is illustrated in Figure 14 [1] that the pore refinement is in progress even after 28 days.

Rice husk ash adsorbs large amount of water due to its high specific surface area. This reduces bleeding water. The high absorption of the RHA has been shown by Hwang [3] (Figure 15). It improves the weakest zone under the aggregate. However, adding the correct amount of rice husk ash is important for achieving high strength. Large amounts of rice husk ash have an adverse effect and reduce strength as shown in Figure 16 [3]. The early strength of

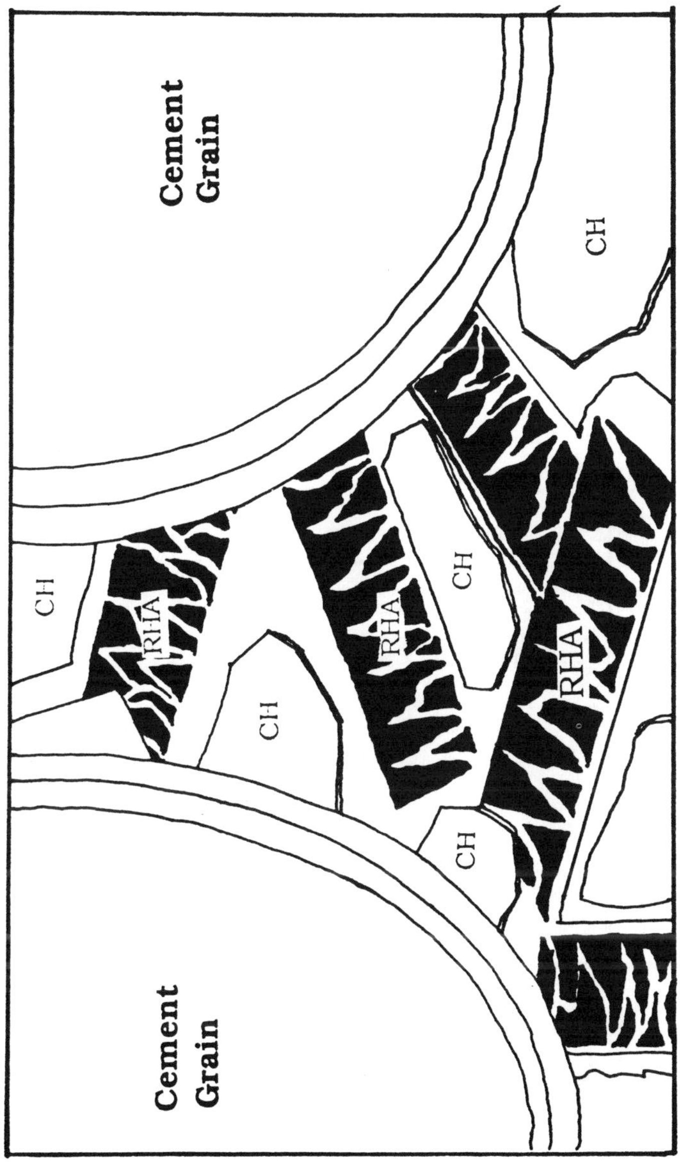

Figure 13. Mechanism of void filling and transition zone strengthening effects of RHA.

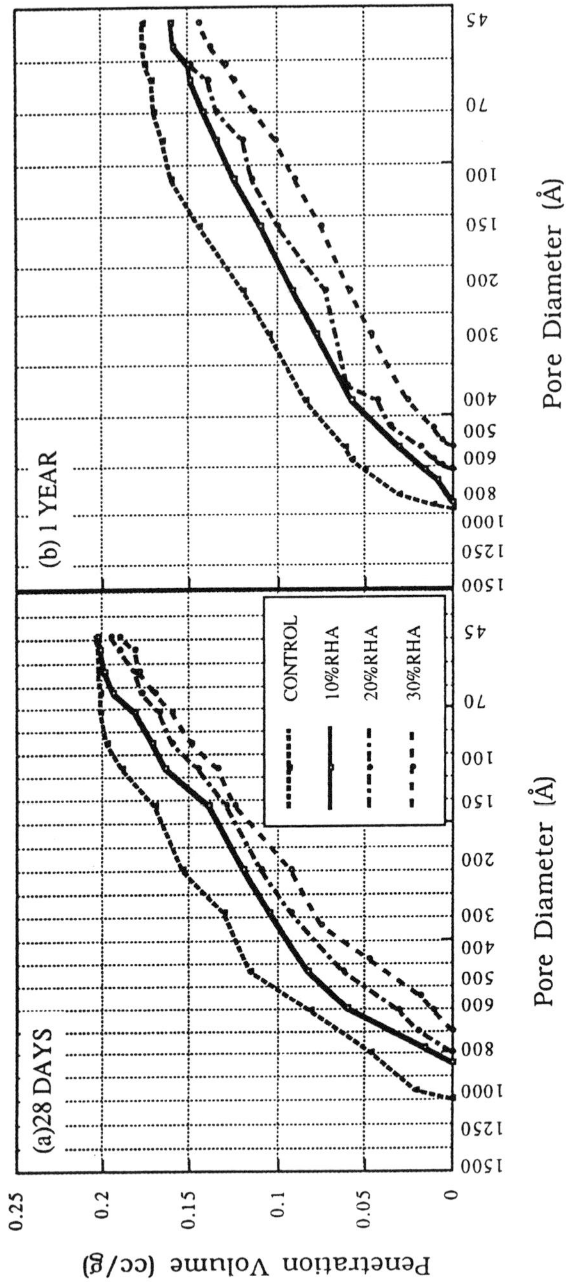

Figure 14. Pore refinement of hydrated cement pastes by RHA. Adapted from Mehta and Monteiro [27].

concrete is a function of water-to-binder ratio. As long as the water-to-binder ratio is kept constant, the early strength of concrete will be similar, but the ultimate strength will be enhanced due to pozzolanic reactions.

It is further seen that above 54 kg/m^3 rice husk ash addition, there is no influence on the strength. However, it decreases the permeability of concrete.

At a high water to cement ratio, the addition of RHA to cement paste will not only reveal a significant effect on strength at early ages, but the strength at later ages also tend to be higher than those with lower water to cement ratios. A higher water to cement ratio also contributes to a lower heat of hydration.

The pore refining effect of rice husk ash has shown a surprising result in permeability of concrete as shown in the last

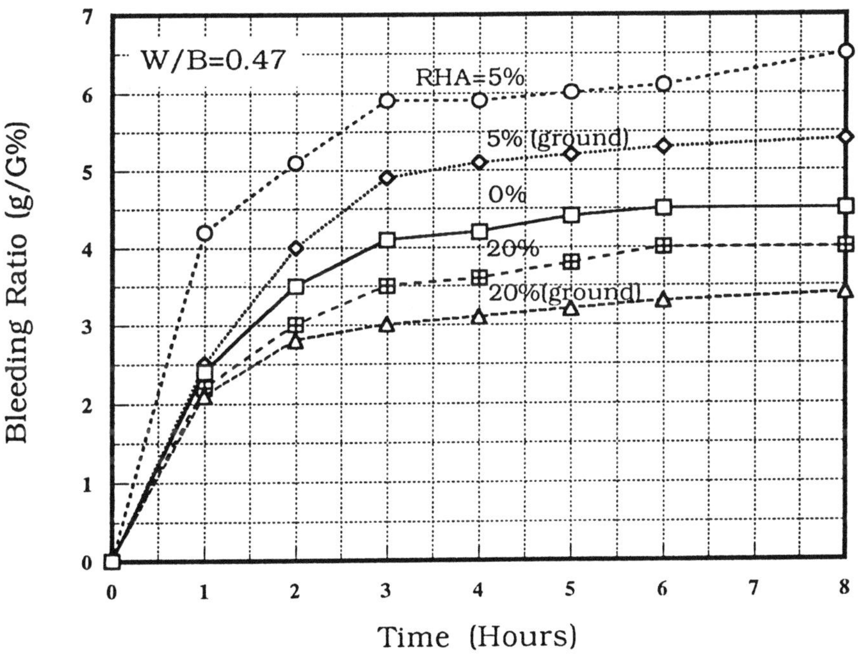

Figure 15. Bleeding ratio of cement paste with RHA. Adapted from Hwang and Wu [3].

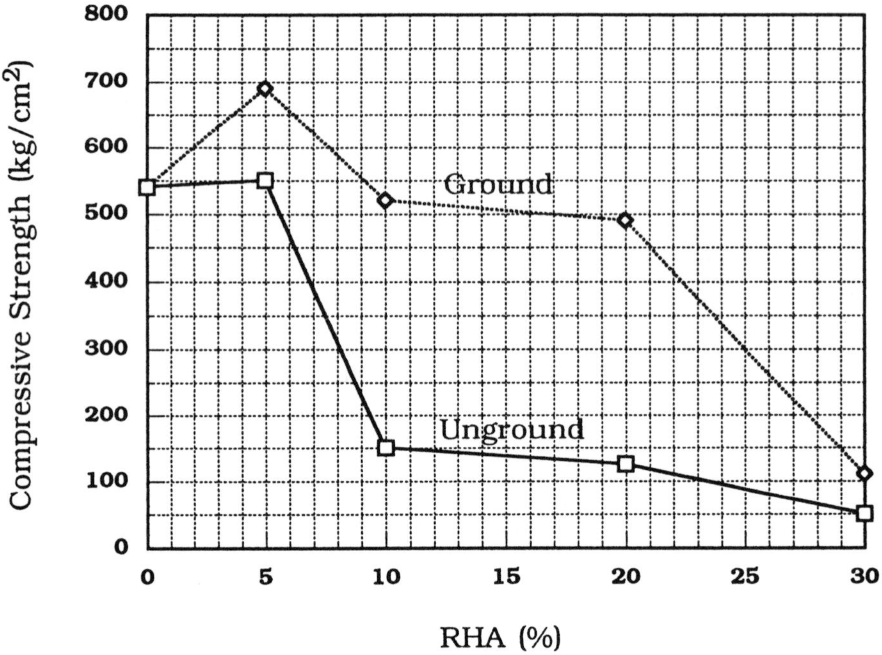

Figure 16. The effect of RHA on the compressive strength of paste. Adapted from Hwang and Wu [3].

column of Table 9 [1]. Each percent of rice husk ash can improve at least 0.6 times of permeability at 1 year. No admixtures or processing techniques in concrete technology are known to yield a concrete product with such low chloride permeability so far. The potential usefulness of rice husk ash, as a cement or concrete additive, for applications where the corrosion of reinforcing steel is a major concern is obvious. Hence, the use of rice husk ash is quite significant for those areas that need water resistance, and good durability like in the marine environment.

Table 9. Mix Proportions and Properties of Concrete Containing RHA. Adapted from Mehta [1].

Mix NO.		RHA (%)	Mix Proportion, kg/m^3					Fresh Properties				Compressive Strength, Mpa				Permeability, Coulombs	
			C	RHA	C.A.	F.A.	W	W/B	W/C	Air,%	Slump, mm	3-d	7-d	28-d	1-yr	28-d	1-yr
1	a	0	392	-	1062	786	128	0.33	0.33	1.0	200	45	56	65	80	3500	2200
	b	9	356	36	1062	786	128	0.33	0.36	1.5	225	42	56	77	86	1260	420
2	a	0	410	-	1044	786	128	0.31	0.31	1.0	240	47	60	66	80	3260	2200
	b	13	356	54	1044	786	128	0.31	0.36	1.5	175	45	60	80	92	870	250
3	a	0	428	-	1026	786	128	0.30	0.30	1.5	225	47	62	70	81	3000	1800
	b	17	356	72	1026	786	128	0.30	0.36	1.5	200	46	65	80	92	390	190

- All mixtures contained a constant amount of a superplasticizer in order to obtain high consistency.
- Coulombs passed in a 6 hours standard test (AASHTO T-277), based on FHWA Report No. RD-81/119, Aug. 1981.

The Modules of Elasticity, Creep and Shrinkage of Concrete with RHA

Modulus of elasticity, creep and drying shrinkage characteristics of concrete are greatly influenced by strength of concrete and stiffness of aggregate. Since ultimate strength of concrete containing pozzolans will result in significant gain in the modulus of elastic and creep will be low after 28 days [24].

Since the addition of rice husk ash reduces bleeding, the constructor needs to carefully protect the concrete surface when conditions for plastic shrinkage cracking prevail [24]. The pozzolanic reaction of rice husk ash refines the pore structure as is indicated in Figure 14, hence at the same water-to-binder ratio the amount of drying shrinkage of concrete with the addition of rice husk ash is slightly higher than that of concrete without rice husk ash [3]. Figure 17 indicates such tendency.

THE DURABILITY PROPERTIES OF CONCRETE WITH RHA

The addition of pozzolans can improve the properties of concrete by modifying the micro- and macro-structure of cement paste. The direct beneficial effect of fine and cellular RHA particles on the bleeding and segregation characteristics of concrete mixtures is due to large water adsorption ability, high internal surface area, as well as microporous and amorphous particles.

The nature of rice husk ash can be viewed in Figure 3. A reduction in bleeding water results in a stronger transition zone between solid matter and cement pastes. This will lead to a more impermeable and durable concrete.

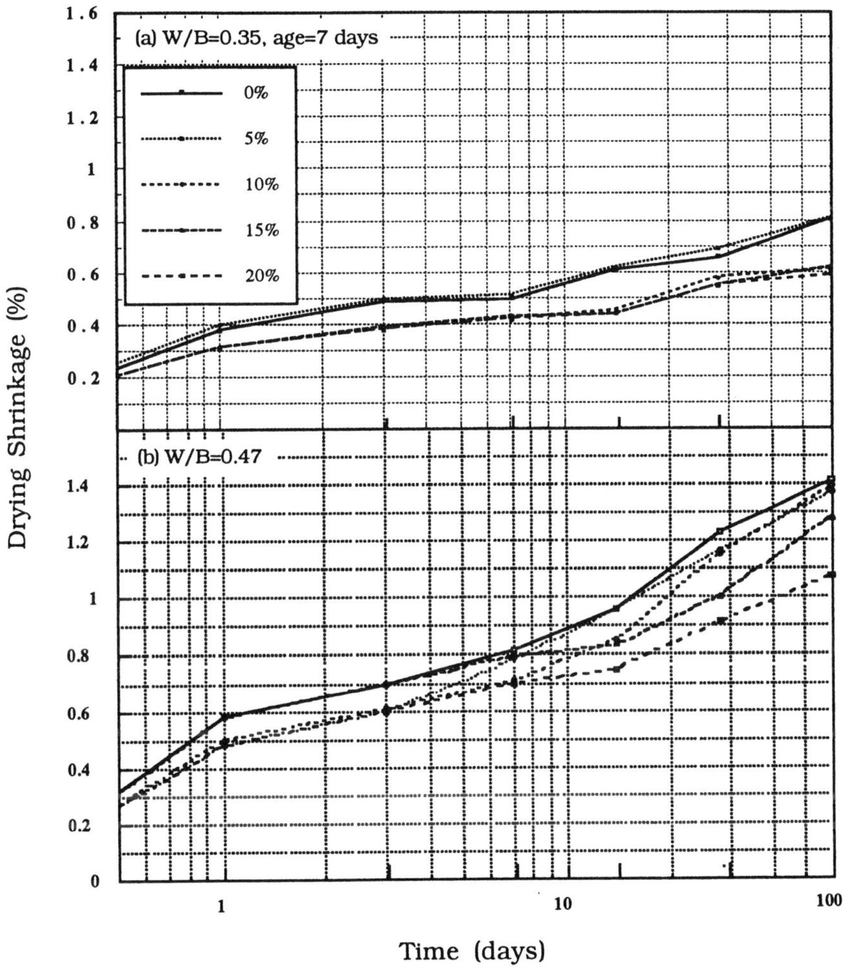

Figure 17. Drying shrinkage of cement paste containing RHA. Adapted from Hwang, Peng and Lin [22].

The Reduction in the Permeability of Concrete

The permeability of concrete plays an important role in durability because it controls the rate of entry of moisture that may contain aggressive chemicals and the movement of water during wetting or drying. The single parameter that has the largest influence on durability is water to cement or water to binder ratio [27]. As the water-to-binder or water-to-cement ratio is decreased, the porosity of the paste is decreased and hence, the concrete becomes more impermeable. The addition of microporous rice husk ash adsorb large amount of water surrounding the solid matter, producing a low water to binder ratio and refining the pore structure. This consequently reduces the permeability of concrete-to-chloride penetration as shown in Table 9 [1]. It has been found that the permeability is significantly reduced after 28-day curing for the cement paste with 10, 20 or 30 percent rice husk ash [1]. The permeability of the cement paste with rice husk ash is believed to be in the range of 1×10^{-11} cm/sec.

Resistance to Thermal Cracking

The mineral admixtures which react slowly at normal temperatures can benefit from the heat of hydration of Portland cement without sacrificing ultimate strength. The use of rice husk ash to replace a portion of the cement directly reduces heat evolution. The possibility of thermal cracking for massive concretes or plastic shrinkage cracking for large flat surface construction will thus be reduced.

Resistance to Alkali-Aggregate Reactions

Alkali-Aggregate reaction is a slow process between the alkali solution from a high alkali portland cement (greater than 0.6% Na_2O eq.) and active aggregate [24,27]. A partial replacement of high-alkali cement by rice husk ash is a direct way to cut down the alkali supply and has been effective in reducing expansion due to the alkali-aggregate reaction. Mehta and Folliard [28] have reported a 95% reduction in expansion for blended cements containing 90% high alkali Portland cement (1.0% as Na_2O) and 10% RHA (Figure 18). This is above the minimum limit of 75% proposed in ASTM C441 for assessment of pozzolans in reducing the alkali-silica expansion. Hence, the addition of rice husk ash plays several roles. First, RHA helps to reduce the supply of alkali from cement. The RHA competes with the slow reactive aggregate and consumes the alkali from cement due to its greater reactivity. Further, it adsorbs the surrounding water and refines the pore structure. This further hinders the diffusion of alkali ions to the surface of aggregate by microporous rice husk ash. Table 8 shows the effect of rice husk ash on the pH value of water.

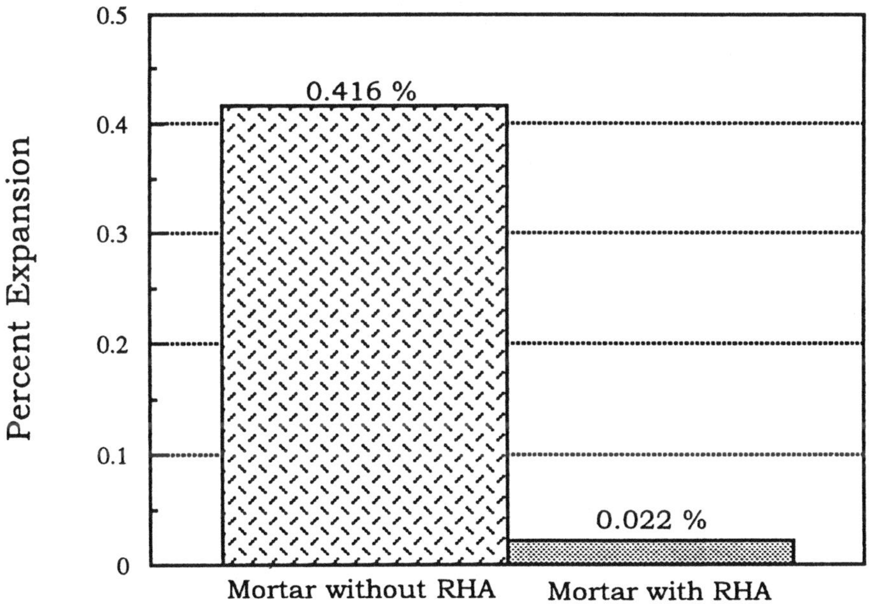

Figure 18. Reduction of alkali-silica expansion according to ASTM C441. Adapted from Mehta and Folliard [28].

Resistance to Acidic Attack

Hydrated cement paste is an alkaline material and therefore, specific attack by alkaline materials is not normally encountered. The situation is quite different for acidic solutions which readily attack basic cementitious material such as concrete and mortar. The addition of rice husk ash has been found to result in excellent resistance of the material to acidic environments [2]. With regard to durability of cement to acidic attack, it may be noted that Portland cement contains 60 to 65% CaO and, upon hydration, a considerable portion of lime is released as free $Ca(OH)_2$, which is primarily responsible for the poor performance of Portland cement concretes in acidic environments [2]. Blended cements with rice husk ash having strength characteristics similar to ordinary Portland cement may contain as low as 20 percent CaO. Upon hydration of the cements, none of the lime would be present in the form of free $Ca(OH)_2$. The products of hydration consist of calcium silicate hydrates and silica gel and, therefore, are more resistant to acidic attack.

Immersion tests of an 8-month period in 1 percent solutions of hydrochloric acid, acetic acid, and lactic acid showed that portland cement mortars were prone to surface softening and disintegration, whereas the mortar cubes made with a rice husk ash cement held their original hard surface and sharp edges. Similar results were obtained when concrete cylinders (W/B=0.4) were immersed in 5% HCl acid solutions, as shown in Figure 19 [6]. Concrete with RHA after immersion test shows less weight loss which indicates the superior feature to resist acidic attack.

Resistance to Corrosion of Steel in Concrete

The corrosion of steel in concrete is a major problem in humid environment. Reduction of alkalinity of the cement paste by carbonation from atmospheric CO_2 and other pollutant gasses like SO_2, Cl^-, and NO_x is generally the first step in the process of corrosion of steel in concrete [24]. In a hydrated Portland cement paste, about

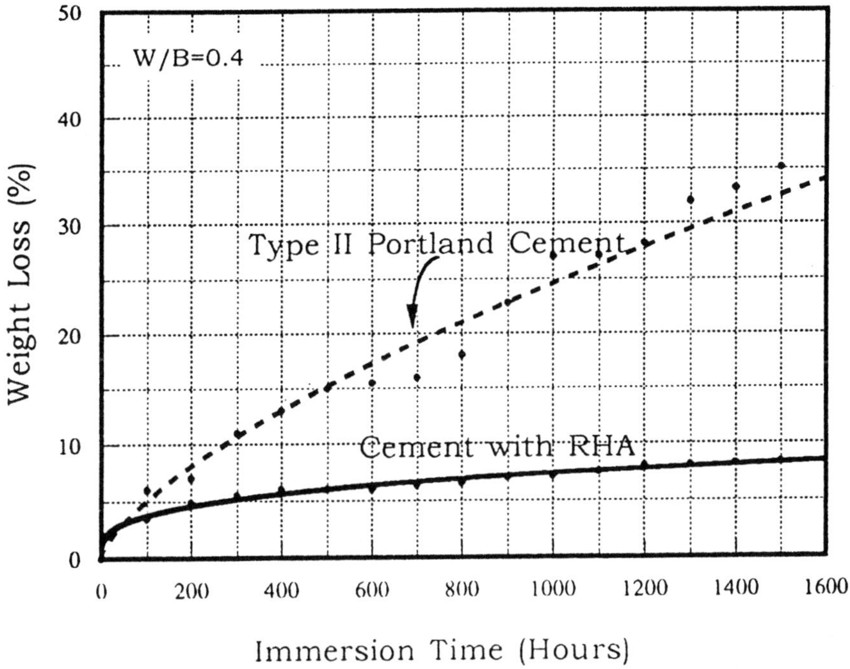

Figure 19. Weight loss of Portland cement with RHA cement concrete cylinders, continuously immersed in a 5% HCl solution. Adapted from Andersson [6].

20% $Ca(OH)_2$ is present to provide the basicity for steel protection. Since pozzolanic reaction consumes $Ca(OH)_2$, theoretically, about 25% reactive silica present in a Portland-pozzolan mixture is sufficient to consume all the available $Ca(OH)_2$ from Portland cement hydration [24]. This has caused some concern among construction engineers who feel that due to the reduced alkalinity of concrete, the addition of pozzolans to reinforced and prestressed concrete would present a danger for corrosion of steel. It is known that the resistance to the carbonation of concrete does not appear to be related to the amount of calcium hydroxide in the pore solution since the pore solutions in mature Portland cement pastes contain little or no calcium. The resistance to carbonation must be thought mainly in physical factors such as permeability of the cement paste. The addition of rice husk ash has shown a considerable reduction in porosity, and has effectively reduced the size of large pores after 28-days curing, and has yielded cement pastes of very low permeability, as shown in Figure 14 [27].

In conclusion, the pozzolanic and cementitious reactions associated with rice husk ash reduce the free lime present in the cement paste on one hand and decrease the permeability of the system on the other hand, thus improving the overall resistance to CO_2 attack. This suggests that even when very active SiO_2-rich pozzolans are used the danger due to carbonation should be of less concern in concretes containing pozzolans [24]. Hence, improved resistance to corrosion of steel in concrete can be anticipated.

Resistance to Sulphate Attack

Sulphate resistance of mortars made with blended cement (10% RHA and 90% high C_3A Portland cement) were tested according to ASTM C1012 [28]. Comparisons were made to a reference mortar with 0.485 water-to-cement ratio without addition of RHA. In making RHA mortars, sufficient water was used to maintain the same flow. Standard bars of 25.4 x 25.4 x 285.8mm were cast and were cured at 38°C for 24 hours when they acquired a strength of 19.6 MPa as required by the test method. The samples were immersed in sulphate solution (50g Na_2SO_4 per liter of water). The results are shown in Figure 20. It was observed that at the end of 16 weeks there was 82% reduction in expansion in RHA mortar compared to the reference mortar. It was further shown that the expansion in the case of reference mortars continued whereas expansion was almost stable for the RHA mortars. The reason for this improvement is attributed to the densification and strengthening of the transition zone which most probably improves the resistance to microcracking. Due to these modifications, penetration of the

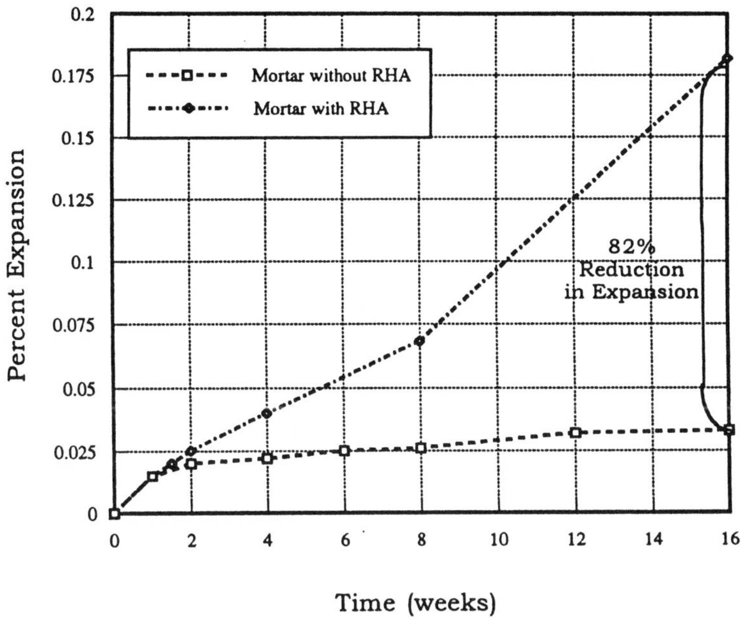

Figure 20. Reduction in sulfate expansion according to ASTM C1012. Adapted from Mehta and Folliard [28].

sulphate solution is reduced, and consequently sulphate resistance is improved.

Resistance to Freezing and Thawing

The development and use of high strength concrete for construction is getting more and more popular. In many parts of the world these concretes are exposed to extreme weather conditions like freezing and thawing. Most of the high strength concretes are made with the use of condensed silica fume. Research has been done to study the freeze-thaw resistance of high strength concrete using condensed silica fume and quite controversial results have been reported. Freeze-thaw resistance of RHA concrete has not been extensively investigated. The only work which the authors came across was done at the University of California at Berkeley [28]. Their work is summarized here.

Prisms (76.2 x 101.6 x 406.4 mm; 3 x 4 x 16 inch) were cast using RHA, CSF (condensed silica fume) and comparisons were made to samples having no mineral admixture. The water-to-binder ratio used was 0.30 and 0.35, respectively. High range water reducer was used for making the concrete specimens. The mix proportions are shown in Table 10. The samples were moist cured for 14 days prior to the test. The freeze-thaw test perform was according to the ASTM 666, Procedure A. The test was performed for 300 cycles unless the durability factor fell below 60, at which time the particular samples were removed from the test. The durability factor after 300 cycles is shown in Figure 21 [28]. The concrete with RHA was only one to clear 300 cycles without reaching the failure criteria (durability factor less than 60). The CSF concrete with water-to-solid ratios of 30 and 35% have shown poor frost resistance with the durability factor less than 10 after 101 and 67 cycles, respectively. It is further seen that

Table 10. Mixture Proportions for Concrete for Freeze-Thaw [28].

Mixture	Cement (kg/m^3)	RHA (kg/m^3)	CSF* (kg/m^3)	Aggregate (kg/m^3)		Water (kg/m^3)	SP ml/kg binder
				Fine	Coarse		
REF30	453	0	0	672	1207	136	33
SF30	376	0	66	672	1207	133	57
RHA30	376	62	0	672	1207	133	53
REF35	420	0	0	672	1207	147	24
SF35	349	0	65	672	1207	144	36
RHA35	349	62	0	672	1207	144	36

* CSF: condensed silica fume

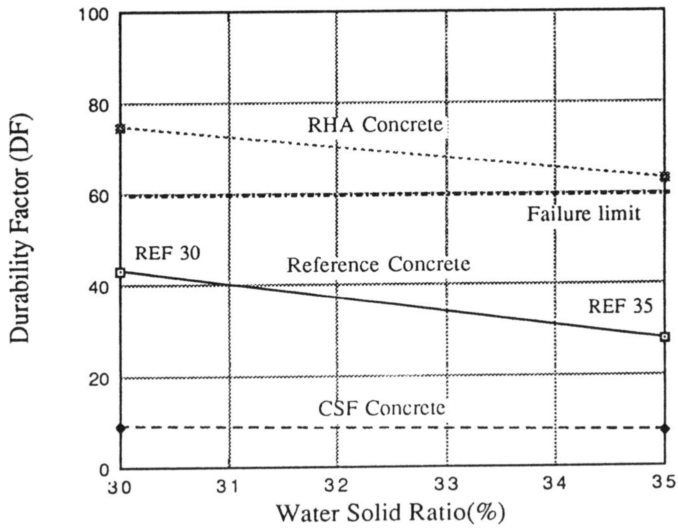

Figure 21. Durability factors of concrete subjected to 30 cycles of freezing and thawing according to ASTM C666, procedure A. Adapted from Mehta and Folliard [28].

the concretes with a water to cement ratio of 0.30 performed better than those with cement ratio of 0.35. The reference concrete has also faired poorly, although its performance was much better than the silica fume concrete. REF30 yielded a durability factor of 43 after 214 cycles and REF35 had a durability factor of 28 after 144 cycles. The poor performance of silica fume concrete is most probably due to the impervious cement/silica fume matrix. The matrix had very small discontinuous pores dispersed randomly throughout. The improved performance of RHA concrete is most likely due to the highly microporous structure. This microporosity provides escape paths for the freezing water inside the concrete. Internal stresses are relieved and subsequently, microcracking is reduced, and freeze-thaw resistance is improved. The freeze-thaw resistance of RHA concrete in a saline environment was also tested. The product performed well [29].

SUMMARY

Annually about 20 million tons of rice husk ash are produced from rice husks which are a waste product from the rice industry. By proper control of the incineration process, the quality of RHA can be assured. The silica content of the ash may be as high as 90-95% with residual carbon as the remaining ingredient. Alkalis ranging from 1 to 3% usually form another impurity. The non-crystalline RHA with 50-100m^2/g surface area behaves reactively and is available in many parts of the world.

To develop pozzolanic activity in the presence of lime or portland cement, the RHA particles need not be very fine. This is because the source of high surface area in RHA is in the microporous structure of individual particles. RHA particles, in the 10-75 μm range, exhibit satisfactory pozzolanic behavior. Also, it may be noted that the impurities normally present in rice husk ash do not cause any adverse effect on the properties of portland cement concrete.

THE RHA in cement performs as a kind of accelerator that shorten the setting time through the adsorption of surrounding water. A lot of C-S-H gel forms inside the microporous RHA and thus tends to strengthen the matrix. SEM micrographs as well as other qualifying measure had shown the effect of RHA on hydration behavior of cement paste.

As a supplementary cementing material, many unique properties of RHA have been reported. With fresh concrete mixtures, pozzolanic additions, such as rice husk ash, have the ability to reduce bleeding and segregation, and thus cause significant improvement in workability. Only RHA has the ability to contribute to strength of portland cement concrete at the early ages of 1 and 3 days. This opens the door of mixing normal pozzolans along with rice husk ash. The latter will act as an accelerator in strength development. Up to 70 percent portland cement replacement has been reported without any adverse effect on strength, 10 to 20% cement replacement even shows a beneficial effect on strength and an extraordinary improvement in the chloride permeability characteristics and other durability properties of concrete. Also, blended portland cements containing 10% rice husk ash have been shown to be quite effective in combating expansion due to alkali-aggregate reaction. Through pore filling effect and pozzolanic reaction of RHA due to high surface area and cellular structure, the permeability of concrete can be significantly reduced. Hence the other durability characteristics of concrete are improved.

The use of RHA has shown superior effect on the properties of concrete. Therefore, the use of RHA should be considered not merely for environmental and economical reasons, but for safety and durability, as well.

REFERENCES

1. Mehta, P. K., Rice Husk Ash-A Unique Supplementary Cement Material, *Advances in Concrete Technology*, Ed. by Malhotra, CANMET, Ottawa, Canada (1992).
2. Mehta, P. K., and Pitt, N., Energy And Industrial Materials from Crop Residues, *Journal Resource Recovery and Conservation*, No. 2, pp. 23-38, Elsevier Scientific Publishing Company (1976).
3. Hwang, C. L., and Wu, D. S., Properties of Cement Paste Containing Rice Husk Ash, *ACI SP-114* (Editor: V. M. Malhotra), pp. 733-765 (1989).
4. Lin, K. M., The Study on the Manufacture of Particle-board Made of China Fir Flakes and Hulls, *A Thesis of National Chung-Hsing University for the Degree of Master*, Republic of China (1975).
5. Tillman, D. A., *Wood as an Energy Resource,* Academic Press Inc., New York, U.S.A., pp. 65-87 (1978).
6. Anderson, L. L. and Tillman D. A., *Fuels from Waste,* Academic Press Inc., New York, U.S.A., pp. 141-159 (1977).

7. Yand, S. J., Exploratory Study of Catalytic Cracking of Cellulosic Biomass (Rice Hull), *A Thesis of National Taiwan University for the Degree of Master*, Republic of China, pp. 5-7 (1980).
8. Pitt, N., Process for Preparation of Siliceous Ashes, *U.S. Patent* No. 3959007. (1976).
9. Longanback, J. R. and Baner, F., *Industrial and Laboratory Pyrolysis,* pp. 476-491 (1976).
10. Tsukishima Kikai Co. LTD, Energy from Solid Waste, *(an unpublished paper in Japanese) (1995).*
11. Schlesinger, M. D., W. S. Scanner and D. E. Wolfson, *Processing Agricultural and Municipal Wastes,* AVI Press, pp. 93-100 (1972).
12. Gieseking, J. E., *Soil Components, Volume 2, Inorganic Components*, Spring-Veraly Berlin Heidelberg, Germany (1975) (a translated paper).
13. Millots, G., *Geology of Clays, Spring-Veraly Berlin Heidelberg*, Germany (1970).
14. Ankra, K., Studies of Black Silica Produced Under Varying Conditions, *PhD Dissertation,* University of California at Berkeley (1975).
15. Yamamoto, Y. and Lakho, S. M., Production and Utilization of Active Rice Husk Ash as a Substitute for Cement, *Proc. Japan Soc. of Civil Engineers,* No. 322, pp. 158-166 (1982).
16. Housten, D. F., Rice Chemistry and Technology, *American Association of Cereal Chemists* (1972).
17. Cook, D. J. and Suwanvitaya, P., Rice Husk Ash Based Cements-A State-of-the-Art Review, *Proc. ESCAP/RCTT Workshop on Rice Husk Ash Cement,* New Delhi (1981).
18. Kapur, P. C., Tube-in-Basket Rice Husk Burner for Producing Energy and Reactive Rice Husk Ash, *Proc. ESCAP/RCTT Workshop on Rice Husk Ash Cement,* New Delhi (1981).

19. Al-Khalaf, M. N. and Yousif, H. A., Use of Rice Husk Ash in Concrete, *J. Cement Composites*, Vol. 6, pp. 241-248 (1984).
20. James, J. and Subba Rao, M., Characteristic of Silica in Rice Husk Ash, *Amer. Ceram. Soc. Bulletin,* Vol. 65, No. 8, pp. 1177-1180 (1986).
21. Boateng, A. A. and Skeete, D. A., Incineration of Rice Hull for Use as a Cementitious Material: The Guyana Experience, *Cem. Concrete Research,* Vol. 20, No. 5, pp. 795-802 (1990).
22. Hwang, C. L., Peng, T. F., and Lin, L. K., The Hydration Mechanism of the Rice Husk Ash Cement Paste, *Journal of the Chinese Institute of Civil and Hydraulic Engineering,* Taipei, Republic of China (To be published) (1995).
23. Hwang, C. L., and Shen, D. H., The Effects of Blast-Furnace Slag and Fly Ash on the Hydration of Portland Cement, *Cement and Concrete Research,* Vol. 21, pp. 410-425 (1991).
24. Mehta, P. K., Pozzolanic and Cementitious By-products as Mineral Admixtures for Concrete-A Critical Review, *ACI SP-79,* Detroit, MI pp. 1-46 (1983).
25. Hwang, C. L., Basic Material Properties and Mixture Proportion of High Performance Concrete, *International Workshop on Civil Infrastructural System-Application of Intelligent Systems and Advanced Materials on Bridge Systems*, Ed. by G. C. Lee and K. C. Chang, Taiwan, pp. 207-224 (1994).
26. Johansen, V., and Anderson, P. J., Particle Packing and Concrete Properties, *Material Science of Concrete,* Vol II, Ed. by Skalny, J., and Mindess, S., The American Ceramic Society, US, pp. 111-147 (1989).
27. Mehta, P. K., Monteiro, J. M., *Concrete Structure, Properties, and Materials,* 2nd Ed., Prentice-Hall, Englewood Cliffs, NJ (1993).

28. Mehta, P. K. and Folliaod, K. J., Rice-Husk Ash-A Unique Supplementary Cementing Material: Durability Aspect, Advances in Concrete Technology, in *Proceedings of Adam Nevill Symposium*, Las Vagas, June 1995.
29. Sandberg, P., Resistance of Non Air-entrained RHA Concrete to Freezing and Thawing in Soline Environment, *Fifth CANMET/ACI International Conference on Fly Ash, Silica Fume, Slag & Natural Pozzolanas,* Milwaukee, June 1995.

5

BLAST FURNACE SLAG - THE ULTIMATE BINDER

Bob Talling and Pavel Krivenko

INTRODUCTION

The dominant role of pure Portland cement is slowly decreasing in the favour of substituted and composite cements. Another very important change is the steadily growing amount of plants specialized in separate grinding and or mixing of components for cements. This makes it possible to optimize the particle size distribution of individual material.

New drying, grinding and mixing technology will make the use of various secondary and waste material as well as several natural materials more feasible and interesting in the nearest future.

The interest in cements containing very high proportions of waste or unprocessed materials and new types of cements designed to utilise specific types of industrial by-products and waste materials is constantly growing. The understanding and the use of alkali-activated binder systems is one of the more promising developments. This chapter introduces the use of alkalis in binder systems, and will bring some very important information about the development of these binders in the former Soviet Union area, especially in the Ukraine. Slag possesses a unique glassy character, the structure and reactivity of which is very strongly dependent upon the chemical composition and cooling conditions. Therefore, it is important that one gains a greater understanding of slag activation before it can be fully commercialized. The hydration of slag is, in principle, easier to control, because the only reactive part of the slag is the glass. Finer grinding of slag strongly promotes hydration and has a very positive effect on the microstructure.

Concrete is a very special composite in which the dry glue (Portland cement) content is from 2 to 20% in volume. Typical glue content is in the range of 7-13% by volume. The volume of water in the fresh concrete is at least as high as the glue volume (w/c=0.31) and in most cases, almost twice the glue volume (w/c=0.62). It is of greatest importance that the glue is of optimum composition and is uniformly distributed in the matrix. It is feasible to develop cements that provide more effective gluing material per added amount of cement.

The future is in combinations of different binders and modifiers into more durable and controllable systems. The target is not only to make the cements and concretes less expensive, but also to give the blend tailored properties suitable for a specified purpose. To be able to do this one must learn how to combine chemistry and material science. There is an increasing demand for a better understanding of material properties and composite materials, as well as better control of the microstructure for development in building materials.

It has been verified that the substituted or blended cements or, the cements based entirely on waste products, can provide performance properties that are better than or not found in Portland cement.

GRANULATED BLAST FURNACE SLAG

General Characterization

A majority of wastes from metallurgical processes are represented by slags. Blast furnace slags are the most widely investigated and have turned out to be the most effective for cement and concrete manufacturing. The main constituents of blast furnace slags are the following oxides: CaO, SiO_2, Al_2O_3, and MgO, i.e., the oxides which are found in the majority of mineral cementitious substances, in particular, in Portland cements. Blast furnace slag is a by-product of iron manufacture and results from the fusion of limestone with minerals from the coke and the siliceous and aluminous residue remaining after the reduction and separation of iron from the ore. The slag, which has a chemical role in removing sulphur from molten iron, rises to the surface and is tapped off from the blast furnace. Blast furnace slag is a versatile material. If it is air-cooled slowly in pits it gives a crystalline product resembling an igneous rock and consisting of a mixture of minerals of which melilites are the main series. Such slag is suitable as dense aggregate for concrete. Alternatively, the slag may be foamed with a limited amount of water to give a lightweight aggregate for concrete. However, the product with the highest economic value is obtained by very rapid cooling using high pressure water jets which leads to a vitrified granular material consisting of glass with a small content of crystalline components. Granulated blast furnace slag is a latent hydraulic material, i.e., it possesses hydraulic properties when suitably activated. It should contain at least two-thirds by mass of glassy slag, preferably, more than 90%. The granulated slag should also consist of at least two-thirds by mass of the sum of CaO, MgO and SiO_2. The remainder contains Al_2O_3 together with small amounts of other oxides. The ratio by mass (CaO + MgO)/(SiO_2) should be greater than 1.0.

Slag contains 30-50 % CaO, 28-40% SiO_2, 8 - 24% Al_2O_3 and 1-18% MgO while Portland cement contains 55-66 % CaO, 20-24% SiO_2, 0-8% Al_2O_3 and up to 5% MgO. This difference in composition, as well as the presence of the compounds of iron, manganese, titanium, sulphur and other elements in slag melts, cause the difference in mineralogical composition between slag and Portland cement clinker as is seen in the phase diagram of CaO-MgO-Al_2O_3-SiO_2 in Figure 1 [1]. Another significant difference is that some of the metals in slag are not fully oxidized, consequently they will have a

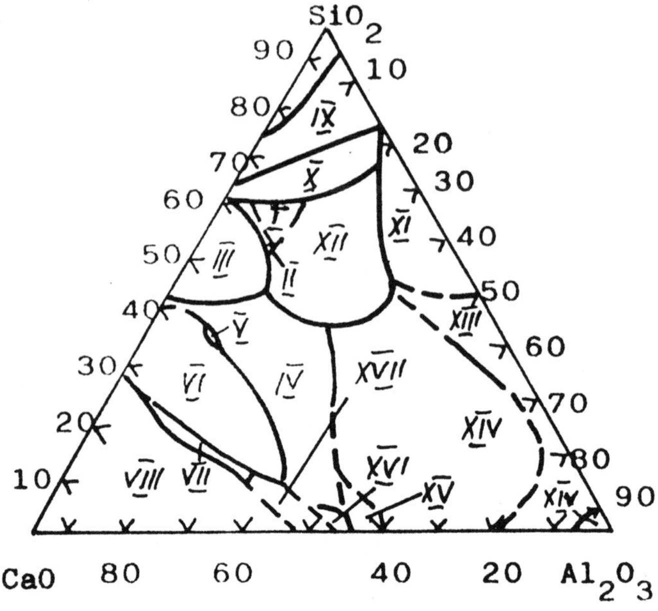

Figure 1. Field of crystallization of a system: CaO- MgO-Al_2O_3-SiO_2. I- diopside; II β- C_2S; III-α-CS; IV- melilite; V- merwinite; VI- C_2S; VII -C_3S; VIII- CaO; IX - cristobalite; X - tridymite; XI - mullite; XII - anorthite; XIII - corundum; XIV - spinel; XV- $C_{12}A_7$; XVI- CA; XVII- C_3A.

different structural role in comparison with the Portland cement clinker. Because of these basic differences in the mineralogical composition, it is clear that the hydration reactions and the hydration products of slag will be different from those of Portland cement.

A set of equations have been worked out for evaluating indirectly the ability of slags to possess hydraulic activity. The ratios between oxides contained in slags serve as a base for this evaluation. Two important equations are the modulus of basicity (M_b) and modulus of activity (M_a).

The modulus of basicity of slag is the ratio between the sum of basic oxides to the sum of acidic oxides by mass [2,3]:

$$M_b = \frac{CaO + MgO}{SiO_2 + Al_2O_3} \qquad (1)$$

Depending on the numerical value of this modulus, slags are divided into two categories: basic for those with a modulus >1 and acidic for those with a modulus < 1.

Aluminium has an amphoteric nature, i.e., it acts as a basic oxide when it is in a sixfold coordination and as an acidic oxide if it is in a fourfold coordination. With this in mind, the modulus of basicity can be calculated using the following equations [4,5]:

$$M_b = \frac{CaO + MgO + 1/3Al_2O_3}{SiO_2 + 2/3Al_2O_3} \qquad (2)$$

or

$$M_b = \frac{CaO + MgO + Al_2O_3}{SiO_2} \qquad (3)$$

The other criterion for evaluating the activity of blast furnace slags is the modulus of activity defined by [2]

$$M_a = \frac{Al_2O_3}{SiO_2} \qquad (4)$$

The hydraulic activity of blast furnace slags trends to increase with increasing modulus of basicity and activity. The impurities of other oxides essentially affect the hydraulic activity of blast furnace slags due to partial replacement of CaO in active compounds and the formation of less active compounds. Iron oxide is found in small quantities and has no proven effect on the hydraulic activity.

The content of sulphide compounds such as CaS, MnS, FeS in slags is usually not essential. Generally, the content of CaS is 2-3% by mass, but even when it reaches 5% by mass, it enhances the hydraulic activity. The CaS interacts with water and hydrolyses:

$$2CaS + 2H_2O = Ca(HS)_2 + Ca(OH)_2 \tag{5}$$

The reaction products promote the activation of the glassy constituent of slag.

Iron sulfide (FeS) and MnS may initiate slag grain disintegration because of hydrolysis according to the following reaction:

$$MnS + 2H_2O = H_2S + Mn(OH)_2 \tag{6}$$

The specific volume of the hydroxide thus formed is higher than that of the starting materials, and is the cause for disintegration of slags. Hydrolysis further promotes disintegration of slag. However, at large volumes, both metal decrease the activity of the slags and retard setting of slag binders.

Titanium is present in slag as Ti_2O_3 and it promotes the rate of polymerisation of silicon-oxygen compounds. The occurrence of complicated silicon-titanium-oxygen complexes and decrease in length of Ti-O bonds as the titanium undergoes transition to four-fold coordination leads to decrease in adsorption of water molecules on the slag particle surface and, hence a decrease in its hydraulic activity [6]. Depending on the total slag composition, a critical value of titanium as TiO_2 is between 1.5 to 2.5%. A higher content has a more pronounced negative effect on slag cements with high slag content.

For evaluating the hydraulic activity of blast furnace granulated slags Budnikov [2], proposed to use a coefficient of quality as defined below:

MgO content up to 10 mass%:

$$K = \frac{CaO + Al_2O_3 + MgO}{SiO_2 + TiO_2} \tag{7}$$

MgO content higher than 10 mass%

$$K = \frac{CaO + Al_2O_3 + 10}{SiO_2 + TiO_2 + (MgO - 10)} \tag{8}$$

Depending upon this coefficient and the chemical composition the slags may be divided in 3 types as shown in Table 1.

Table 1. Characterization of Slags.

Characteristics	Limits For Types		
	First	Second	Third
Coefficient of Quality: not less than	1.65	1.45	1.2
*Al_2O_3 not less than	8	7.5	no limit
MgO not more than	15	15	15
TiO_2 not more than	4	4	4
MnO not more than	2	3	4

*The oxide content mentioned in the table are in mass%.

Structure and Hydraulic Activity of Slags

The oxide composition of blast furnace slag forms various minerals. Analyzing the phase diagrams of the corresponding oxide systems suggests the existence of up to forty double and triple compounds. A majority belongs to silicates, aluminosilicates, aluminates and ferrites [8]. As seen in Figure 1, the mineralogical composition of the slag is characterized by the presence of compounds with a lower basicity as compared to the minerals of Portland cement clinker. The granulated blast furnace slags consist mainly of a glass with crystal inclusions of larnite β -Ca_2SiO_4 and melilite, which is itself a solid solution of Ca_2AlSiO_7 - $Ca_2MgSi_2O_7$ in basic slags and mainly melilite in acidic slags. The crystals of larnite, rankinite, $Ca_3Si_2O_7$, pseudowollastonite, α-$CaSiO_3$, and melilite dominate over glass in non-granulated slowly-cooled basic slags, while anorthite, $Ca_2Al_2(Si_2O_8)$, dominates in acidic high alumina slags. An increased MgO content gives merwinite, $Ca_3Mg(Si_2O_8)$, monticellite, $CaMgSiO_4$, and diopside, $CaMg(SiO_3)_2$ [7].

The hydraulic activity of slags may be evaluated by the content and ratio between minerals contained in the corresponding crystalline slag. These minerals, under thermal treatment hydrate according to the sequence mentioned in the reference [9]: C_3S > calcium alumoferrites >β -C_2S > slag glass of basic composition > slag glass of acidic composition > melilite > γ -C_2S > merwinite > monticellite > low basic calcium alumosilicates and silicates (anorthite > rankinite > pseudowollastonite) > fayalite. The hydraulic properties of the slags are determined by the structural irregularities of the constituent phases, the state of which is predetermined by set of chemical, physical and technological conditions involved in their formation.

As a rule, the phase composition and structure of the slags depends on their chemical composition and cooling conditions. Freely cooled blast furnace slags with modulus of basicity M_b=1 contain approximately 50-70 % by mass of crystalline phase, whereas the high basic slags with M_b=1.5 are mainly crystalline [1]. In acidic slags that contain essential amounts of alumina and are characterized by high viscosity, even with slow cooling, a glassy structure is formed [7]. Common in the existing structural theories is the existence of micro-inhomogeneity and the existence of multi-phases within the structure of a glassy matter along with the existence of microfields with structures close to that crystalline materials (ordered crystal lattices) referred to as crystallites. The formation of crystallites (which may be high basic minerals such as melilites and orthoclases) lowers the basicity of the rest of the melt [10] and results in a more acidic composition of the glassy phase of partially crystallized basic slags. This correlates well with the data that CaO, MgO, FeO are concentrated predominantly within the crystalline phases, and the slag glass is found to have an alumino-silicate composition [1]. When slag granulation begins at lower temperatures, the resultant granulated slag has a low reactivity. In some slags, it has been observed that the glass has separated into two different phases each with a distinct composition. This explains certain variations in the reactivity of slags. It is important that the slag producer optimizes the granulation process to provide the best possible reactivity and stability.

According to the modern views [11] a liquid slag is itself a micro-heterogeneous melt consisting of elementary cations (Ca, Mg, Na, Mn, Fe, etc.) and elementary (oxygen, sulphur) and complex anions of different complexity KxOy (where K is Si, P, Al, Ti, B, V, Cr, etc.).

The complex anions play an important role in the hydraulic hardening ability of slags. The highest hydraulic activity is characteristic of the groups consisting of elementary anions such as $(SiO_4)^{4-}$ and $(AlO_4)^{5-}$ isolated with calcium cations much like the structure of orthosilicates [1]. The similar micro-heterogeneous ion groups in the same or another form may be found in the rapidly quenched slag glass. At slow cooling rates, these groups may be the structure-forming units and by rearrangement and aggregation may occur as more complicated inclusions. The degree of condensation of the structural units in the cooled slag depends upon the content of calcium and magnesium available to keep them in a separated state.

According to a theory of discrete anions [12, 13], a content of 10-33 mole% of MeO in the molten system MeO-SiO_2 stabilises discrete circular anions $(Si_{2n}O_{6n+3})^{6-}$ or $(Si_{4n}O_{8n+4})^{4-}$ that divide a space network from silico-oxygen tetrahedron into separate groups. From 33 to 50 mole% MeO a slag consists of circular anions$(Si_8O_{20})^{8-}$, $(Si_6O_{15})^{6-}$, $(Si_4O_{12})^{8-}$, $(Si_3O_9)^6$. Between 50-66 mole% MeO there occur short ($n \leq 3$) chains $(Si_nO_{3n+1})^{(2n+2)-}$ along with the elementary circular ions. At MeO-content above 66 mole-% the structure of the slag destroys increasingly until the isolated tetrahedron $(SiO_4)^{4-}$ dominate. The silicates of the system MeO-SiO_2 are represented in the slag by pseudowollastonite, wollastonite, and larnite. The structure of pseudowollastonite α-$CaSiO_3$ consists of double layers of tetragonal rings $(Si_3O_9)^{6-}$ separated by layers of octahedrons, (CaO_6). The structure of wollastonite, β-$CaSiO_3$, consists of the chains $(SiO_3)^{2-}$, in which a recurrent periodicity equal to three tetrahedra. The Ca^{2+} - ions are between the chains surrounded by six oxygen atoms forming the irregular/disordered octahedron groups. The larnite, β -Ca_2SiO_4, structure consists of separate tetrahedra $(SiO_4)^{4-}$, where each one is surrounded by eight atoms of calcium. Among the above listed minerals, only larnitepossesses hydraulic properties.

The introduction of Al_2O_3 in acidic melts of the CaO-SiO_2 system, in which the silicon-oxygen network dominates, does not lead to its destruction [15]. Rather, it leads to transition of the structure into a field of hydraulically inert anorthite with a framework structure or, gehlenite.

The anorthite and gehlenite are formed when separate groups with the characteristics of the two polymeric forms are combined in the presence of oxygen and Me^{2+}. When aluminum atoms dominate over the silicon in the melt, gehlenite occurs as $Ca_2Al_2SiO_7$ and a network structure is formed by the tetrahedra $(SiO_4)^{4-}$ and $(AlO_4)^{5-}$ united in a six-member ring. The networks are united by the calcium and oxygen ions to form a laminated structure. The structure of anorthite, $Ca_2Al_2Si_2O_8$, consists of a threefold framework of alternating tetrahedra of $(SiO_4)^{4-}$ and $(AlO_4)^{5-}$, in the cavities of which the Ca^{2+} cations are located. The aluminosilicate and the siloxane chains may be exposed, as the amount of aluminium atoms is not sufficient to surround all the silicon atoms, as shown below.

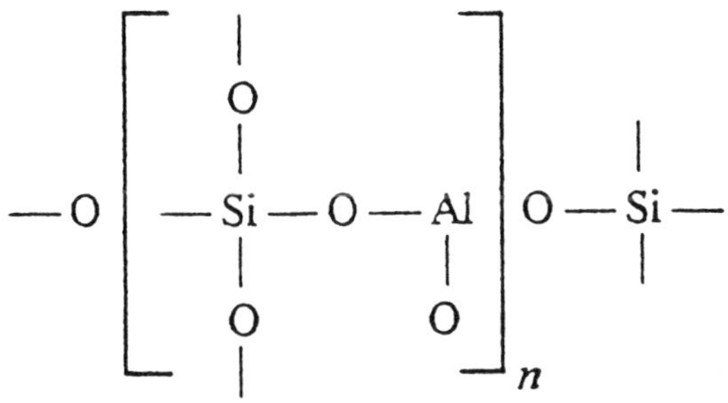

The lack of binding properties of anorthite and gehlenite may be attributed to a high degree of condensation of structural tetrahedra and the presence of high-strength polyalumino-siloxane or siloxane bonds. Breaking the bonds is difficult due to an insufficient ion force of the mixing medium. In this case, the aluminium atoms are situated in tetrahedral coordination and behave like silicon atoms. This is the reason for treating both silicon and aluminum (in fourfold coordination) as a glass network formers when calculating the modulus of basicity and other coefficients. Fourfold aluminium cannot be determined precisely, and is commonly estimated to be 2/3 of the total content of Al_2O_3.

The structure-forming role of MgO in blast furnace slags is closely related to the CaO and Al_2O_3 content [5,17]. For example, when one-half of the Ca^{2+} ions are replaced by Mg^{2+} ions, a diopside ($CaMg(SiO_3)_{2)}$, occurs in the CaO-SiO_2 system. The diopside has a pyroxene structure consisting of chains of SiO_3^{2-} that are joined by bonds with the Mg^{2+}-ion in octahedral coordination. Four of the oxygen ions are situated on one side and are combined with one Si atom and one Mg atom while the four oxygen ions on the opposite side are combined with two Si-atoms.

The ratio of Al_2O_3 to MgO affects the areas in which the solid solutions such as Ca_2AlSiO_7 - $Ca_2Mg(Si_2O_7)$ exist. This determines the degree of hydraulic activity of the slags. For example, at low Al_2O_3 content (<5 %) an increase in MgO greater than 5 % causes MgO to act as a glass-forming agent [7]. This lowers the activity of the slag. Increasing the Al_2O_3-content and simultaneously replacing CaO by MgO the activity enhances and is maximized at 20% Al_2O_3.

The glass content requirement for granulated blast furnace slag is not supported by specific test methods and the wide range of results that can be obtained diminish the value of this specific requirement.

By data reported in [6], analyzing the akermanite and gelhenite structures in the process of transition from a crystalline to glassy state demonstrated that the electron paramagnetic centres on the tetrahedra $(SiO_4)^{4-}$ are not found in the crystalline minerals, whereas in the glassy minerals these centers occur resulting in the higher activity of latter.

Thus, an important characteristic of hydraulic activity of the slag with a mineralogical composition, is the degree of condensation of structure-forming elements ($(SiO_4)^{4-}$ and $(AlO_4)^{5-}$), determining a formula of anion complex, a number of non-bridge atoms of oxygen in the tetrahedron $(SiO_4)^{4-}$ and the type of structural elements [7]. The most general expression for the correlation between these values was found to be the ratio of Si to O defined according to the following relation:

$$f_{Si} = \frac{\nu SiO_2}{(\nu Me_2O + \nu MeO + 3\nu Me_2O_3 + 2\nu MeO_2 + 5\nu Me_2O_5)} \qquad (9)$$

where:

f_{Si} = the degree of binding (or condensation) of the silicon-oxygen tetrahedra in silicates

ν = the oxide content in the glass, mol.% [7, 16].

At Si/O ratios between 0.25 to 0.286, the $(SiO_4)^{4-}$ tetrahedra are found isolated as well as in pair form surrounded by cations. At Si/O ratios between 0.286 to 0.333, the tetrahedra form groups as di-orthogroups and chains and rings. Finally at Si/O ratios between 0.333 to 0.4, the tetrahedra form chains and rings, and occasionally space framework with the cavities. Because of this, the slags are divided into four main structural types described below[7]:

1. Orthosilicate-type slags where the dominate structure consists of silicon-oxygen tetrahedra isolated by Ca^{2+} cations.

2. Melilite-type slags where the dominate structure consists of silicon-oxygen tetrahedra partially combined with each other to form the diorthogroups $(Si_2O_7)^{6-}$ and with $(AlO_4)^{5-}$ -tetrahedra.

3. Wollastonite-type slags which do not contain large circular or chain groupings of silicon-oxygen tetrahedra. At SiO_2 + $2/3Al_2O_3$ less than 50 mol%, separated tetrahedra and diorthogroups were found to exist, when greater than 50 mol%, restricted space groupings.

4. Anorthite-type slags where the dominate structure consists of space framework consisting of tetrahedra $(SiO_4)^{4-}$ and $(AlO_4)^{5-}$ with Ca^{2+} cations in the cavities.

Figure 2 presents the disposition of the indicating points of commercially produced slags based on the classification by a degree of binding f_{Si} [7].

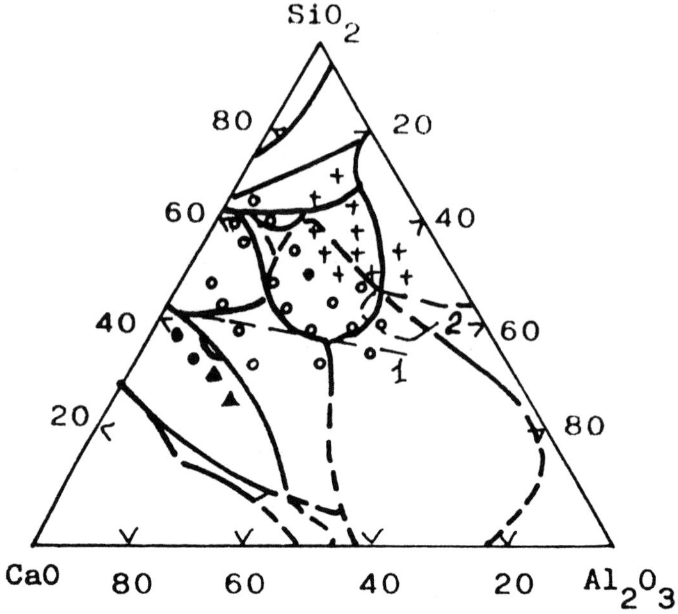

Figure 2. Disposition of indicating points of synthesized slags according to their classification by IR-spectra [7]. 1- 990 cm^{-1}, 2- 1050 cm^{-1}) and ratio (Si + 2/3Al)/O (s -0.286; l -0.287 to 0.333; m- 0.333 to 0.4; + -0.4 to 0.444).

PROCESSES OF HYDRATION AND HARDENING OF SLAG CEMENTS

Hydration of Slags

The interaction of crystalline and glassy silicates and alumosilicates with water or solutions involves a set of complicated processes such as adsorption, ion exchange, hydration, hydrolysis, and others resulting in the change of the structure of initial slags and occurrence of new phases.

Among potential crystalline phases occurring in slag, the β - Ca_2SiO_4 is found to react easily with water, even at room temperature. Rankinite, $Ca_3Si_2O_7$, reacts weakly [18], whereas gehlenite, melilite, akermanite, anorthite are practically inert under these conditions [19]. Of the potential glassy phases of slags, the finely divided gehlenite glass composition hydrolyses weakly at room temperature [20] and melilite glass is less reactive than the gehlenite [21]. Akermanite-rich glass hydrolyses rather intensively [22]. The wollastonite and anorthite glasses do not react with water at room temperature [23].

It is known that, when in contact with water, slag glasses form a thin film of silicic gel and alkali earth hydrosilicate preventing further hydrolysis on their surface [24,25]. When activators like calcium sulphate, lime, caustic soda and others are introduced, no retardation of the hydrolysis process occurs.

With sulphate activation, calcium sulphate interacts directly with the alumina, calcium hydroxide and water to form hydrosulphoaluminates $3CaO\text{-}Al_2O_3\text{-}CaSO_4\text{-}12H_2O$ along with other new pH phase formations during the hardening processes [2].

With alkaline activation, the introduction of calcium hydroxide, sodium hydroxide, soda and others in an aqueous solution, leads to the formation of the corresponding silicate hydrate. The calcium silicates are known to be the structure-forming phases whereas the sodium silicates are soluble. In the work of Glukhovsky, the alkali metal compounds were incorporated in larger quantities. The study showed the alkali metal compounds did not serve as activators, but rather as the major constituents promoting the formation of alkaline and alkaline-alkali earth hydroalumosilicates along with alkali-earth silicates [26].

A mechanism of hydration was proposed for these systems by which the process of hydrolytic destruction of slags is caused by the break up of the covalent bonds (Si-O-Si, Al-O-Al, Al-O-Si) [27,28]. It is similar to the transformation of silica and alumina to hydrate forms. Alkaline cations provide the course for hydrolytic destruction of low-active, low-basic phases as the first step of structure formation by acting as a catalyst for destruction. As soon as the condensation and crystallisation processes develop, the alkali cation participates actively with Ca and Mg in the formation of new compositions initiating the formation of alkaline and alkaline-alkali earth hydroalumino-silicates.

New Hydrate Formations in Slag Cements

The qualitative and quantitative compositions of the products resulting from the hardening of concrete based on slag cements are determined by both chemical and mineralogical compositions of slag, nature of activator and curing conditions. When slag is activated under normal conditions with 5-10% by mass of lime or Portland cement, low basic laminated calcium hydrosilicates of composition (0.8-1.35)CaO-SiO_2-2.5H_2O or CSH(B) are formed (after Bogg), along with dicalcium hexagonal metastable hydroaluminate 2CaO-Al_2O_3-8H_2O or hydrogehlenite 2CaO-Al_2O_3-SiO_2-8H_2O [2]. Generally, under such conditions, the concentration of calcium oxide in an aqueous solution does not exceed 0.2-0.3 g/l.

The mixing of larger quantities of lime (up to 25-40%) or Portland cement (up to 50-65%) with the basic blast furnace slag initiates the formation of hydrosilicates or hydroaluminates of increased basicity. During thermal curing, when heated to 373K a calcium hydrosilicate with one molecule of water is formed, whereas at heating up 553K, the corresponding hydrate contains only 0.5 molecule of water.

When activating the slag by alkali-metal compounds, the composition of new phase formations is determined by the quantity of these compounds. In the $Ca(OH)_2$-SiO_2-H_2O-NaOH system, with $Ca(OH)_2$:SiO_2 equal to 1, the following sequence of phase formation occurs: at NaOH:SiO_2<1 tobermorite 1.13 nm with a degree of condensation of silicon-oxygen anions intermediate between chain and laminated: at NaOH:SiO_2=1-5 tobermorite 1.13 nm with chain silicon-oxygen anions; at NaOH:SiO_2>5 - Na-replaced tobermorite with isolated orthoions [29].

In slag alkali blends with a high content of alkali metal compounds, conditions are created first, for the formation of predominantly low-basic compounds and secondly, for the foramtion of mixed alkaline-alkali earth compounds.

The phase composition of the hardening products of slag alkaline binders are represented by the tobermorite-like low basic calcium hydrosilicates of the CSH group (B), hydrogarnets of variable composition, silicic acid, alkaline hydroalumosilicates of zeolite and mica types as well as mixed alkaline- alkali-earth compounds (Table 2). Their composition depends primarily upon the basicity of the slags and varies with the quantity of amphoteric oxides.

Table 2. New Hydrate Formations Slag Cement.

Mineral	Stoichiometric Formula	Solubility, kg/m^3
	Slag Alkaline Cement	
CSH(B)	$5CaO\text{-}6SiO_2\text{-}nH_2O$	0.050
Xonotlite	6CaO-6SiO2-H2O	0.035
Riversidite	$5CaO\text{-}6SiO_2\text{-}3H_2O$	0.050
Plombierite	$5CaO\text{-}6SiO_2\text{-}10.5H_2O$	0.050
Gyrolite	$2CaO\text{-}3SiO_2\text{-}2.5H_2O$	0.051
Calcite	$CaCO_3$	0.014
Hydrogarnet	$3CaO\text{-}Al_2O_3\text{-}1.5SiO_2\text{-}3H_2O$	0.020
Sodium-calcium hydrosilicate	$(Na,Ca)SiO_4\text{-}nH_2O$	0.050
Thomsonite	$NaO\text{-}CaO_5\text{-}Al_2O_3\text{-}10Si_2O_5\text{-}12H_2O$	0.050
Hydronepheline	$Na_2O\text{-}Al_2O_3\text{-}2SiO_2\text{-}2H_2O$	0.020
Natrolite	$Na_2O\text{-}Al_2O_3\text{-}3SiO_2\text{-}2H_2O$	0.020
Analcime	$Na_2O\text{-}Al_2O_3\text{-}4SiO_2\text{-}2H_2O$	0.020
	Clinker cement	
Calcium hydroxide	$Ca(OH)_2$	1.300
C2SH2	$2CaO\text{-}SiO_2\text{-}nH_2O$	1.400
CSH(B)	$5CaO\text{-}6SiO_2\text{-}nH_2O$	0.050
Quadri-calcium hydroaluminate	$4CaO\text{-}Al_2O_3\text{-}13H_2O$	1.080
Tricalcium hydroaluminate	$3CaO\text{-}Al_2O_3\text{-}6H_2O$	0.560
Calcium hydro sulphoaluminate	$3CaO\text{-}Al_2O_3\text{-}3CaSO_4\text{-}31H_2O$	high

The thermodynamic calculations performed for the system: "granulated blast furnace slag + sodium hydroxide" manifest the formation of the calcium zeolite named gismondite, CaO-Al_2O_3-SiO_2-$4H_2O$, along with low-basic calcium hydrosilicates during hardening of the binders for a slag with $M_b > 1$, whereas when using a slag with $M_b < 1$, the alkaline hydroalumino-silicate of analcime composition Na_2O-Al_2O_3-$4CaSO_2$-$2H_2O$ is formed [30].

CEMENTS USING BLAST FURNACE GRANULATED SLAGS

Clinker-Based Slag Cement

Generally, slag cement is manufactured through the compound grinding of the individual components in the required portions. However, as granulated blast furnace slag is more difficult to grind than clinker, the clinker component of the slag cement accumulates in the finer grain size fraction, while the granulated blast furnace slag concentrates in the coarser fraction. If the quantity of granulated blast furnace slag is larger than the clinker component, it cannot participate sufficiently in the hydration reaction during the compound grinding process. This causes a reduction in the cement strength which can only be compensated by finer grinding or separate grinding of the raw materials. The separate grinding of granulated blast furnace slag and subsequent mixing with the ground Portland cement is therefore, a viable alternative to compound grinding. It is always more beneficial to grind mono-mineral material or glass to a higher fineness than finely crystalline multi-mineral materials such as clinker. Separate grinding of slag allows the producer to optimize milling for slag grinding. Granulated slag is usually moist and has to be dried. Separate grinding allows free selection of the grinding temperature, because no concern has to be taken regarding dehydration of the sulphate carrier as in cement grinding. Modern grinding plants for separate grinding of granulated blast furnace slag employ tube mills, tube mills with high-pressure grinding rolls and roller mills.

By using separately ground slag at the concrete plant, the concrete producer gains access to a range of cements containing from 0 to 95% slag according to the new cement standards. Only two silos are needed. In many cases, it is beneficial to have one normal-hardening and one rapid-hardening cement along with the slag. It may be beneficial to have an additional silo for slag with high fineness, as well. Very finely ground slag has a very positive effect on the microstructure of the Portland slag matrix. One reason is that the finely ground slag, which is roughly about the same size as CSH-agglomerates and calcium hydroxide crystals, strongly influences the morphology of the hydrated phases. The calcium hydroxide crystals become less orientated and less flaky when introducing finely ground slag in the paste.

The European standard for cement (ENV 197-1:1992) identifies 25 types of cement. Among them are two varieties of clinker-based slag cement, as specified below:

➢ Portland slag cement , which is sub-divided into 2 types: the first consisting of 80- 94 mass% clinker and 6-20 mass% granulated blast furnace slag, and the second consisting of 65-79 mass% clinker and 21-35 mass% granulated blast furnace slag;

➢Blast furnace cement, which is sub-divided into 3 types: the first type consisting of 35-64 mass% clinker and 6-20 mass% granulated blast furnace slag, the second of 20-34 mass% clinker and 66-80 mass% granulated blast furnace slag, and third which consists of 5-19 mass% clinker and 81-95 mass% granulated blast furnace slag.

In general, clinker-based slag cements must meet the same requirements for physical and mechanical properties as Portland cements. Depending on the type and quantity of slag introduced to cement, the density is in the range of 2.8-3 g/cm^3, bulk density varies between 0.9- 1.2 g/cm^3 and in compacted state, 1.4- 1.8 g/cm^3. The normal plasticity and water demand of the clinker-based cement is equal to, or somewhat lower than that of Portland cement, whereas water bleeding at equal W/C is somewhat higher when compared to Portland cement. To reduce water bleeding and enhance the density of the cement stone, it is favorable to replace up to 10% of the slag by a pozzolanic additive in the grinding process. Slag is most suited as a ternary component in blended cement.

Slag material with the appropriate particle size can be blended with different cements, fly ash, silica fume, or very finely ground limestone to create cements with performance and function superior to normal one- or two-component cements. The need for multi-functional cements is steadily growing as structures become more complex in design and larger in size with advances in civil engineering, construction techniques and diversification of the social structure. As a result, multi-component blended cement has rapidly gained attention. The correct choice of components and their fineness makes it possible to tailor a concrete for a specific need.

During the last decade, a great deal of information and data have been published about the durability of slag when used in conjunction with Portland cement. Results vary, primarily because of a lack of understanding of the compositional effects of slag and also because correct proportions were not used in the concretes. However, there are many accepted data concerning the effects of slag, in general. For a complete evaluation, it would be necessary to sub-divide all the results according to replacement level, clinker composition, physical characteristics, curing conditions and fines included by the aggregate. With low replacement levels, the effects are not very dramatic, but at higher replacement levels, the binder system becomes more sensitive. All general characteristics can easily be found in any textbook or proceeding concerning cement replacement material.

Clinker-based slag cement is more effective if its fineness is higher when compared with that of ordinary Portland cement [32, 33]. The most effective way is to separately grind the slag to a high fineness, or even to an extremely high fineness. Fine grinding makes it possible to involve a great quantity of slag without decreasing the main physical and mechanical properties of the resulting cement. Grinding the slag finer involves higher costs, but the substantial improvements in the quality of the resulting concrete, and particularly in its pore structure and water tightness, can greatly offset the higher initial costs through longer and more durable service life and lower repair costs. Proper combination of slag fineness, water-binder ratio and the right choice of aggregate and superplasticizer results in no, or only marginal, increases in cost. One important reason is that the amount of superplastisizer can significantly be reduced without affecting workability because the slag particles absorb less. A substantial reduction in total pore volume as well as a significant improvement in impermeability to water is also possible. Finely ground slag is very competitive with silica fume for certain concretes.

Non-Clinker Slag Cement

Lime-Slag Cement. Lime-slag cement is a hydraulic binder produced by the combined or separate grinding of blast furnace granulated slags with lime (air-hardening or hydraulic, lime hydrate or quick lime), either with gypsum or without it. The lime content in the binder varies from 15% to 30% and gypsum does not exceed 5%.

The lime-slag cements are themselves slow- setting and hardening, but after long-term hardening in water the cements gain considerable strength. These cements are intended for producing low-grade concretes and mortars to be used in making underground and submarine structures under the influence of soft, or weakly mineralized water. The cements can also be used in ground structures but require special care during initial periods when the surface of the object must be covered to prevent it from drying.

Some recent developments in this area are slag and desulphurisation waste blends, slag oil shale ash blends and high-lime fly ash blends. Slag also lowers the expansion in these systems and gives a rather high final strength to these blends when properly optimized.

Sulphate Slag Cement. Sulphate-slag cement contains dihydrate, 1.5 hydrate gypsum and anhydride. Fired dolomite, lime or Portland cement may be introduced additionally to enhance its activity. For example, the ordinary sulphate- slag cement contains 80-85% of slag, 10-15% of anhydride or gypsum dihydrate, and up to 5% Portland clinker or 2% lime [34]. The binder activity after 28-days approaches 30-40 MPa. It hardens more intensively with slightly increased temperature. This cement differs from Portland cement and clinker-based slag cement due to its increased resistance to soft and sulphate waters, organic acids and oils. The highest efficiency is attained when using sulphate-slag cement for massive concrete underground and submarine structures, especially those subjected to attack of aggressive waters and leaching. One rather new and very interesting combination is created with desulphurisation products.

High Alumina Cement Slag Cement Blends. Blends of high alumina cement and ground granulated blast furnace slag can effectively be used in rapid hardening mortars and plasters as well as in high temperature applications. These blends also have a good chemicals resistance. A 50% replacement level is said to be enough to suppress the strength loss of high alumina cement. This field of application is relatively new, and will most likely be an important field for research for several years.

Slag Alkaline Cements

A slag alkaline cement (SAC) is composed of ground slag and alkaline components which may be introduced in the process of grinding the slag or preferentially as solutions in the concrete mixer [26,35]. To regulate technological and physico-mechanical properties additives are introduced, among which Portland clinker is the most commonly available [36]. Depending on the additive content, the Ukrainian standards for slag alkaline cement identify four types as shown in Table 3 [37].

Table 3. Slag Alkaline Cement Types.

Types	Portland Cement Clinker Mass%	Modulus of basicity of ggbfs
SAC 0	-	M_b - 0.6
SAC 2	2-1	M_b -1.05
SAC 4	4-1	M_b - 0.95
SAC 6	6-1	M_b -1.05

Open-hearth furnace, converter, steel-refining dumped slags, belite slimes and high-calcium ashes can be introduced as long as their content does not exceed 25% of binder mass.

Depending upon cement type and nature of alkaline component the slag alkaline cements are divided into strength classes as specified in Table 4.

A long-term and systematic research study work has demonstrated that the physico-mechanical and technological properties of the SAC are easily regulated by changing the slag and alkaline component basicity and by introducing additives of mineral and organic origin [28,36,38]. For example, the additives include: clays, rocks, glasses, glassy rocks and industrial wastes, cakes and clinker substances such as Portland cement clinker. This permits introduction into the production line a wide range of substances of both man-made and natural origin that have compositions different from that of blast furnace slags. Secondly, it allows the production of a wide range of materials with special properties: quick hardening, shrinkage-free, expanding, stressing, oilwell, heat- and corrosion resistance to minerals and organic media with pH values greater than 3 and less than 7.

Table 4. Slag Alkaline Cement Strength Classes.

Brand			
SAC 0	SAC 2	SAC 4	SAC 6
Soda			
300	300	300	300
400	400	400	-
-	500	500	500
Soda-alkali-solution			
300	300	300	300
400	400	400	-
500	500	500	500
-	-	-	600
Sodium metasilicate, $M_s = 1$			
300	-	300	300
400	400	-	-
500	500	500	500
600	-	-	-
-	700	-	700
-	-	800	-
900	-	-	-
-	1000	-	1000
-	-	-	1100
1200	-	1200	-
Sodium disilicate, $M_s = 2$			
300	300	-	300
400	-	400	400
500	500	500	500
600	-	-	-
700	700	700	700
800	-	800	800
900	900	900	-
-	1000	-	-
Sodium silicate, 2 - M_s -3			
300	-	300	300
400	400	-	400
500	-	500	-
600	600	-	600
700	700	700	700
800	-	800	-
900	900	-	-

CONCRETES CONTAINING GRANULATED BLAST FURNACE SLAGS

Clinker-Based Slag Cement Concretes

The main properties and technological use of clinker-based slag cement concrete do not differ essentially from those of Portland cement. Nevertheless, due to considerable variations in the slag content and variations in its chemical and mineralogical composition the differences and peculiarities may exist. This was demonstrated by large-scale studies held during a few years of research at the Research Institute on Reinforced Concrete (NIIZHB), Moscow [39] and by other researchers.

When designing the concrete mix the following should be taken into account: the hydraulic activity of the slags with $M_b>1$ is 90% (maximum) of that of Portland cement [40].

The water requirements of concrete mixes depends insignificantly upon the slag content in a cement. The introduction of the slag leads to an increase in normal plasticity, the standard cone slump remains constant and even increases in some cases. When using a high replacement level, a reduction in water content is highly recommended.

For compositions with increased slag content, especially 60-80%, steam curing is used. The optimal heating temperature was found to be 363-368 K. A prevailing influence on strength is attributed to the intensive hydration of the slag constituent, the strength of the clinker-based slag concrete with 80% slag may exceed a standard (28-day) strength by 20% [43]. After steam curing, 80-90% of the standard strength should be achieved.

A series of additives introduced in the grinding process has been found to activate hardening and promote a wider use of the clinker-based slag cement [44-46]. Waste from titanium- magnesium production (KEK) is recommended as one of the best [45]. The main constituents of the KEK are non-hydrated silicon and aluminium oxides (more than 30%), mixed iron and titanium oxides (Fe_2O_3 -

TiO_2, FeO-TiO_2), $CaCO_3$, and $CaCl_2$ present in insignificant quantities (0.15-0.2% of a cement mass) as shown in Table 5.

Some of the samples were steam cured. The steam curing schedule was; 2hours setting, 3 hours heating, 6 hours holding the temperature, 2 hours cooling. In the first column the compressive strength shown is just after the steam curing, in the second column 28 days after steam curing and in the third column after 28 days of normal curing. Reference material in Table 5 has no additives.

Sodium silicate decreases strength by promoting the encapsulation of the binder grains with the products of Na_2SiO_3 hydrolysis; 0.5 % NaOH enhances the strength of the concrete after steam curing correlating well with the data on acceleration of cement hydration in the presence of hydroxides and carbonates of alkaline metals [27,28,47].

The efficiency of alkaline additives increases in the compositions with a high slag content (60-80%) by accelerating the speed of dissolution of the slag glass. These additives initiate the strength gain by 10-30% after steam curing and by 5-10% after 28 days.

The activating influence on the clinker-based slag cement is rendered also by calcium nitrite, nitrite-nitrate NNC), and nitrite-nitrate-chloride [44,46]. The influence of $CaCl_2$ on clinker based slag concrete is shown in Figure 3.

The introduction of 2-3% additives initiates the strength increase by 12-25%, frost resistance by 50-100 cycles and water permeability decline over 2.1- 2.8 times. The strength of the concrete with 70% acidic slag and the NNC was found to be at the same as Portland cement concrete, and in the absence of the additives it was decreased by about 20-25% [44].

The effect from the plasticizers in the concrete mixes is maintained, and their content may be considerably decreased due to less adsorption of surfactants on the slag grains. The properties of concretes with two slag levels are illustrated by the data given in

Table 5. Strength of Clinker-Based Slag Concrete with 50% Slag (W/C=0.52)

Additive	Additive %/ wt. of cement	Compressive Strength, MPa/ % ref. past treatment		
		Steam 2+3+6+2hrs	28days steam	28 days normal
	0	18.0/100	31.0/100	25.1/100
KEK	5	19.5/109	34.2/110	27.1/108
Na_2CO_3	0.5	18.7/104	29.6/95	25.9/103
	1	16.8/93	29.2/94	19.6/78
NaOH	0.5	21.8/121	30.9/100	27.0/108
	1	19.4/108	30.8/99	24.4/97
$CaCl_2$+ $NaNO_2$	2	23.4/130	37.8/122	29.6/118
	4	24.3/135	39.7/128	32.8/131

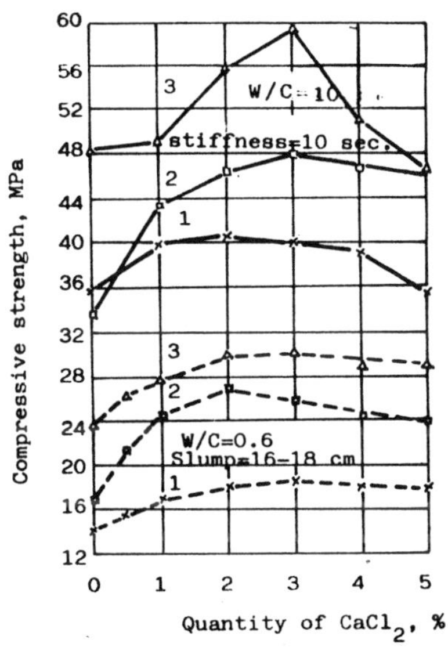

Figure 3. Influence of $CaCl_2$ on strength of clinker-based slag concrete with standard strength 35 MPa. 1) steam cured under loading 0.2 N/cm^2 at T=369K-373K; 2) 28 days of normal hardening; 3) 28 days of normal hardening after steam curing:—— low consistency mix; ------ high consistency mix.

Table 6. Compressive strength was tested on 150x150x150 mm cubes as well as on 40x40x160 mm mortar prisms. These are mentioned as R_{cub} and R_{pr}, respectively. Bending strength was tested on 100x100x500 mm prisms, referred to as R_{rf}.

The results show that the steam curing improves the strengths in concrete with a binder of high slag content, while the effect on the E-modulus is insignificant.

Table 6. Properties of Clinker-based Slag Concrete (W/C =0.5) (Basic Slag, M_b=1.13)

Curing Conditions	Slag Content in cement mass%	R_{cub} MPa	R_{rf} MPa	R_{pr} MPa	E-10^3 MPa
28 days normal hard. past steam curing	40	36.5	3.77	22.5	28.2
	80	33.5	4.32	23.6	32.2
28 days, normal hard.	40	41.0	4.24	32.7	38.1
	80	27.1	3.72	19.6	33.3

The frost resistance of Portland slag cement concrete depends upon the basicity of slag and the curing conditions as shown in Figure 4. Freezing was carried out at -15°C for 16 hours and thawing at room temperature +20°C for 8 hours.

The clinker-based slag cements are used most advantageously in concrete structures situated in aggressive media (mineralized and sulphate waters), concrete elements produced under heat treatment, and in monolithic construction in hydrotechnical structures with restricted requirements to frost resistance [54].

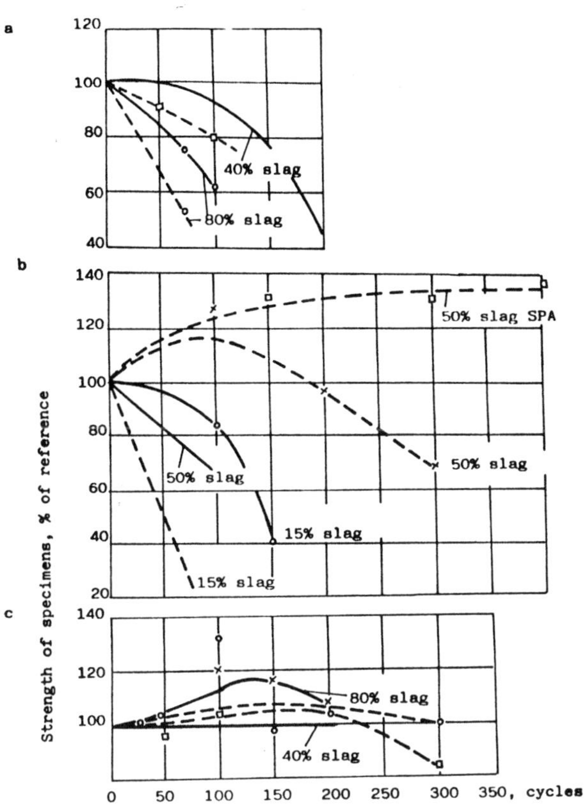

Figure 4. Frost resistance of concretes from cements with different slag content: a) weakly acidic slag (M_b=0.95; M_a=0.38); b) basic slag (M_b=1.13; M_a=0.21); c) basic slag (M_b=1.13; M_a=0.26). ________28 days normal hardening; ---- 28 days of normal hardening after steam curing.

Slag Alkaline Concrete or Alkali Activated Slag Concrete

The slag alkaline concretes are specified according to the following main features: structure, density, kind and grain size distribution of aggregates used, binder composition and curing conditions.

Slag alkaline concrete is in itself an artificial stone produced by moulding and subsequent hardening of designed and compacted mixes of coarse and fine aggregates, ground slag and alkali metal solutions and in some cases, corrective additives.

The heavy weight concretes refer to normal constructional concretes from slag alkaline concrete. The lightweight concretes are divided into constructional, constructional and heat insulating. From the guaranteed values of compressive strength the following strength classes for heavy weight concretes are specified (B10=10MPa):

Class	Grade	Class	Grade
B10	M200	B55	M800
B15	M250	B60	M900
B20	M300	B70	M1000
B25	M400	B80	M1100
B30	M500	B90	M1200
B40	M600	B100	M1300
B50	M700	B110	M1400

Depending upon the service conditions of the objects and structures the following grades for the slag alkaline concretes are specified:

- frost-resistance (cycles): F200, F300, F400, F500, F600, F700, F800, F900, F1000;
- water impermeability: W4; W6; W8; W10; W12; W14; W16; W18; W25; W30.

The constructional lightweight slag alkaline concretes of classes B15-B50 (M200-M600) with mean densities ranging from 1400 to 1800 kg/m^3 are prepared using expanded clay, blast furnace granulated slag, artificial porous aggregates, shell rocks, slag pumice, wooden wastes. The same aggregates are used in the production of constructional heat-isolating concretes with a mean densities ranging from 500 to 1400 kg/m^3 which are characteristic of compressive strengths from 3.5 to 40 MPa and heat conductivity from 0.17 to 0.4

W/m K. The heat-insulating slag alkaline concretes with mean densities from 170 to 450 kg/ m^3 and compressive strengths from 0.25 to 2 MPa and heat conductivity 0.059 to 0.13 W/m K have been designed with the use of expanded perlite.

The slag alkaline concretes are manufactured using a conventional chemical additive-containing cement technology [27, 37]. The main physico-mechanical characteristics of the slag alkaline concretes are associated with the special characteristics of slag alkaline cements.

The strength characteristics of the slag alkaline concretes and their stability in time are attributed to the composition of the slag alkaline cement and aggregates used. The highest compressive strength (116- 140 MPa) features the concretes from slags with M_s=1-1.05 and soluble glasses. When using a carbonate aggregate the strength reached is 45-80 MPa. Figure 5 shows the relation between strength gain and slag content of a concrete with a slag content from 300 to 600 kg/m^3 .

The dependence between strength of slag alkaline concrete and solution/slag ratio (S/S) as seen from Table 7 is less essential as compared with that of Portland cement concrete [56].

The strength of slag alkaline concrete is also influenced by curing regimes. Differing from a Portland cement concrete, which gains a lower final strength when accelerated curing is used, the slag alkaline cement being optimally cured approaches 130% of its standard strength.

The slag alkaline cements are characterised by long term strength development after gaining a standard strength. The rate of this increase depends upon the composition of the slag alkaline cement used (Figure 6) and curing conditions (Figure 7).

Examination of the slag alkaline concrete behaviour at a multiple dynamic loading by flexural loading [58] indicates that by an endurance criterion, the slag alkaline concretes are not inferior in durability to Portland cement concrete of the same strength. The

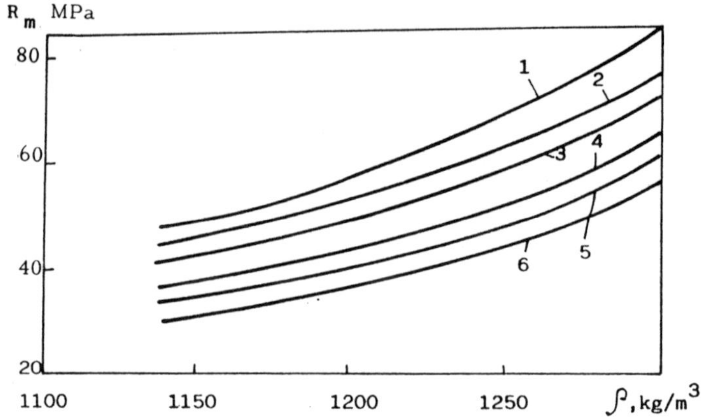

Figure 5. Strength gain of slag alkaline concrete of normal hardening from slag alkaline cement No 2(slag with M_s=1.16) depending upon density of sodium disilicate and slag content: 1 -6- 550, 450, 425, 350, 300 and 250 kg/m^3, respectively.

Table 7. Dependence between Slag Alkaline Cement Strength and S/S.

Alkaline Component	Compressive Strength, MPa at S/S						
	0.25	0.27	0.28	0.29	0.30	0.36	0.40
Sodium disilicate	112	134	126	118	117	106	102
Sodium metasilicate	128	121	118	113	103	94	90
Soda-alkali melt	-	74	82	71	66	53	45
Soda ash	-	-	52	65	57	44	40

Note: Concrete mix proportions, kg/ m^3: ground basic slag- 550, sand - 500, crushed stone- 1150. Curing regime: 3 hours heating, 6 hours holding at T=368 ± 5K and 3 hours cooling. 3+6+3 hours.

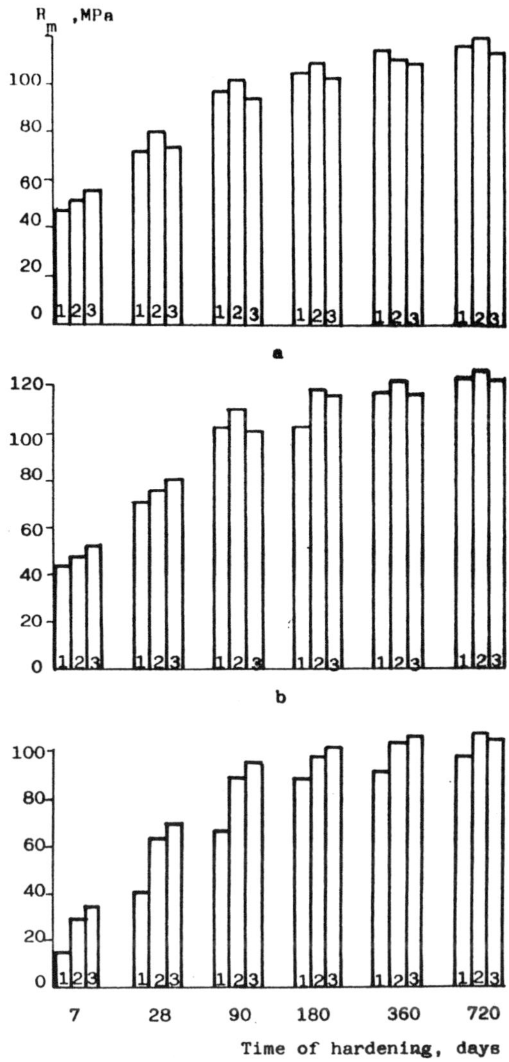

Figure 6. Histogram of changes in time of compressive strength of slag alkaline concrete with slags of $M_b>1$(a), $M_b=1$ (b), and $M_b<1$ (c) hydrated in the presence of sodium disilicate: 1) without additives; 2 and 3) with additives of Portland cement clinker 1% and 3%, respectively.

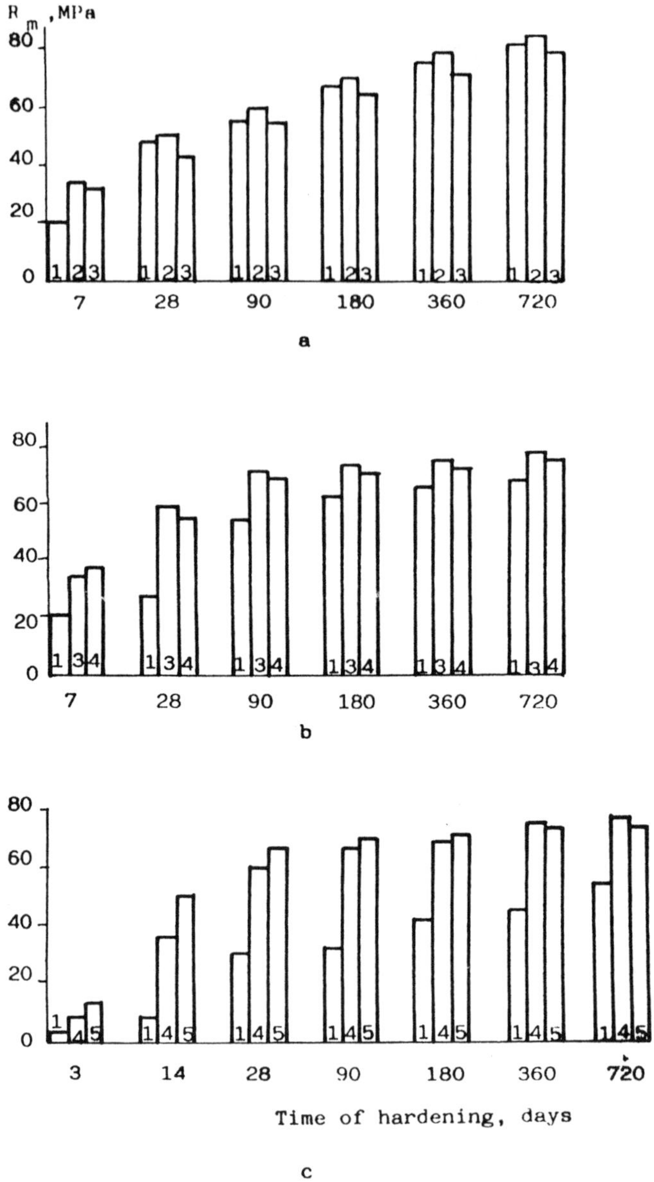

Figure 7. Character of strength change of slag alkaline concrete with slag of M_b>1 and soda ash: 1, 2- steam cured and hardened in normal and air-dry conditions; 3 - hardened in normal conditions at T=291 ± 2K.

alkaline component pore solution has an ability to repair the micro-defects and cracks in the slag alkaline concrete when the structures "rest." This gives increased service reliability to the slag alkaline concrete.

The shrinkage deformation of the slag alkaline concrete are caused by the nature of the alkaline component and type of slag, as well as curing conditions [38,55]. The deformability of the slag alkaline cement stone can be improved by accelerating the formation of crystalline hydro aluminosilicate compounds. The occurrence and crystallisation of the aluminosilicate phases take place very slowly. The introduction of highly basic and in alkaline medium quickly hydrating minerals to the binder initiates this process [28,36,59] and decreases the shrinkage deformations (Figure 8).

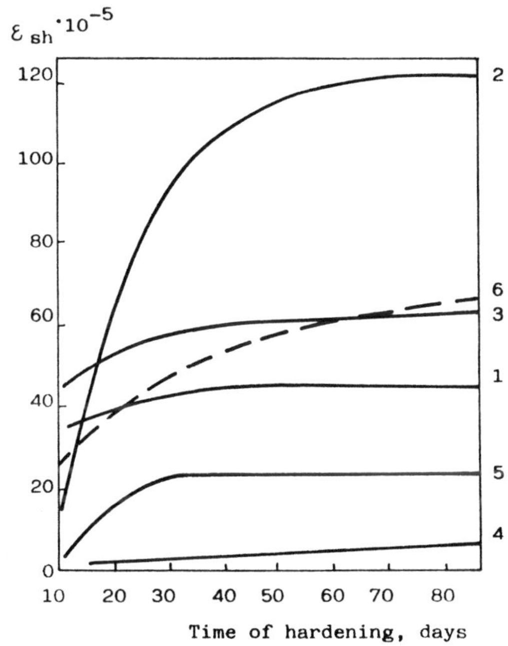

Figure 8. Shrinkage deformation of cement-sand concrete from blast furnace slag (M_b=0.99) and soda solution (ρ=1200 kg/cu m) depending upon an additive introduced: 1) 6 references for comparison: additive-free slag alkaline cement and OPC, respectively; 2- C_3S; 3- β-C_2S; 4- C_3A; 5- C_4AF.

The resistance to shrinkage crack is closely related to shrinkage and swelling deformation. It is seen from data in Table 8 that after 50 cycles of moistening and drying, the cracks were found only on the surface of the Portland cement concrete.

Table 8. Crack Resistance of Slag Alkaline Concrete at Shrinkage Swelling Deformations.

Nos	Material content, kg/ m³			Quantity of solution ***	Coefficient of crack resistance in conditions, cycles,					
	Slag	Sand	Aggregate(fraction 10 mm)*		Moistening-drying		Air-dry		Humid	
					50	100	50	100	50	100
1	505	303	1310	255	1.21	1.11	1.14	1.16	-	-
2	495	260	1250	265	1.12	0.97	1.00	0.88	2.5	2.27
3	495	360	1230	300	1.28	1.27	0.58	0.43	-	-
4	495	360	1250	265	2.70	2.15	1.05	1.05	1.1	1.15
5	495	360	1230	300	3.93	2.67	0.66	1.56	-	-
6	485	370	1190	285	1.62	0.75	1.06	1.08	-	-
7	495**	360	1250	280	0.50	0.59	0.52	0.58	-	-

Note: * - burnt rocks are used as aggregate;

** - Portland cement is used as binder;

*** - soda- alkali solution is used.

This demonstrates the higher tensile properties of slag alkaline concrete. The maximum tension allowed may also be increased by introducing plasticizing additives, decreasing aggregate size or improving adhesion resistance of the "cement stone-aggregate" contact zone [38].

The deformations at short-term loading are to a large extent, determined by the curing conditions. As is seen from Table 9, during the first days after steam curing the initial modulus of elasticity of the high-strength slag alkaline concrete, as well as of Portland cement concrete, is lower when compared to concrete with normal hardening. During subsequent hardening, under normal conditions the modulus of elasticity does not change significantly. By 28 days it reaches a value of 35×10^3 MPa and by 360 days $41 \cdot 10^3$ MPa.

Table 9. Time Dependence of Strength and Deformation Characteristics of High Strength Slag Alkaline Concretes.

Concrete Age, days	Normal Hardening			Normal Hardening (Past steam cure)		
	*Strnth cube MPa	Strnth prism MPa	MOE 10^{-3}Mpa	Strnth cube MPa	Strnth prism MPa	MOE 10^{-3}MPa
1	-	-	-	128.5	75.7	30.2
28	90.5	60	36.1	115.3	77.7	34.7
90	113.2	76.4	39.2	126.2	83.5	35.1
180	117.5	81.7	41.3	129.7	92.5	37.5
360	119.6	85.7	45.1	130.1	93.5	41.0

Note: Stiffness of concrete mix - 10sec., slump - 4 cm.
* Strnth = strength; MOE = modulus of elasticity

While hardening the specimen (after steam curing) in air-dry conditions, the initial modulus of elasticity decreases somewhat, but the cube strength increases.

It is possible to eliminate the negative influence of air-dry conditions on performance of the steam-cured concrete by introducing additives initiating the crystallization of a gel phase, and by decreasing the amount of the alkaline components. Additives which can be successfully used are: finely ground clinker, various forms of lime, cement or lime kiln dust, and slags containing free lime.

Under long-term influence of loading, creep values of the slag alkaline concrete depend primarily on the concrete mix proportions, type of a slag and alkaline component, as well as curing conditions.

It has been established that if the slag alkaline concrete is formulated using a soda prepared from the nepheline raw material, sodium hydroxide and sodium metasilicate, the above properties improve [27,37]. The relationship between the specific creep of concrete and the type of alkaline component is shown in Figure 9.

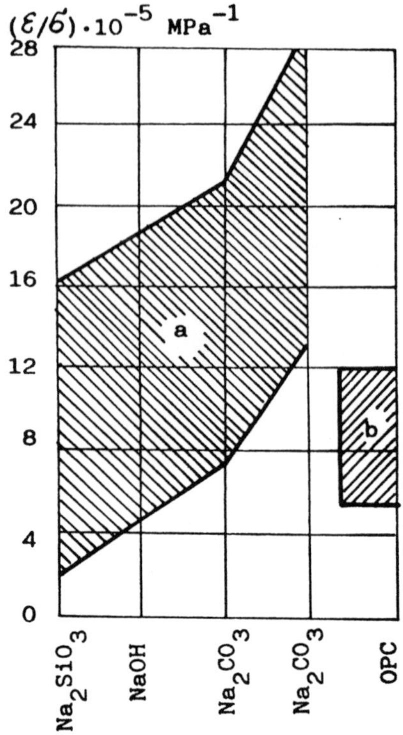

Figure 9. Dependence between specific creep of concrete and type of alkaline component: "a"- field of slag alkaline concrete; "b"- field of Portland cement concrete.

By regulating the rate of crystallization of new phase formations, it is possible to govern the creep of the slag alkaline concrete [28,36] and attain, if necessary, the values corresponding to the values of creep of a Portland cement concrete [37].

The weather resistance depends mainly upon the pore structure of concrete. Better is the pore structure, higher will be the weather resistance of concrete. Slag addition modifies the pore structure, resulting in much higher weather resistance. It is seen from Figure 10 that the alternate drying and moistening does not affect on the compressive strength of the slag alkaline concrete.

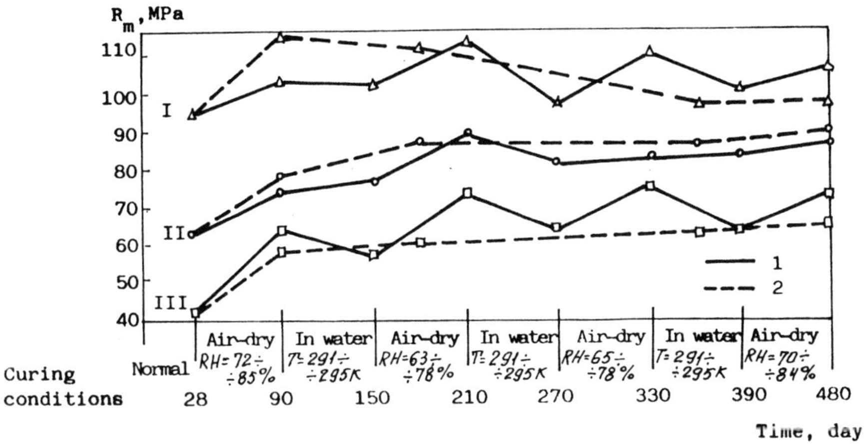

Figure 10. Compressive strength development of concrete in conditions: 1) alternate moistening-drying; 2) normal hardening; I- slag alkaline concrete (sodium soluble glass with M_s=2); II-slag alkaline concrete (alkaline carbonate as alkaline component); III-concrete from road Portland cement.

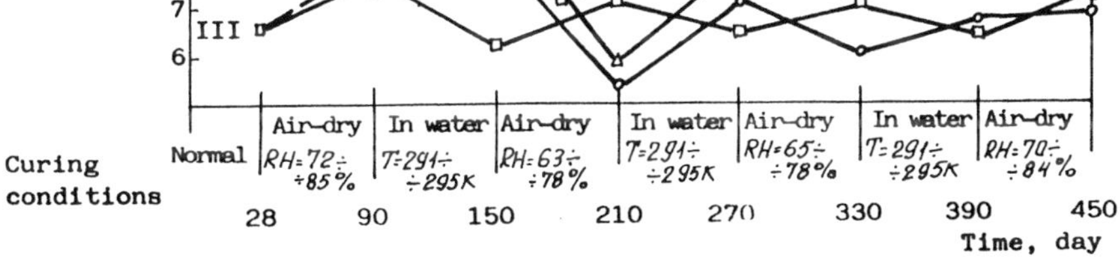

Figure 11. Flexural strength development of concrete in conditions: 1) alternate moistening-drying; 2) normal hardening; I- slag alkaline concrete (sodium soluble glass with M_s=2); II- slag alkaline concrete (alkaline carbonate as alkaline component); III-concrete from road Portland cement.

Flexural strength is more sensitive to humidity conditions (Figure 11). When cycled (14 days) with alternate moistening and drying the slag alkaline concrete shows considerable variations [60, 61]. A similar phenomena is also observed while examining Portland cement concrete [62].

The reasons for the decline in flexural strength is the low crystallinity of the hydrates of young slag alkaline concrete and corresponding high content of water saturated gel phase [64]. The loss of flexural strength in slag alkaline concrete is recovered in under water storage within 60 days after additional 360 days water storage the flexural strength has increased by 15-40%. The corresponding period for recovering of flexural strength loss for Portland cement concrete is more than 60 days [63].

It has been shown that optimization of the slag alkaline concrete structure with the aim of stabilizing its physico-mechanical and deformative characteristics can be accomplished by intensifying crystallization of the gel phase, or by modifying the pore structure (Figure 12) [38,64].

The freeze-thaw resistance of the slag alkaline concrete depends strongly on the type of alkaline component and its density. Therefore, the concrete with optimal composition from soda-alkali solution is characterized by a frost resistance of 200-700 cycles and is inferior to a Portland cement concrete of thc same class. The slag alkaline concrete from soluble silicates withstands 1000 cycles and more. The freeze-thaw resistance was tested by freezing the concrete to -15°C for 16 hours and thawing at 20°C.

The high freeze-thaw resistance of the slag alkaline concrete is attributed to a lower pore liquid freezing temperature as compared with that of Portland cement concrete. The pore liquid in the capillaries of the slag alkaline concrete freezes gradually. A slag alkaline cement concrete cooled to T=243°K has no residual deformations; when cooled down to 7T=233°K the residual deformation is 0.01 mm/m and at further cooling down to T=213°K 0.15 mm/m (Figure 13).

The corrosion resistance of the slag alkaline concrete also exceeds that of Portland cement concretes while freezing and thawing in salt solutions . Slag alkaline concrete with sodium metasilicate tested in 5%-solutions of the following salts: NaCl, $MgSO_4$, and $CaCl_2$ withstood 45, 60, and 100 cycles without visual deterioration of the structure [27,37,38].

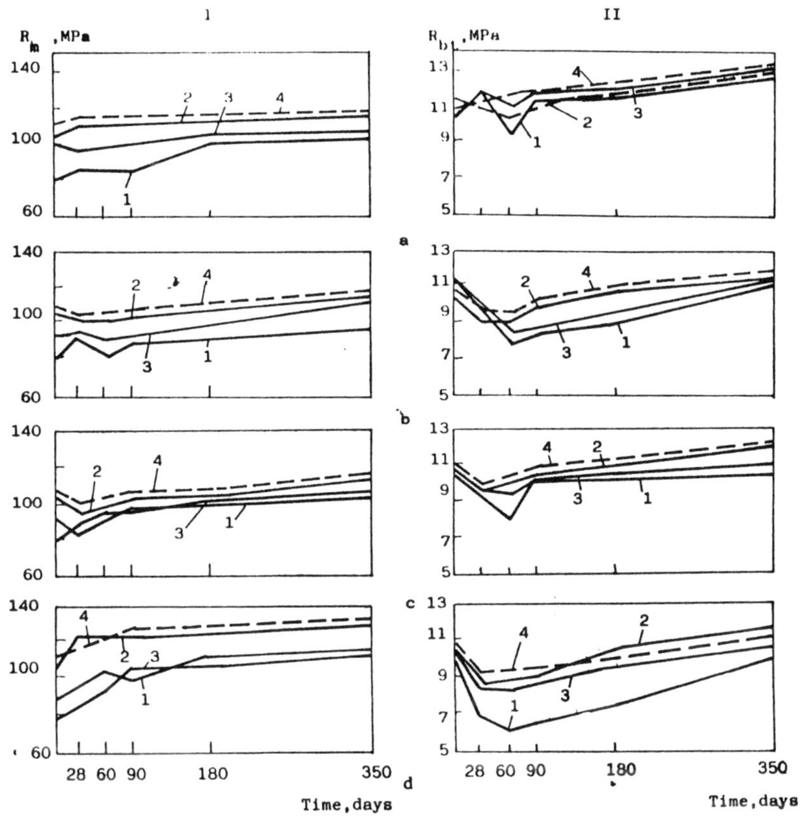

Figure 12. Compressive (I) and flexural (II) strength development of fine aggregate slag alkaline concrete when specimens were cured under: a) normal conditions; b) water; c) air-dry; d) alternate moistening-drying; 1) slag alkaline cement; 2) slag alkaline cement + 10% dehydrated zeolitic rock; 3) slag alkaline cement+2% petrolatum emulsion; 4) slag alkaline cement+10% dehydrated zeolitic rock +2% petrolatum emulsion.

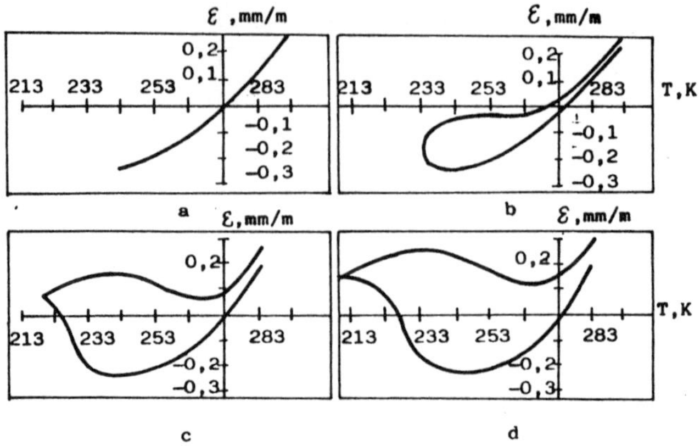

Figure 13. Deformations of slag alkaline stone (moisture content in specimens- 19.3 %) during thawing and freezing: a) 243°K; b) 234°K; c) 219°K; d) 212°K.

The corrosion resistance of the slag alkaline concrete as well as the Portland cement concrete is caused, first of all, by the density and impermeability of cement stone. However, at the same densities, slag alkaline concrete is characterized by higher resistance attributed to the lack of hardening products of calcium hydroxide, calcium high-basic hydrosilicates and hydroalumino-silicates [38,65].

A number of studies have demonstrated that the corrosion resistance of the slag alkaline concrete depends upon such variables as chemical and mineralogical composition of the slag used, nature and alkaline component content and curing conditions [27,36,37,66].

It has been established that the macro- and microstructures of cement stone may be regulated by optimizing the type and quantity of alkaline components, initiating synthesis of stable new formations or modifying their composition and structure, preventing development of exchange reactions between the ions of the corrosive medium and cement stone phases [28,38].

By using finely ground modifying additives (like calcined clays, certain zeolites, calcined magnesium silicate hydrates, certain metal oxides etc.) or cetrtain metal salts or solutions, the hydration products of slag alkaline cements concrete can be tailored for chemical-resistant applications [28,38,67].

The heat resistance of the slag alkaline concrete depends upon various factors [28]. While designing and formulating the compositions of heat-resistant slag alkaline cements, it should be taken into account that their thermo-mechanical characteristics are determined by the following: *(i)* the ability of the hydrated phases to topotaxial recrystallization into anhydrous new formations; *(ii)* the formation in the fired stone structure of epitaxial intergrowths as a result of the crystallo-chemical similarity of the occurring phases; *(iii)* the presence of 5-10% eutectic melts (of cement mass) providing a flow of the process of structural self-reinforcing; *(iv)* highly fragmentary structure determined by the optimal ratio of glassy to crystalline phases in the fired stone.

The given models of synthesis of thermo-mechanical properties have been used while producing a range of heat-resistant materials [38,68], which by their efficiency, exceed the traditional heat-resistant concretes, including those from high-alumina cement.

FIELD EXPERIENCE AND SPECIAL USE OF SLAG ALKALI CEMENTS

The results of 35-years of use of the slag alkaline cements in construction for different fields of application such as: hydrotechnical, mining, agricultural, industrial and residential construction may be considered as the proof of their effectiveness and high service performance. Pilot testing followed by commercial scale production of the slag alkaline concretes was initiated in 1958-1964.

The improved performance characteristics of slag alkaline concrete in comparison with Portland cement concrete made it possible to use them under severe environmental conditions and to replace them with structures produced from different varieties of Portland cement concretes. The results of long-term observations [69] show that strength continues to increase by 1.5- 2 times and can exceed designed strengths. The microstructure of the concrete provides a high frost resistance for the structures. Hence, the frost resistance, which is one of the major factors determining the durability of a concrete, exceeds the standard requirements of Portland cement concrete under the same conditions.

The structural densification of the concrete decreases capillary porosity and improves water impermeability with time. For example, at early ages a cast-in-situ concrete with manufactured channel facings, had a water impermeability of 0.6 MPa, and after 12 years of service, 1.8 MPa. In another case, the water impermeability of concrete designed for an irrigation system had increased by 2MPa after 9 years service.

The dense structure of the concrete and high-alkalinity medium in the hardened cement stone prevents corrosion of reinforcements. After 6-12 years in service, steel reinforcements in irrigation systems, road pavement, channel facings, tubes and soil anti-slipping systems had no visual signs of corrosion, even when the protective layer was only 3 mm thick [70].

Observations of several structures in 1978-1979, 1985 and again in 1987 showed that after 15-25 years of service, the strength of alkali-activated slag concrete remained stable. The constructions were characterized by good appearance and no failures had occurred.

The high durability of alkali-activated slag concretes and articles is verified by experience of their service under severe conditions. After 8- 12 years of service under the complex effects of mechanical loading from track and wheeled vehicles, the aggressive influence of organic media, as well as exposure to heat and cold, road pavements, agricultural yards, warehouses for fertilizers (mineral and organic), aerodromes for agricultural purposes and decontamination areas prepared from the cast-in-situ concrete were found to be in satisfactory condition. Deterioration and scaling of the surfaces was not present. Structures produced from a Portland cement concrete placed side-by-side for comparison, had remarkable deterioration of the surface with exposed deep layers of aggregates and reinforcement.

The alkali-activated slag composite materials turned out to be very effective for use in special construction and non-construction engineering. Concretes of increased acid resistance based on a alkali-activated slag with additives of dehydrated magnesium silicate or zeolite-containing rocks have been used as an underlayer for the floors in a sulphuric acid pumping station warehouse and slabs used in the facing of drainage channels in the areas where chemical water is purified.

An experience gained from the use of heat-resistant slag alkaline compositions [72] in the linings of magnetodynamics weighers and metal feeders proposed for pumping of non-ferrous molten metal testifies that the residual strength after interaction with aluminium melt (T=1073°K) was 93- 95 %, and the metal resistance determined in accordance with a standard procedure met the requirements imposed on them. The linings of the magnetodynamic pump after 6 months in service were found to be undamaged while those from heat-resistant high-alumina cement concrete mix had been damaged after 1-2 months of service.

A highly interesting application is in making oilwell cements by proper selection of slag and utilizing its thermal resistant properties. A nickel slag was used as a basic component for the cements with increased adhesion and thermal resistance. A soluble glass with density 1100-1300 kg/m^3 was used to produce an impermeable oilwell cement concrete . The hardened cement stone possesses high adhesion to metal and natural rocks. This property may be characterized indirectly by a limit of shearing stress when a cylindrical cement specimen is pressed out through a metallic ring. The limit of shearing stresses increases with an increase in a density of the alkaline solution and hardening temperature and may exceed 9 MPa. The gas permeability of the cement stone depends upon mix proportions, density of soluble glass and temperature and duration of hardening. Fully impermeable stone may be produced under optimum conditions. The optimized cements allow one to synthesize a cement stone characterized by high strength, low gas permeability, increased adhesion to casing pipes and rocks providing reliable fixing of oilwell pipes at a working temperature of 358-473°K.

High strength and thermal resistant compositions for the filling of a reverse side of galvanoplastics copies of the dies of machine-tool attachments, and electroconducting concretes have been formulated [73]. It was established that in steel yokes dies of a slag alkaline concrete were undamaged even at a pressure of 250 MPa. The concrete is resistant to temperatures exceeding 573°K. An important advantage of the alkali-activated slag concrete is its high rigidity promoting a good support of the galvanoplastics copy that do not depend on service temperature. Experimental studies using 35 mm thick specimens showed that the coefficient of rigidity, determined as the ratio of the specific pressure to compressive deformation attained by a concrete, is 0.8-1.2 x10^4 MPa/m for alkali-activated slag concrete, whereas for the state-of-the-art metal-polymer composition, even at normal temperatures, has a much lower value, 1.7-1.9 x 10^3 MPa/m. At a temperature rise up to 373-393 K the rigidity of the concrete decreases. It is worth mentioning that for steel, the coefficient of rigidity is 1.8-2 x 10^4 MPa, these values are close to those for this concrete. A properly tailored slag alkaline concrete is almost shrinkage-free. This property results in high volumetric stability and accuracy of the dies, but its heat conductivity is low. It is possible to eliminate this disadvantage by using metallic fillers and inserts contacting with a copy. A very effective method is to use the alkali-activated slag cements in the production of the dies for cold plastic metal working [74] utilizing such special properties as high impact elasticity and hardness. Research and development studies on slag alkaline concretes continue in the direction of how to create the materials that meet the requirements for impact elasticity, hardness and wear resistance. This work is intended in full substitution of two-layer construction of a shape-forming die by one-layer construction from monolithic concrete.

The possibility of manufacturing slag alkaline concretes with regulated special properties like logarithmic decrements, shrinkage deformation and others extends the perspectives of their use in the machine-tool industry for base members of machine tools [75].

A fine aggregate slag alkaline concrete reinforced with dispersed basalt fibers characterized by enhanced vibrodamping over a range of medium and high frequencies from 500 to 1000 Hz was used as base members for precision machine tools. The slag alkaline concretes with required deformation characteristics have been used as base members of machine tools for flexible industrial modules [76, 77]. The reference specimens of the parts of the base members are characterized by the compressive and flexural strengths R_m = 85 MPa and $R_{b,t}$= 7.5 MPa, respectively, and the logarithmic decrease of 0.024. The enhancement in vibrodamping of the test specimens over a range of medium and high frequencies (500-10000 Hz) as compared to that of base members from vibration resistant iron, allows one to increase the accuracy of the treated surfaces of the objects. The use of a slag alkaline concrete with regulated damping characteristics provides an increase in service period of cutting tools used in flexible industrial modules.

Slag alkaline cements has been successfully used for localization and long-term disposal of radioactive wastes. An effective way to solve the problem of radioactive waste decontamination is based on modelling the natural processes of mineral formation occurring at temperatures and regimes conventional for building materials production. By proper selection of slag alkaline cements, it is possible to produce an artificial stone of high chemical resistance, low rate of leaching (in water), thermal and radiation resistance [78-81].

The development of properly tailored compositions for low- and intermediate-level radioactive waste processing, in particular, by solidifying the waste into artificial mineral-like compounds for long-term storage appears to be of interest. The mixes consist of a blend of calcium low-temperature zeolite of $CaO\text{-}Al_2O_3\text{-}(2\text{-}4)SiO\text{-}2nH_2O$, kaolin and calcium hydroxide that can encapsulate 20-25% of the radioactive alkali metal carbonates of the total mass of the material to be processed. The alkali metal ion leach rate does not exceed 0.71 exp^{-6} g/cm^2 per day. Experience gained in the processing of the radioactive wastes from alkali metal liquid heat bearers of nuclear power plants with the goal of long-term and environmentally safe storage is of great importance. An artificial stone produced in accordance with the technology developed contains alkaline radio-nuclides, including Cs-137, which are chemically bound into the compounds similar to natural minerals of zeolite type. The composite materials proposed for decontamination of liquid radioactive wastes from nuclear power plants are characteristic of high content of alkali metals and high water stability. The rate of Cs-137 leaching is 3×10^{-6} g/cm^2 per day.

High performance of the slag alkaline cements, which in many aspects surpasses the known state-of-the-art systems and vast experience gained in terms of composite materials and concretes in constructional engineering and in special fields of applications testifies to their high technical potential. Some recent publications provide a more detailed information about alkali-activated (slag) binders and their potential fields of applications [86,87]. The experience gained in the Ukraine and former Soviet Union is now available and can be used as a base for future development. By introducing a part of the binder system in liquid form, it is possible to control and tailor the properties of the alkali activated systems. This short review is only meant to put forward some thoughts about the potentials for inorganic binders in general. While a concrete engineer is fighting with the durability of a 25MPa concrete, it is possible to use the same binders in production of high-tech products.

CONCLUSIONS AND RECOMMENDATIONS

Future binders will be based on local sources of natural, secondary and waste materials. Cement and concrete blends are tailored to fulfill local needs. A better understanding of material properties and processes involved in hardening reactions leads to successful combinations of materials of different origin.

Slag-based cements will find a very large application area in environmental applications like solidification of waste materials and contaminated soils. The use of slags for civil engineering and geotechnical applications is widely recognized. They can also be successfully used as ion-exchange substrates in fresh and waste water treatment. The principles of alkali activation will be more widely accepted as an alternative to Portland cement activation.

New drying, grinding and blending techniques have made it possible to increasingly use secondary materials more effectively as binders in concrete. One continues to learn how to benefit from all material-specific properties. The synergistic effect in blends will become a new major branch in concrete science. By selecting the right proportion and particle size distribution for all components it is possible to attain synergistic effects which give improved durability properties. The interaction with fine particles present in the aggregate and filler is of greatest importance when designing the optimum blend. It is suggested that a non-profitable association for the benefaction of alternative binders is formed.

REFERENCES

1. Botvinkin, O.K., About Multiplicity in Glasses Structure, Gosizdat Stroit.Materialov, Moscow (1955).
2. Volzhenskiy, A.V., Burov, Yu.S. and Kolokol'nikov, V.S., Mineral Cementitious Substances, Stroiizdat Publish., Moscow (1973).
3. Satarin, V.I., Slag Portland cement, in: *Proceed. VI Int. Congress on the Chemistry of Cement*, 1974, vols 1, 2(1), 2(2) and 2(3), Russian with English preprints, Stroiizdat, Moscow (1976).
4. Parker, T.W. and Ryder, .F., *J. Iron Steel Inst.*, 156, 21 (1942).
5. Coale, R.D., Grinding technologyWolhuter, C.W., Jochenes, P.R. and Howat, D.D., Cementitious Properties of Metallurgical Slags, *Cement and Concrete Res*. 3, 81-82 (1973).
6. Royak, S.M., P'yachev, V.A. and Shkolnik, Ya.Sh., A Structure of Blast Furnace Slags and Activity, *J.Tsement* 8 (1978).
7. Govorov, A.A., The Processes of Hydrothermal Hardening of Slag Dispersions, Naukova Dumka Publish., Kiev (1976).
8. Cavalier, G., Sandrea-Deudon M., *Rev.Metall.*85 (1960).
9. Vinogradov, B.N., Raw Materials for Manufacturing Autoclave Slag Concretes,Stroiizdat Publish., Moscow (1966).
10. Kramer, V., Blastfurnace Slags and Slag Cements, in: *Proceed.of IV Int. Congress on the Chemistry of Cement*, vol. II, 957-73 (1964).
11. Baidov, V.V., Ultra-acoustic Studies and Microstructure of Silicate Melts, in: *Properties and Structure of Slag Melts*, Nauka Publish., Moscow (1970).
12. Bockris, J.O.M., Mackenzie, J.D., Kitchener, J.A., Trans. Farad. Soc. 51 12: 1734 (1955).

13. Bockris, J.O.M., White, J.L., Tomlinson, J.M., Trans. Farad. Soc. 52: 299-310 (1956).
14. Midgley, C.M., Acta crystall 5 (1952).
15. Bregg, U., Klaringboul G., Crystalline Structure of Minerals, Mir Publish., Moscow (1967).
16. Cherkinsky, Yu.S., Polymineral High-temperature Transitions during Synthesis of Cements, in: *Proceed. Meeting on the Chemistry and Technology of Cement*, Stroiizdat Publish., Moscow (1967).
17. Taylor, H.F.W., The Chemistry of Cements, London (1964).
18. Flint, E.P., Wells L.S., *J.Res.Nat Bur.Stand.*6 (1934).
19. Butt Yu.N., Varshak, B.G., Mayer, A.A., *J. Stroitelnye Materialy* 2 (1961).
20. Cleret de Langavant, *J., Rev. Mater.Constr.*:401-402 (1949).
21. Lapin, V.V., in: *Blastfurnace Slags in Construction*, ,p.p.60-77, Gosstroiizdat Publish., Kiev (1956).
22. Keil, F. and Gille, F., Zement-Kalk-Gips 2:229-232 (1949).
23. Budnikov, P.P., Pankratov, V.L., in: *Scientific Reports of NIITsement* 11(42), 28-32 (1961).
24. Mylius, F., Groschuff, E., Zeitschr.Anorg.Chemie 55, 101-108 (1907).
25. Cleret de Langavant, *J., Rev. Mater.Constr.* 330, 49 (1937).
26. Glukhovsky, V.D., A New Building Material, in: *Bulleten Gosstroya*, Gosstroiizdat Publish.,15 (1957).
27. Glukhovsky, V.D., Krivenko, P.V., Pashkov, I.A., et al, Slag Alkaline Fine Aggregate Concretes (V.D.Glukhovsky ed.), Vischa Shkola Publish., Kiev (1981).
28. Krivenko, P.V., Syntesis of Binders with Required Properties in a System "Me20-MeO-Me_2O_3-SiO_2-H20", Dissertation for Dr.Sc. Degree, Kiev Polytechnic Institute (1986).
29. Timashev, V.V., Pekunova, E.V., Nikonova, N.S., On Rate of Condensation of Silicon-oxygen Anions in Calcium Hydrosilicates and Lime- siliceous Compositions, in: *Proceed.Seminar on Calcium Hydrosilicates and Their Use*, pp.128-132, Kaunas, USSR (1980).

30. Raksha, V.A., Investigations of Influence of Chemical Composition of Slag on Properties of Slag Alakline Binders and Concretes, Dissertation for PhD, Kiev Civil Engineering Institute (1975).
31. Japanese Standard JISR 5211-1979 "Blastfurnace Slag Cement".
32. Akimova, T.N., Shestoperov, S.V., About Properties of Slag Portland Hardened in Steam Curing Conditions, *J.Beton i Zhelezobeton* 10:13-19 (1969).
33. Schroder, F., Slag and Slag Cement,in: *Proceed.of V Int. Symp. on the Chemistry of Cement IV*, p.171, Tokyo (1968).
34. Budnikov, P.P., The Chemistry and Technology of Silicates, Naukova Dumka Publish., Kiev (1964).
35. Glukhovsky, V.D., A Binder, The Author's Certificate of USSR No 449894, priority of 19.07.1958.
36. Krivenko, P.V., Regularities of Formation of Structure and Properties of a Cement Stone and Slag Binders, in:*Proceed.II National Conference on Slag Alkaline Cements, Concretes and Structures*, pp. 10-16, Kiev (1984).
37. Glukhovsky, V.D., Krivenko, P.V. et al, Manufacture of Concretes and Structures from Slag Alkaline Binders, Budivelnik Publish., Kiev (1988).
38. Krivenko, P.V., Pushkaryeva, E.K., Durability of Slag Alkaline Concrete, Budivelnik Publish., Kiev (1993).
39. Vysotskiy, S.A., Malinina, L.G. Concretes from New Kinds of Low-energy Cements, in:*A Review, TsNIIS*, Moscow (1987).
40. Wesche, K.H., Schubert, P., Zumahlstoffzement and Betonzusatzstoffe, *J. Betontechnik* 3: 80-84 (1982).
41. Vysotskiy, S.A., The Efficiency of Steam Curing of Concretes from Slag Portlands of Various Compositions, PhD Dissertation, Moscow (1977).
42. Mironov, S.A., Kurbatova, I.I. et al, The Extraction Method for Defining Content of Blastfurnace Slag in Cements, *J.Tsement* 12: 11-12 (1976).
43. Mironov, S.A., Kurbatova, I.I. et al, The Influence of Hydration Activity of Slag Portland Cement on Physico-mechanical Properties of Concretes, *J.Tsement* 6: 8-9 (1978).

44. Ratinov, V.B., Shestoperov, S.V. et al, The Protective Properties of Concretes from Slag Portland Cements, *J.Beton i zhelezobeton* 7 (1974).
45. Shokotova, B.G., Pinchukova, V.F. et al, An Additive Accelerating Hardening of Slag Portland Cement, *J.Tsement* 11:11 (1983).
46. Maiko, V.P. The Perfection of Manufacturing the Reinforced Concrete Articles from Slag Portland Cements, Dissertation for PhD, NIIZHB, Moscow (1982).
47. Jawed, I., Skalny, I., Alkalies in Cement, *in:A Review, Cem. and Concrete Res.*, 8 1, 37-52 (1978).
48. Batrakov, B.G., Silina, E.S, High -strength Concretes from Slag Portland Cements, *J.Stroit.Mater. i Konst.* 3:15-16 (1985).
49. Schorr, K., Frost-und Tansalzbestandigkeit von Zementein aus unterchiedlicher Zementarten mit verschiedenen Fluga-schebeimengungen, *J. Betonwerk + Fertigteil-Technik* 1, 26-22 (1983).
50. Vysotskiy, S.A., Bykova, I.V., The Influence of Kind and Content of Slag in a Cement on Frost-resistance of Concrete, in: *Technology, Calculations and Design of Concrete and Reinforced Concrete Structures*, pp. 50-54, NIIZB, Moscow (1979).
51. Shestoperov, S.V., Fedner, L.A., Shestoperov, V.S, The Plasticized Concretes during Steam Curing, in: *Works of MADI*, Issue 38 (1972).
52. DIN 1164 Portlandzement, Eisenportlandzement, Hochofenzement Blatt I, (1970).
53. Mitrovic, N., Miletic S., Moskovbjevic M. Blended Hydraulic Cement Based on the Clinker with High Alkali Content and Addition of Blastfurnace Slag, *J. Silikates industries*, 49 2 (1984).
54. Stol'nikov, V.V., Fominyh, V.A., Slag Portland Cements for Hydrotechnical Applications, Energiia Publish., Leningrad (1972).
55. Firsov, N.N., A Directed Structure Formation of Slag Alkaline Concretes with the Aim to Regulate Their Properties, Dissertation for PhD, Kiev Civil Engineering Institute,Kiev (1985).
56. Butt, Yu.M., Kolbasov, V.M., Berlin, L.E., The Influence of W/C ratio on Structure, Strength and Frost Resistance of a Cement Stone, *J.Beton i Zhelezobeton* 11:9-10 (1974).

57. Gots, V.I., The Influence of Temperature on Structure Formation Processes and Properties of Slag Alkaline Concretes Dissertation for PhD, Kiev Civil Engineering Institute,Kiev (1983).
58. Krivenko, P.V., Rumyna, G., V., Nagaichuk, V.M. et al,The Processes of Structure Formation in a Contact Zone "Slag Alkaline Cement- Aggregate," *J.Tsement* 11-12:64-70 (1991).
59. Glukhovsky, V.D., Krivenko, P.V., Gelevera, A.G, The Influence of the Processes of Structure Formation on Deformations of Slag Alkaline Cement Stone, in:*Proceed. II Regional Conference on Physico- chemical Mechanics of Dispersive Systems and Materials*,p.p.92-94, Naukova Dumka Publish.,Kiev (1983).
60. Kutti, T., Malinowski, R., Influence of the curing conditions on the flexural strength of alkali activated blast-furnace slag mortars, in: *Works of Chalmers University* 10, pp. 9-18, Chalmers University Publication , Goteborg, Sweden (1982).
61. Malolepzy, J., Deia, J. The Influence of Curing Conditions on the Mechanical Properties of Alkali Activated Slag Binder*, J. Swedish Cement and Concrete Research Institute*: 18-24 (1988).
62. Craf, O., Die Eigenschaften des Betons, Lweite neubearbeitete Auflag.-Berlin: Springer-Verlag, Gottingen, 10-16 (1960).
63. Milenkovskaya, K. M., The Influence of Water Saturation of Concrete on Its Strength and Deformability, *J.Hydrotehnicheskoe stroitel'stvo* 4:13-16 (1960).
64. Krivenko, P.V., Pushkaryeva, E.K., Scherbina, L.V., Enhancement of Stability of Physico-mechanical Characteristics of Slag Alkaline Binders in Conditions of Alternate Moistening-drying, *J.Tsement* 11-12:9-15 (1991).
65. Krivenko, P.V., Physico-chemical Bases of Durability of a Slag Alkaline Cement Stone, *J.Tsement* 11:2-5 (1990).
66. Kurdowski, W., Duszak, S., Trybalska, B., Corrosion of Slag Cement Paste in Strong_Chlorides Solution, in: *Proceed.I Int. Conference on Slag Alkaline Cements, Concretes and Structures*, pp. 961- 970, Kiev (1994).
67. Krivenko, P.V., Pushkaryeva, E.K., Brodko, O.A., Acid-resistant Slag Alkaline Binders of Hydraulic Hardening, *J.Tsement* 11-12:16-23 (1991).

68. Pushkaryeva, E.K., The Heat Resistant Materials from Alkaline Cementitious Systems, Dissertation for Dr.Sc. Degree, Kharkov Civil Engineering Institute (1995).
69. Krivenko, P.V., Physico- chemical Bases of Durability of a Slag Alkaline Cement Stone,*J.Tsement* 11:2-5 (1990).
70. Petropavlovsky, O.N., Chirkova, V.V. and Demianova, L.E., The Experience of Manufacturing, Exploitation and Perspectives of Development of Raw Materials Base for Slag Alkaline Binders, Concretes and Constructions, *J.Tsement* 6:20-22 (1990).
71. Pushkaryova, E.K. and Brodko, O.A., Physico-chemical Bases of Synthesis of Fire-Corrosion Resistant Slag Alkaline Materials, *J.Tsement* 11:16-18 (1990).
72. Krivenko, E.K, Pushkaryova, E.K. and Brodko, O.A., An Experience of Exploitation of Slag Alkaline Concretes in Acidic Aggressive Media, in: *Proceed. Scientific Research Conference "New Materials and Technologies in Construction,"* AGMI Publish., Alchevsk,Ukraine, pp. 180-183 (1992).
73. Pushkaryova, E.K., Synthesis of an Artificial Stone with Required Properties Based on Alkaline-alkali-earth Cementitious Systems, Dissertation for PhD, Kiev Civil Engineering Institute (1985).
74. Brainman, B.M., Use of Slag Alkaline Concretes in Galvanoplastic Dies, in: *Proceed.III National Conference on Slag Alkaline Cements, Concretes and Structures*, pp. 261-263, Kiev (1984).
75. Belitsky, I.V.,Tkalenko, S.A. and Tkalenko, A.A., Perspectives of Use of Slag Alkaline Concretes in Press Tools, in: *Proceed. III National Conference "Slag Alkaline Cements, Concretes and Structures*, Kiev (1984).
76. Krivenko, P.V., The Slag Alkaline Polymer- impregnated Concretes in Machine-Manufacturing, in: *Proc. 6th Int.Congress on Polymers in Concrete* p.p.24- 27, Tongji University, Shanghai, China (1990).
77. Krivenko, E.K, Pushkaryova, E.K., Brodko, O.A. et al, Binders for Base Members of Machine-tools, USSR Author's Certificate N 1721034 (1992).

78. Krivenko, E.K, Pushkaryova, E.K., Brodko, O.A. et al, Concrete Mix for Base Members of Machine-tools, USSR Author's Certificate N 1708794 (1992).
79. Glukhovsky, V.D., Krivenko, P.V., Skurchinskaya, J.V. et al, Composites for Solidification of Alkali Metal Carbonate-Containing Radioactive Waste, USSR Author's Certificate N 1471571 (1987) (unpubl.).
80. Glukhovsky, V.D., Krivenko, P.V., Skurchinskaya, J.V. et al, Processing Technique for Alkali Metal- Containing Radioactive Wastes Solidification, USSR Author's Certificate N1625252 (1989) (unpubl.).
81. Krivenko, P.V., Skurchinskaya, J.V., Lavrinenko, L.V., Environmentally Safe Immobilization of Alkali Metal Radioactive Waste, in: *Proc. Int. Confer. "Concrete 2000,"* pp. 1579-1587, Dundee, Scotland, UK(1993).
82. Krivenko, P.V., Skurchinskaya, J.V. and Lavrinenko, L.V., Environmentally Safety Localization of Alkali Metal Radioactive Waste within Alkaline Binding Systems, *J.Tsement* 3:31-33 (1993).
83. Krivenko, P.V., Skurchinskaya, J.V., Pushkaryova, E.K. and Chislitskaya E.V.: Synthesis of an Artificial Stone Based on Alkali-Containing Wastes, in: *Proc. Scien. Pract. Conference "Use of Industrial Wastes in Building Materials Production,"* pp. 123-126, Volgograd (1989).
84. Krivenko, P.V., Skurchinskaya, J.V., Lavrinenko, L.V. et al, Utilisation of Electroplating Sludges in Slag Alkaline Binders Manufacturing, *J.Tsement*, 3:37-39 (1993).
85. Krivenko, P.V., Skurchinskaya, J.V. and Samoilenko, J.I., Slag Alkaline Compounds for Electroplating Sludges Utilisation, *J. Stroit. mat.i konstr.* 4:29 (1991).
86. *Proceedings of the First International Conference on Alkaline Cements and Concrete held in Kiev; Ukraine 11-14 October 1994".*
87. Talling, B., J. Brandstetr.: Clinker-free Concrete Based on Alkali Activated Slag. *Progress in Cement and Concrte; Mineral Admixtures in Cement and Concrete, Vol 4 (Ed. S.L.Sarkar).* pp. 296-341, New Delhi (1993).

6

RED MUD AND PHOSPHOGYPSUM AND THEIR FIELDS OF APPLICATION

Wieslaw Kurdowski and Francois Sorrentino

INTRODUCTION

Red mud and phosphogypsum belong to the family of wastes produced in large quantities like blast furnace slag (BF slag), fly ash or municipal waste incineration ash (MWA). Red mud is a residue obtained by the Bayer process during the manufacture of alumina. Phosphogypsum is a by-product from the production of phosphoric acid by the wet process.

Many technologies have been designed to utilize red mud or phosphogypsum and have been found to be successful. In spite of this, their utilization has been very limited. Whereas, both BF slag and fly ashes have found worldwide development. To determine the cause for this lack of acceptance, many parameters have to be considered from purely technical factors to economics. The economics of the system must be balanced by environmental factors.

The stabilization of MWA is an example of a non-economical process required by governmental regulations under the pressure of environmental lobbies. The toxicity of the waste was the primary motivation for development and utilization.

Waste by-products are found worldwide with several concentrated areas for red mud and phosphogypsum. Figures 1 and 2 show the distribution of bauxite and phosphates ores.

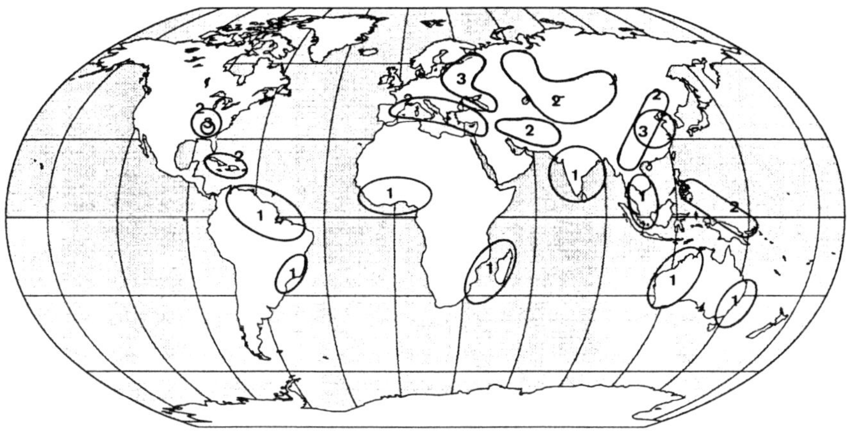

Figure 1. Geographical distribution of bauxite deposits: 1) laterite bauxite, and 2) Karst bauxite [4].

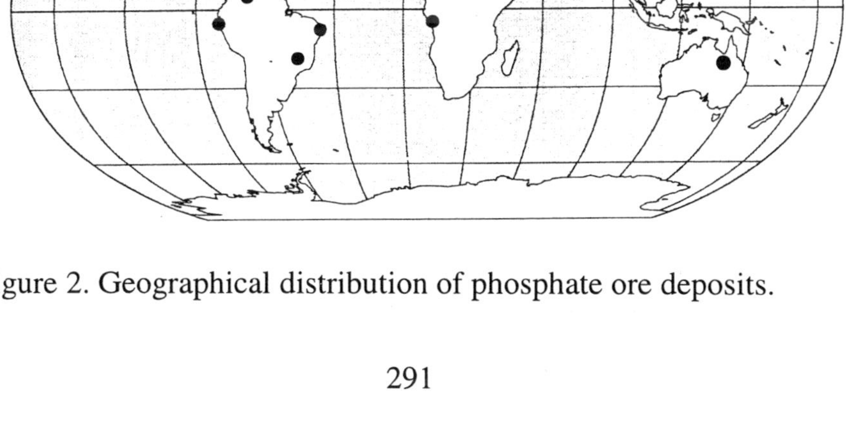

Figure 2. Geographical distribution of phosphate ore deposits.

In spite of many possible venues for utilization of these wastes, they are usually land filled or disposed of at sea. The first reason for this is that red mud and phosphogypsum are not considered to be toxic materials and the producers are under less pressure by environmental or governmental regulations. The second reason is that the characteristics of the red mud and phosphogypsum (chemical, mineralogical composition, physical state) are not readily amenable for economical use in large-scale processes.

Nevertheless, because of the large production of these by-products, research must be done to find an economical use for these wastes. In this chapter, the authors describe the most recent findings for the utilization of red mud and phosphogypsum, especially for the construction industry.

RED MUD UTILIZATION

Introduction

Red mud is the by-product of the manufacture of alumina from bauxite by the Bayer process. World production of aluminum is 15 millions tons per year (mt/y) provided from about 30 mt/y of alumina produced from 50 mt/y of bauxite which give approximately 30 mt/y red mud [1-3]. The production rates of red mud depend on the origin of the bauxite. From Surinam bauxite, 0.33 tons of red mud are produced per ton of alumina (Al_2O_3), while the production of red mud from Jamaican bauxite is one ton per ton of Al_2O_3 and from Arkansas bauxite it is two tons per ton of Al_2O_3 [4].

Bauxite is an heterogeneous material composed mainly of aluminum hydroxide minerals (gibbsite which is the trihydrate, diaspore and boehmite which are the monohydrates). The principal impurities common to nearly all known deposits of bauxite are aluminum silicates (clays), iron and titanium oxides. The quantity of impurities varies from one deposit to another as does the proportion of trihydrate to monohydrates. Bauxite deposits are essentially old soils and occur in a variety of geological and geomorphic situations. They are classified as: 1) lateritic bauxite deposits underlain by alumino silicate rocks, they are products of "in-situ" tropical weathering 2) Karst bauxite deposits cover the uneven, karstified surface of limestones and dolomites 3) Tikhvin-type deposits. They overlay the eroded surface of aluminosilicate rocks, but have no relationship to them. These deposits (close to Tikhvin in Russia) are products of the erosion of lateritics-bauxite deposits, are transported to the area, and are then overlain on the eroded surface of alumino-silicate rocks. The most important reserves of bauxite are Guinea, Australia, Brazil , Vietnam, India, Jamaica [5].

Although a large number of methods to produce alumina have been developed on a laboratory scale, most of the commercial production of alumina is obtained by the Bayer process or a modified version. Patented by Karl Bayer in 1888, this process consists of leaching bauxite with caustic soda at elevated temperatures and pressure. The products of this reaction are a solution of sodium aluminate and an insoluble "red mud residue" which is removed by decantation and filtration. The pure sodium aluminate solution is then cooled and seeded with crystals of alumina hydrate. Although retains some characteristics of the mother bauxite, red mud is somewhat different because of the chemical and physical treatments.

The alumina industry is facing an important problem regarding the settling/separation of red mud during the production process. The settling/separation of red mud not only affects the productivity of alumina but also its purity. Quick-settling red mud permits an effortless filtration process for producing high purity alumina [4].

The red mud leaves the main circuit as a slurry which contains typically 20/40% solids, which means that every 0.2/0.4 tons of red mud solids produced and separated contains between 0.6 -0.87 tons of solution which at this stage, typically contains 180 to 220 g/liter of caustic soda. Obviously, if this separated red mud sludge is simply discarded, there would be a considerable loss of caustic soda. Given the fact that the process usually consumes up to 3-4 tons of caustic soda per ton of alumina produced and that one ton of caustic soda is about the same price as one ton of alumina, the entire process is quite not economically feasible unless the bulk of the entrained soluble caustic soda in the red mud can be recovered. For many potential applications, the presence of alkalies is undesirable and processes have been studied to remove alkalies from red mud by treatment with lime and humic acid [6].

The directions of the researches undertaken to utilize red mud are based on their chemical, mineralogical and physical properties. Red mud contains alumina, iron and titanium oxide and it is thermodynamically possible to recover the metals. Red mud can be added to blast furnace as poor iron ore or reduced to titanium recovered by chlorination. Titanium can also be obtained by precipitation of hydroxide and aluminum as sulphate [7,8]. Other processes allow minor elements such as V, Zr and Mo to be extracted. The reduction process to recover iron produces a slag which can be used in cement directly [9,10] or, after recovery of soda and alumina by a leaching process [11-15]. Until recently, none of these processes were economically feasible.

The production of absorbents to remove sulphur dioxide from industrial effluents has been proposed as an application for red mud [16].

Various methods of using red mud have been suggested and reviewed in the literature [17- 29]. Ceramic materials can be made from the mud, but the quantity of this utilization is limited [30-32]. Red mud is used as a filler in mortars based on bitumen and other resins. A filler was developed for use in bituminous road construction and in plastics. A road surface of 600,000 m^2 with 2500 tons of red mud filler have been constructed [33,34].

Red mud is also used in rubber and plastics products, as an artificial aggregate in concrete and bricks [35-37] or, by addition to the raw mix of cement to modify its properties [38] particularly, its color [39]. The production of cements by addition of lime, fly ash, gypsum has also been proposed [40,41]. Red mud has been recommended as an additive to cement to improve its strength and modify its setting behavior [42,43].

In the following schema (Figure 3), the authors have attempted to summarize the potential applications classified by the initial properties and treatment (wet or dry, with or without additives, firing, etc.). In this chapter, attention is focused on building applications for red mud rather than potential metallurgical or ceramic uses. The chapter was not written to provide an exhaustive review of literature for all applications, just the uses that have the greatest likelihood of success in an industrial setting. The following items have been considered: characteristics of red mud, production of bricks and building elements, addition to cement mix and manufacture of special cements [44], red mud as a pozzolanic material, pigment for concrete, addition to concretes, aggregates and lightweight elements, special treatment to manufacture alumina and Portland cement raw material [45] and the environmental impact of red mud disposal.

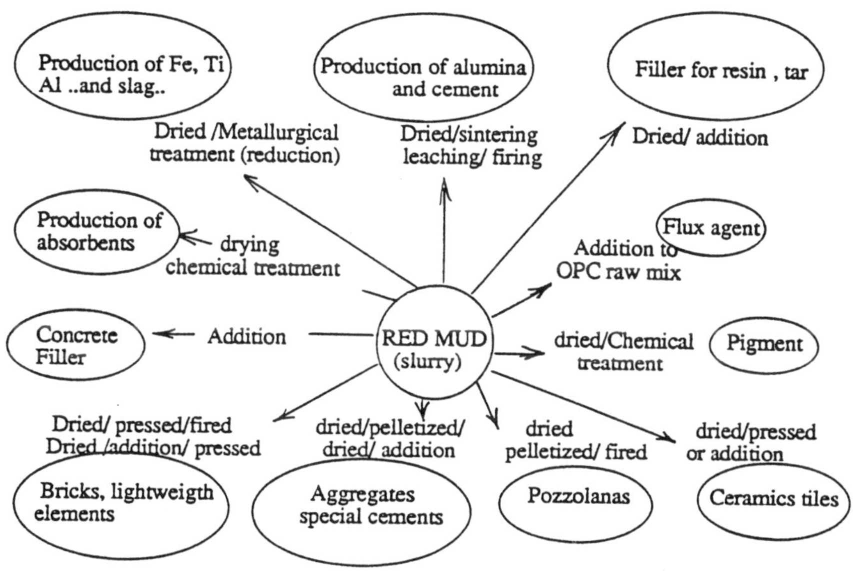

Figure 3. Schematic showing possible uses for red mud.

Characteristics of Red Mud

The chemical composition of red mud varies with the quality of the bauxite and processing procedures [4]. Examples of chemical compositions from different countries are given in Table 1. The references mention the author and the country of treatment or the origin of the bauxite.

Table 1. Chemical Compositions of Several Red Muds.

Refs.	Fe_2O_3	Al_2O_3	TiO_2	Na_2O	SiO_2	CaO	CO_2	LOI
7	33 -56	11-24	3.6-7	3 -8	4-14	3.50	n.g	n.g
52 Egypt	31	20	21	6	8	4	n.g	n.g.
77 France	19.40	27.3	1.90	5.2	10.2.	8.6	11.4	16.5
40 France	40-43	18	5	8	12-14	6 -11	0	0
77 Guinea	56	21.1	2.3	0.40	3.7	0.2	0.8	15.7
77 Soviet	41.6	14.2	3.8	3.3	9.2	12	6.6	11.88
74 Japan	49	19	9	3	9.14	2.5	0	8.5
37 India	26.4	26.24	22.1	4.20	6.55	3.5	0	10.88
99 US	41.6	16.40	5.5	6.80	1.30	6.7	n.g	n.g
70 US	46.25	14.80	5.86	1.85	4.17	8.8	n.g	n.g
34 Germany	28	25.8	n.g	9.70	n.g	1.1	n.g	n.g
55	30- 35	20- 35	20-35	4-5	6-8	n.g	n.g	n.g
46 Jamaica	53.8	10.33	8.58	8.25	8.25	n.g	n.g	n.g
46 Brazil	43.7	16.76	4	9.80	9.80	n.g	n.g	n.g
46 Greece	41.9	14.80	5.80	10.02	10.02	n.g	n.g	n.g
69 Australia	32.82	25.29	8.68	16.77	16.77	2.11	n.g	7.66

The specific surface (BET) and the density are in the range of 10/21 m^2/g and 2.5/ 2.7 g/cm^3, respectively. When mixed with water, red mud releases OH- and the pH is increased to a value of 11. Red muds have 12.3 % of matter that is soluble in hydrochloric acid and 1.4% of matter that is soluble in water.

The mineralogical composition, as well as the chemical composition, depends on the mother bauxite and on the production process. The mud is composed of minerals that come directly from the bauxite (anatase, boehmite, calcite, gibbsite, goethite, hematite, calcium ferrite, kaolinite) and of compounds produced by the process. Typical red mud phases which are not contained in bauxite can be divided into three groups:

1). "NAS" phases: $3(Na_2O\ Al_2O_3\ 2SiO_2)Na_2X$
($X = CO_2^{2-}\ 2OH^-$, SO_4^{2-}, $2\ Cl^-$)

2). "CAS-CFS" phases: $3CaO(Fe_2O_3)_x\ (Al_2O_3)_{1-x}\ kSiO_2$
$(6\text{-}2k)H_2O$

3). "NT-CT" phases: $Na_2Ti_3O_7 \bullet 3H_2O$, kassite
perovskite, portlandite.

Three techniques are commonly used to characterize red mud: differential thermal analysis, (DTA), infrared spectroscopy (IR) and x-ray diffraction. A specific x-ray method was developed to determine the qualitative and quantitative analysis of red mud [47].

Figure 4 shows typical traces of red mud, analyzed by these techniques. The x-axes represent: temperature for the DTA curves, cm^{-1} for the IR trace and 2-theta for the x-ray diffraction diagram. The y-axes represent: the intensity of the transformation, the absorption or, reflection (arbitrary units).

From infrared techniques, the hydroxyls bound in gibbsite are detected by the presence of 3617/3524/3446/3390/3092 cm^{-1} and hydroxyl bound with kaolinite by 3690/3638/3612 cm^{-1}. Free water corresponds to 1641/1631cm^{-1} and 1425 cm^{-1} corresponds to the vibration CaO in calcite. The wavelengths 1035/1110/ 1000 /978 are attributed to Si-O in kaolinite and 1021/921/914 /875/801/744/662 to the bound Al-O. The presence of MgO in the silicate structure is detected by the wavelength 659/623/560 cm^{-1} and 680/ 556 /538/471 represents Fe-O.

From differential thermal analysis, 3 endothermic peaks are detected. They correspond respectively, to gibbsite at 314°C, kaolinite at 538°C and carbonates at 891°C.

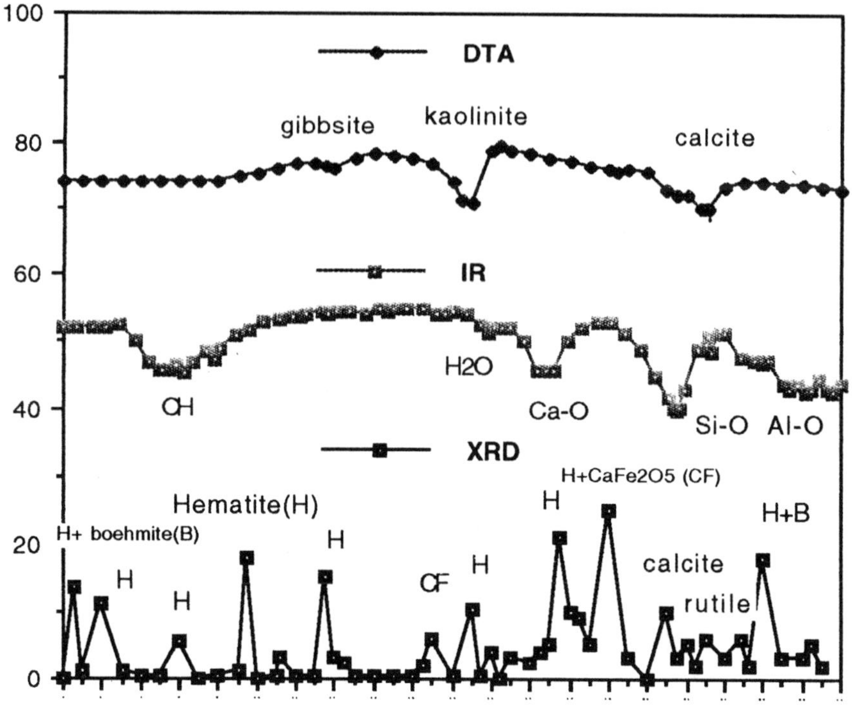

Figure 4. Red mud characteristics.

The preceding description corresponds to an average of different red muds. As previously stated, the exact composition depends on the origin of the bauxite. For example, some red mud contains large quantities of kaolinite while others are rich in TiO_2 [48].

One problem which can occur in certain bauxite and thus, red mud products, is the level of radioactivity. For example, Jamaican bauxite and the corresponding red mud contains levels of U^{238} and Th^{232} that are significantly greater than those found in most other local soils. When used as building materials, concerns regarding radiation hazards to building occupants have lead to studies on the radiation hazards to building occupants have lead to studies on the radiation characteristics of the red mud. Using Jamaican red mud for constructing walls in homes is likely to result in an effective gamma dose; equivalent increments above background levels on the order of ≤0.72 mSv/y to individual occupants. The actual dose would depend on the amount of time the residents remain within the house on a daily basis [49]. The mean specific activity of Ra^{226}, Th^{232} and K^{40} in red mud is 477/705 and 153 Bq/Kg and 202/211/140 Bq/Kg for red mud cement. The equivalent absorbed dose for residents of houses built with red mud cement is 2.65/3.49 mSv/yr [50].

For a room, constructed entirely of conventional materials, calculations provide an effective equivalent background dose to the body of 0.24 mSv/y. The uncertainty in these calculated values is on the order of ±10%. Such a level of radioactivity is not common. This effect is not the same for all red mud or bauxite and specific studies have to be carried out for each case.

Utilization in the Building Industry

Production of Bricks and Building Elements. Using red mud for the production of bricks is a simple and economical method to utilize this waste material [51]. There are two basic processes. The first requires that the red mud be pressed into blocks and fired at high temperatures. The second process mixes the red mud with binders and makes use of either hydraulic hardening, or firing at low temperatures.

The blocks are fired in electrical muffle furnace at 900°C after initially drying at room temperature for two days. The brick mix that consists of pyrophyllite, fly ash and clay shows an improvement in final strength. Bricks made with red mud have a compressive strength of 1.9MPa and 32% water absorption, while bricks of pyrophyllite have a compressive strength of 3.8 MPa and water absorption value of 27.97 %. Bricks consisting of a 50/50 mix of red mud and pyrophyllite have even higher compressive strengths: 16.5 MPa and lower 19.8% water absorption [52].

In a similar process, bricks of a suitable mix of clay, red mud and fly ash have been molded or extruded and fired at 1000-1100°C in commercial kiln. These bricks have very high compressive strengths on the order of 40-70 MPa. Because of their characteristic color, texture and density, the bricks are particularly suitable for facing purposes in the construction of multi-story buildings [36,37].

In another process, bricks are made by the addition of lime to red mud and are pressed into the desired shape using 4MPa of pressure. The strengths developed after 28 days are 4.75 MPa with 5% lime and 4.22 MPa with 8% lime. Bricks specimens after one year storage in natural weathering showed up no apparent damage. The bricks can be used to construct walls for low-cost shelters.

A mixture of equal quantities of red mud and lime is fired at about 1220°C produce 50% of C_4AF and 20% of C_2S. The addition of 2% gypsum gives a mixture that can be briquetted [53].

Sintering of alumina doped with phosphoric acid occurs at low temperatures (<250°C) [54]. The method consists of mixing orthophosphoric acid and red mud, compacting the mix in a steel die at a pressure of 0.5-1.0 ton/cm^2 followed by firing at low temperatures (180-300°C) in air for 1-2 hours. A phosphoric acid content of 0.5-2% was required to achieve compressive strengths in the range of 4-7MPa [55].

Other possible uses include: molding building matcrials with red mud and residues from aluminum plants [56], plastic waste, Ca or Zn stearate as heat stabilizers and antioxidants have also been described, but their applications are very limited [57].

Building elements can also be produced by mixing red mud with cement and different types of calcium sulphate [58,59] or with cement, sand and polymer dispersions [60].

Cement Raw Mix Additives. Red mud is used in two ways, either in the production of ordinary Portland cement as a source of iron oxide or alumina, or in the raw mix of special cements. Red mud is a good source of Fe_2O_3 for cement raw mix [61], but modifies somewhat the microstructure of the clinker [62].

The use of red mud must be limited to keep the clinker liquid and the moduli under control, a maximum of 8-10 % red mud can be incorporated in the raw mix when no clay is used. Its use lowers the temperature of liquid formation by approximately 100-150°C. The alkali content of red mud also limits the amount that can be added [63,64].

The chemical composition of red mud allows its use in special-cement manufacture. Cement clinker is produced by mixing calcareous materials with red mud, press molded and fired at 1300-1450°C in a tunnel kiln [65].

Red mud has been proposed as a raw material for producing Ferrari cements [66-69]. An equal weight of red mud and silica is mixed with lime and fired three times at 1450°C in order to obtain a typical Ferrari cement composition (Al_2O_3 = 4.41%, SiO_2=21.40% Fe_2O_3=5.59% and CaO=66.12%). The 3, 7, 14, 28 days compressive strengths are respectively 7.4, 13,15 and 29 MPa.

In an other case, a mixture containing 20-50% cement kiln dust (CKD), red mud (20-50%), calcium oxide (5-25%), calcium sulfate (10-25%) and alumina (0-25%) is mixed and fired between 1200-1450°C to give a cement having the following chemical composition: 20-70%CaO, 5-35% Al_2O_3 and 5-25% Fe_2O_3, 0-20%SiO_2 and 4-25%SO_3. The compressive strengths after 1,7 and 28 days are 23, 30 and 36 MPa, respectively [70].

Red mud (5-30%) is melted in liquid steel slag. The molten material is rapidly cooled forming a glass and C_2S, C_2MS. The ground powder is mixed with about 3-5% plaster, cement or lime to produce a hardening agent for weak soil or sludge [71,72].

A mix of lime and red mud (85/15), fired at 1400°C, ground to 88 microns is added to concrete mix and autoclaved at 180°C. The bending strength is 16.3 MPa (7.5 MPa for the standard) [73].

Red Mud as a Pozzolanic Material. Red mud contains components that, in combination with lime, produces calcium aluminate hydrates. These reaction products are able to develop strength. The reaction of alumina from red mud to calcium sulfate leads to ettringite formation which also develops strengths under certain conditions. Several processing scenarios are possible such as 1) red mud used without any special treatments except drying, 2) red mud is filtered at a defined treatment prior to being mixed with lime or sulfates.

A simple mixing of 70% red mud and 30% CaO gives a product with compressive strength of 7MPa. The hydrates formed after 4 days are $Ca(OH)_2$, C_4AH_{13} and C_4AH_{11}. By addition of 10% gypsum a compressive strength of 15 MPa is obtained after 7 days. At later ages, ettringite expansion will occur [40].

These results have been confirmed by more recent experiments. Cements prepared by mixing red mud with and or Portlandite shows a decrease of the flexural strength as function of time [74].

The addition of sulfate prevents crack formation in mixtures of lime/red mud and slow setting time. Ettringite is formed and C_4AH_{13} begins to form gradually when the content of Portlandite increases. The instability in strength is improved by adding fly ash and an additional 7% of portlandite (20MPa after 90 days). The instability in strength is caused by the decomposition of ettringite in presence of CO_2 atmosphere. Ettringite decomposes into gypsum, gibbsite and calcite. The solution is found by a mixture of fly ash, Portland cement and red mud (25, 25 and 50%, respectively) which develop strengths of 27MPa after 7 days. Red mud acts as an accelerator of Portland cements and fly ash contributes to long-term strength by a pozzolanic effect. Portland cement is used because it contains the silica necessary to rapidly obtain high strengths. The addition of aluminous fly ash allows the diminition of portlandite [40]. The addition of blast furnace slag and plaster is an alternate solution [75].

After removing the sodium hydroxide by burning at 950°C with lime, the red mud provides a good mortar binder when mixed with gypsum and Portland cement [76].

By comparing the thermal activation of kaolinite and clays, Pera [77,78] has developed a promising solution. Red mud was calcined for 5 hours at different temperatures between 600°-800°C.

Calcination modifies the specific surface and the mineralogical composition of red mud. Blended Portland cements containing 30% or 50% of the calcined red mud were studied for the hydration products, strength and durability. Red mud attains maximum reactivity when calcined at 600°C because the portlandite produced during the hydration of the cement is consumed through the pozzolanic effect of the red mud. The calcined red mud, when used in combination with Portland cement contribute to the formation of hydrated alumino silicates and hydrogarnets. Good compressive strengths were obtained with the blended cement containing 30% calcined red mud. Mortars cast with these blended cements were placed in solutions of sea water and acetic acid. The results indicated good stability of mortars to these environments.

Pigments for Concrete. The red color of the mud, essentially due to the high iron and titanium oxide content is used to stain concrete to imitate the color of bricks, or to produce special architectural effects. Several processes exist to enhance or modify the basic color of the red mud. The processes are based on a chemical treatment by controlling the pH, or a mixing with other oxides, or with moderate firing. A few examples are mentioned here.

➢Red mud, mixed with a weak acid is stirred and de-watered to 40%. The pH is adjusted with $Ca(OH)_2$. The product is filtered and dried to obtain a reddish brown powder, which is used as a pigment for manufacture of building materials such as, colored stone plates and concrete products [79,80].

➢Red mud with 30% water is mixed with 20% dry, ground, yellowish iron oxide. This mixture(3-10%) is added to cement and produces concrete with clay brick color [81].

➢After drying, calcining and grinding, red mud is used to stain aggregates [82].

Addition to Concretes. The addition of red mud to concrete leads to modification of the properties which are described below.

The addition of red mud, replacing 30% of sand exhibits higher strengths than concrete with quartz aggregates for in built-in place piles [83]. Red mud concretes have a better durability than standard concrete in defined conditions. They exhibit less penetration by sodium aluminate solution after 3-4 years immersion. Cubic C_3AH_6, and $CaCO_3$ were found in red mud concrete instead of thermonatrite [84]. The stability of concrete used in fluor media is increased. This increase is attributed to the formation of HF resistant C_4AF during normal aging and C_3AH_6 during steam curing [85,86]. The strength of Portland cement began to decrease when the red mud additions exceeded 15-20% [87].

Red mud can be associated with a water-reducing agent [88], with SiO_2 fly ash which results in a compressive strength increase and can be used to fill trenches or repair roads [89]. Autoclave curing (180°C at 10 atm. for 5 hrs) of the specimen was effective in further increasing the compressive strength [90].

Red mud (1-15%), cement (15 -25%) and filler (65-84%) are used for mortar and colored concrete blocks, filling depleted mine shafts and dam structures. The blocks exhibit good weather resistance [91].

The use of red mud slurry is an economic solution because it saves the cost of drying. A composition containing Portland cement (20-25%), calcium sulfate semihydrate (3-5 %), red bauxite sludge (20-40%) and blast furnace slag has been proposed and studied [92,93].

Organic ammonium salt solidifies red mud slurries and the result can be used as fillers for construction materials, cement plastics, or for land fill materials [94].

Sludge from municipal sewage treatment plant is treated by a mixture of lime, BOF dust, lime kiln dust and red mud. The final composition hardens under atmospheric conditions [95].

The addition of 10% red mud to calcium aluminate cement allows one to control the setting time and the rate of heat evolution to any desired degree without diminishing the long-term strength of the cement. The one day strengths are, however, reduced by the addition of red mud which acts as a retarder [96].

Aggregates and Lightweight Elements

Aggregates. Aggregates for concrete, bitumen or resin are obtained by drying, pelletizing and firing. With the addition of red mud, there are several additional steps, as outlined below.

Red mud is dried, granulated and calcined at 1000-1400°C to obtain a chemically inert, moderately dense, hard and abrasion-resistant material which is compatible for use with cement, tar, asphalt and other building materials [97].

Because of its high iron content it will be heavier than the average natural rock. The color changes during firing from red to black or grey black as a result of the crystallization of the iron oxide (hematite, magnetite or a mixture of the two) [99]. The aggregates produced with red mud are pelletized and fired at 1100-1300°C and have a density, crushing strength and water absorption of 2.76 g/cm^3, 26 MPa and 1.44%, respectively. In comparison, the values of the same properties for gravel are 2.65 g/cm^3, 25 MPa and 1.94%, respectively. The concrete made with red mud aggregates has a 28 day compressive strength of 32 MPa versus that of 27 MPa for the standard [100].

Steam-cured mixtures of red mud, sand and white cement produce aggregates [98]. Steam-cured concrete prepared with an aggregate containing 20% sand and 80% red mud with water /cement ratio of 1.27 exhibits a 28 day compressive strength of 13.3 MPa or, 10 MPa with 100% red mud aggregate W/C=0.5 [101].

The crushing strength depends on the cooling rate 27/25/27 MPa at 50°/min 100°/min 260°C/min to 500°C. The sample must be cooled slowly [102,103]. The red mud is pelletized, dried and fired at 1200°C for 2.5 hours.

Pellet formation is improved by addition of water soluble sodium alginate which is dissolved in red mud slurry to form spherical pellets by addition of 5% $CaCl_2$ and fired at 1250°C. The pellets have a crushing strength of 7.5 MPa [104].

Addition of slags to red mud and firing improves the crushing strengths of the aggregates and decreases the sintering temperature. These slags can be fluor containing wastes from aluminum industry [105] converter slag [106,107], or spinel from chromite extraction residue [108].

Lightweight Elements. A great demand exists in the construction industry for lightweight aggregate (density less than 0.88 g/cm^3).

Red mud containing 50% iron oxide with a density of 5.25 g/cm^3, the remaining constituents have a density of 2.7 g/cm^3. To attain a density of 1g/cm^3 , the aggregate would have to contain 72% voids. This would be difficult to achieve without reducing the strength. On the other hand, most of the lightweight aggregates are formed by expansion of gases generated internally while the product is in a plastic state. The raw material must not melt sharply, which is the case for red mud. To obtain lightweight aggregates from red mud, it is necessary to add a material which decreases the melting point, broaden its softening range and generate the necessary gases within the softening range. Many materials have been tested and include: carbonaceous materials (oil, bituminous coal, graphite phyllite), chemical products such as sodium carbonate, sodium silicates, sodium phosphates, calcium carbonates or minerals such as perlite, clay schiste alunite, talc, epidote or mixture of them [99]. To obtain lightweight aggregates, the quantity of additives must be higher than 10%. Many solutions have been proposed [109,110].

Slurries of red mud are mixed with clay and perlite containing foaming agent and stabilizers , the resulting foam is dried and fired at 1190°C to produce building blocks and slabs with a density of 0.48-1.12 g/cm^3 [111].

Lightweight aggregates with a density of 1.25 g/cm^3 and 0.71% water absorption are obtained by mixing 80% red mud and 20% micaceous liparite followed by pelletizing and firing at 1200°C [112].

Lightweight aggregates have been produced by firing iron rich red mud and fly ash in a 50/50 proportion [113,34] and by mixing red mud with tailings of coal dressing [34], or by mixing with recovered oil and expendable clay that are the granulated, uniformly dried and fired [114].

Red mud burned at 900-1000°C with lime gives a product which is mixed with four times the quantity of Portland cement, silica, lime and a gas forming additive can be steam cured to produce lightweight elements [115].

Lightweight building elements are also produced by mixing red mud, an amphoteric metal (Al), sodium silicate and red mud. The foam is produced by the attack of aluminum by sodium silicate. Hardening occurs at low temperature but can be accelerated by heating at 100°C [116].

During manufacturing of lightweight concrete, addition of red mud decreased the consumption of cement from 230-270 to 170-195 kg/m3 concrete. An increase in the elastic modulus of the concrete, as well as an improvement in its deformation characteristics resulted [87].

Simultaneous Production of Alumina and Portland Cement. In this technology patented by Grzymek [117], red mud is mixed with limestone and sintered in a rotary kiln. The resulting product which contains calcium aluminate and belite disintegrates because of the transformation of beta to gamma belite into dust with a 20 micron particle size. The powder is then treated by a soda solution which decomposes calcium aluminate into calcium carbonate and sodium aluminate which remains in solution. The extracted powder (based on gamma belite) is mixed with the Portland cement raw meal to accelerate clinkering reactions. An increase of productivity of 20-40% has been estimated. The equipment required for the production of alumina and cement is the same as that normally used in the cement factory. This technology has been industrially tested by the Groszowice cement factory which produced 50,000 tons of alumina and a 500,000 tons of cement.

Red Mud Disposal. In spite of the effort described previously to utilize red mud, most of this residue remains unused and the disposal of the discarded red mud is a great environmental problem for two main reasons. The first is the toxicity due the high pH (caustic soda). The second reason is the volume of the residue. Every ton of alumina produced leaves about one ton of solid residue in suspension and 4 tons of slurry. The red mud slurry is usually discharged into sealed or unsealed artificial or natural areas [118,4]. The impoundment areas include deep ponds, tailing dams and dykes, stacking, 5% slope mud stacking by solar drying. Most of the storage areas are lined with impervious earth or plastic liners, but the possibility of seepage or leaking still exists (i.e., Lake Hidalco at Renukoot in India). The environmental impact of red mud disposal could come from the seepage of the alkaline solution into the groundwater which may result in contamination affecting industrial, domestic and agricultural water suppliers. It could also come from the vast area of land used for mud disposal, or from dust pollution generated from the dried surface of the pond.

PHOSPHOGYPSUM UTILIZATION

Introduction

Phosphogypsum is a by-product originating from the production of phosphoric acid by the wet process. In this process, the raw phosphate is treated by sulfuric acid and, besides the main product which is phosphoric acid, gypsum and a small quantity of hydrofluoric acid are obtained. The chemical reactions occurring in this process are represented by the following formula:

$$1/2Ca_{10}(PO_4)_6F_2 + 5H_2SO_4 + 10\ H_2O \rightarrow 3H_3PO_4 + 5\ CaSO_4.2H2O + 1/2H_2F_2$$

Corresponding molecular weights are as follows:

½ x 1008.68 + 5 x 98.08 + 10 x 18.02 →3 x 98 + 5 x 172.17 + ½ x 40

For each ton of anhydrous P_2O_5 about 5 tons of phosphogypsum are obtained. The world production of phosphoric acid is about 22 millions tons (Mt) of P_2O_5 and thus the quantity of phosphogypsum corresponds to about 110 Mt [119]. This is more than twice the quantity of natural gypsum of which about 50 Mt is used for the manufacture of gypsum building materials.

In the 1960s and 1970s, a great number of studies were undertaken concerning the production of gypsum building materials from phosphogypsum [120]. Recent studies have also been published, but current research in the area is not prevalent because no practical solution for the radioactivity problem has been found [120-125].

Several technologies were devised and introduced into industrial practice. The most popular were those of Knauf and Giulini in Germany, Rhône Poulenc in France, Imperial Chemical Industries in England and Yoshino and Nitto in Japan. In the past, several plants used phosphogypsum. For example, in France of Rhône Poulenc in Rouen and of CDF Chimie Air Industry at Douvrin [126] and in Great Britain of Imperial Chemical Industries, but they were closed in 1970s [119]. The Giulini plant in Ludwighafen was also closed. Currently, only two European plants using phosphogypsum from phosphoric acid remain in operation today [127].

There are two main reasons for the plant closures. First was the presence of radioactive elements in the phosphogypsum and second, the relatively high moisture content in the phosphogypsum, which is 20-25% higher than natural gypsum (normally 1-2%). In contrast to this situation, there is widespread use of phosphogypsum in Japan both in the gypsum and cement industry due to the lack of natural gypsum resources. To reduce the level of radioactivity, the phosphogypsum is mixed with flue gas gypsum [119].

Technologies of Wet Phosphoric Acid Production

There are four wet phosphoric acid processes which differ in technology and the type of calcium sulphate obtained [119].

- Single-stage conventional dihydrate process (Prayon, Dorr, Oliver). Calcium sulphate dihydrate is obtained directly with a free-water content of about 25%.

- Two-stage hemihydrate-dihydrate process (Nissan). Hemihydrate is obtained in the first stage. In the second stage, the hemihydrate is converted to dihydrate, after which it is filtered to a free-water content of about 20%.

- Single-stage hemihydrate process (Veba, Fisons). Calcium sulphate hemihydrate is obtained directly with a moisture content of about 20%.

- Two stages dihydrate-hemihydrate process (Central Prayon, Central Glass). The dihydrate obtained in the first stage is converted to hemihydrate in the next stage. The most widely used process in the single-stage conventional dihydrate processes (84%) [119].

Based on the $CaSO_4$-H_3PO_4 system, Dahlgren [128] established the equilibrium curves that result from the modification of calcium sulphate (Figure 5) [128].

The AB curve corresponds to the equilibrium $CaSO_4\,2H_2O \rightarrow CaSO_4$ II. The CD curve corresponds to the metastable equilibrium $CaSO_4\,2H_2O \rightarrow CaSO_4 \bullet \frac{1}{2}H_2O$. Above the EF curve, the transformation of hemihydrate to anhydride II is rapid and occurs over several hours. Below the curve, the reaction is slow and requires several days for completion. The hemihydrate is the most stable

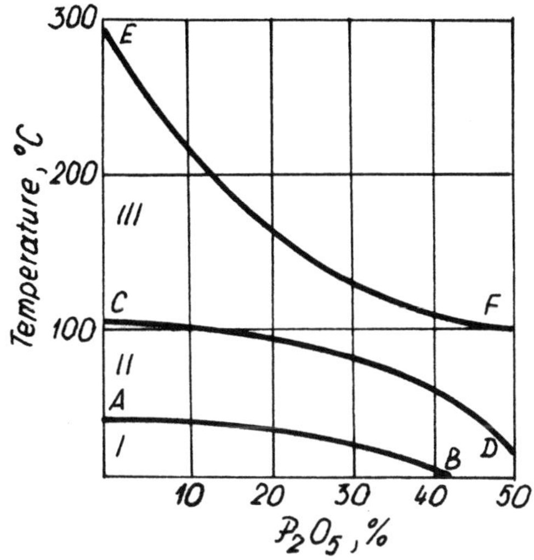

Figure 5. Phase diagram of the system $CaSO_4$-H_3PO_4. I-stable dihydrate, metastable anhydrite, unstable hemihydrate; II-stable anhydrite, metastable dihydrate, unstable hemihydrate [128].

below the EF curve and the stability increases with decreasing temperature. Below the AB curve, precipitation of $CaSO_4{\bullet}2H_2O$ occurs. In the presence of free acid (H_2SO_4) hemihydrate is formed, but quickly decomposes. The transformations are as follows:

$$\underset{\text{unstable}}{\alpha CaSO_4{\bullet}\tfrac{1}{2}H_2O} \xrightarrow{\text{rapid}} \underset{\text{stable}}{CaSO_4\ 2H_2O} \xrightarrow{\text{slow}} \underset{\text{metastable}}{CaSO_4\ II}$$

Between the AB-CD curves, anhydride II is stable but calcium sulphate precipitates as dihydrate:

$$\underset{\text{unstable}}{\alpha CaSO_4{\bullet}\tfrac{1}{2}H_2O} \xrightarrow{\text{rapid}} \underset{\text{metastable}}{CaSO_4\ 2H_2O} \xrightarrow{\text{slow}} \underset{\text{stable}}{CaSO_4\ II}$$

Above the CD curve, hemihydrate is formed and the transformations are as follows:

$$CaSO_4 \bullet 2H_2O \xrightarrow{\text{rapid}} CaSO_4 \bullet 1/2H_2O \xrightarrow{\text{slow}} CaSO_4\,II$$

unstable — metastable — stable

In practice, there is rarely a pure solution of phosphoric acid. Frequently, calcium sulfate crystallizes from the mixture of phosphoric and sulfuric acids causing displacement of the equilibrium curves, as shown in Figure 6. When the calcium sulfate crystallizes as dihydrate, the liquid phase contains 28-30% P_2O_5 and for the crystallization of hemihydrate the liquid phase contains 40-50% P_2O_5.

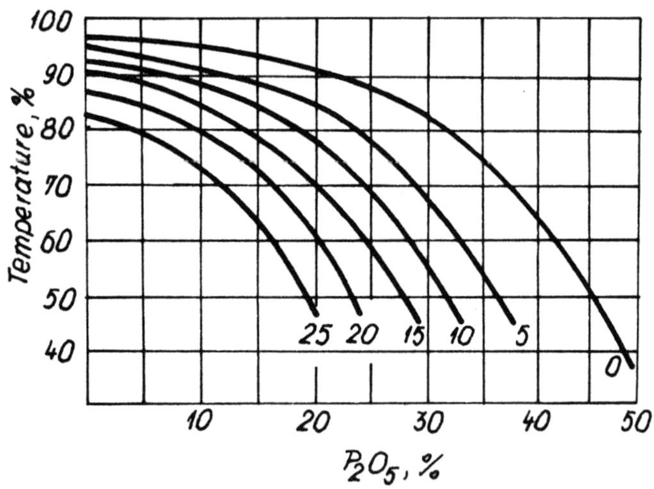

Figure 6. The equilibrium dihydrate-alpha hemihydrate in the mixture of H_3PO_4 and H_2SO_4 [128]. The figures near the curves represent % of sulphuric acid.

Impurities in Phosphogypsum

The impurities in phosphogypsum are a result of phosphoric acid production and the type of raw materials used. In principle, hemihydrate phosphogypsum is much purer than dihydrate[119].The amount of moisture in hemihydrate is less than 10% and for dihydrate it is between 20-30% [129].

Phosphogypsum contains free acids such as, phosphoric and sulfuric acid and in smaller amounts hydrofluoric and silicic-fluoro acid (H_2SiF_6) and its pH is ≤ 2.5. The acids act to increase corrosion between the plaster and structural elements in the building or the equipment used to make the pre-fabricated plaster products. Free acids tend to retard setting. The physical form of phosphogypsum is a fine grained powder composed of lamelli forms or lath-shaped crystals of very small particles up to approximately 200 microns in size. The specific surface area can range from 200-600 m^2/kg [130]. The most prevalent impurities are phosphates (mono- and dicalcium phosphate), fluorine, silica (coarse quartz and fluorosilicate), sodium compounds (major components) which are readily soluble in water and organic matter, whereas alumina and iron are soluble in smaller amounts. In Table 2, some analysis is given. Slack [131] overviews the compounds, which can crystallize in wet phosphoric acid production when the concentration of phosphoric acid is 28-32% . The compounds are shown in Table 3. The phosphogypsum from magmatic phosphates from Kola contain some rare earth elements [132].

The impurities can exist in several different forms. These include admixtures such as, organic matter or silica adsorbed by phosphogypsum particles, or as solid solutions in gypsum crystal lattice. In coarse admixtures, there are always some grains of unreacted phosphoric ore, which can be separated by screening. Lutz states that the cut-fraction for the wet screening process is at approximately 170 microns [129].

An important problem is the content of natural radioactive elements, principally Ra_{226}, which is present in all sedimentary

Table 2. Composition of Various Phosphogypsums From the Dihydrate Process.

Components	Phosphate rock		
	Kola	Florida	Morocco
Crystal water	19.5	20.0	20.0
CaO	32.0	31.0	32.0
SO_3	45.0	44.0	44.0
SiO_2	1.1	2.4	1.3
Al_2O_3	0.28	0.12	0.06
Fe_2O_3	0.08	0.02	0.02
P_20_5 total	1.6	0.78	1.3
P_20_5 soluble in water	0. ?	0.27	0.3
F	1.7	0.6	0.7
Na_2O	0.3	0.18	0.18
Organic matter	--	0.24	0.15
Ln_2O_3	0.56	--	--

Table 3. Compounds Formed During the Wet Acid Process (Concentration 28-32 %) [131].

$AlH_3(PO_4)_2$ $2H_2O$	$Fe_3(H_3O)H_8(PO_4)_6$ $6H_2O$
$AL_2K(PO_4)_2(OH)2H_2O$	K_2SiF_6 (cubic mod.)
$CaSO_4$ $2H_2O$	$MgSiF_6$ $6H_2O$
$CaSO_4$ 0.5 H_2O	Na_2SiF_6(hexagonal)
$CaSO_4$ (anhydride)	$Na_xMg_xA_{12-x}(F,OH)_6$ H_2O
CaF_2	Na_3AlF_6
$CaSO_4$ $SiF_6AlF_6(OH)12H_2O$	SiO_2 xH_2O
$Fe(H_2PO_4)_2$ $2H_2O$	

phosphates ores. About 20-40% of these radioactive elements remain in phosphogypsum [133]. For example, phosphogypsum dihydrate from Florida phosphate contains approximately 15-30 mg of uranium per kilogram and hemihydrate seven times more [133]. About 60% of the uranium in hemihydrate cakes can be readily leached after hydration of the cake, but the residual amount is very difficult to remove economically ([133]. Only phosphogypsum from magmatic phosphate ores contains a very small quantity of radioactive matter and can be accepted for use in building materials. The purification process of dihydrate cake, which works by separating fine particles in hydrocyclones, has only a limited influence on the content of radioactive elements.

Some impurities are present as solid solutions in the gypsum crystal lattice and as such, are very difficult to separate. These impurities include calcium orthophosphate $CaHPO_4•2H_2O$ and monosodium orthophosphate, $NaH_2PO_4•2H_2O$, which are isomorphic with $CaSO_4•2H_2O$. These impurities can be separated only by re-crystallization during the autoclaving process. The $(AlF_5)^{2-}$ and $(FPO_3)^{2-}$ ions can substitute for the $(SO_4)^{2-}$ anions in the gypsum crystal lattice. Usually, a portion of the sodium ions are present as solid solutions in gypsum crystal lattice. Finally, the uranium U(IV) ions substitute for the calcium ions in the $CaSO_4$ xH_2O crystals [133].

Fine fractions, which contain organic matter and very high levels of impurities within the phosphogypsum particles in the form of solid solutions, can be separated by flotation or by application of hydrocyclones. As a rule, the particle size cut for hydrocyclones varies between 10-15 microns [129]. Where the organic impurities have foaming properties, the addition of a flotation agents are not required [129]. The content of water soluble compounds, principally orthophosphoric acid, H_3PO_4, can exceed 1%. Also phosphates, primarily, $Ca(H_2PO_4)_2$, and sodium compounds such as sulphate and Na_2SiF_6, are soluble in water.

All these water soluble impurities are very harmful for the production of plaster because they cause longer setting times (Phosphor ions), cause efflorescence (sodium compounds) , hinder the adhesion of gypsum to the paperboard production and influence on deformation of plaster blocks. The impurities should be removed or transferred into water insoluble compounds.

To remove the soluble impurities, the phosphogypsum must be washed and treated with calcium hydroxide. The addition of calcium hydroxide is very important for the conversion of soluble phosphor ions into water insoluble calcium phosphate $Ca_3(PO_4)_2$. Also soluble fluorine ions are converted into water insoluble CaF_2. In some processes, instead of calcium oxide or carbonate, potassium carbonate is added, which according to Efimov [134], provides better results.

The particle structure of phosphogypsum, which ranges from lamelli form to lath-shaped particles, results in thixotropic gypsum pastes of poor flowability and cannot be used without modification [119]. These defects can be eliminated only by very fine grinding or re-crystallization [119].

Utilization of Phosphogypsum

There are several areas of utilization for phosphogypsum. The two main fields are

- conversion of phosphogypsum to plaster and plaster products and

- replacement of natural gypsum in cement production.

Other applications include:

- production of sulfuric acid and cement clinker. This application has been recognized since its inception in 1886 by Cummings and Lesh. The method was further developed by Müller and Kühne during World War I.

- manufacture of ammonium sulfate,

➢ gypsum for road foundations. Phosphogypsum for this purpose has been used in France as an activator in fly ash and lime mixtures. The proportions are 91% fly ash, 4% lime and 5% gypsum [135].

➢ soil stabilization. In this technology, gypsum-lime [136] can be used. Gypsum-lime is a mixture of phosphogypsum and quick lime. Water is added and calcium hydroxide and hemihydrate are obtained. The mixture has been used for soil stabilization with good results and Nieminen [136] found that ettringite forms after one year, as a reaction product of clay minerals and stabilizing agents.

Another promising use for phosphogypsum is the preparation of ash phosphogypsum material for filling voids in rocks mass (mining backfill) [137]. Further uses may be as a mineralizer for cement clinker production [138], as a raw material component for expansive clinker manufacturing [139] and finally, as a raw material for special cement production [140].

Conversion of Phosphogypsum to Plaster and Plaster Products. Many scientific works and many patents have been published on the conversion of phosphogypsum to plaster, but only a few of them were put into practice. In this chapter, the authors will address several practical methods.

Knauf's method is described below [119,141,142]. Knauf developed three processes: SI , SII, SIII. In the SI process, the dihydrate obtained as filter cake in wet phosphoric acid production is used. The phosphogypsum is purified by flotation and washing out the water soluble and solid admixtures impurities [141]. Both treatments are carried out immediately after discharge from the phosphoric acid filter by remashing and repeating filtration. The relative small quantity of washing water that is involved contains very little solid matter and can be returned to the phosphoric acid circuit thus, saving water. The purified phosphogypsum is dried in directly heated units with fluidizing equipment. This is followed by very fine grinding to improve the flowability of the plaster. As mentioned earlier, the elongated forms of phosphogypsum particles results in thixotropic gypsum paste of poor flowability. It cannot be used in that form and this is the reason for the grinding step. The beta-calcium sulphate obtained is of very high purity so that the products are not affected by efflorescence and no cardboard damage occurs when used in plasterboard. A flowchart of the Knauf phosphogypsum processes is presented in Figure 7.

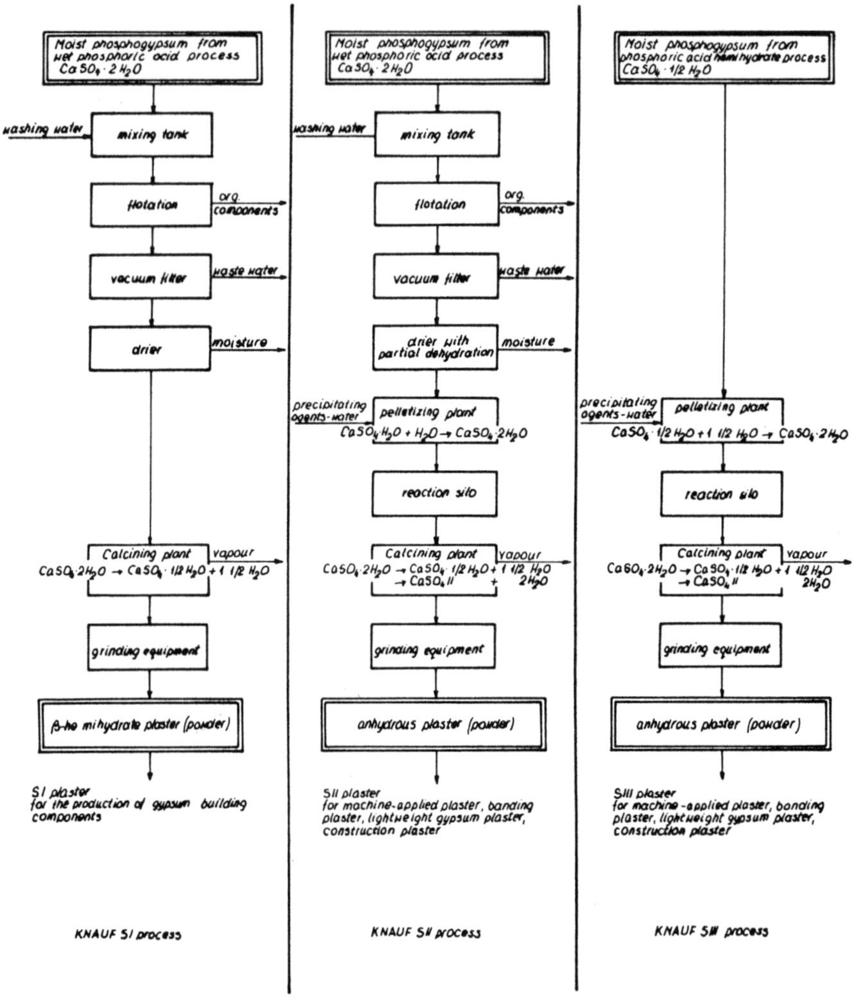

Figure 7. Basic flow chart of the phosphogypsum process [141].

The product of the Knauf SII process is anhydride II used in construction plasters. It is coarsely ground with 20% of the particles larger than 200 microns. Since phosphogypsum has very fine particles with an average particle size of about 50 microns for the production of construction plaster, it is necessary to enlarge the grains. For the purification of phosphogypsum, the same equipment used in the SI is used in the SII process. The purification requirements are lower than those for the SI process. Grain enlargement is carried out after purification. In this process, after drying and simultaneous partial dehydration, the material is pelletized in conjunction with the incorporation of additives that precipitate the phosphate that affect the setting time.

This procedure takes a relatively long time and is carried out in reaction silos. At the same time, a re-crystallization takes place whereby the thixotropic properties are eliminated and a high degree of plasticity is obtained. The matured pellets can then be calcined to anhydride II in a rotary or grate kiln [141]. Following calcination, the pellets undergo grinding to the optimum granulometric composition required for construction plaster.

Anhydride II, the product of S II process, can also be used in the cement industry as setting retarder and for production of gypsum building components. The construction plaster produced in the SII process can be applied by hand on the building site as conventionally practiced. But it can also be converted to special building plasters, such as machine-applied plaster, bonding plaster and pre-mixed lightweight plaster.

The Knauf SIII process has been specially developed for one- and two-stage wet phosphoric acid processes in which hemihydrate is obtained. These phosphoric acid processes are the Central-Prayon and Veba-Chemie process. The Knauf SIII process is almost as economical as natural gypsum technology for the production of construction plaster, but is linked to the phosphoric acid hemihydrate process [141].

The process requires no purification of the initial cake hemihydrate. The hemihydrate is derived directly from the phosphoric acid filter and is then recrystallized. This procedure eliminates the thixotropic properties of phosphogypsum. After intermediate storage, the gypsum can be calcined and ground to construction plaster in the same production facilities as those used in the SII process. The product can be applied directly by hand or the previously described techniques can be used. In the cement industry, the lumpy, recrystallized phosphogypsum, can also be used directly as a setting time regulator.

The gypsum plant of Gebr. Knauf Westdeutsche Gipswerke at Castrop-Rauxel (Ruhr District) has used the Knauf SI and SII processes since 1963. The gypsum plant Net B Knauf et Cie at Eheim near Liege in Belgium is using the Knauf SIII process and has a capacity of 0.2 Mt /year of calcined plaster. In Table 4, the properties of these plasters produced by the Knauf processes are shown.

Table 4. Properties of Plaster Produced by Knauf Processes [141].

	Beta hemi hydrate Plaster	Construction Plaster	Machine Applied Plaster	Bonding Plaster	Light-wt. Gypsum Plaster
water/plaster ratio	0.70	0.55	0.48	0.66	0.64
Initial set (min)	5	5	75	40	45
Final set (min)	16	21	150	75	85
Gips-element density (kg/m^3)	1050	1340	1200	840	920
Flexural strengths (MPa)	4.2	6.8	2.2	1.4	1.7
Compressive strengths (MPa)	10.5	20.4	6.0	3.5	3.9
Coverage m^2/100 kg plaster	--	--	9.8	13.0	12.0

The technology developed by CDF Chimie and Air Industry was applied in the plaster plant at Douvrin near Arras in France [126]. The process is initiated in a mixing tank (Figure 8), then the coarse

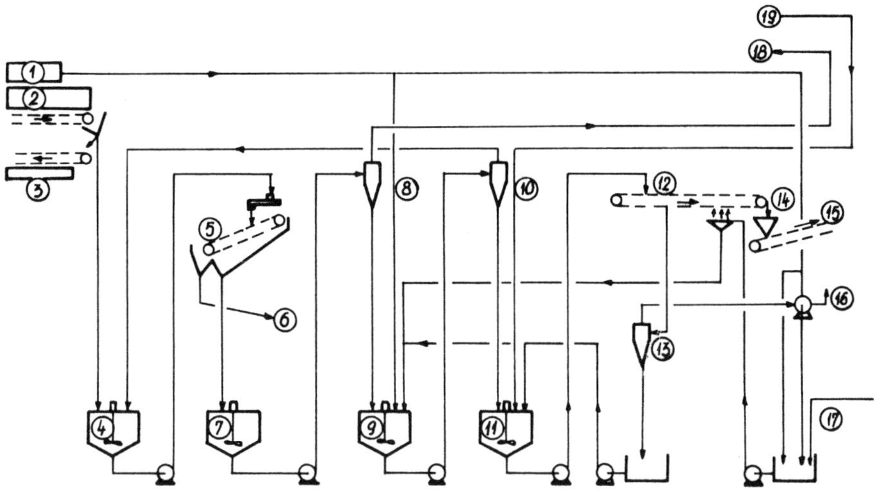

Figure 8. Flow sheet of CDF Chimie and Air Industry Process [126]. 1) fresh water, 2) phosphogypsum supply, 3) towards barrows, 4) pulper tank, 5) sifting screen, 6) clarification station, 7) washing tank, 8) hydrocyclones, 9) washing tank, 10) hydrocyclones, 11) neutralizing tank, 12) continuous vacuum filter, 13) filtrate separation tank, 14) clean water phosphogypsum, 15) thermic treatment section, 16) vacuum pump, 17) scrubber water, 18) contaminated water (to clarification station) and 19) lime solution.

fraction is separated by screening and the material is washed and the fine fraction is separated in hydrocyclones. Once completed, the entire process is repeated. In another tank, lime for neutralization is added and the material is filtered using a continuous vacuum filter and fed to a pneumatic dryer and calcining kiln. If necessary, the calcination may be continued in a second installation. The hemihydrate is then cooled and stored in silos prior to use.

Very similar technology was applied by Rhône Poulenc. Initial purification of phosphogypsum is carried out by storing the material as a suspension and removing the larger particles by screening. Also, lime is added to neutralize the acidity. If the phosphogypsum has a low level of impurities, as is frequently the case for high-grade apatite ore, further purification may be unnecessary, as the main impurities are soluble and are removed by water during filtration of the suspension prior to calcining. Using this process, 80-90% of the soluble impurities are removed and recovery of phosphogypsum is greater than 97%.

When further purification is necessary, it may be carried out in one of two ways, depending on the properties of the phosphogypsum [143]. In a phosphogypsum in which the crystals agglomerate, as is commonly obtained from the acidification of Florida rock, there is usually a high concentration of soluble and organic impurities. To remove these impurities, the phosphogypsum must be washed three times in a hydrocyclone. Gypsum recovery is between 70-90% and over 95% of the soluble impurities are removed [143]. The efficiency of the organic matter removal depends on the original concentration but is apparently adequate in most cases [143]. When the phosphogypsum is particularly finely divided, or in regions where water is scarce, flotation is substituted for hydrocyclones. This provides satisfactory extraction of organic material, a removal of soluble impurities between 85-90% and a gypsum recovery of 90-95%.

Rhône Poulenc also used several drying and calcination techniques. In the plant at Rouen, wet phosphogypsum was introduced into a pneumatic dryer and dry material was then calcined to beta-hemihydrate in a fluidized bed, but the majority of heat was supplied by a steam-heated heating coil immersed in the bed [143]. Schematic representation of the process is shown on Figure 9. Some average data for Rhône Poulenc process are described in Table 5 [143].

The Cerphos process is based on the assumption that a large portion of the impurities in phosphogypsum can be found in particle

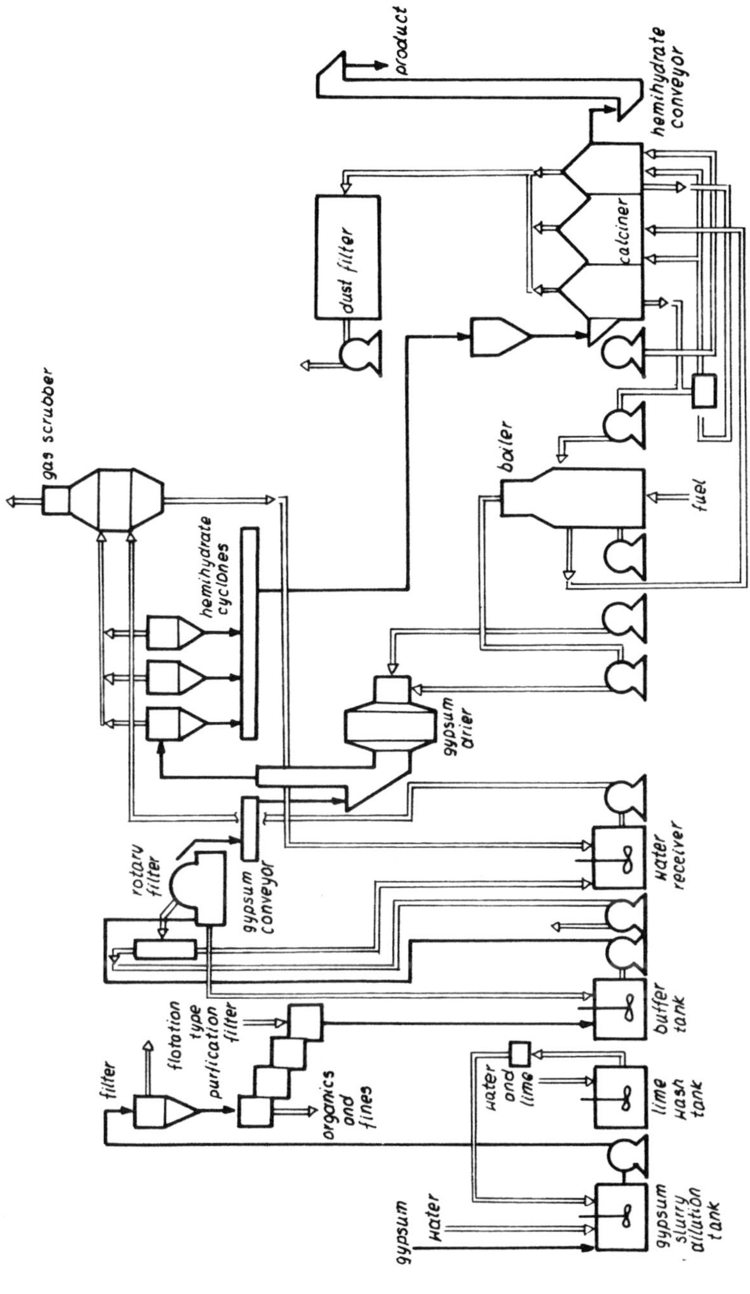

Figure 9. Rhône Poulenc process, incorporating flotation and two-step calcination [143].

Table 5. Average Utilities Consumption Per Ton of Plaster for Rhône Poulenc Process [143].

Phosphogypsum (kg)	1.5
Water (m^3)	3
Lime (kg)	5-10
Fuel oil (kg)	45
Power (kWh)	50
Labor (manpower shift)	2

size fractions larger than 160 microns and less than 25 microns. Phosphogypsum is cleaned by removing coarse and very fine particle fractions. In a phosphogypsum from Morocco phosphate, the particle size fractions larger than 170 microns are heavily enriched with fluorine, silicon and sodium salts [129]. On the other hand, the particle size fractions less than 25 microns are rich in organic substances and in phosphate combined in the gypsum crystal lattice [129].

Approximately 75% of the starting material can be recovered by this method [129]. The basic principle of the Cerphos process is shown in Figure 10. The phosphogypsum is slurried with water and the gypsum suspension in water is then pumped to an elevated wet vibratory screening machine to remove the coarse size fraction. The cut size of this wet screening process lies at approximately 160-170 microns. The coarse material, the screen overflow, still contains some unreacted raw phosphate from the phosphoric acid. This pumpable suspension is returned to the phosphoric acid plant. The screen underflow goes directly into a multi-stage flotation system [129]. The very fine gypsum particles containing a high level of impurities are entrained in the foam which has been formed. The flotation underflow is fed to a belt filter then washed and neutralized with a limestone suspension. The filtrate and wash water from the filter is

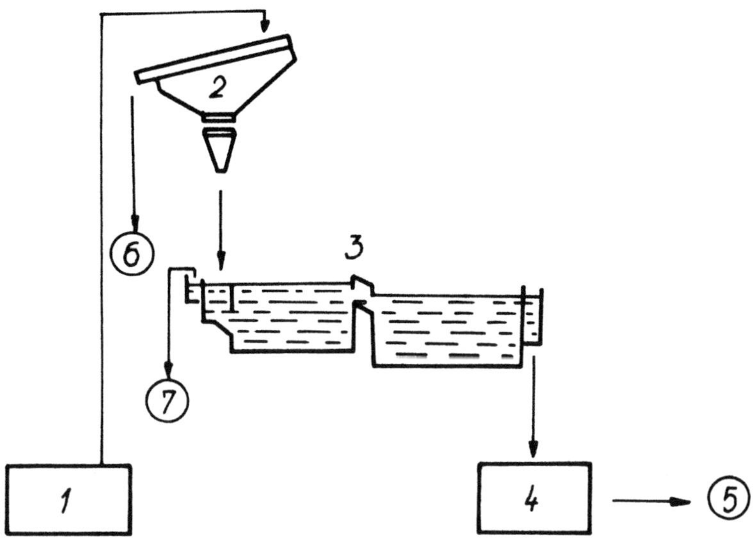

Figure 10. Schematic representation of the Cerphos process [129]. 1) slurrying the phosphogypsum with water, 2) wet screening, 3) flotation, 4) filtration, 5) calcining, 6) oversize (> 168 microns), 7) foam: size fraction <25 micron.

returned and reused for gypsum slurring or can be returned to the phosphoric acid factory.

Depending on the origins and wet phosphoric acid digestion process, it is not always possible to make a clean cut between coarse and fine particles or else this is associated with large losses of raw gypsum [129]. A cleaning effect can be achieved by a simple displacement of washing process on the filter, as the water soluble impurities are rinsed out by the displacement effect of the fresh water

[129]. The use of hydrocyclones appropriately located before the solids/liquid separation stage, has been shown to be effective in cleaning the material [129].

The process with washing into the filter is shown on Figure 11 [129]. Before washing, the fine fractions are separated using hydrocyclones. For drying and calcining, a pneumatic dryer and a classic rotary calciner by Büttner-Schilde-Haas AG are used.

The conversion of phosphogypsum into alpha hemihydrate involves a totally different technology. This process was particularly developed by Giulini in Germany [144]. Phosphogypsum filter cake coming from phosphoric acid plant is slurried and diluted to such extend that the slurry contains about 450g of solid matter per liter. This slurry is then fed through an intermediate tank, where a number of chemicals and flotation additives are added, into a flotation unit, where organic impurities and some of the soluble impurities are

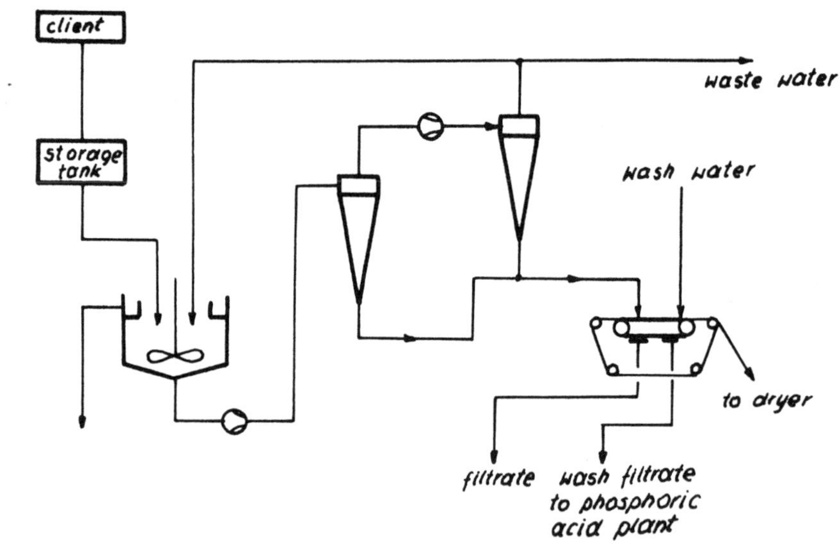

Figure 11. Gypsum cleaning using hydrocyclones [129].

removed (Figure 12). Then the underflow from flotation is pumped into a thickener, to which water and steam are added in order to remove the remaining soluble impurities, a portion of the insoluble impurities and to raise the temperature of the slurry which is then transferred at a controlled rate continuously to the autoclave. The temperature is 105-125°C and the pH of the mother liquid is 1-3 [145]. The size and shape of crystals may be controlled during autoclaving by addition of chemical additives and regulating the pH and temperature and by varying the solids concentration in the suspension. A re-crystallization of calcium sulphate, which takes place in the autoclave, permits the releasing of impurities locked in the crystal structure of hydrate, mainly sodium and calcium phosphates. Alpha hemihydrate is continuously removed from the autoclave and after centrifugation and can be directly used for production of prefabricated plaster products. It can also be dried and stored as construction plaster. Hemihydrate from Giulini process has been used for a number of applications, including partition blocks, jointing cement, plasterboard, and various building plasters [145]. The utilities consumed per ton of alpha hemihydrate in Giulini process are given in Table 6.

Table 6. Utilities Consumed Per Ton of Alpha Hemihydrate in the Giulini Process [145].

	Wet alpha hemi-hydrate for wall block production	Dry,alpha-hemi hydrate powder	Dihydrate pellets for the cement industry
LP steam hot water 3bars, (tons)	0.4-0.6	0.4-0.6	0.4-0.6
90°C , (m^3)	0.5	0.5	0.5
Raw water (m^3)	2	2	2.1
Electric power(kwh)	25	35	28
Fuel gas or fuel oil, (kcal)	-	200000	-
Cost of chemicals (DM)	1.0	1.0	1.0

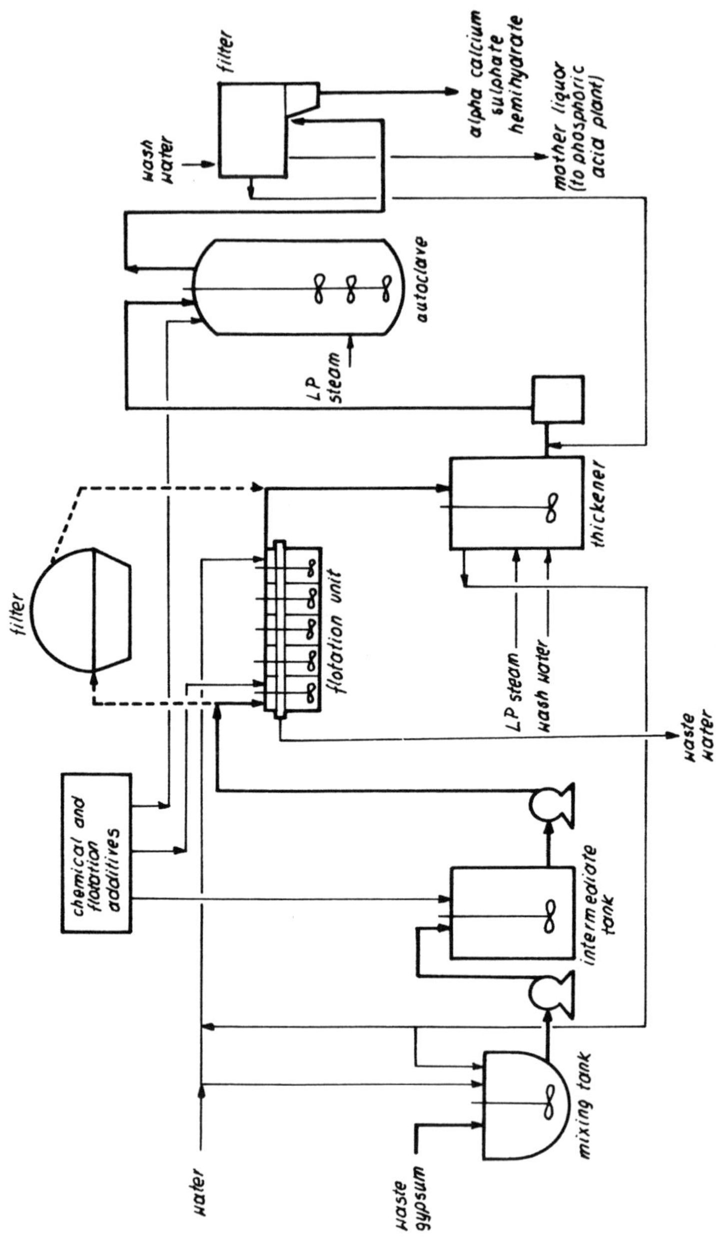

Figure 12. The Giulini hemihydrate process [145].

The Giulini process for producing alpha calcium sulphate hemihydrate was the only wet phase process to have been used in commercial plants until 1978 [145]. The content of impurities in the alpha hemihydrate after the Giulini process are very low and in Type 2 alpha hemihydrate, there is approximately 0.10% F , 0.08 P_2O_5 and 0.04 Na_2O [145].

Until about 1970, the Imperial Chemical Industries Ltd. (ICI) of England was one of the largest users of phosphogypsum [143]. The purification system was simple and only dilution in water using a vortex mixer with further neutralization was applied [143]. However, in the mid-1970s, ICI was obliged to close the two factories that used this technology and developed a new process designed to produce alpha hemihydrate by a wetphase dehydration process in an autoclave [146]. To prevent the formation of anhydride scale on indirect heating surfaces, ICI developed a direct heating process using steam injection into a stirred vessel. Optimum temperatures (150-160°C) were attained without the formation of anhydride. In the technological line (Figure 13) two stirred reactors in series were adopted [146]. In this way the total conversion time was less than 3 minutes. When the phosphogypsum is used in solution-phase conversion, the hemihydrate crystals vary in shape according to the impurities present, which in turn largely depend on the source of the phosphate rock used in the phosphoric acid plant. One of the impurities often contained in phosphogypsum and released during wet-phase dehydration is aluminum. Under reaction conditions, this forms a series of Al-OH complexes whose equilibria are determined by pH [146]. By controlling the reaction conditions, it is possible to use of this modifier to ensure the formation of crystals of a desired habit .

Addition to Cement. The use of phosphogypsum as a retarder for cement is minimal. Worldwide its consumption is no more than about 1.7 million tons per year, most of which is used by countries lacking natural gypsum resources such as Japan (0.9 Mt/year) and Korea (0.5Mt/year), with some minor quantities consumed in Brasil and central Europe [119].

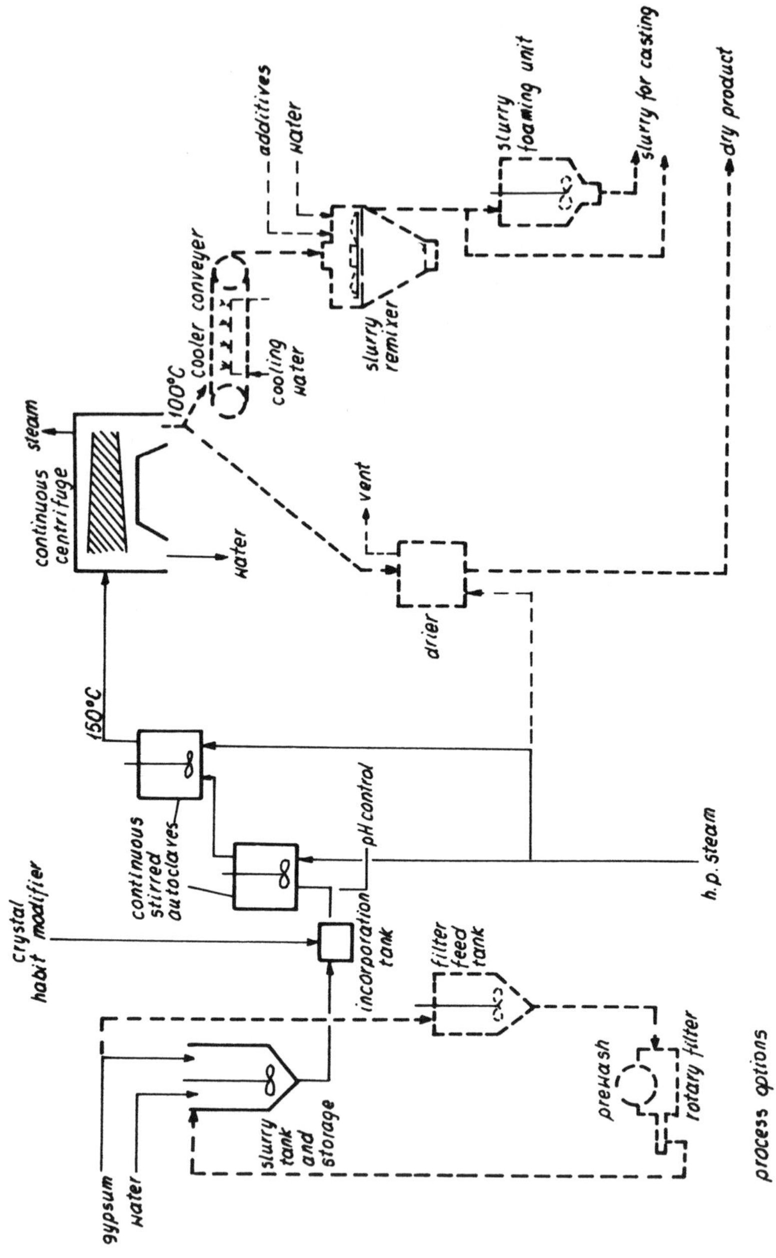

Figure 13. The ICI continuous alpha hemihydrate process [146].

According to Wirsching [119], the main reason for so low consumption of phosphogypsum by the cement industry are its high moisture content and its impurities, which influence on the strong retardation of cement set. Wet phosphogypsum is difficult to handle because it does not easily flow from the silos, is difficult to feed and proportion correctly, and freeze into icy lumps in cold weather. Moreover, the acid phosphate impurities (H_3PO_4, $Ca(H_2PO_4)_2$ and $CaHPO_4$) interfere in an unpredictable way with setting time of cement and retard the early strength development of concrete [119].

Tabkikh and Miller [147] established that the influence of the fluorine ions is stronger than that of phosphoric ions and although present simultaneously act probably independently (Figure 14).

Phosphogypsum before used in cement industry needs purification and pelletizing. The degree of purification required will naturally depend on the phosphoric acid process and the phosphate rock used; for example, if the by product is in the hemihydrate state, it will have undergone a recrystallization in which the co-crystallized impurities have been released, and so only neutralization as the hemihydrate re-hydrates to gypsum is necessary [145]. On the other hand, phosphogypsum in the dihydrate state needs considerably more purification. This can be accomplished by purification processes similar to those used for beta-hemihydrate plaster. Conversion of the relatively pure dihydrate produced by hemihydrate-dihydrate processes into a material suitable for cement additives, while not as simple as using the by-product hemihydrate, is easier than using phosphogypsum from conventional dihydrate processes [145].

Phosphogypsum produced by the New Nissan hemihydrate - dihydrate process is comparatively pure and can be used to manufacture cement additive in a process developed by Nissan Chemical Industries Ltd [145]. A flow diagram is shown in Figure 15. Phosphogypsum is first dried in a rotary drier and neutralized with lime before it is transferred to a rotary kiln for calcination. The conditions are adjusted so that a mixture of soluble anhydride and hemihydrate is obtained. From the kiln, dehydrated gypsum is fed to

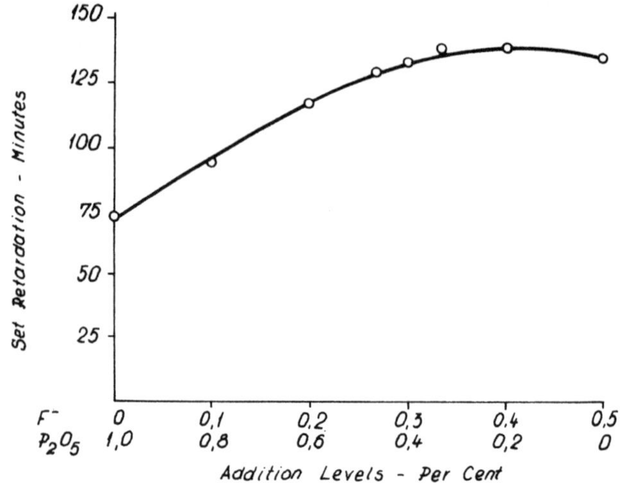

Figure 14. Setting time versus composition: cement + Na_2SiF_6 + $Ca(H_2PO_4)_2$ H_2O [147].

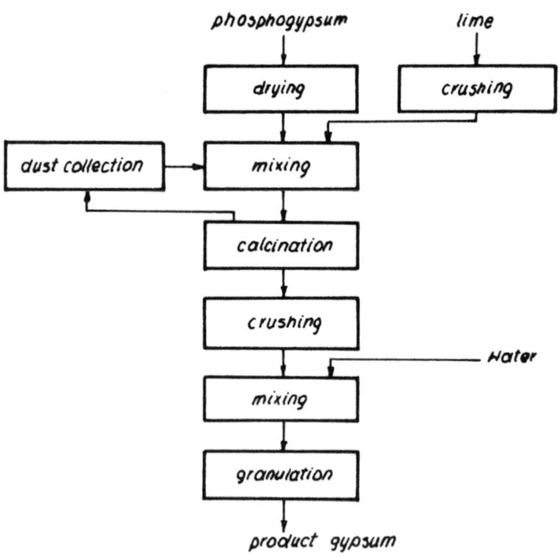

Figure 15. Nissan process for manufacturing cement additives [145].

a pugmill, where it is transferred to a mixing unit, in which water is added, into a pan granulator. In this machine, dehydrated gypsum is rehydrated and granulated simultaneously. The utilities consumed per ton of products are given in Table 7 [145].

Table 7. Utilities Consumption of Nissan Gypsum Cement Additive Process (per ton of gypsum).

Phosphogypsum (27% moisture) (kg)	1216
Lime (85% CaO) (kg)	30
Water (kg)	700
Steam (kg)	45
Heavy fuel oil (kg)	60

(gypsum product contains 10% moisture)

Portland Cement and Sulfuric Acid. From a chemical standpoint in traditional Portland cement production, calcium carbonate can be replaced with calcium sulfate producing CaO and sulphur dioxide, from which sulfuric acid can be manufactured. It is not possible to decompose calcium sulfate using only thermolysis because of its chemical stability. Coke as reducing agent must be used, but even so, higher calcination temperatures are required and the energy consumption of the process is much larger than that of the ordinary lime and clay processes. A process was developed during World War I by W.J. Müller and H. Kühne and a prototype plant was built at Leverkusen and remained in operation until 1931 [148]. Then Müller helped Imperial Chemical Industries Ltd. to set up a plant at Bilingham in 1929. This plant was replaced in 1972 by the twin sulphur-burning Lurgi Double Catalysis plants, as part of a modernization program. Further plants were erected at Miramas near Marseilles (France) in 1937, and in Wolfen (Germany) in 1938, which was rebuilt in 1954 [148]. Another plant at Coswig was also constructed. Other plants were built in Poland at Wizow and in Austria at Linz, as well as two additional plants in the United

Kingdom [148]. All these plants were successively closed or converted to burn sulfur which is a much more economical process.

The flow diagram process of Müller-Kühne is presented on Figure 16. Instead of anhydride, gypsum can also be used, but it must be calcined before use [148]. In the kiln, two main reactions take place which involve calcium sulfate and coke. Between 900-1100°C, a portion of the calcium sulfate is reduced to calcium sulfide:

$$CaSO_4 + 2C \rightarrow CaS + 2\,CO_2 \qquad (1)$$

As solids pass into the hotter part of the kiln, the calcium sulfide reacts with the remaining calcium sulfate:

$$CaS + 3CaSO_4 \rightarrow 4CaO + 4SO_2 \qquad (2)$$

There may also be some direct reaction of calcium sulfate and coke to produce sulfur dioxide:

$$2CaSO_4 + C \rightarrow 2CaO + 2SO_2 + CO_2 \qquad (3)$$

The stoichiometry of reaction 3 is the same as the combination of reactions 1 and 2 and it can be seen that one mole of carbon is required for every two moles of calcium sulfate processed. In practice, however, a slight excess of coke over the stoichiometric quantity is used to compensate for side reactions such as shown in equations 4 and 5 [148]

$$3CaS + CaSO_4 \rightarrow 4\,CaO + 4S \qquad (4)$$

$$CaS + 2\,SO_2 \rightarrow CaSO_4 + 2S \qquad (5)$$

Other constituents of kiln bath introduced with clay, also react with anhydride accelerating its decomposition with simultaneous formation of silicates, aluminates and ferrites. As an example, the reaction with silica can be given:

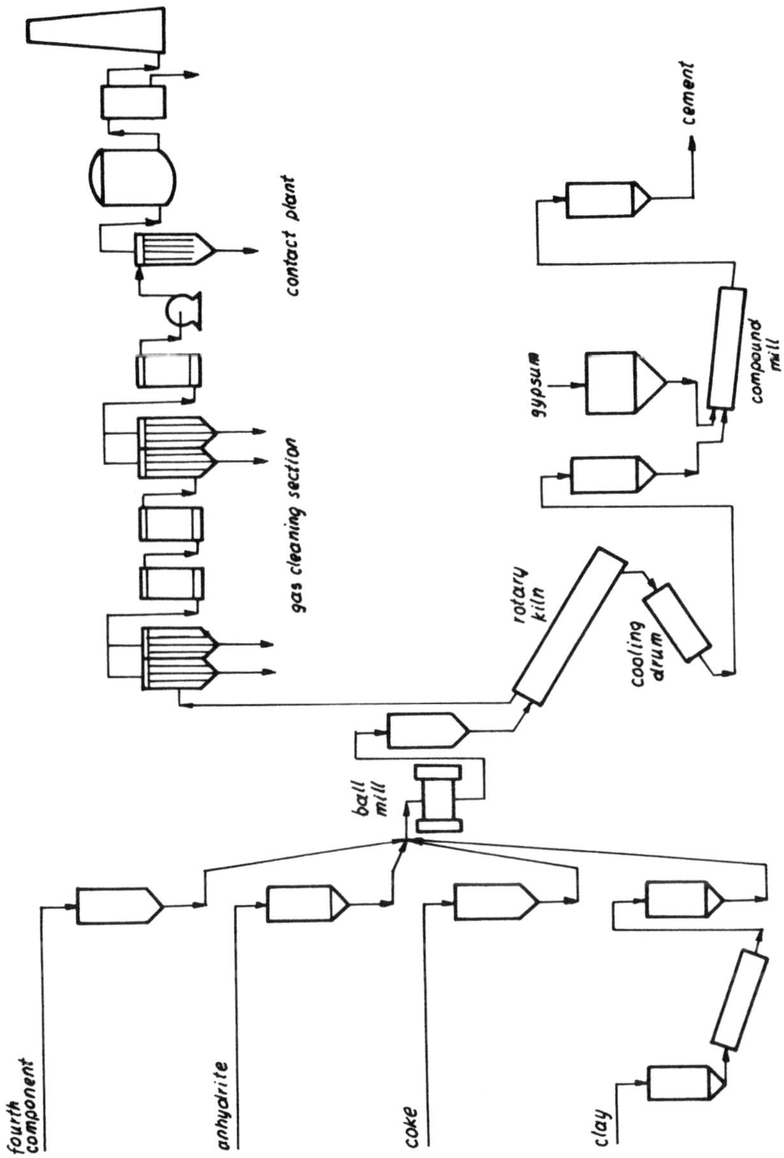

Figure 16. Basic cement/sulfuric acid process (Müller-Kühne process) [148].

$$2CaSO_4 + SiO_2 \rightarrow Ca_2(SiO_4) + 2SO_2 \qquad (6)$$

The hot gases leaving the kiln inlet are subjected to very thorough cleaning to remove solid before entering the sulfuric acid plant [148]. A typical sequence of dust control equipment is composed of cyclones,a dry electrostatic precipitator, wet scrubber/ cooler and a wet electrostatic precipitator immediately preceding the drying tower.

In the use of phosphogypsum for the production of Portland cement and sulfuric acid, three factors must be taken into consideration: moisture, phosphates and fluorides impurities.

It has been already mentioned that the raw materials for the Müller-Kühne process have to be dried before use and gypsum must be first calcined to remove water of crystallization. The calcium sulfate hemihydrate produced in some phosphoric acid processes such as the Fisons Hemi hydrate process or Central Prayon process has lower total moisture content and is more suitable for association with a cement/sulfuric acid plant than conventional dihydrate process. Also phosphogypsum from hemihydrate, dihydrate-hemihydrate or hemihydrate-dihydrate type of phosphoric acid process contains considerably less phosphate than the by-product from the conventional dihydrate process, and is therefore less likely to cause problems in the cement/sulfuric acid process.

Since very little of the phosphate content of phosphogypsum is volatilized in the kiln, the main effects of the cement/sulfuric acid process are realized on the properties of the cement [148]. The phosphates have a negative effect on clinker formation because, as Nurse [149,150] and Gutt [151]have shown, the solid solution of C_3P in C_2S is formed by the decomposition of alite. This solid solution has the decomposition product consisting of 7% P_2O_5, 66.5% CaO and 26.5% SiO_2, which is close to Nagelschmidtite $7CaOP_2O_5\ 2SiO_2$, but with a smaller content of C_2S. Nurse [149] elaborated the formulae for calculation of the clinker mineralogical composition in the presence of phosphorous. For each percentage of P_2O_5, 9.9% of tricalcium silicate is decomposed and the content of belite and belite solid solution is increased by 10.9%. Another approach presented by Gutt [151], states the P_2O_5 forms up to a 2% solid solution in tricalcium silicate and its primary field of crystallization is extended to 13% of P_20_5. However, both authors give a very similar limit for phosphorous content in cement clinker, (Nurse 2% and Gutt 2.5%). Indeed for 2% P_2O_5 content in clinker forms a cement of good quality, but strength development is somewhat slower [152]. For these reasons 1% of P_2O_5 was adopted as the limit by Chemie Linz AG [148]. In contrast, it was established that small quantities of phosphorous have a beneficial influence on the mortar strength and enhance the final strength [150,153,154]. The increase is very small and is about 0.2% [154]. In contrast to the phosphate, the fluorine content of phosphogypsum is lowered during calcination [148]. The exact proportion varies according to the origin of the phosphate rock and the conditions used in the phosphoric acid process, but it has been observed that up to 40% of the fluorine is released as gaseous fluorine compounds and mixes with the sulphur dioxide kiln gas [148].

Since fluorine is extremely damaging to the sulfuric acid plant catalysts, attacking both the silica support and the active vanadium component, the fluorine must be removed as completely as possible by wet scrubbing in the gas cleaning section [148]. The presence of fluorine in the clinker has a beneficial effect because it counteracts the negative influence of phosphate during the sintering process [155]. However, there seems to be some correlation between the fluorine content and the residual sulphate content of the clinker, and fluorine and sulphate together appear to be capable of adversely affecting the hydraulic properties of the cement in a somewhat unpredictable manner [148,156]. For this reason, it is considered desirable to restrict fluorine levels in the phosphogypsum feed to about 0.15 [148,156,157]. As phosphogypsum frequently contains 1-2% fluorine, some purification is usually required. Various techniques have been proposed, primarily based on the calcination of phosphogypsum which is necessary in any case for elimination of water- with or without the addition of sulfuric acid [148].

Among the industrial processes, the best known is the Chemie Linz AG, in cooperation with Krupp as far as kiln systems are concerned, especially heat exchangers. Chemie Linz adopted the shaft exchanger, but with extensive modifications when compared to the type of exchanger used in cement industry [148].

Because Chemie Linz advises that the concentration of phosphates in the phosphogypsum be kept at 0.5 % P_2O_5 or less, some treatment of the phosphogypsum is required. Similarly, the fluorine content must be reduced to meet the company's specification of 0.15% fluorine. Chemie Linz has developed two procedures for the control of these impurities [148]. In the first procedure, wet phosphogypsum is mixed before calcination, with sulfuric acid and hydrated silica. Fluorine is released and enables the phosphate to be washed out of the calcium sulfate with water [158]. The second procedure controls only the fluorine content and adds silica to the reaction system of the phosphoric acid plant [148].

Unfortunately the cement/sulfuric acid process cannot compete economically with other methods of sulfuric acid production. Especially now, when the demand for sulfuric acid is rather low and the price of fuel is high, the comparison is very disadvantageous for the cement/sulfuric acid process. With capital costs currently in the neighborhood of five times that of a sulfur-burning plant of equivalent capacity and an energy input requirement of approximately twice that of the total energy output of the sulfur-burning process, the cement/sulfuric acid process is hardly competitive. High interest rates and energy costs are not offset even if the phosphogypsum can be charged in at zero cost [148].

Other Directions of Phosphogypsum Utilization

Ammonium Sulfate Manufacture. The ammonium sulfate is used as a fertilizer especially in rice -growing areas [119]. It can also be used in the production of other fertilizers formulations such as ammonium nitrate or certain grades of compound fertilizer or it can be used as a source of sulfate for precipitating calcium in nitrophosphate process [145]. The Leuna-Merseburg process was re-designed in 1965 for the application of phosphogypsum by Chemie Linz [119,145]. Phosphogypsum used in this process must be purified. The moist by product is reslurried and filtered using a rotary filter to remove most of the harmful impurities, including free acid, compounds of fluorine and silica, organic matter and fines (Figure 17) [145]. Prepared phosphogypsum is then transferred to the reaction section of the plant, which is the same whatever the original form of the calcium sulfate [145]. Here, by reaction with an excess of ammonium carbonate, ammonium sulfate and calcium carbonate are formed as follows:

$$(NH_4)C0_3 + CaSO_4 \rightarrow CaCO_3 + (NH_4)_2 SO_4 \quad (7)$$

An alternative method of performing the conversion of calcium sulphate to ammonium sulfate and calcium carbonate is exemplified by the process developed in sixties by Continental Engineering of the Netherlands [145,159].

According to Wirsching [119], the demand for ammonium sulfate could adequately be met by large quantities of it arising as a by-product in the chemical industry, so the possibility of using phosphogypsum from Chemie Linz Process becomes non-existent.

Phosphogypsum as Soil Conditioner; Gypsum is used to combat alkalinity and salinity and is particularly effective on soils having a high sodium concentration [145]. Specifically, the physical properties especially permeability of these soils are improved by using gypsum. Cationic species such as sodium in the soil tend to become adsorbed in the clay and the build-up of sodium, in particular, can cause an increase in alkalinity, stickiness and can also cause crust formation which results in the soil becoming impermeable. By adding gypsum, the sodium ions are displaced by calcium and the salts formed may be leached from the soil by irrigation. The crust is broken, alkalinity is reduced and the permeability of the soil restored [145]. Phosphogypsum can be used as a soil conditioner, and

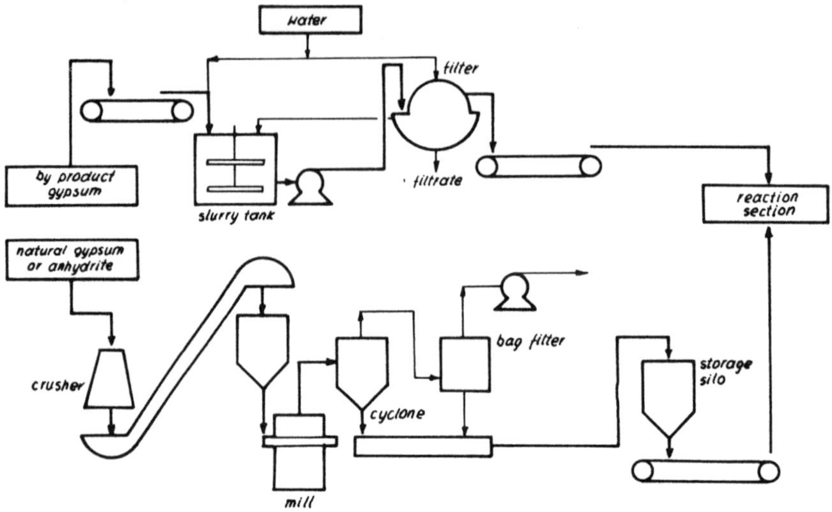

Figure 17. Preparation of gypsum for ammonium sulphate production [145].

furthermore, unlike other uses in which it replaces natural gypsum, no purification is necessary and it can be used directly [145].

Rare Earth Recovery from Phosphogypsum. Production of phosphoric acid by the dihydrate process from Kola apatite gives phosphogypsum, which can be a source of lanthanides and contains about 0.5-0.6 Ln_2O_3 [132]. Kijkowska et al [160] described a method which produces a solid containing 20-25% Ln_2O_3. About 50% of the lanthanide content of the phosphogypsum can be recovered in a concentrated form. In the first stage, phosphogypsum is leached using sulfuric acid. The optimum conditions for leaching phosphogypsum dihydrate are as follows [160]:

Temperature: 50°C
Sulfuric acid conc.: 10-15%
Mass ratio, solution/solid: 2.5-3
Time of leaching: 45-60 min
Recovery efficiency: 50%

The leachate slurry is stirred for 60-120 minutes at 60°C, after which, the solid is filtered off and washed with water. The filtrate contains the rare earths and is subjected to the second stage, in which it is evaporated at 100°C where the lanthanide-rich solid calcium sulphate hemihydrate crystallizes. The solid is filtered off and washed with water. The concentrated filtrate from the second stage contains P_2O_5 from the phosphogypsum.

The main component of the lanthanide-rich solid was calcium sulfate hemihydrate, however, when the solid was sampled at a lower concentration of H_2SO_4 (<30%), small amounts of dihydrate were found, and at higher concentrations (> 45% $H_2SO_{4)}$, anhydride appeared sporadically. Although the concentration of rare earths reached 25% Ln_2O_3 on occasion, no lanthanide salts apart from calcium sulfates were detected by x-ray diffraction [160].

The process of leaching lanthanides with sulfuric acid purifies the phosphogypsum to such an extent that 70% of the P_2O_5 and 60-70% of the F content are removed and it is suitable for further application in the construction industry [132,160].

Two different purification processes approaches have been tested. The processes are illustrated in the flowchart shown in Figure 18 [132]. In the approach shown in Figure 18a, phosphogypsum was purified by leaching out the rare earths with sulfuric acid and was then dried. The product was beta-hemihydrate and contained 0.6 of P_2O_5 and 0.1% of F. In the approach shown in Figure 18b, the filter cake of purified gypsum consisting primarily of di-hydrate was immediately re-slurried with sulfuric acid and anhydride seed crystals were used to promote the transformation from dihydrate to anhydride. The temperature was 50°C and the sulfuric acid concentration from 40 to 50%. The solid/solution ratio was 1 and the time for re-crystallization was 2 hours. The slurry was filtered and washed with water. The filter-cake was then dried at 450°C and after drying, the anhydride contained 0.04 of F and a very small amount of phosphate (<0.10%) [132].

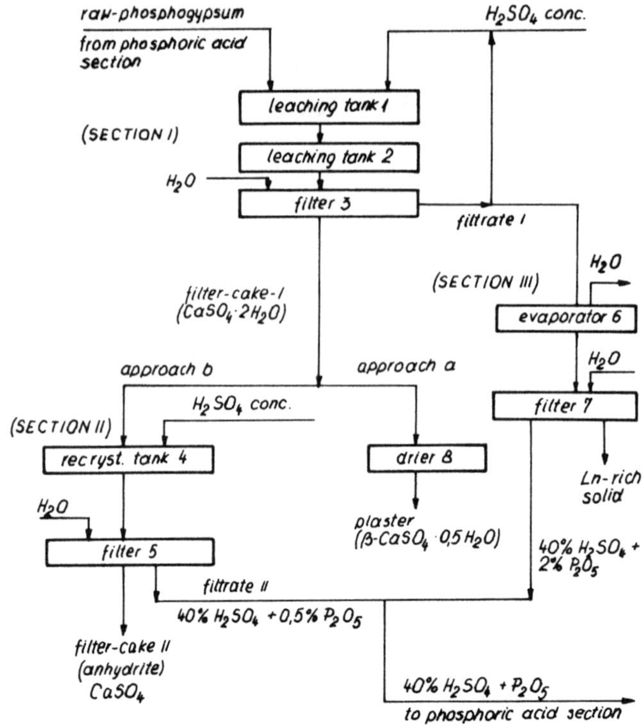

Figure 18. Purification of Kola phosphogypsum [132].

In a recent paper, Jarosinski [161] describes the properties of anhydride II obtained in this process. The anhydride II was washed to pH 7 and dried at 180°C. Re-crystallization of dihydrate into anhydride II, which occurs in the liquid phase, makes possible rare earth recovery and the elimination of admixtures adsorbed on the crystal surface and those substituted in the crystal lattice of dihydrate. The anhydride contains little phosphate and about 0.01 insoluble in water F [161]. The anhydride was ground to 400m^2 /kg and K_2SO_4 was added as an accelerator for hydration [161]. The possible utilization of purified phosphogypsum combined with recovery of lanthanides and phosphates could lower the costs of lanthanide production to such an extent that the process may be viable, particularly if environmental costs are taken into account [132].

SUMMARY AND CONCLUSIONS

As can be seen from this review, there are many potential uses for phosphogypsum that have been investigated and tested on an industrial scale over the past the last few years. Unfortunately, because of economic, high energy consumption, and radioactivity problems the utilization of phosphogypsum is limited. Wirching [119] claims that currently only about 4.5-5 millions tons of phosphogypsum are used annually. In the building materials industry 3.6Mt /year are used, 2.1 Mt/year in the gypsum industry and 1.45 Mt/year in the cement industry [119]. The principal consumers are Japan, South Korea, Germany and Benelux Countries. Other directions, especially the chemical industry (cement/ sulfuric acid, ammonium sulfate) as well as, soil improvement and road construction consume together about 1Mt/year [119].

Under these circumstances, the majority of phosphogypsum must be subjected to dispoal options. There are two disposal methods: wet and dry stacking. The Oresund site near Landskrona in Sweden is an example of wet stacking [119]. In the course of 10 years from 1980 to 1990, approximately 2.3Mt of phosphogypsum have been accumulated on an artificial island [119]. For dry stacking, normally gravel pits or lignite coal mines serve as disposal areas, or there are specially designated areas above ground [119]. The last scenario frequently results in severe pollution of surrounding areas. Moist phosphogypsum, in form of filter cake, is also taken out to sea in coastal barges for discharge [119].

It appears at present that there are no positive prospects for increasing phosphogypsum utilization. Also, the quantity of very pure gypsum from gas desulphurisation (flue-gas gypsum), is growing rapidly. Several plants have replaced phosphogypsum with flue-gas gypsum.

REFERENCES

1. McMichael, B: *Industrial Minerals,* 12, 19-37 (1989).
2. Benbow, J: *Industrial Minerals*,9, 67-83 (1988).
3. Jarvis, D.A. : *Industrial Minerals,*12, 38-42 (1991).
4. Li, L.Y: Ph. D. Thesis, Quen's University Kingston (Can) (1993).
5. Bardossy, G. et al: Lateritic Bauxite, Elsevier Scientific Publishing Co. Amsterdam, Oxford , New York (1990).
6. Senbanyak, T.: *German Patent*, DE 2159584 derwent 7239613 (1972).
7. Murat, M. : *Valorisation des Déchets et des sous Produits Industriels*, Publ. by Masson (France) (1981).
8. Fursman, O.C. et al: *U.S. Dept. of the Interior, Bureau of Mines Report of Investigation,* R.I. 7454, 1-32 (1970).
9. Visnyovsky, L.: *Vasipari Kutatoint Evkonie,* 1, 24-35 (1963).
10. Hemman, A.O.: *German Patent*, 1 143 529 (1963).
11. Miller, Ya. et al: *Tsvetn. Metal,* 36, 2, 45-49 (1963).
12. Horvath, H. et al: *German Patent,* 2218464 (1962)
13. Horvath, Gy: *Acta Technica Academia Scientiarum Hungaricae,* 79 (3-4), 413-449 (1974).
14. Horvath, Gy et al: *Aluminiumpari Troszt.* (Budapest) (1975).
15. Dobos, G. et al: *Banyasz Kohasz Lapok Kohasz*, 105, 9, 417-426 (1972).
16. Mayers, J.G. et al: *US Patent,* 3580702 (1971), Chem. Abst. 75,52, 527 (1971).
17. Thakur, R.S. et al: *J. Sci. Ind. Res.*, 33, 408-16 (1974).
18. Thakur, R.S. et al: *J. Sci. Ind. Res.* 42, 87-108 (1983).
19. Yanagida, H. et al: *Kagaku kogaku*, 41, 369-70, Chem. Abst. 087-122381 (1977).

20. Battacharjee, S.C. et al: Bauxite Tailings,in *Proc. of aqn Int. Conf. Kingston, Jamaica,* Publ. by the Jamaican Bauxite Institute , Kingston (1987) .
21. Building Research Institute: *BRI Tech. Notes*, N°4 (1985) .
22. Gerhard, H.: *Neue Hutte*, 33, (11) 424-429, (1988).
23. Knight, J.C. et al: *Journal of Material Science*, 21, 2179-2184 (1986).
24. Logomerac, V.G.: Bauxite Tailings, in *Proc. of aqn Int. Conf. Kingston, Jamaica*, Publ. by the Jamaican Bauxite Institute, Kingston (1977).
25. Puskas, F.: Bauxite Tailings, in *Proc. of aqn Int. Conf. Kingston, Jamaica*, Publ. by the Jamaican Bauxite Institute127-130 (1986).
26. UNEP "Environmental Aspects of Alumina Production" - A Technical Review (1985) see ref 4 .
27. UNIDO: *Report of Project* N° SI/JAM/81/802 (1983).
28. Wagh, A.S.: *Final Report to IDRC* II 121-143 (1986).
29. Mc Leod, D. et al: Bauxite Tailings, in *Proc. of aqn Int. Conf. Kingston, Jamaica* Publ. by the Jamaican Bauxite Institute, 131-136 (1987).
30. Riedel, R.: *German Patent*, DE 2739493, Chem. Abst. 090-208944 (1979).
31. Amat di San Filipo, P. et al: *Ziegelindustrie International*, 2/3, 67-74, 4 , 133-139 (1988).
32. Amat di San Filipo, P.: *Italian Patents* 1154799 (1987), 1154800 (1987).
33. Satalkin, A.V. et al: *Chem. Abst.*, 59, 6121 (1963).
34. Wargalla, G. : *Forshungsber. Bundesminist Forsh. technol.* (BFTEAJ, 03407608) Vereinigte Aluminum werke A-G Luenen, Chem. Abst. 096-109627 (1980).
35. Wilkening, S.: *German Patent*, DE 2210007, Chem. Abst. 080-051836 (1973).
36. Dass, A. et al: *Materials and Structures*, vol 23, 136, 252-255 (1990).

37. Dass, A. et al : *J. Nat. Building Orgn.* (India), 24,29-33 (1979).
38. Rybakova, I.S. et al: *Stroit Mater Konstr.* 24, 2 (1987).
39. Foerster, H.: *German Patent*, DD57543, Chem. Abst. 070-11738 (1967).
40. Vaquier, A.: *Annales de l'institut Technique du Batiment et des Travaux Publics*, 391,57, 98-103 (1981).
41. Sugimoto, S. et al: *Japanese Patent* 8313 (1961), Chem. Abst. 59, 11020D (1963).
42. Horiguchi, Y. et al: *Rika Gaku Kenyusho Hokoku*, 37, 101-105 (1961).
43. Brodko, A.S. et al: *USSR Patent,* 417 386 (1974) .
44. Ferrari, F.: *Il Cemento,* 65, 11, 221-233 (1968).
45. Grzymek, J. et al: *Cement Wapno Gips,* 26/38, 164-171 (1971).
46. Zambo, J.: *Travaux* 15, 265 (1979).
47. Sasvari, J. et al: *Materials Science Forum,* vols 166/69, 695-98 (1994).
48. Momtazi, S. A.: Ph. D. Thesis; *Inst. Nat. Si. App.*, Lyon (1991).
49. Pinnock, W.R.: *Health Physics*, vol 61, 5, 647-651 (1991).
50. Wang, K.: *Huajing Kexue* , 13, 90-3, Chem. Abst. 118-197023 (1992).
51. Shmorgunekoet, N.S. et al., *Epitoanyag*, 31 (11) 436 (1979).
52. Ampritphale, S.S. et al., *Silicates Industriels,* 3/4, 31-35 (1987).
53. Skelety, I., *Proceedings of the 6th Conference on Silicate Industry*, Budapest, 399-408 (1961), Herausgegeben von Becke and F. Tamas Budapest (1963).
54. Mathur, J.P. et al., *Ceramic Bulletin* vol 55 , 3, 301 (1976).
55. Murthy, J.S. et al., *Chem. Eng. World,* 28, 10, 59-60 (1993).
56. Kacker, K.P. et al., *Indian Ceram.* 19, 11, 712-723 (1977).
57. *Chinese Patent*, CN 1068836 A (1993), Chem. Abst. 120-108718.

58. Nakanishi, K., *Japanese Patent*,74- 45892 (1974).
59. Mori, S. et al., *Japanese Patent*, 74- 69759 (1974).
60. Horioka, K., *Japanese Patent,* 74-99719 (1974).
61. Tonak, T. et al., *Cimento Bul.* 31, 1-6 (1994), Chem. Abst. 121-236986.
62. Akhmetov, I.S. et al., *Isv. Vyssh. Uchebn. Zaved Stroit.* 3, 66-9, Chem. Abst. 118-130630 (1992).
63. Mullik, A.K., *Proceeding of the First Symposium on Cement Industry Solutions Waste Management* (Calgary) 489-512 (1992).
64. Wargalla, G., *Erzmetall*, vol 26, 1,18-20 (1973).
65. *Chinese Patent,* CN 1070388 (1993), Chem. Abst. 119:25521.
66. Mauvelian, M.G. et al., *Doklady Akad. Nauk Armyan S.S.R*, 26, 33-36 (1958).
67. Fulda, W. et al., Tonerd und Aluminum, *Ergebnisse und Erfahrungen aus der Betriebs Praxis 1920-1952*, band 1Tonerde S. 123 De Gruyter W. Co Berlin (1953).
68. Szekelety, I., *Epitoanyag,* 12, 28-37 (1960).
69. Amat di San Filipo, P., *Zem. Kalk Gips*, 11, 584-586 (1989).
70. Kunbargi, H., *International Patent,* WO 86/05773 (1986).
71. Tormari, M. et al., *US Patent*, 4652310 (1988), derwent 87-101081.
72. Yoshizawa, *German Patent*, 2308851 (1974) derwent 74-34915.
73. Ogura, H. et al., *Japanese Patent,* 7812527 (1978),Chem. Abst. 089-079289.
74. Ikeda, K.O., *Symposium on Cement Chemistry*, Rio , 6 , 389-394 (1986).
75. Siberian Automative Transportation institute: *Russian Patent,* 1691337 (1991), Chem. Abst. 116-240783.
76. Sugimoto, S. et al., *Japanese Patent*, 61-8315 (1961).
77. Pera, J. et al., *Am. Concr. Inst.*, SP 132 , 749-761 (1993).
78. Pera, J. , *Mat. Res. Soc. Symp. Proc.* vol. 245 , MRS , 55-61 (1992).

79. Kawai, M., *Japanese Patent* 93139727 (1993), Chem. Abst. 119-145310.
80. Yokota, A. et al., *Japanese Patent,* 79130628 (1979), Chem. Abst. 092-095698.
81. Kneidinger, W., *Austria Patent,* AT 381299 (1989), Chem. Abst. 105-213393.
82. Autran, L.M.A., *French Patent* 2688796 (1993) derwent 93-370947.
83. Buraev, M.I. et al., *Komplsn. Ispol'z Miner.,* Syr'ya 10, 66-9, (1983) Chem. Abst. 100-038710
84. Buraev, M.I., *Beton Zhelezobeton* (Moscow) 2, 41-2 (1986) Chem. Abst. 105-010974.
85. Buraev, M.I., *Beton Zhelezobeton (Moscow)* 7, 44-5 (1986) Chem. Abst. 105-157864.
86. Shpynova, L.G et al., *Izv. Vyssh. Uchebn. Zaved. Stroit.* Arkhit. 5, 68-72 (1986), Chem. Abst. 095-137289.
87. Tyan, V.A. et al., *Izv. Vyssh. Uchebn. Zaved Stroit.* Arkit. 7, 56-60 (1990), Chem. Abst. 115-014146.
88. Kanazawa, H. et al., *Japanese Patent* 89261253 (1989) Chem. Abst., 112-239695.
89. Baryla, J.M., *French Patent* 2512440 (1983) derwent 83-35106.
90. Kohno, K. et al., *Semento Gijutsu Nenpo,* 42, 136-9 (1988), Chem. Abst. 111-083124.
91. Klein, M. et al., *German Patent*, 3633413 (1988), Chem. Abst. 108-191683.
92. Popov, V.A. et al., *Russian Patent,* 691337 (1991) derwent 92-282967.
93. Serban, L. et al., *Mater. Constr.* 21, 2/3, 81-4 (1991).
94. Kane, J., *US Patent,* 4146573 (1979) derwent 79-29434.
95. IU conversion system: *US Patent*, 4028130 (1976) derwent 77-43201.
96. Scripture, E.W. et al., *US Patent,* 2099176 (1937).
97. Rivera, F. G. et al., *Spanish Patent* ES 2040640 (1993), Chem. Abst. 120-115129.

98. Unemura, K. et al., *Japanese Patent*, 75-91626 (1975).
99. Blank, H.R., *Journal of Testing and Evaluation JTEVA* , 4, 5,355-358 (1976).
100. Sato, N. et al., *Japanese Patent*, 7607024 (1976) Chem. Abst., 084-139929.
101. Bakulina, E.I. et al., *Beton Zhelezobeton (Moscow),* 1, 36-7 (1989), Chem. Abst.110-218040.
102. Ishihara, M.I. et al., *Japanese Patent,* 51007023 (1976) Chem. Abst. 084-139928.
103. Ishihara, M.I. et al., *Japanese Patent,* 7607022 (1976) Chem. Abst. 084-139927.
104. Seto, H. et al., *Japanese Patent,* 8614159 (1986) Chem. Abst. 105-047963.
105. Nippon Light Metal, *Japanese Patent,* 54048813 (1979) Chem. Abst. 091-162275.
106. Ishibashi, Y. et al., *Japanese Patent,* 53054230 (1978), Chem. Abst. 089-116802.
107. Nagasaki, M. et al: *Japanese Patent,* 734819 (1973) Chem. Abst. 080-051814.
108. Albright. M. et al., Wilson LTD : *Belgium Patent,* BE 759522 BE 759522 (1971) derwent 71-38263.
109. Hayashi, S. et al., *Japanese Patent,* 7452213 (1974), Chem. Abst., 081-095724.
110. Matsubara, Y., *Japanese Patent,* 7903831 (1979), Chem. Abst. 090-141463.
111. Nakamura, H.H. et al., Final Report from Illinois Institute of Technology, Research Institute on Contract 14-09-0070-392, *U.S. Bureau of Mines*, Chicago III, 23 (1968).
112. Ishihara, M. et al., *Japanese Patent,* 4906324 (1974), Chem. Abst., 081-175240.
113. Bayer, G. : *Erzmetall,* vol 25 N°9 154-157 (1972).
114. Zueblin, E. AG., *Belgium Patent*, BE 898589 (1984) derwent 84-120901.
115. Hashahi, S. et al., *Japanese Patent*, 74-52213 (1974).
116. Casanova, J., *French Patent,* 70 31781 (1972).

117. Grzymek, J., *Cement Wapno Gips,* 8, 229 (1967), *Polish Patent* N° 43443, 43444 (1959) 44998 (1960).
118. Robinsky, E.I., *CIM Bull.* 68, 47 (1975).
119. Wirsching, F., *Müll- und Abfallbeseitigung*, 67,8581,1-15 (1982).
120. *The 8th Congress of Eurogypsum,* Lucern (1970).
121. *Proc. 1st Int. Symp. on Phosphogypsum*, Lake Buena Vista, Florida Inst. Phosphate Res., Florida, 5-7 Nov (1980).
122. *Proc. 2nd Int. Symp. on Phosphogypsum*, Florida Inst. Phosphate Res., Bartow, Florida, 2 (1988).
123. *Proc. 3rd Int. Symp. on Phosphogypsum,* Florida Inst. Phosphate Res., Ed. by W.Chang, Orlando, Florida, December (1990).
124. *Second Workshop on By-Products of Phosphate Industries* (1984).
125. *Third Workshop on By-Products of Phosphate Industries* (1985).
126. Baron, J. et al., *Zem.-Kalk-Gips*, 30, (8) (1977).
127. Boullenois, G., *Private Communication* (1995).
128. Dahlgren, S.E., *J. Agr. Food Chem.* 8, 411(1960).
129. Lutz, R., *Zem.-Kalk-Gips*, 47, 690 (1974).
130. Osiecka, E., *Phosphogypsum, Plaster and Building Elements,* (in Polish) Arkady, Warsaw (1980).
131. Slack, *Phosphoric Acid,* Part 2, 652-3 (1968).
132. Jarosinski, A. et al., *Phosphorus & Potassium*, 164, 24 (1989).
133. Hurst, F.J. et al., *Hydrometallurgy,* 9, 69 (1982).
134. Efimov, V.T. et al., *Stroit. Mater. Konstr.,* 20, 1, 31 (1978).
135. *French Standard,* NF P 98-124 (1992).
136. Nieminen, P., *Conference Int. By Products and Wastes in Civil Engineering*, 229, Paris , see ref 127
137. Janiczek, S. et al., *Przegl. Gorn.* 36,10, 503 (1980).
138. Mehta, P.K. et al., *Cem. Conc. Res.*, 7, 537 (1987).
139. Valentini, G. et al., *Thermochemica Acta*, 113, 269 (1987).
140. Abdul-Maula, S. et al., *Mat. Res.Soc. Symp. Proc.*, 245, 315 (1992).

141. Wirsching, F., *TIZ*, 105, 6, 383 (1981).
142. Wirsching, F., *Rev. Mat. Const.* 671-2, 234 (1970).
143. Getting Rid of Phosphogypsum III, *Phosphorus & Potassium,* 94, 24 (1978).
144. Haerter, M., *Tonind. Ztg,* 95, 1, 9 (1971).
145. Getting Rid of Phosphogypsum: *IV Phosphorus & Potassium*, 96, 30, July/ August (1978).
146. Allen, M., *IV Phosphorus & Potassium*, 78, 42 (1975).
147. Tabikh, A.A. et al., *Cem. Conc. Res.,* 1, 663 (1971).
148. Getting Rid of Phosphogypsum: *II Phosphorus & Potassium,* 89, 36, May-June (1977).
149. Nurse, R.W., *J. appl. Chem.*, 2, 708 (1952).
150. Nurse, R.W., *Proc. of the 7th Conf. on the Silicate Industry,* Budapest, 279, Akademiai Kiada (1965).
151. Gutt, W., *5th Int. Symp. on Cem. Chem.,* Tokyo , 1, 93 (1968).
152. Kurdowski, W. et al., *Cem. Wapno Gips (in Polish),* 24, 12, 349 (1969).
153. Welch, J.W. et al., *4th Int. Symp. on Cem. Chem.,* Washington (1960).
154. Ierchov, L.D., *Cement (in Russian)*, 4, 19 (1955).
155. Gutt, W. et al., *Cement Technology,* 2, 2, 41, and 3, 91 (1971).
156. Gutt, W. et al., *Cement Technology*, 2, 1, 9, and 3, 91 (1971).
157. Belting, K. W., *Zem.-Kalk-Gips*, 8, 387 (1972).
158. *Austrian Patent Specifications* N° 273042 (1969), Chem. Abst. 071-083079
159. *Sulphur,* 85, 29, Nov/Dec (1969) see ref 145.
160. Kijkowska, R. et al., *Phosphorus & Potassium*, 163, 22, 9/10 (1989).
161. Jarosinski, A., *Cem. Conc. Res.*, 2, 99 (1994).

7

USE OF LIGNIN-BASED PRODUCTS IN CONCRETE

Ajoy Kumar Mullick

INTRODUCTION

Lignin is one of the principal constituents of wood. The main sources of lignins are the wood-based industries such as paper and pulp. Because of the inherent structure and some intrinsic properties, lignin-based products find applications as binders, surface-active agents and dispersants etc. The principal use of lignin-based products in concrete manufacture is as chemical admixtures.

Chemical admixtures often form a vital "fourth" ingredient of concrete, in addition to cement, water and aggregates. These admixtures are used to modify the properties of concrete (or mortar or grouts) so that these become more suitable for the intended applications. The properties in question cover the concrete in the fresh as well as hardened states. These include controlling setting times, improving workability, pumpability and prevention of segregation etc. in fresh concrete. Air-entrainment, strength, permeability and enhancing durability are some of the properties of concrete in the hardened state that can be modified by the use of chemical admixtures. Chemical admixtures usually have certain organic compounds or mixes of organic and inorganic compounds.

The types of such chemical admixtures and characteristics of performance are defined in various national Specifications and Committees' reports [1-3].

One important class of such admixtures are the water-reducing, set-controlling admixtures, which are used to reduce the water requirement of concrete mixes for a given workability or to modify the setting properties, or both [3]. Some air-entrainment also take place. Of late, high-range water-reducing admixtures referred to as 'superplasticizers' have become available, which render much greater workability to concrete [1,2]. Historically, lignosulphonate salts have formed the basis of water-reducing admixtures for the last fifty years [4]. The other major applications of lignin-based concrete admixtures include as pumpability aids, integral water-proofing compounds and superplasticizers [5], either on its own or in combination with other chemical compounds.

Scope of the Report

The origin and the nature of lignins; processing required for development of concrete admixtures; and performance of lignosulphonate salts as chemical admixtures in concrete; their effects on hydration, microstructure and resultant properties of concrete including durability are discussed in this chapter.

THE NATURE AND SOURCE OF LIGNINS

Much of the description of lignins and lignosulphonates given in this and the next Section are based on the information contained in Reference 6.

Next to cellulose, lignin is the most abundant constituent of the woody structure of higher plants. It constitutes 24 to 28 percent of the mass of dry wood. As a result, lignin is the second most abundant organic substance available in nature. Lignin acts as a cementing medium to bind the matrix of cellulose fibers together in a rigid woody structure.

Being closely associated with cellulose, most of the lignin is bound to plant polysaccharides. It is rather difficult to separate lignin from other components of wood, without altering it's form to some extent. The exact chemical nature of lignin as it occurs in wood is not known. The nature of isolated lignins are relatively more well established. In any case, it is now widely accepted that lignin in wood is a polymeric substance resulting from random combination of products of polymerization of coniferyl alcohol and sinapyl alcohol, depending upon the type of the tree [6]. Nuclear magnetic resonance (NMR) analysis of lignin in wood indicates that it is a polymer with an aromatic nature. It does not indicate any observable crystalline form under the electron microscope. X-ray diffraction (XRD) studies indicate that lignins are amorphous polymers. Experimentally isolated or commercially available lignins are friable solids, usually powders, having densities 1.3 to 1.4 g/cc and a refractive index of 1.6. The color is usually brown.

The low specific viscosity of lignin solutions suggests that lignin is a branched-chain polymer with a lower degree of polymerization than cellulose. It is believed that there are many lignins with varying properties and composition, depending on the source and method of isolation [6]. Functional groupings such as hydroxyl, methoxyl and carbonyl have been identified in the polymer. A typical elemental analysis for wood lignin is shown in Table 1 [6].

Table 1. Average Elemental Analysis of Wood Lignin [6].

Species of Lignin	Amount, %			
	C	H	O	OCH_3
Coniferous	63.8	6.3	29.9	15.8
Deciduous	59.8	6.4	33.7	21.4

The main units which are contained in wood lignin are shown in Figure 1 [6]. Although wood lignin has a definite structure, it cannot be represented by a single formula. Possible substructures and linkages occurring in lignin are described in Reference 6.

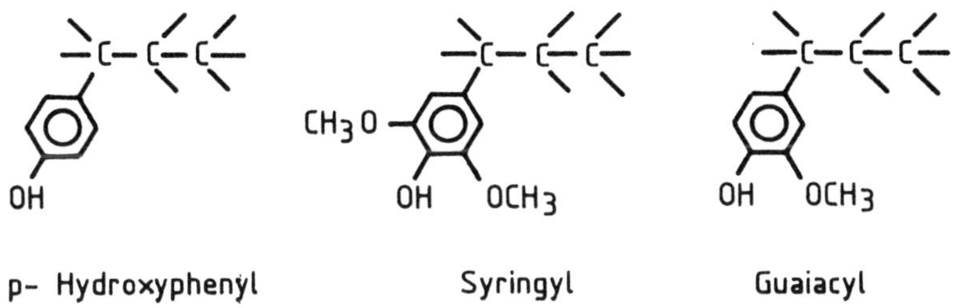

Figure 1. Units in wood lignin [6].

As the above description implies, the most abundant source of lignin are the trees. The wood wastes and bark wastes from the lumber industry and wastes from agricultural operations can provide extremely large amounts of lignin, not even a fraction of which is currently utilized. Paper and pulp industry provide a more accessible source of lignin; spent liquors from the wood-pulp industry being the principal commercial source. Lignin in spent liquors is usually called 'waste' lignin. This is no longer an accurate description given the possible applications for lignin-derived products and its potential value as a fuel. Lignin is sometimes used to denote the entire non-cellulosic portion of wood and much of the application of lignin-based products depend upon separating the carbohydrates, tannins, rosins and others from the non-cellulosic portion of wood, so that purified and, at the same time, modified lignins are obtained. While unmodified lignins can be used as binders, fillers and resin extenders, chemical modification of isolated lignin is required to improve its performance when used in adhesives and structural polymers, coatings and dispersants. For the present report, processing and modification of lignins as required for applications as dispersing and surface-active agents are of particular interest.

LIGNOSULPHONATES

Spent pulping liquors from wood pulping processes, which give commercial lignins as byproducts, contain lignin sulphonates as the principal component. The commonly used term for these, and also for the total solids in the spent liquor, is lignosulphonate [6]. In the sulphite-pulping process of the paper industry, wooden logs as well as chips and sawdust from lumber mills are used. The pulp is made by cooking in a pulp digester the wooden chips, sometimes mixed with sawdust, under pressure in a solution of sulphurous acid and either calcium, magnesium, sodium or ammonium bisulphite. During the cooking, the lignin and other encrusting substances of the wood are dissolved, leaving the cellulose fibers still containing lignin. The composition of a typical pulping liquor is given in Table 2 [6].

Table 2. Composition of Pulping Liquor [6].

Constituent	Percent
Lignosulphonate	55
Hexose	14
Pentose Sugars	6
Sugar acids and residues	12
Resins and extractives	3
Ash	10
TOTAL	100

Use of lignosulphate as adhesive binders was well known even before World War II. They were also used as soil binders and linoleum adhesive. The sulphite spent liquor is applied as binder and dust cover on roads, playground, and parking areas. Lignosulphonates are used as adhesive for plywood and particle board. Lignins find use in ceramics, carbon black, dyes, pigments and many other products. Lignosulphonates are also used as dispersants and foaming agents in the manufacture of gypsum board. They are used as conditioning agents in oil-well drilling mud, where they control viscosity and gel characteristics and cause a strong fluid-impervious cake along the sides of the well bore. This area of application has received relatively greater attention for development [4]. Because of the surface activity property, lignosulphonates are used as emulsifiers to stabilize emulsions in the manufacture of asphalt emulsions and oil-based drilling mud. Other uses dependent on the surface activity include, ore floatation, electrolytic refining and industrial water treatment [6]. Such applications will not be discussed in any detail in this chapter.

The major application of lignosulphonate-based products in concrete are those which exploit their dispersing and surface-active properties. Such admixtures are usually based on either the calcium or sodium salts of lignosulphonic acid, derived by pulping wood, using lime and sulphurous acid mixtures. Water-reducing admixtures are usually described as calcium lignosulphonates and may or may not contain a proportion of sugars [7]. For superplasticizing operations, sodium lignosulphonates have been found to perform better than calcium lignosulphonates [4]. Modern-day concrete admixtures based on lignins are formulated from purified products, from which the bulk of the sugars, low molecular weight fractions and carbohydrate contaminates have been removed [4,8].

Calcium lignosulphonate is prepared by precipitation from solution by the Howard process, by adding calcium hydroxide to the spent liquor in three stages [6]. The first precipitate is calcium sulphite which is recycled to the pulping liquor. The second step separates an insoluble basic calcium lignosulphonate. The third-stage precipitate, which consists of a small additional amount of ligninsulphonate together with excess lime, is added to the lime of the first stage of a succeeding batch.

Lignosulphonates of various bases may be prepared by base exchange or in the pulp digestion itself. For example, adding lime to ammonium lignosulphonate liquor liberates ammonia and gives calcium lignosulphonate. Various methods have also been developed for separating sugar from lignosulphonate and for separating different molecular weight fractions. Fermentation and yeast growth remove much of the sugars, whereas specialized lignosulphonate products are produced by techniques such as ion exchange, ion exclusion, gel adsorption, electrodialysis, ultrafiltration, and reverse osmosis [6].

Typical chemical composition of calcium and sodium lignosulphonates, commercially available in India and used in admixture manufacture, are shown in Tables 3 and 4, respectively [9]. The chemical formula of lignosulphonate is very complex, but can be

Table 3. Specification of Calcium Lignosulphonate [9].

Color	Yellowish Brown
Bulk Density	0.20 to 0.25 kg/lit.
pH (3% Solution)	3 ± 0.3
Calcium	$4 \pm 0.3\%$
Total Sulphur (As S)	5.55%
Sulphate (As S)	0.2%
Sulphite (As S)	0.35%
Sulphonate Sulphur	5.0%
Reducing Sugars	22%
Moisture	5%
Ash	9.5%

Table 4. Specification of De-sugared Sodium Lignosulphonate [9].

Color	Dark Brown
Bulk Density	0.35 - 0.40 kg/lit.
pH (3% Solution)	8 - 9
Sodium	7.5 - 8.5%
Calcium	0.4%
Reducing Sugars	2% - 3.5%
Iron	0.1%
Moisture	5%

said to be a polymeric sulphonate containing sufficient groups in the polymer backbone to allow adsorption on to cement particles without any significant interference with the hydration reactions [8]. Lignosulphonates are considered to be foremost among natural long-chain surface-active agents which can impart a dispersing action on to the cement particles in water; others belonging to this class of surfactants are (poly-) carbohydrates and products of hydrolysis of animal proteins. A specific molecular structure is required for such a dispersing action on cement-water systems and the presence of hydrophilic and hydrophobic groups in a preferred distribution has been considered to be important [10].

Ramachandran has described the basic units of lignosulphonate based superplasticizers [11]. He stated that lignosulphonate molecule consist of substituted phenyl propone units with hydroxyl, methoxy, carbonyl and sulphonic acid groups; molecular weight may vary between a few hundred to 100,000 units. The chemical structure of sodium lignosulphonate superplasticizer is depicted in Figure 2 [11].

Infrared spectroscopy (IR) has been used to characterize lignosulphonate-based admixtures for concrete [12]. Typical infrared spectra shown in Figure 3 showed characteristic bands at 2950 cm^{-1}, a broad band due to sulphonate at 1200 cm^{-1} and a weak band at 870 cm^{-1}. The absorption band at 1200 cm^{-1} characteristic of sulphonate

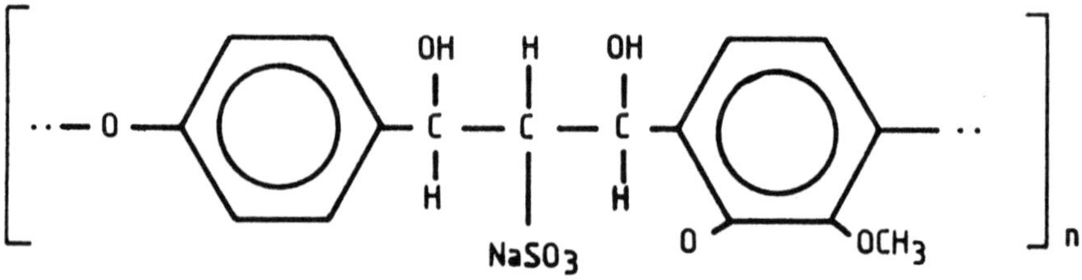

Figure 2. Chemical structure of sodium lignosulphonate superplasticizers [11].

was a common feature in the IR spectra of condensates of sulphonated melamine formaldehyde and sulphonated naphthalene formaldehyde also, besides other bands being present in each case [12].

USE OF LIGNOSULPHONATES IN CONCRETE

The desirable property of concrete in the fresh state is that it should have adequate fluidity without segregation and bleeding, placeability, compactability and finishability appropriate to the conditions of placement. These properties are collectively called 'workability.' Workability of concrete depends upon the mix proportions, aggregate characteristics and environmental conditions.

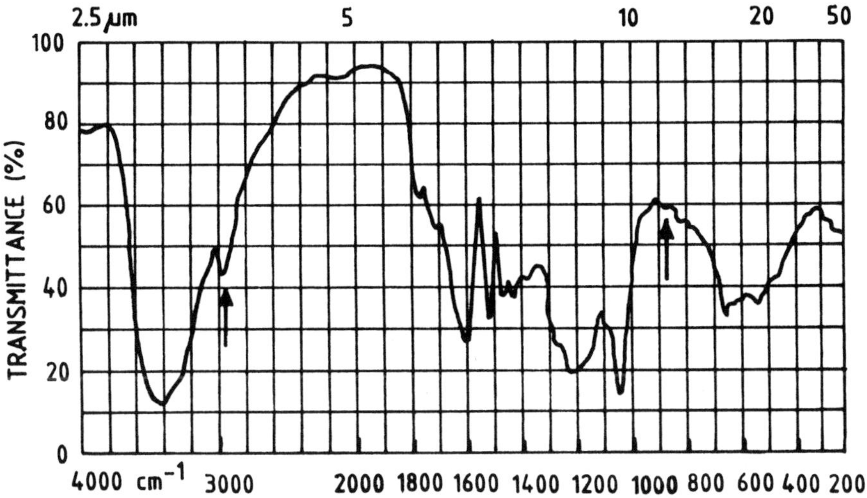

Figure 3. IR spectra of calcium lignosulphonate [12].

With all other parameters remaining the constant, the workability of concrete increases with the water content in the mix. The important properties of concrete in the hardened state i.e., strength, deformation characteristics and durability improve, the lower the water-cement ratio, i.e. the lower the water content for a given cement content in the mix. Thus requirements of water content in concrete in the two states may appear to be contrasting.

The role of chemical admixtures, which can bring about improvement in workability of concrete without increasing the water content, or allow the water-cement ratio to be lowered without impairing the workability, are thus obvious. These admixtures are called 'water-reducing agents' or 'plasticizers;' superplasticizers also belong to the same functional group [3]. The water-reduction is achieved due to surface-active property of the chemicals. It has been noted that many long-chain (or even short-chain) polymers of either natural or synthetic origin can be surfactants, which can improve the workability of concrete, provided they have the specific molecular structure [10].

Modified lignosulphonates are among such natural long-chain surfactants which improve workability of concrete. For the last few decades calcium- and sodium- lignosulphonates have been widely used as the base of water-reducing agents and of late, of superplasticizers [4, 13-15]. Expressing the efficiency in terms of relative increase in slump of mortars due to addition of admixture as related to the slump of mortar without the admixture, the efficiency of lignosulphonates (0.33) have been reported to be comparable to that of sulphonated melamine formaldehyde condensates (0.33) and somewhat lower than that of naphthalene formaldehyde condensates (0.58) [10]. The latter two synthetic surfactants form the base of a majority of superplasticizers produced currently [3,13]. Hereinafter, these types of superplasticizers will be denoted as M-type and N-type, respectively. Lignin also forms the base of integral water-proofing compounds and pumpability aids for concrete, in which advantage is again taken of its surface-activity [5].

Mechanisms of Action

The surface-active property of lignosulphonates gives rise to the following actions in the cement-water system [5,11]:

I) reduction in surface tension of water
ii) adsorption on C_3A* and C_3S* phases of cement
iii) electrical repulsion (large negative zeta-potential)
iv) dispersion
v) deflocculation

**Note :* *Usual cement notation,* $C = CaO$, $S = SiO_2$, $A = Al_2O_3$ $F = Fe_2O_3$, $\bar{S} = SO_3$, $H = H_2O$ *etc. have been used.*

The surface activity of lignosulphonates is due to an unbalanced charge of electricity carried by these. When such a surface-active agent is put into water, the 'active' charged end of the material sticks into the water and the other end is out in the air

(Figure 4) [7]. Plasticizing water-reducers, due to their ability to migrate to the water-air boundary, reduce the surface tension at the boundary. The charged ends of the surface-active agents protruding into the water in the manner shown in Figure 4 results in attraction between these molecules being less than the attraction that occurs between the water molecules, which are dipoles and normally have a strong affinity. The net result is a lowering of surface tension at the air-water interface [5]. Such lowering of surface tension aids better wetting and hydration reaction. In case of some detergent-derived products used as admixtures, this is one mechanism by which the workability of concrete mix was improved [16].

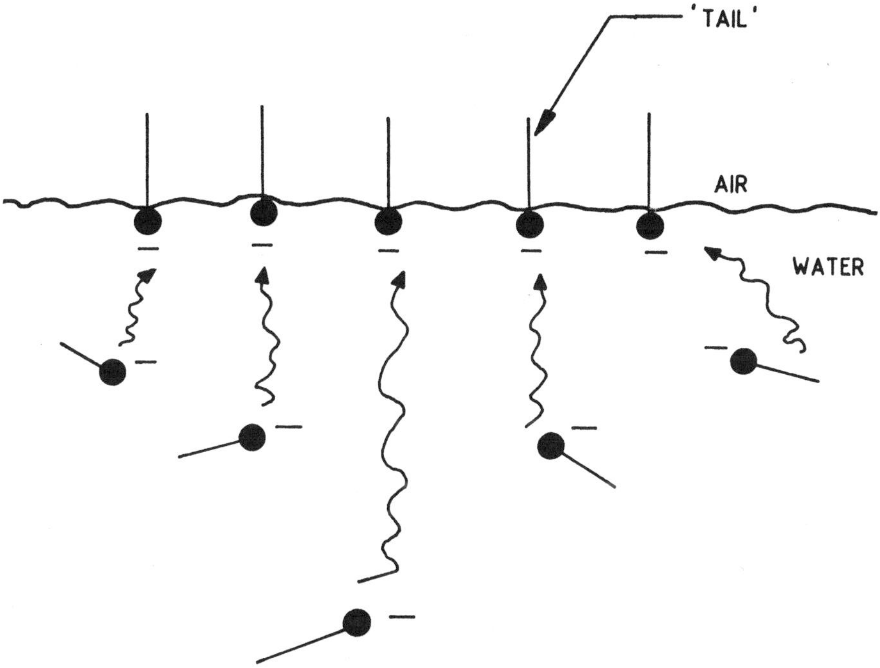

Figure 4. Migration of water-reducing agents to the surface of the water [7].

Reduction in surface tension also results in some air-entrainment in concrete [7].

In cement-water dispersion, the 'tail' end of the surfactant gets adsorbed onto the cement particles. These fine particles of cement in water, which under normal circumstances would tend to floc together because of physical and chemical forces of attraction between them, would now be dispersed. This results in deflocculation, which leads to greater surface area of cement particles being exposed to water. At the same time, the amount of water which may have been trapped inside such a floc of cement particles gets released (Figure 5) [7]. The net effect is a greater mobility of the system, resulting in higher workability.

The dispersing and deflocculating action is shown schematically in Figure 6 [17]. Fresh cement paste, even at reasonably high water content, forms a flocculent structure, the intergranular bonds being derived from Coulomb attraction and van der Waals forces. Addition of a dispersing agent like lignosulphonate salts disperses the particles. As a result of this, for the same number of cement grains per unit volume, intergranular distances increase and the size of the largest pores in the configuration decreases (Figure 6). The effect is also explained in Figure 7, which shows the particle size distribution of cement, cement+water system, as well as when a plasticizing admixture like lignosulphonate is added to the latter [11].

The cement has a higher percentage of fine particles than in cement paste. This indicates that cement in contact with water forms surface products which tend to coalesce or floc. Significantly enough, the particle size distribution in case of cement paste with a plasticizing admixture is finer than in case when no admixture is added (Figure 7); showing the dispersing or deflocculating action.

Lignosulphonate admixtures, when added to cement-water system, are immediately adsorbed on to the C_3A phase of cement and substantial amounts are adsorbed within a few seconds. Adsorption on C_3S phase starts after 4 and 5 hours and continues thereafter [11]. A film is formed around the cement grains which prevents or reduces the hydration reactions with water. This is one explanation of set

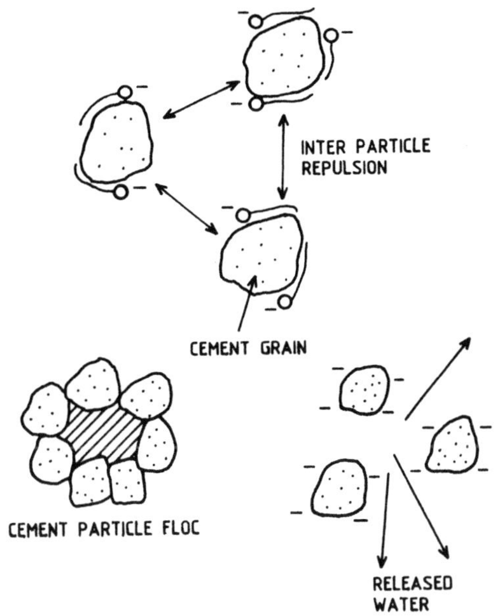

Figure 5. Effect of surface-active agents on cement particle flocs [7].

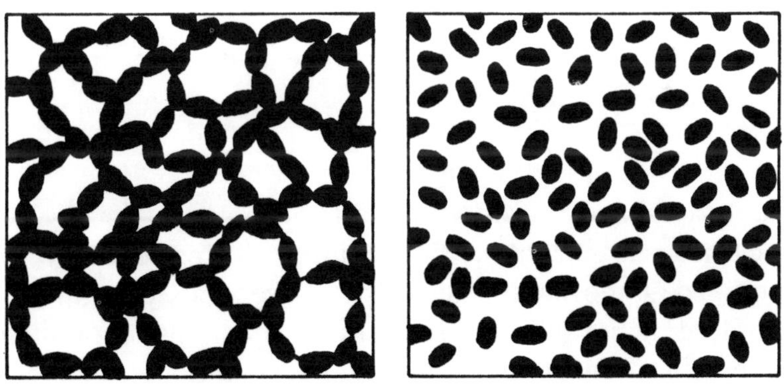

Figure 6. Schematic model of cement paste in flocculated {left} and dispersed {right} states [17].

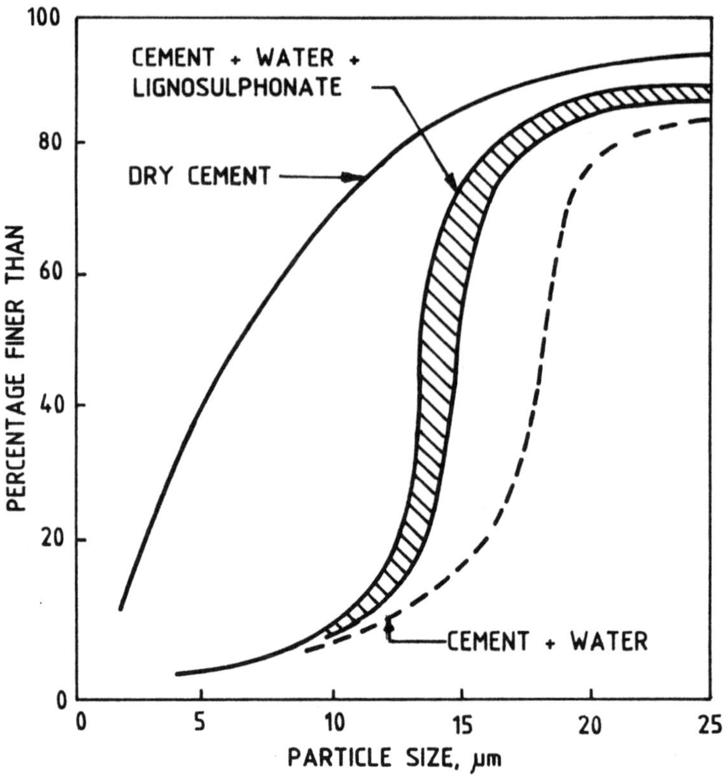

Figure 7. Effect of lignosulphonate admixture on particle size distribution in cement [11].

retardation that is caused by addition of some plasticizers, as will be discussed subsequently. In addition, the presence of sugars in unrefined lignosulphonates also cause set retardation. Such set-retarding capabilities are useful in situations involving delay in placement of concrete due to long hauls, in hot weather conditions and where 'cold joints' in the structural units have to be avoided [3]. Set-retarding water-reducing admixtures help in maintaining the increased workability longer i.e., in alleviating what is called 'slump loss' [18].

Adsorption on Cement Particles and Electrical Repulsion

The dispersive action of lignosulphonates and deflocculation of cement particles is a physico-chemical effect. The stability of colloidal particles in suspension in water is usually due to development of charge caused by adsorption of ions [11]. As explained in Figure 5, due to adsorption of lignosulphonate ions on them, the cement particles develop similar charges and repel each other. On the basis of colloidal chemical principles applied to cement-water-admixture system, Ramachandran [11] has explained that the potential gradient is high within a distance of one ion of the dispersed phase of a colloid. The potential gradient becomes lower, further into the dispersing medium. The first layer of adsorbed ions is tightly bound. The difference in potential between the outer fixed layer of the adsorbate and the bulk of the dispersing medium constitute the electro-kinetic potential or zeta-potential [11].

Techniques of measurement of adsorption of admixture and change in zeta-potential of cement-water system have been described [19,20]. The amount of admixture adsorbed by cement is calculated by the differences in concentration of the admixture in the aqueous phase before and after adsorption, by UV absorption spectroscopy [11,19]. Cement, water and the admixture are mixed for about five minutes. The liquid phase is separated by vacuum filtration and is diluted to the optimum concentration for the measurements. Zeta-potential measurements are based on electrophoretic technique. The zeta-potential of colloidal particles is measured by determining the rate at which these particles move in a known electric field. The liquid phase of cement-water-admixture is separated by vacuum filtration as in adsorption tests. Small portions of the solid are then added to the filtered liquid phase. A suspension is obtained having the same ionic strength as the original paste and diluted enough so that it can be observed microscopically [19].

The stability of a colloidal system like cement-water-lignosulphonate admixture depends upon the zeta-potential; the suspension is more stable, the higher the zeta-potential. The dispersing action of such admixtures is explained in terms of lowering of zeta-potential, that results upon adsorption of the admixture in the cement paste phase [11,19,20]. Cement in water has a negative zeta-potential which enables it to adsorb both anionic and cationic surfactants [21]. The zeta-potential of the cement-water system without any additives is about -8 millivolt (mV) [19]. Incorporation of surfactants or superplasticizers has been found to decrease the zeta potential to a value on the order of -25 mV to -30 mV [19,21]; values on the order of -40 mV or lower have also been reported [22].

Exhaustive investigations on the effects of lignosulphonate-based admixtures and superplasticizers on the adsorption characteristics and zeta-potential on the cement as well as various cement minerals have been reported [20,23-26]. The majority of the work reported is on the C_3A phase and its hydration products, both with and without gypsum, because of the major influence of this cement phase on the setting, workability and general rheological characteristics of the cement paste.

Costa and Massazza have reported the adsorption characteristics of sodium lignosulphonate-based superplasticizers on synthetically prepared β-C_2S [23]. Both β-C_2S and C_3S phases were reported to adsorb calcium- and sodium-lignosulphonate admixtures to similar extents.

Typical results on the amount of sodium lignosulphonate admixture adsorbed on β-C_2S paste after 30 minutes of contact with lime-water (water/solid ratio 0.4) as a function of the equilibrium concentration is shown in Figure 8. At saturation, the amount of admixture adsorbed was about 0.2%, corresponding to equilibrium concentration of 3 to 4 g/l. Figure 9 gives the zeta-potential value of the particles as a function of the amount of adsorbed lignosulphonate superplasticizers [23]. In the presence of the superplasticizer, the zeta-potential decreased regularly up to -25 mV as adsorption increased. The maximum adsorption in case of β-C_2S paste was slightly lower than that found in solid-water suspension. Rheological measurements (discussed in a later section) carried out on β-C_2S

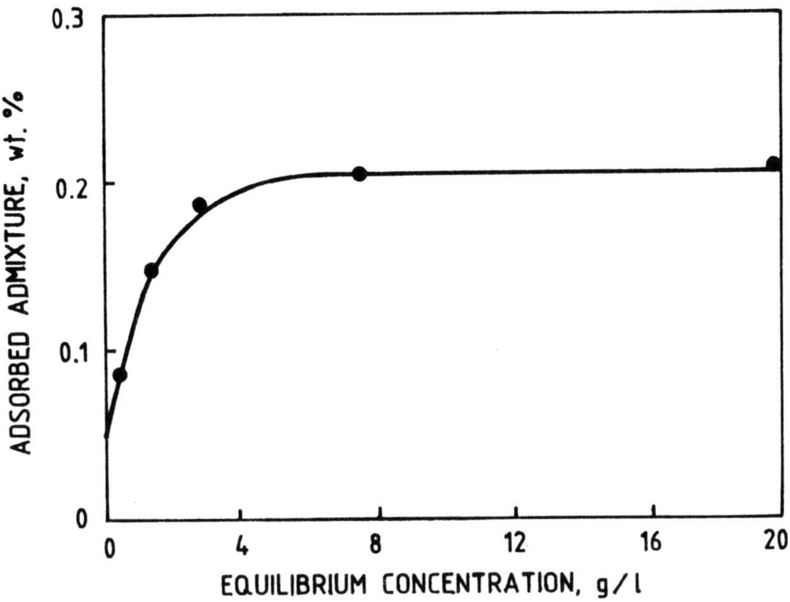

Figure 8. Adsorption isotherms of β-C_2S mixed with lime water (w/s =0.4) containing different concentrations of lignosulphonate superplasticizers [23].

paste showed that plastic viscosity attained the minimum value and the yield value became zero when the lignosulphonate admixture saturated β-C_2S and zeta-potential reached the lowest (highest absolute) value. These results were taken to indicate a correlation existing among the fluidization effect exerted by lignosulphonate admixtures, their adsorption on the solid and the changes in the zeta-potential [23].

Ramachandran has investigated the role of lignosulphonates on the hydration of C_3A and studied adsorption-desorption isotherms of system containing C_3A, and the hydrates C_4AH_{13} -C_2AH_8 and C_3AH_6

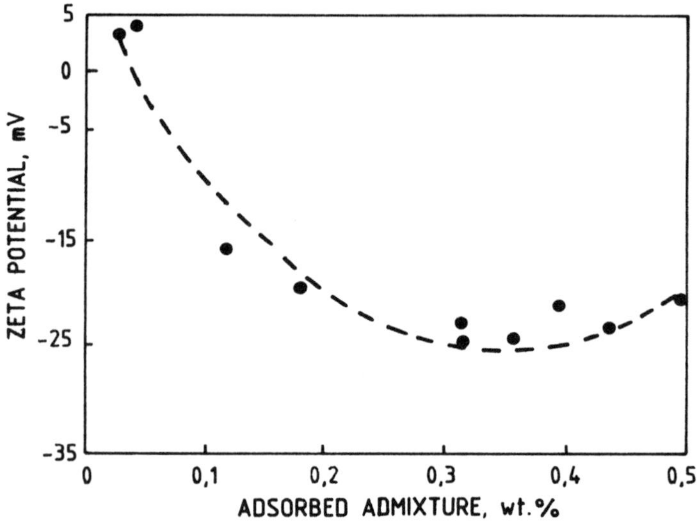

Figure 9. Zeta potential of β-C_2S dispersed for five hours in lime-water containing sodium lignosulphonate superplasticizer, as a function of the amount of adsorption [23].

in the presence of Ca- and Na-lignosulphonate [24]. In aqueous medium, both hexagonal and cubic aluminate hydrates were found to adsorb lignosulphonates irreversibly (Figure 10). Substantial amount of lignosulphonate are adsorbed on C_3A. Lignosulphonate salts play an important role in set retardation and both sugar-free calcium- and sodium-lignosulphonates were found to retard the hydration of C_3A as efficiently as commercial lignosulphonates [24]. This support the contention that set retardation effect of lignosulphonate admixture is due to adsorption on C_3A phase, in addition to the effects of sugar and carbohydrate [7].

Massazza and Costa have reported the effect of sodium lignosulphonate superplasticizer on the hydration of C_3A + $CaSO_4$,$2H_2O$ mixture [25]. The admixture was found to retard the hydration of C_3A and formation of ettringite. In the absence of the admixture, gypsum in the system was consumed in hydration

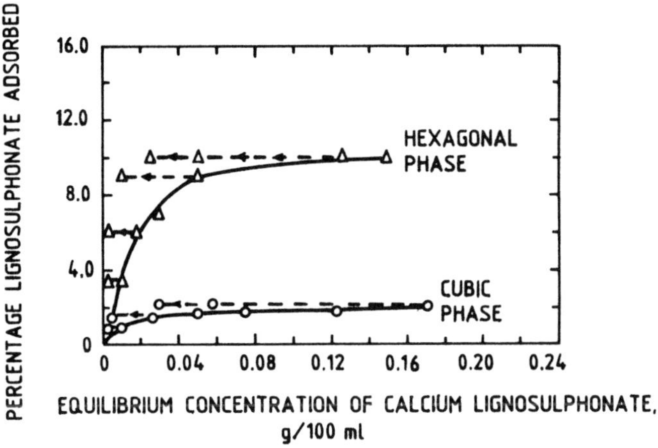

Figure 10. Adsorption-desorption isotherms of calcium lignosulphonate on the hexagonal and cubic phase of calcium aluminate hydrate [24].

reactions within four hours. However, in the presence of 1 percent sodium lignosulphonate, gypsum was found to persist for up to 36 hours [25].

Massazza, Costa and Barrila had reported adsorption characteristics and zeta-potential measurements carried out on calcium aluminate monosulphate hydrate ($C_3A{\bullet}C\overline{S}{\bullet}H_{12}$) and C_3A dispersed in solution containing sodium lignosulphonate [20]. Adsorption isotherms in the aqueous medium are shown at the right-hand side of Figure 11. The lignosulphonate admixture was found to be completely adsorbed when present in low concentrations. The left-hand side of Figure 11 shows the zeta-potential measurements corresponding to the amount of adsorbed admixture. The zeta-potential was found to decrease linearly as the adsorption increased and stabilized to a constant value when the isotherm reached values approaching the maximum adsorption [20]. The saturation values of zeta-potential were on the order of -30 mV, which were quite negative and are capable of creating sufficient repulsive forces.

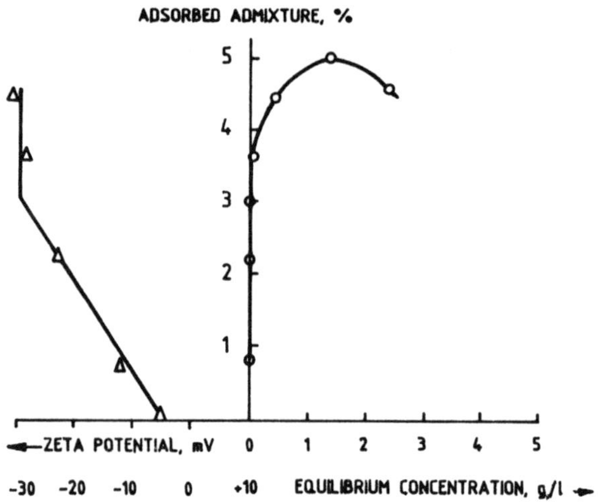

Figure 11. Adsorption of sodium lignosulphonate and change in zeta potential of monosulphate aqueous suspension [20].

Collepardi et al. have investigated the influence of sodium lignosulphonate admixture in combination with Na_2CO_3 on the hydration of C_3A + calcium hydroxide system [26]. Addition of these two admixtures retarded the hydration of C_3A. Initial hydration of C_3A were substantially blocked for five minutes by the addition of sodium carbonate and sodium lignosulphonate and similarly the hydration of C_4AF phase was substantially blocked for about 30 minutes. Significant changes in the zeta-potential were observed in case of simultaneous use of both the admixtures. In C_3A + H_2O system (i.e., without admixtures) the value was +37.4 mV and by the addition of the admixture combination, it reduced to a value of -41.5 mV. The changes in zeta-potential were less remarkable when either of the admixtures were used alone. It was concluded that the large increase in the flowability due to the action of sodium carbonate + sodium lignosulphonate admixture depends on zeta-potential changes as well as retardation of the initial stages of reaction of C_3A [26].

Rheological Aspects

Mechanisms of action of lignosulphonates in concrete have also been examined in terms of the rheological characteristics of the fresh cement paste [27,28]. The parameters can be conveniently explained in terms of tests with rheometers [29]. In such tests, an impeller is immersed in fresh concrete which is rotated at different speeds and the corresponding torque is measured. As shown in Figure 12, the relationship between the speed of rotation and the torque gives a measure of the yield point i.e., the amount of work input needed to enable the concrete to flow, and the slope indicates the plastic viscosity i.e., the resistance to flow [4]. Some other test methods employed the movement of a steel sphere in mortars or pastes [27]. The lower the values of yield point and plastic viscosity, the greater is the workability of concrete or fluidity of the cement paste.

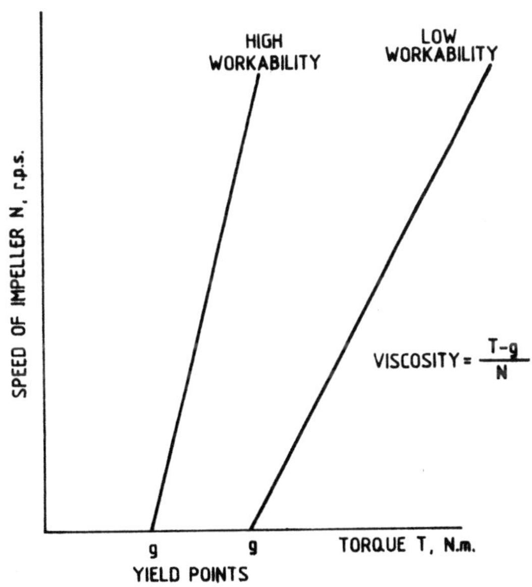

Figure 12. Rheological characteristics of concrete [4].

The schematic model of flocculated cement paste shown in Figure 6 indicates that the paste, on mixing, will have some mechanical strength and a yield value [17]. The dispersing function of a water-reducing admixture or superplasticizer leads to modification of development of such a floc structure or partial, or even total dispersion of pre-existing flocs. Such a fluidization of the paste results in marked lowering of both the yield stress and the plastic viscosity [11,30]. A typical relationship between workability of concrete measured by flow test and the yield value is shown in Figure 13 [4]. Addition of sodium lignosulphonate based superplasticizer was found to decrease the yield value of concrete. In this series of tests, increase in the flow from 50cm to 60cm corresponded to decrease in yield value from 2.562 to 1.579. The trend of effect of addition of the admixture on change in viscosity was less clear [4].

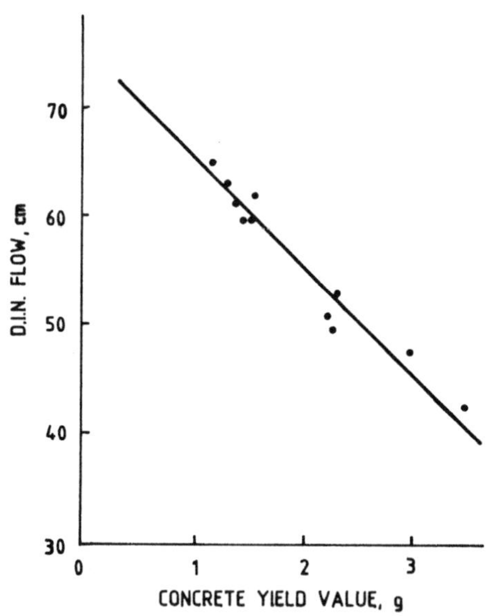

Figure 13. Typical relationship between workability (flow) and yield value of concrete [4].

Kishitani et al. [27] and Nishibayashi et al. [28] measured relevant rheological constants of cement paste or mortars incorporating admixtures and employed test methods, in which the resistance to the movement of a steel sphere through the sample of cement paste or mortar placed in a container was measured. In the latter case, the container was fixed to a vibrating table and the test was carried out both under vibration and non-vibration conditions [28]. Kishitani et al. measured the shear stress and shear rate of movement of the sphere when a lignosulphonate based air-entraining water-reducing admixture was incorporated in the cement paste at the mixing stage. The air content was in the range of $4 \pm 1\%$. In Figure 14, yield value and plastic viscosity of the cement paste, calculated from shear rate - shear stress relationship, due to addition of the admixture, are indicated as ratios of those of plain cement paste having the same water-cement ratio. The yield value was increased compared to the plain paste, the ratio being 1.3 ± 0.1. Such an increase was presumably due to air-entrainment, which is believed to increase the stiffness of the cement paste. The plastic viscosity was decreased, the ratio being of the order of 0.3 to 0.5. Compared to these, addition of sulphonated melamine formaldehyde condensate superplasticizer decreased both the yield value and plastic viscosity substantially. The yield value could be lowered to one-tenth or less of that of the plain paste. Such extreme reduction in yield value was taken as the rheological indicator of superplasticizers, contrasted to normal water-reducing admixture [27].

In the tests of Nishibayashi et al. [28] the performance of a calcium lignosulponate based water-reducing agent was measured on 1:3 cement mortars in which the admixture was added one hour after the initial mixing. The flowing behavior of fresh mortar with the delayed addition of admixture represents plastic flow. Typical results are summarized in Figure 15. Addition of admixture reduced the yield value (the intercept of the curves on the abscissa in Figure 15) compared to the plain mortar. Vibration conditions further reduced the yield value. The slope of the curves in Figure 15 indicative of plastic viscosity, did not show much change with variation in

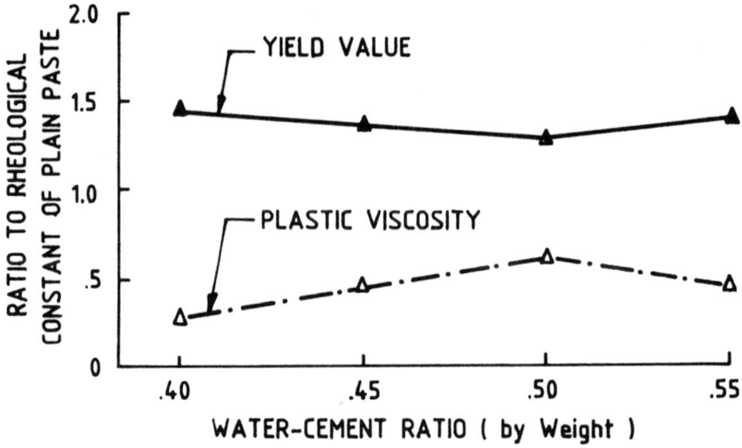

Figure 14. Change in rheological constant of cement paste due to lignosulphonate admixture [27].

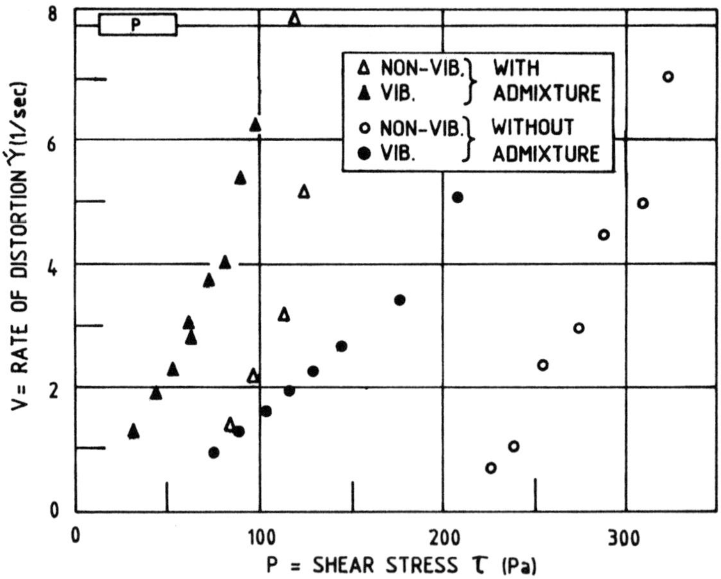

Figure 15. Effect of lignosulphonate admixture and vibration (4g, 120 Hz) on P-V curve for cement mortar [28].

vibration conditions (frequency, acceleration). However, vibration tended to increase the plastic viscosity and decrease the yield value compared to non-vibration conditions [28]. With an elapse of time after the addition of the admixture, yield values increase [4,28]. The tests above show that rheological parameters can represent the effects of lignosulphonate-based admixtures on the workability status of fresh concrete on mixing and thereafter.

EFFECTS ON HYDRATION OF CEMENT AND MICROSTRUCTURE OF CONCRETE

Hydration Characteristics of Cement

The effects of lignosulphonate-based admixtures on the hydration characteristics of cement have been investigated.

In general, it is held that these admixtures may alter the rate of hydration reaction, but the products of hydration are unaltered [18,31]. In one investigation, the effect of calcium lignosulphonate based water-reducing and set-retarding admixture produced by the sulphite process with subsequent removal of sugar by fermentation, on the initial hydration of low and high C_3A cements were investigated [31]. The effects of hydration on cement were investigated by measuring the heat of hydration in an isothermal conduction calorimeter and non-evaporable water content. The latter was determined by loss of ignition at 1000°C from the state of immersion in liquid nitrogen to terminate the hydration reaction at the required age. Typical results on heat of hydration for the two types of cement are shown in Figures 16 and 17, respectively. From the results of heat evolution (h) as well as rate of heat evolution and acceleration (dh/dt and d^2h/dt^2 respectively), both with the admixture in two dosages and without, it is found that induction and acceleratory periods were extended with the addition of the admixture as well as by increasing its dose. The effectiveness of the admixture in delaying

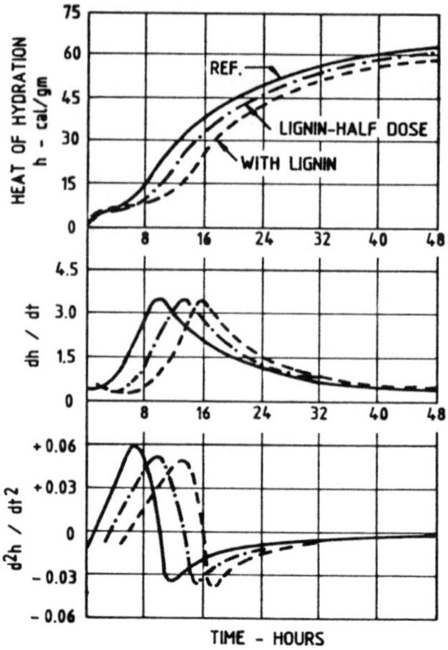

Figure 16. Effect of lignin admixture on hydration of Type I cement [31].

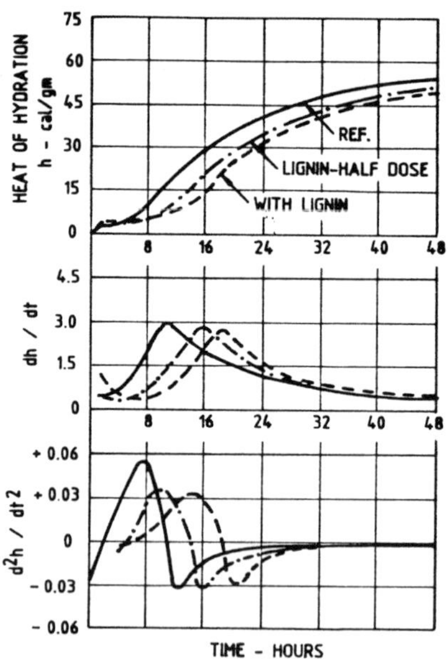

Figure 17. Effect of lignin admixture on hydration of Type V cement [31].

the occurrence of the peak hydration rate and reducing the one-day heat of hydration was greater for the cement low in C_3A content.

The changes in non-evaporable water content with time are shown in Figure 18. It is found that, in a manner similar to the evolution of heat of hydration, the non-evaporable water content in the initial period is somewhat reduced in the presence of calcium

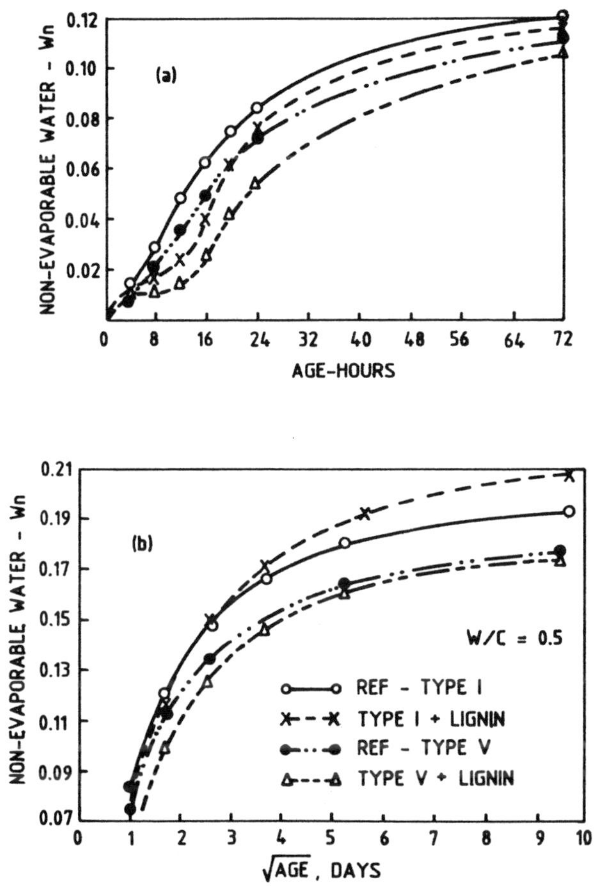

Figure 18. Influence of lignin admixture on non-evaporable water content for Type I and Type V cements. a) early age; b) later ages [31].

lignosulphonate admixture, but at 7 days and later ages, it is slightly greater than for plain mixes. This implies that a retardation in the early stages of hydration process is more than compensated by higher rate of hydration than the control at later stages.

The relationship between the heat of hydration and non-evaporable water content for the different cements, with and without calcium lignosulphonate admixture is shown in Figure 19.

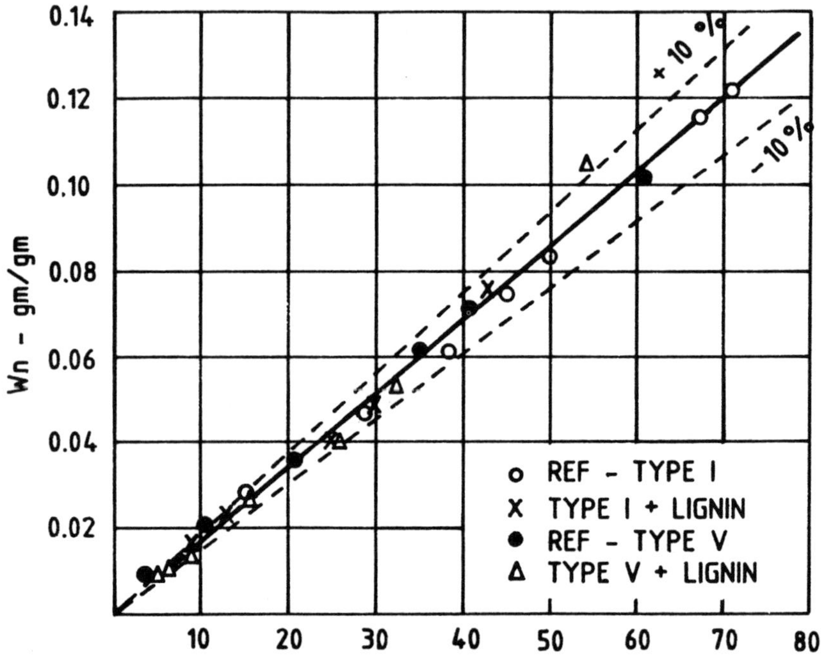

Figure 19. Relationship between h and w_n for Type I and V cements, admixture-lignin based [31].

It is found that a similar relationship exists between the two parameters, irrespective of cement composition and presence or absence of lignosulphonate admixture. This suggests that the same kind of hydration products are being formed at all stages of hydration and that the presence of admixture can be only related to an alteration in the rate of reaction rather than any substantial changes to the products of hydration.

Ramachandran [18] has reported the action of various types of retarders/water-reducing admixture on the rate of hydration, strength and properties of mortars containing a Sulphonated Melamine Formaldehyde Superplasticizer (SMF). Calcium lignosulphonate admixture retarded the setting time, when used alone or with SMF. The rate of hydration was followed by a differential scanning calorimeter; typical results, at the time of initial set are shown in Figure 20. Differential scanning calorimeter (DSC) curves

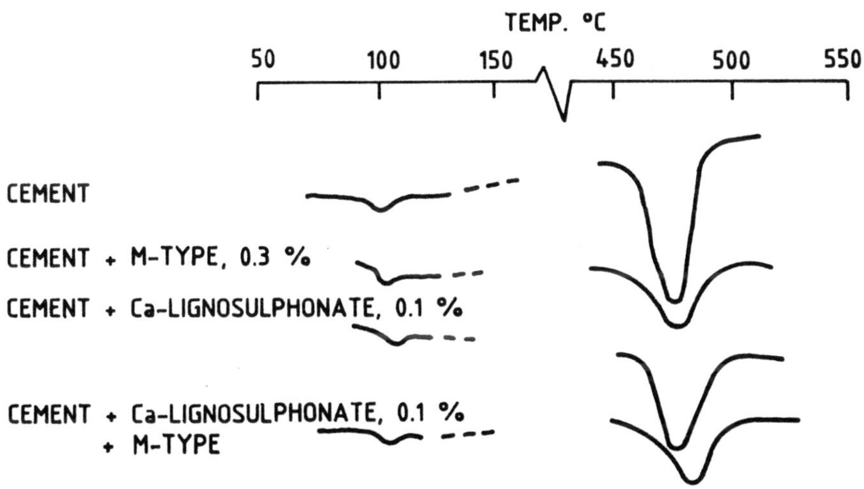

Figure 20. The DSC curves at time of initial set cement mortars containing various admixtures [18].

of initially set mortars showed small endothermal peaks at about 100°C representing C-S-H and/or ettringite phases, followed by another endothermal peak of larger intensity between 450-500°C, indicative of decomposition of $Ca(OH)_2$ formed by hydration of C_3S [18]. Results indicated that at the time of initial set, which is delayed in the presence of calcium lignosulphonate and/or SMF, some hydration reaction have taken place, but the extent of hydration is lower than in case of reference mortar without any admixture. Accordingly, the peak intensity for $Ca(OH)_2$ is reduced in the presence of these admixtures (Figure 20).

Microstructure

The effects of chemical admixture and superplasticizers on the microstructure of hardened concrete in general, and with the use of lignosulphonate-based products in particular, would depend upon the specific propriety of concrete which is targeted to be tailored. If, a water reducer or a superplasticizer is used to increase the workability of concrete without lowering the water/cement ratio, the microstructure can be expected to be of one type, than when these admixtures are employed to allow the use of low water/cement ratio concrete, either for obtaining high strength or increased durability. If lignosulphonate based admixture entrain significant amount of air, microstructure would show the feature of typically air-entrained concrete i.e., the system of discrete air bubbles of required sizes and spacing. If supplementary cementitious materials like fly ash, slag or condensed silica fume are added with a water reducing agent or a superplasticizer, the microstructure of yet another type can be expected. More published information on the effects on microstructure of concrete are available with N-type and M-type superplasticizers than with lignosulphonate based products. Since the effects on the microstructure can be expected to be similar in either cases, the broad features are described below.

The main water-reducing action of a plasticizer results from deflocculation and dispersion effects on the cement paste. In high or normal water-cement ratio concrete, containing a water-reducing agent or a superplasticizer to produce a high workability or 'flowing' concrete, the microstructure would show a more dispersed and an orderly arrangement of cement hydration products. In case of low water/cement ratio mixes, the microstructure would show typically more dense features. The study of hydration reaction already discussed have shown that lignosulphonate based salts may alter the rate of hydration of cement but the ultimate extent of hydration and type of microstructure formed may not be different.

Ramachandran's investigation on the hydration reaction in concrete containing different types of superplasticizers as well as lignosulphonate based admixtures was mentioned above [18]. In this investigation, the comparison of the progress of hydration was made by differential scanning calorimetry (DSC) on mortar specimens corresponding to the initial setting time, which is determined by ASTM C-403 (Proctor Needle) method. In this method, a set (initial or final) corresponds to a certain resistance to the penetration of the test needle offered by the hydrated system. DSC curves indicated lower endothermal effect i.e. lower extent of hydration till the time of set in mortars containing calcium lignosulphonate salts. In other words, a smaller amount of hydration product formed in the presence of lignosulphonate salt offered the same resistance to the penetration of the needle, as did the reference mortar containing no admixture and which had a larger amount of hydration product (Figure 20). This indicates that a mortar containing calcium lignosulphonate admixture is better dispersed and forms a more compact network [18].

Similar trends were observed in another investigation on the influence of N-type superplasticizer on the rate of gain of strength of concrete at very early ages [17]. Typical results on microconcrete made with cement content 400 kg/m^3, w/c=0.5 and aggregate of 2.5 mm maximum size are shown in Figures 21 and 22. It was pointed out that the initial strength development is due to the intergranular bonds in the contact areas of cement grains in a floc. In the very early ages up to 2-3 hours, the presence of admixture delays development of strength due to deflocculation (Figure 21). Deflocculation also

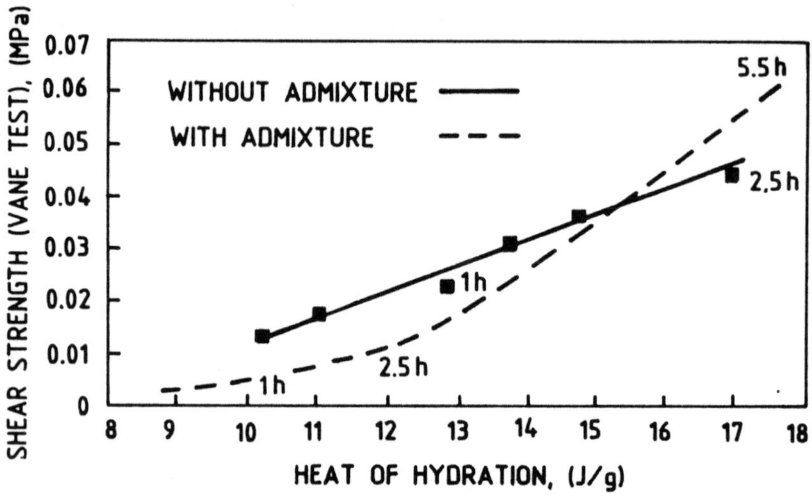

Figure 21. Compressive strength as a function of heat of hydration before and during setting [17].

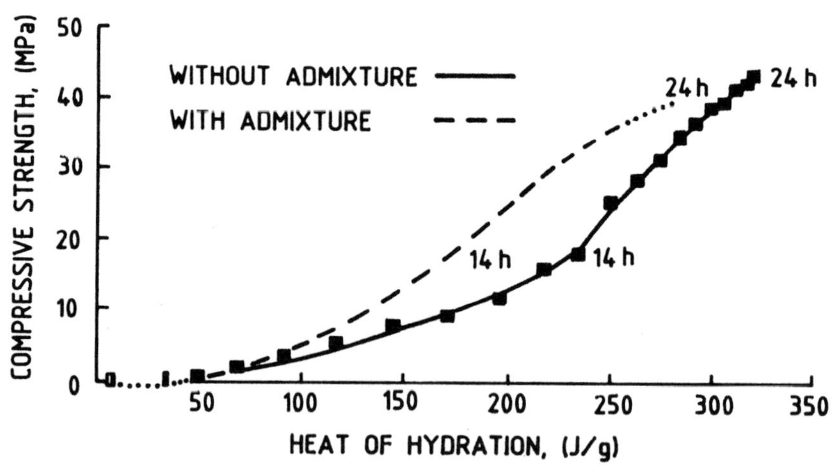

Figure 22. Compressive strength as a function of heat of hydration during the first 24 hours [17].

results in reduction in the size of the largest pores (Figure 6). Accordingly, the pores need less volume of hydrates to fill up. Once sufficient hydrates are formed, the volume of filled up pores is large enough to offset the initial effect of deflocculation and the strength of concrete up to 24 hours is greater for the same degree of hydration, in the presence of the admixture (Figure 22) [17].

It has been suggested that $CaO:SiO_2$ ratio of the C-S-H product is somewhat increased in the presence of some superplasticizers and, in the hydration of C_3A, the hydration product may form as a gel coating the C_3A surface, instead of hexagonal plate formed initially in non-admixed system [11]. At a water-cement ratio of 0.6, comparison of SEM photographs of C_3S-hydrate formed in the presence of M-type superplasticizer showed a more compact structure and seemed to be less porous than the reference specimen. Although no substantial differences have been observed in the morphology of cement pastes hydrated with and without superplasticizers, finer particles with a denser structure can form with superplasticizers [11].

The beneficial effects of addition of silica fume or pozzolana results from both filler action in altering the pore structure, as well as by lowering the water cement ratio as a consequence of incorporating water-reducing agent or a superplasticizer, which results in a denser microstructure. In particular, the use of silica fume and of water-reducing agent or superplasticizer go hand-in-hand, and the resultant advantages in the concrete are due to the effects of both types of additions. Berke et al., have described the microstructure of concrete containing silica fume and/or fly ash, which were in addition to the cement and not as its replacement [32]. In these investigations, one series of concrete mixes, containing a nominal cement content of 360 kg/m^3 with water-cement ratio of 0.43 had a slump of 100-150 mm and 6% air by incorporation of a superplasticizer. Examination of the microstructure with scanning electron microscopy (SEM) revealed the paste structure to have similar morphology in each case, with the primary constituents being a tightly woven network of C-S-H gel. There were, however, differences in the porosity and degree of crystalinity exhibited by each paste structure. Generally, the structure of the fly ash concrete exhibited less porosity than that of portland cement concrete. In case of the silica fume concrete, there was no evidence of capillary pores and only very small (< 0.5 um) gel pores were visible at high magnification [32]. Concrete containing silica fume exhibited a tighter bond between aggregate particles and cement paste as compared to the fly ash or Portland cement concrete. Based upon these microstructural observations, it was concluded that the improved performance of concrete containing silica fume at the early ages were only partially due to pozzolanic reaction. Much of the improvement in properties of concrete was attributed to the development of the tighter, less porous paste structure, as a result of a capillary and gel pore spaces being filled by very small silica fume particles. At the later ages, the improved performance is also due to the increased contribution from pozzolanic reaction [32].

PERFORMANCE IN CONCRETE

Among the chemical admixtures for concrete, which were commonly in use prior to development and widespread use of superplasticizers, lignosulphonate products were the basic continuents of water-reducing agents, pumpability aids and integral water-proofers [5]. Out of these, use in water-reducing admixtures of various types, e.g., normal, retarding, accelerating types [1] were more prominent and consisted of the basic lignosulphonate salts [3,4]. As a result, majority of published information on the performance of lignosulphonate based admixture in concrete are in the area of water-reducing or plasticizing agents [5,14].

Water-reducing, set-controlling admixtures are used in three basic modes of application. They are used to increase the workability of concrete for the same w/c ratio and cement content; to reduce the water content in the concrete mix for the same workability without altering the cement content, thereby lowering the w/c ratio and resulting in higher stregnth; or to reduce both the water and cement contents to maintain the same w/c ratio, workability and stregnth but effect economy in cement content [2,3]. These are also used to modify the setting time of concrete. A common side-effect of many water-reducing admixtures is retardation of setting of concrete. Incorporation of other retarding or accelerating materials in required dosage can result in a water-reducing agent which is also retarding, accelerating or normal (i.e., non-retarding) [3].

The effect of a lignosulphonate-based admixture on the properties of fresh concrete (workability, setting time, air content) and hardened concrete (strength, durability) depends on the intended mode of action. Standard specifications [1,2] prescribe the performance requirements and a summary is shown in Table 5. The actual performance in concrete is often better than the minimum prescribed in Table 5. For example, although standard specifications lay-down a minimum 5% reduction in water content with use of a water-reducing agent, reduction up to 10 percent is commonly expected, in order to justify its use [8,14]. In addition, the performance is influenced by the cement characteristics, properties of the aggregates, concrete mix proportions and the ambient environmental conditions [3,5,14], which are discussed later. Under such situations, no generalized description of performance of these admixtures is possible; what is summarized below should be taken only as indicative of the general trend.

Setting Time

Intrinsic to the action of delaying the initial hydration reaction as mentioned before, lignosulphonate based admixtures generally prolong the setting times of concrete. This is more so when sugar is present. Ramachandran has reported an investigation on the performance of calcium lignosulphonate-based water-reducing admixtures in comparison to other type and in combination with M-type superplasticizers [18]. All the admixtures which were investigated increased the setting time of concrete. Calcium lignosulphonate based water-reducing set-retarding agent, at a dosage of 0.4% by weight of cement in mortars increased the initial setting

Table 5. Summary of Performance Requirements of WR Admixtures as per IS:9103 and ASTM-C-494 (Type A).

Item	Value
I) Water content, Max. % of control	95
ii) Allowable deviation in setting time from control sample, (hr-min)	
Initial	1*, 1 - 30+
Final	1*, 1 - 30+
iii) Compressive strength, Min., % of control, at ages,	
3, 7, 28 days	110
6 months, 1 year	100
iv) Flexural strength, Min., % of control, at ages	
3, 7, 28 days	100
v) Bleeding, % increase over control, max	5
vi) Length change, % increase over control sample, max	0.010
vii) Relative Durability factors, min.	80

* IS ; + ASTM

time from four and a half hours to nearly double the value (Figure 23). In Figure 23, the effect of M-type superplasticizers, as well as the combination of M-type superplasticizer with calcium lignosulphonate admixture, are indicated for comparison. It is seen that the combination increased the setting time more than the superplasticizer alone [18]. Lignosulphonate based superplasticisers were also found to increase the setting time of concrete in another investigation [4], which will be discussed later.

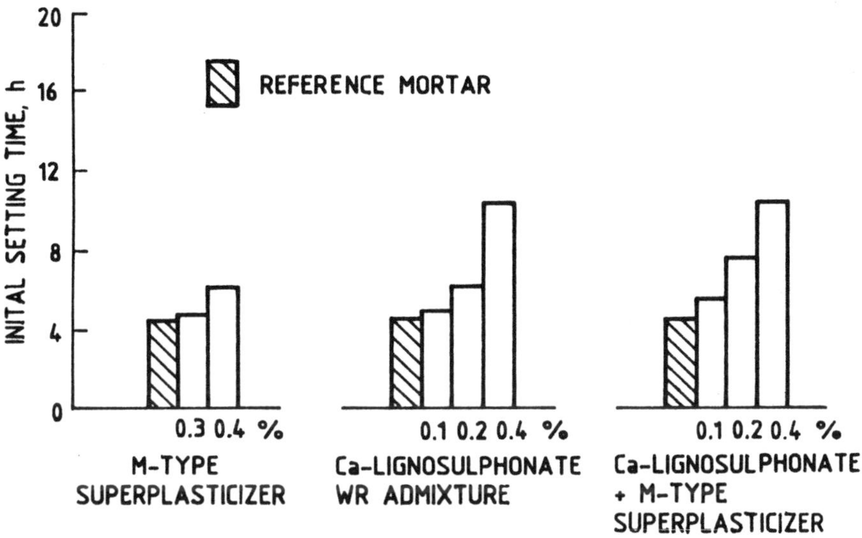

Figure 23. Influence of calcium lignosulphonate admixture and M-type superplasticizer on initial setting times of mortars [18].

El-Rayyes has described the performance of a non air-entraining, set-retarding water-reducing admixture (ASTM C-494, Type D) of lignosulphonate-base on concrete [33]. The water-cement ratio in concrete mixes varied from 0.40 to 0.65 and the admixture dosages were 0.4% (normal), 0.5% (above normal) and 0.6% (high). The initial temperature was 23-24°C, but the subsequent exposure temperatures were 36-45°C. This investigation being concerned with the effect of ambient air temperatures on the set behaviour of fresh concrete and the effect of admixtures in retarding the set, the results will be presented in greater detail later. For the present, it can be summarized that lignosulphonate based admxiture retarded the setting of concrete and more than offset the accelerating effect of higher ambient temperature on initial and final setting [33].

Air Content

The mechanism of action of lignosulphonate based admixtures in concrete includes lowering the surface tension of air-water interface [7]. Hence, some air-entrainment in concrete is always expected, as is with surface-active agents in general [5]. The amount of air entrained is usually small; in the range of 0.2 to 0.5% [8]. However, higher values in the range of 2-6% have also been reported [3]. It has been suggested that the line dividing the mechanism of action of a concrete plasticiser (e.g. based on lignosulphonates) and air-entraining agents (e.g. vinsol resins) is rather thin, in that initially both can produce large numbers of small air bubbles [5]. The difference is in the stability of the foam thus formed and, therefore, in the air content of concrete in the hardened state.

For a satisfactory air void system to result in the hardened hydrated cement paste, combined use of water-reducing agents and air-entraining agent are also reported [34,35]. Fukuda et al., have reported the use of lignosulphonate based water-reducing admixture alongwith an air-entraining agent in a concrete mix of w/c = 0.65 and cement content 268 kg/m^3 of concrete; air content of 4.5% resulted and was maintained up to 60 minutes after mixing [34]. Comparable performance was obtained with the use of a superplasticiser based on β-naphthalene sulphonate-formaldehyde and containing a reactive polymer, along with the air-entraining agent. For similar slump and air content, the dosage of the air-entraining agent in the later case could be reduced to about half of that with the lignin-based water-reducing agent [34].

Kobayashi et al., have reported the performance of calcium lignosulphonate based concrete admixture, with and without air-entraining agent (sodium resin acid base) [35]. Water-cement ratio of concrete mixes incorporating lignosulphonate based water-reducing agents were 0.42 and 0.55. The slump ranged from 90 to 110 mm. The cement content was 440 kg/m^3 of concrete in mix with w/c = 0.42 and 300 kg/m^3 in case of water-cement ratio 0.55 and somewhat lower when air-entraining agent was added. Air of the order of 3.0 and 3.7% were entrained in fresh concrete with the use of lignosulphonate based admixture, which increased up to 4.3-4.8%, respectively with the addition of air-entraining agent [35].

Effect on Workability and Slump Loss

As mentioned earlier, the most widespread use of lignosulphonate based admixtures have been to increase the workability of concrete. When added to concrete mixes without lowering the water content, typically, a 75 mm slump concrete with w/c = 0.55 can have the workability enhanced to approximately 135 mm [8]. Alternatively, a concrete mix having 90 mm slump can retain the same workability or even higher, although the water content in the mix is reduced to the extent of nearly 18% [7].

The initial value of the workability of a concrete mix is gradually reduced with time. This is due to continued hydration reactions, which commences immediately after mixing with water specially on the C_3A phases; and due to atmospheric evaporation of mix water, which depends on the ambient temperature. Typical data on loss of workability with time after mixing for plain concrete mixes made with different types of cement are shown in Table 6 [36]. The

Table 6. Typical Slump Loss in Plain Concrete Mixes [36].

Cement Type	Initial Slump,	Slump (mm) after Time Interval (min)						
	mm	15	30	45	60	75	90	120
Rapid hardening cement	225	197	168	153	124	102	76	34
Ordinary Portland	225	215	193	178	159	147	131	114
Portland slag cement	225	218	206	188	173	160	146	131
OPC + Flyash (15-25%)	225	214	204	195	178	157	140	108

same trend is observed in concrete mixes containing water-reducing admixtures and superplasticizers [18,34,37]. The increase in workability of concrete, that is obtained with the addition of admixture is only transient and not maintained beyond 30 to 60 minutes (Figure 24) [37]. In an investigation, the reference concrete had an initial slump of 57 mm, which reduced to 13 mm in two hours time [18]. Addition of calcium lignosulphonate based admixture resulted the initial slump to increase to 127 mm, which decreased to about 40 mm in two hours (Figure 25) [18]. These values were, nevertheless, more than double that of reference concrete.

Because of such loss of workability taking place with use of admixtures, some specifications and recommendations stipulate the workability of concrete mixes containing plasticizers to be measured not earlier than 15 minutes and not later than 30 minutes after mixing with water [16,37]. Delayed addition of lignosulphonate-based admixtures, a few minutes after mixing with water, as indeed also in

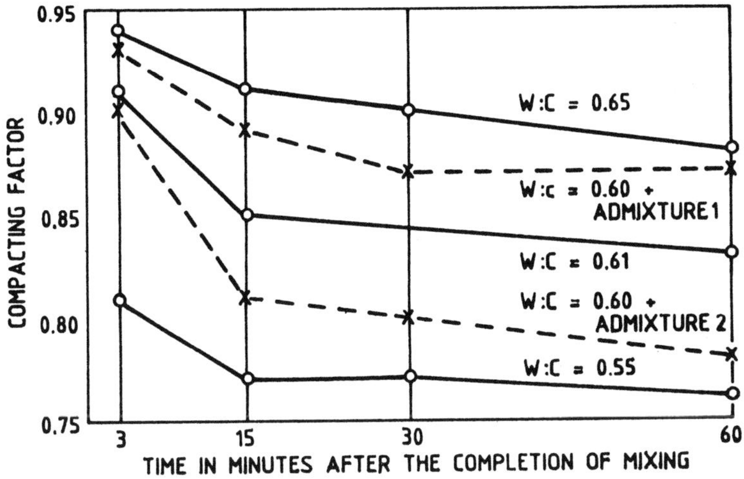

Figure 24. Change in workability (compacting factor) with time in concrete mixes with or without lignosulphonate admixture [37].

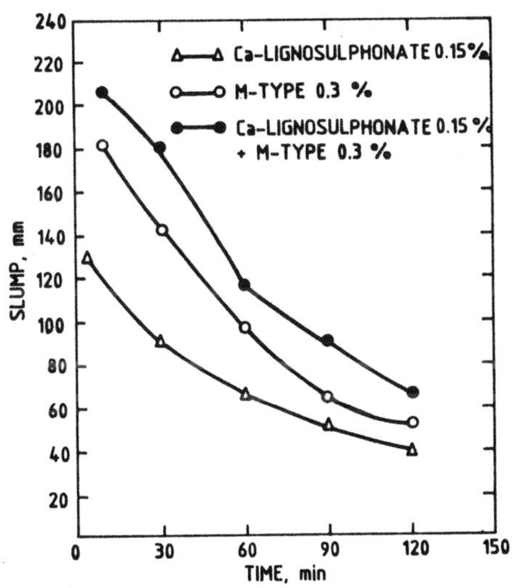

Figure 25. Effects of lignosulphonate admixture and M-type superplasticizer on slump loss of concrete [18].

water-soluble dispersant in the highly alkaline medium in hydrating case of many N- and M-type superplasticizers, has often been recommended for better performance [18]. In case of M and N-type superplasticizers, repeated dosing, or simultaneous use with a retarding admixture have been found to reduce the slump loss [13,18]. Fukuda et al have described the effect of a reactive polymer, along with an N-type dispersant. The reactive polymer is converted to a cement-water system and acts as a precursor to the conventional dispersing action of the superplasticizer [34]. Compared to a lignosulphonate based water-reducing admixture, where the slump decreased from 183 mm to about 130 mm in one hour, the above described superplasticizer enabled slump of the order of 180 ± 20 mm to be maintained up to one hour (Figure 26) [34].

In Figure 25, it was shown that the slump of concrete was higher, both initially and after two hours, when an M-type superplasticizer was used than when calcium lignosulphonate based admixture was used. The combined addition of the two, however, gave better results in that both the initial slump as well as slump after two hours were higher than in case of the superplasticizer alone [18]. Among the many factors involved in the slump loss phenomena, accelerated formation of ettringite by the reaction of C_3A with the SO_3 phase is an important cause. Accordingly, an admixture, which can retard the hydration reactions and also have a dispersive action, e.g. a set-retarding water-reducing admixture, can help in retarding slump loss [18]. Similar results were reported with an N-type superplasticizer which was modified with a retarding agent added, which provided longer slump maintenance in hot weather (32°C) conditions [38].

Lowering of Water-Cement Ratio and Increase in Compressive Strength

When advantage is taken of the improved workability and consequently possibility of reducing the water content in concrete by addition of water-reducing agents, the compressive strength of

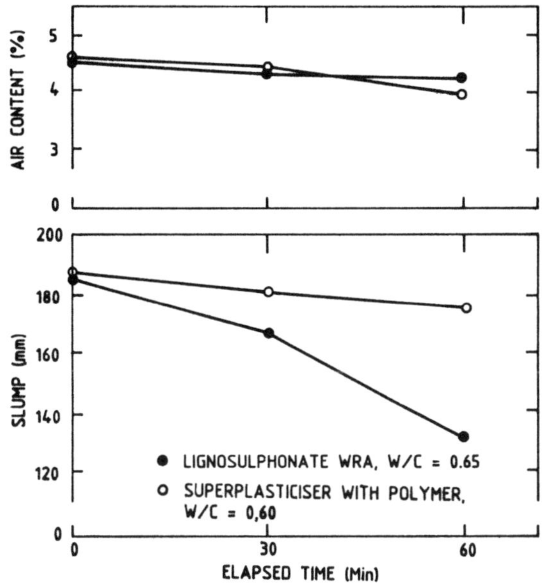

Figure 26. Effects of admixtures on variation in slump and air content of concrete with time [34].

concrete is increased primarily due to reduction in the water/cement ratio of the mixes. For a typical concrete of water/cement ratio 0.55 and a slump of 75 mm, if the water content can be reduced to correspond to a water/cement ratio 0.5, an increase in 28 days compressive strength of the order of 25-30% can be expected [8]. For the given cement characteristics and the mix proportions adopted, the increase in strength follows typical strength - water/cement ratio relationship. The increase in compressive strength is expected to be valid at all the ages, unless excessive set retardation and increase in air content takes place, as can be sometimes expected with lignosulphonate-based admixtures. In one investigation, the water/cement ratio of concrete could be reduced from 0.48 to 0.47, 0.46 and 0.45 respectively by addition of a modified lignin type water reducing agent at dosages of 0.06, 0.12 and 0.18% solids on cement [39]. As a result, the early age (at 18 hours) compressive strength of concrete increased from 14.8 MPa to 21.3 MPa at curing temperature of 22°C [39]. In another instance, the one-day compressive strength of mortar containing calcium lignosulphonate based retarding type water-reducer, decreased from 8.9 MPa (control) to 8.0 MPa, and the strengths up to 90 days were generally equal or lower than the control. This was due to initial retardation of setting and entrainment of air in the presence of lignosulphonate [18]. Another set of typical results on the use of carboxylic polymer modified lignosulphonate based admixture (designated A) is shown in Table 7, in which the performance is also compared with an N-type (C) and M-type (E) superplasticizers [40]. Using two different types of cement and two sets of concrete mixes of medium workability (80 to 90 mm slump) as well as high workability (180 to 200 mm slump), it was found that the reduction in the water/cement ratio with lignosulphonate type admixture was comparable to those with the M- and N-type superplasticizers. Although lignosulphonate based admixture entrained somewhat higher air content than the superplasticizers, the compressive strength at all ages, 1, 7, 28 and 90 days were generally higher [40]. These investigations were concerned with the effect of the lignosulphonate based admixtures as well as the superplasticizers on the strength and durability of concrete including the effects of incorporation of condensed silica fume [40,41]. The results will be discussed in the following sections.

FACTORS INFLUENCING THE PERFORMANCE OF LIGNOSULPHONATE ADMIXTURES

The performance of water-reducing set-controlling admixtures, like those based on lignosulphonate salts, depend upon the cement used, the concrete mix proportions, the temperature and humidity conditions and workmanship including the sequence and timing of addition of the admixture [3]. The mechanism of action of lignosulphonate admixtures in concrete were discussed in an earlier

Table 7. Effect of Lignin-Based Admixture (A) and M and N-type Superplasticizers (C,E) on Properties of Concrete Mixes [40].

Cement	Mix	Admix. and dosage	W/C	Air %	Slump mm	Compressive Strength MPa, after days 1	7	28	90
II-Z/35A	8	--	0.50	2.2	80	9.1	18.5	21.4	24.9
	9	A,0.4	0.44	3.5	80	12.2	23.6	27.8	31.1
	11	C,1.4	0.45	2.6	80	11.2	20.5	25.6	30.1
	13	E,0.7	0.46	2.7	80	9.7	19.0	22.3	26.1
	21	--	0.63	1.9	185	8.3	17.4	20.0	23.4
	22	A,0.8	0.54	4.2	190	7.75	22.3	27.8	32.4
	24	C,2.4	0.54	2.6	200	10.0	20.9	25.1	28.0
	26	E,1.2	0.53	2.1	190	9.7	19.5	22.3	27.3
I/45A	41	--	0.48	2.5	80	16.0	25.1	29.6	35.5
	42	A,0.4	0.45	3.2	90	20.3	32.5	37.7	41.7
	44	C,1.4	0.45	2.7	90	19.8	28.1	35.1	38.2
	46	E,0.7	0.45	2.9	80	16.7	26.5	30.5	34.3
	54	--	0.58	2.2	180	13.4	22.1	27.8	28.5
	55	A,0.8	0.54	3.4	180	17.4	28.1	33.9	39.0
	57	C,2.4	0.53	2.6	180	16.9	28.1	31.4	35.2
	59	E,1.2	0.52	3.0	180	15.3	23.3	28.1	35.0

section in terms of physical and chemical actions on the cement-water system. It is, therefore, possible to visualize that any factor, which can have a bearing on these physico-chemical reactions, will influence the performance of the admixture in concrete.

Cement Characteristics

Among such factors, the role of cement characteristics have been examined and reported in sufficient detail [14,42]. Generally, two types of influences have been recognized. The first is the efficiency of the admixture to effect water reduction in concrete mixes and increase in compressive strength. It has been reported that the effect of water-reducing admixtures, including lignosulphonate based plasticizers, on strength and related properties of concrete are modified by the C_3A and alkalies present in cement. The performance of the admixture tend to decrease as the proportions of these constituents in cement increase [14]. While the reduction in water demand due to addition of water-reducing agents were found to be comparable, the improvements in compressive strength were greater, the lower the alkalies and C_3A content in the cements [42]. Thus, the chemical composition of cement has a large influence on the effectiveness of a given water-reducer, which cannot be explained by the reduction in water-cement ratio alone.

The other aspect is the influence on setting time and sometimes abnormal setting behavior, which is caused due to cement-admixture incompatibility. Severe stiffening may arise in false set, which can be overcome by extended mixing time or remixing of the concrete; quick setting, which is accompanied by heat generation and which cannot be overcome by remixing, can also be caused [14]. Most important among the properties of cement that relate to such cement-admixture incompatibility are the kind and amount of calcium sulphate added to cement and the content of readily available alkalies [43-45]. Polivka and Klein had reported that set-retarding admixtures were more effective with cements of low alkali and low C_3A content

[42]. Tam had also reported that an increase in either C_3A or the alkali content in cement reduced the effectiveness of lignosulphonate-based admixtures to delay the initial and final setting times [43]. In particular, it was found that increase in alkali content in cement from 0.53-1.30% had very little effect on the setting times of plain mortar mixes, but the effect on mixes containing lignosulphonate based admixture was to substantially reduce the setting time, i.e., shortening the delay time over plain mixes [43].

Apart from C_3A content, the fineness of cement is also important in affecting the fluidizing effect of plasticizers and superplasticizers; the higher the fineness of cement, lower is the efficiency of the admixture [15]. It has also been found that, for the same chemical composition of cement clinker, the form of calcium sulphate added as set regulator has an influence on the efficiency of plasticizers; the fluidizing effect is more in the presence of calcium sulphate dihydrate ($CaSO_4$,$2H_2O$) than in case of hemihydrate ($CaSO_4$,$1/2H_2O$) (15,44). Hemihydrate liberates sulphate ions much faster than gypsum, which, in turn, does faster than anhydrite. The alkali sulphates from the cement clinker (K_2SO_4,Na_2SO_4 or sodium-potassium sulphate) dissolve faster than any sulphate based on calcium. It has been explained that lignosulphonate molecules, having a selective adsorption effect on $CaSO_4$-surface, decrease the rate of solution and reduce the SO_4^{--} concentration in the liquid phase. If the anhydrite/gypsum ratio is higher than the optimum, the lack of quickly available sulphates required to form ettringite around C_3A phases, may hasten C_3A hydration and quick stiffening of the paste [44].

Guanlun and Ping have reported rapid setting of concrete mixes made with cement containing fluorogypsum as set controller, when a calcium lignosulphonate based water-reducing, set-retarding admixture was used [45]. This is because of swift adsorption of calcium lignosulphonate on fluorogypsum which is a strong adsorbent, thereby restricting the release of SO_4^- ions in the liquid phase and formation of ettringite. The problem was solved by delayed addition of the admixture after an initial mixing period of about one minute (Figure 27) [45]. Such a delay allows more SO_4^{--} ions to dissolve and some ettringite to form before the lignosulphonate film is adsorbed on to the fluorogypsum surface. Delayed addition of plasticizers and superplasticizers is favored from the consideration of increase in workability also, as it allows more of the admixture ions to be available to disperse the hydrating cement particles, rather than most of it being adsorbed on to the C_3A- gypsum system [11,15,18,30].

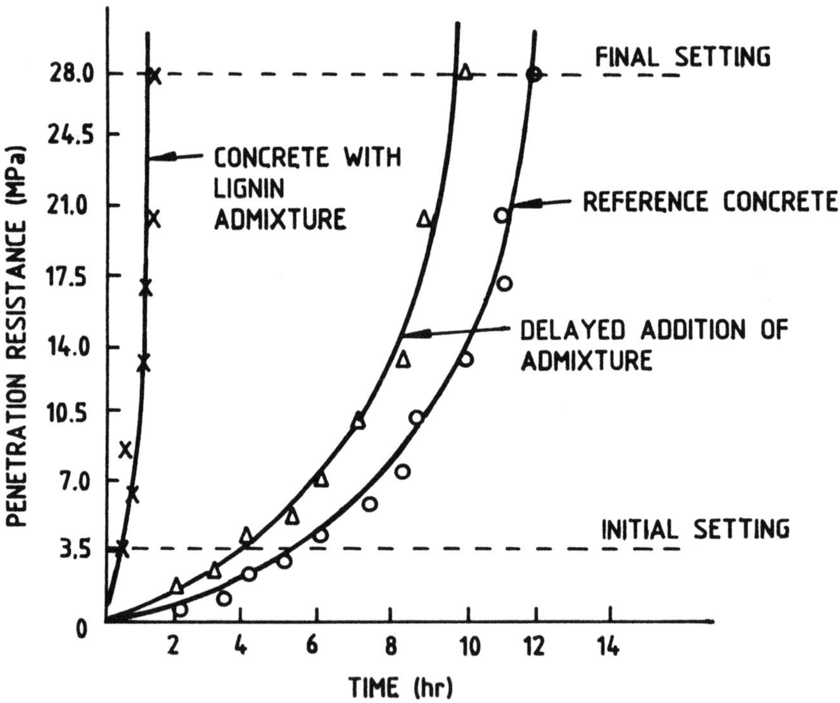

Figure 27. Effects of time of addition of lignosulphonate admixture on rapid setting of cement containing fluorogypsum [45].

Quon and Malhotra have reported the performance of a modified lignosulphonate-based superplasticizer on concrete mixes made with high-alumina cement [46]. In spite of higher dosages of the admixture, required high slump in low water-cement ratio (0.36) mixes were not obtained and there was extremely rapid slump loss, with concrete reverting to the original slump in less than 20 minutes. The degree of conversion of monocalcium aluminate decahydrate to tricalcium aluminate hexahydrate after moist curing at 21°C for 180 days was not affected by the incorporation of the superplasticizer (Figure 28) [46].

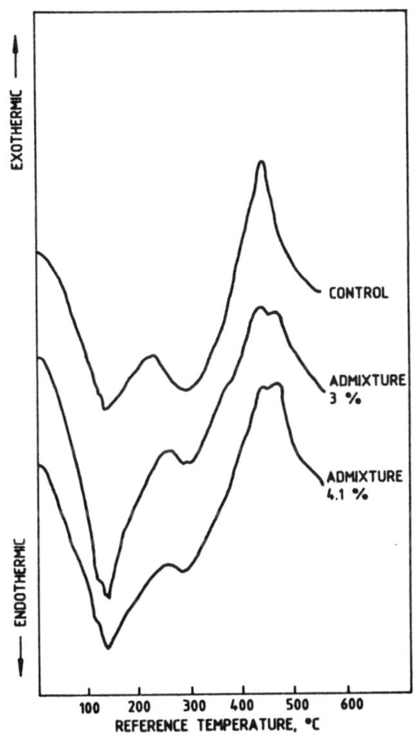

Figure 28. Thermogram of high-alumina cement concrete with and without lignosulphonate superplasticizer [46].

Concrete Mix Proportions

The influence of concrete mix proportion on the performance of an admixture manifest through the quantities of cement and water available. In high strength, low water-cement ratio concrete mixes, there is relatively greater quantity of cement particles and lesser water to accept sulphates. Thus there are more C_3A sites available for immediate reaction in contact with water, but less SO_4^{--} ions in the solution phase, because of quantity of water being limited. Therefore, when plasticizers or superplasticizers are used, the sulphate must dissolve in water more rapidly to prevent rapid slump loss. Accordingly, the type of sulphate addition becomes critical and the optimum amount of SO_3 and alkalies in cement for use with a particular admixture and in particular mix proportion may be different than what is adopted for cement manufacture, on the basis of acceptance tests performed on standard mortar or cement pastes of higher w/c ratio and pure water [47].

In normal strength concrete mixes, for particular ambient conditions and dosage of lignosulphonate-based water-reducing, set-retarding admixture, the initial and final setting times of concrete mixes were found to be greater, the higher the water-cement ratio [33]. In view of so many causative factors being involved, it is imperative that the performance of any admixture must be evaluated before use with the particular cement and aggregate, in trial mixes with actual mix proportions likely to be adopted in construction, and keeping the ambient field conditions in view [2,3,14].

Ambient Temperature

Setting, hardening and workability of concrete take place concurrently with the hydration reaction of cement. Like many other chemical reactions, these processes can be thermally activated. Accordingly, ambient temperature and humidity conditions influence the setting, strength development and workability of concrete, and performance of admixtures in concrete. Higher ambient temperature accelerate the setting, reduce the workability and increase workability loss of concrete due to faster hydration reactions as well as loss of mix water due to evaporation. Lignosulphonate-based water-reducing, set-retarding admixtures can be used in hot weather conditions to delay the setting time and alleviate excessive slump loss [3,33].

In an investigation concerned with use of lignosulphonate base water-reducing, set-retarding admixture to offset rapid setting of concrete in hot weather, it was found that use of admixture delayed the set when the ambient temperature was 40-42°C beyond the values of unadmixed concrete at 23-24°C (Figure 29) [33]. When the unadmixed concrete was also allowed to set at the same high temperature, use of admixture delayed the set up to 104 percent, the effect being relatively greater, the lower the water-cement ratio [33].

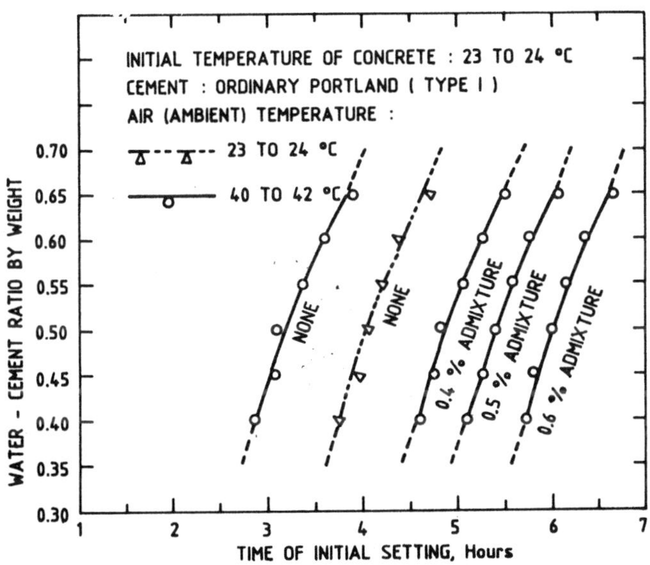

Figure 29. Effect of w/c ratio, air temperature and dosage of lignosulphonate admixture on initial setting time of concrete mortars [33].

USE AS SUPERPLASTICIZERS

As the name indicates, superplasticizers, also called high-range water-reducing admixtures (HRWRA), impart much greater workability to concretes of normal mix proportions and water content than what can be expected with the use of water-reducing agents described earlier. Typically, a concrete mix designed for normal workability of 25 to 50 mm slump, may attain, by incorporation of superplasticizer, a slump of about 200 mm, and remain cohesive without excessive bleeding, segregation or abnormal retardation of setting. This class of application is called 'flowing concrete' [3] and is used in large volume placements, in congested locations, where great amount of reinforcement is present or when concrete is required to be pumped. Superplasticizers are also used for producing high-strength concrete, because reduction in water content up to 30 percent for the same workability is possible. This will allow a low w/c ratio (of the order of 0.25) to be used to attain a 28-day compressive strength of the order of 80 MPa or higher [11,13,15]. Another distinct application is in concrete mixes of low w/c ratios as required from considerations of durability.

Development

The more widely used superplasticizers are based on sulphonated melamine formaldehyde condensates (M-type) and sulphonated naphthalene formaldehyde condensates (N-type), although development and performance of superplasticizers based on modified lignosulphonates have also been reported [4,15]. Rixom and Waddicor had observed that, in spite of general improvement in the purification and separation processes, lignosulphonate materials had played only a minor part in the area of superplasticizers, and their applications have been mainly restricted to incorporation as cheaper dilutants to N-type and M-type formulations. Requirement of increased dosage levels in order to attain high workability, excessive set retardation and air-entrainment due to the presence of undesirable impurities were cited as the main reasons [4].

Development of lignosulphonate materials to overcome these deficiencies, so that these become suitable for superplasticizing admixtures, were reported [4]. The variables in the lignosulphonate molecule, which influence relevant properties of fresh concrete e.g. air-entrainment, set retardation, slump and slump loss with time, were identified. A range of lignosulpholnate products of varying molecular weight, degree of sulphonation, sugar content and cationic type, which were prepared by chemical, ion exchange and ultrafiltration techniques, were investigated [4]. This resulted in identification of a lignosulphonate material, which had been purified by ultrafiltration to remove both the carbohydrates and lignosulphonates of low molecular weights. Sodium lignosulphonate salt was prepared by ion exchange and the degree of sulphonation was kept as high as possible as per the manufacturing process [4]. As is well known, sugar content has to be minimized so that no prolonged set retardation takes place [4,15].

Another example of synthesizing lignosulphonoate products to obtain superplasticizing admixture has been reported, in which spent liquor was subjected to heat treatment to remove sugar content and the resultant product was subjected to ultrafiltration, so that materials of lower molecular weight and sugars were removed [48]. The purified lignosulphonate products having molecular weight in the range of 20,000 to 50,000 were reported to have satisfactory performance as superplasticizer. It was concluded that although such tailored lignosulphonate products may perform as a superplasticizer, the problems of air-entrainment, delayed setting and foaming have to be studied further [15].

In addition to lignin-based superplasticizers as described above, lignosulphonate salts have also been used as auxiliary admixtures in conjunction and M-type and N-type superplasticizers. Their performance in concrete are described below.

Performance in Concrete

Although modified lignosulphonates are recognized as one of the possible bases for high-range water reducing admixtures as well as admixtures for flowing concrete [3,13], information on the performance of such superplasticizers are not as widely reported as those based on sulphonated naphthalene or melamine condensates. General observations like 'the setting time of superplasticized concrete is sometime delayed, especially when lignosulphonate-based superplasticizers are used, but this is of no serious consequence in concrete operations' have been made [13]. Excessive retardation and air-entrainment may arise when lignosulphonate based water-reducers are used in very high dosages to attain 'flowing' workability or significant water-reduction to allow low water-cement ratio in the mix. On the other hand, there are instances when lignin-based water-reducing agents have been used to result in high workability of about 200 mm slump, or compressive strength of about 50 MPa with 90 mm slump and high strength concrete of 80 MPa with or without condensed silica fume; in such cases, the performance have been comparable to those of conventional M- or N-type superplasticizers [40,49,50].

Performance of an effective superplasticizer, developed of a fractionated lignosulphonate raw material of high average molecular weight and low in carbohydrates, has been reported [4]. Comparison was made with conventional melamine and naphthalene-based superplasticizers in terms of extension of workability and compressive strength development of concrete mixes, and similar results were obtained. The reference concrete mix had cement content 350 kg/m^3, water-cement ratio 0.65 and 20 mm MSA gravel aggregate. Assessment of effect of superplasticizers on workability of concrete was carried out in terms of yield value and viscosity characteristics measured on a rheometer [29]. A summary of important results corresponding to the lowest and the highest dosages of the superplasticizers are shown in Table 8. The results of yield value tests indicated that sodium lignosulphonate were more efficient than M-type and N-type superplasticizers. Therefore, a smaller dosage were required to produce a concrete of required workability (e.g. 60 mm flow as per DIN-1048). It was observed that lignosulphonate based formulation resulted in better workability retention than either M-type or N-type superplasticizers. Figure 30 indicates the rate of change of slump with time. The time before the concrete reversed to its original slump was greater with lignosulphonate based superplasticizer and was 1.7 and 1.45 times greater as compared to M-type and N-type superplasticizers respectively [4].

All the admixtures retarded the initial setting time but such retardation was within reasonable limits. The air content was higher with the addition of lignosulphonate than in the control and was also greater than with M or N type superplasticizers (Table 8). While for moderate dose (0.5 percent) of the admixtures, the compressive strength at 1 day was comparable, at the highest dose of 1.5% it was lower for sodium lignosulphonate; with addition of chloride free accelerator (CFA) the initial strength was similar to the control mix. The compressive strengths after 28 days were comparable in all cases.

Investigations of Kobayashi et al., on the performance of calcium lignosulphonate based admixture in concrete in comparison to M and N type superplasticizers, used along with an air-entraining admixture, were discussed earlier. Results summarized in Table 9 show that, for typical concrete mixes of water/cement ratio 0.42, all the three types of admixture or superplasticizers used in appropriate dosages resulted in comparable slumps, and air content was higher in case of lignosulphonate based admixture [35]. Investigations of Artigues et al, which were discussed earlier indicated that a carboxylic polymer-modified lignosulphonate based admixture was as efficient and water reducer as M- and N-type superplasticizers (denoted as admixtures A, C and E respectively) and resulted in greater compressive strength up to 90 days (Table 7) [40]. Results of Ramachandran discussed earlier have indicated that calcium ligno-sulphonate based water-reducing admixtures gave substantially higher values of slump than the reference concrete, but less than M-type

Table 8. Comparison of Performance of Superplasticizers in Concrete [4].

Super-plasticizer	Dose @ 30 w/w%	Yield value, g	Plastic viscosity h	Air Content %	Compressive strength, MPa, 1d	7d	28d
Control	--	--	--	1.1	11.5	33.3	38.5
Na Ligno-sulphonate	0.5	2.229	1.074	2.0	10.5	33.2	38.5
Na Lig+CFA	0.5	2.249	1.104	2.2	12.8	34.4	40.5
N-type	0.5	2.966	1.286	1.1	11.5	31.8	39.5
M-type	0.5	3.524	1.498	1.6	12.0	32.5	40.5
Control	--	--	--	1.5	10.3	33.3	41.0
Na - Lig	1.5	1.280	1.051	2.1	6.0	32.3	39.0
Na-Lig+CFA	1.5	1.234	1.152	2.0	9.8	35.5	43.0
N-type	2.0	0.998	1.073	1.4	11.4	32.0	40.8
M-type	2.67	1.091	1.632	1.2	11.2	33.5	41.0

Note: Ordinary Portland Cement @ 350 kg/m^3 of Concrete
w/c = 0.65
Initial Slump - 50 mm
CFA = Chloride-Free Accelerator

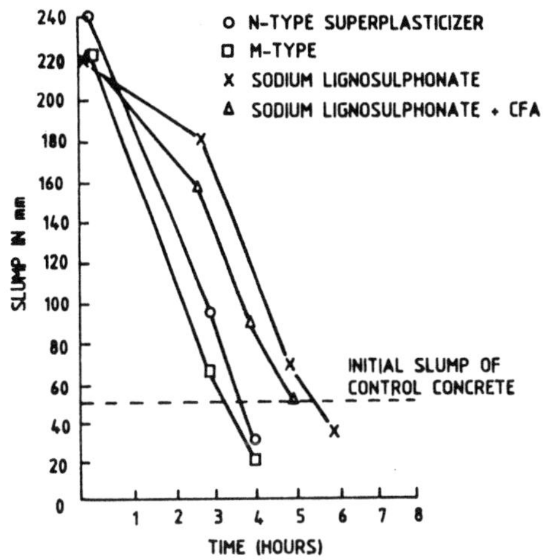

Figure 30. Variation of concrete slump containing different superplasticizers with time [4].

superplasticizer; the latter also resulted in higher compressive strength in mortar [18]. Similarly, slump and slump retention with calcium lignosulphonate admixture were somewhat less efficient than with the M-type superplasticizer; however, the best results were obtained when both were used in combination (Figure 25) [18].

In Combination with Superplasticizers. Other results of the performance of lignosulphonate-based admixtures in combination with M- or N-type superplasticizers are reported by Kisitani et al. [50] and Roberts[39]. In the former investigations, conventional high slump concrete mixes from ready mixed concrete plants were of water-cement ratio 0.55 and cement content 349 kg/m^3 with a nominal 180 mm slump; another mix of nominal 210 mm slump had cement content 333 kg/m^3 and water/cement ratio 0.57. These mixes contained lignosulphonate-based air-entraining water-reducing agent. Flowing concrete of 210 mm slump could be achieved with further addition of N-type superplasticizer, with reduced cement contents (324 and 295 kg/m^3 respectively) and at the same w/c ratios. In general, both types of admixtures behaved comparably in that high

Table 9. Effect of Lignosulphonate Admixtures and M- and N-type Superplasticizers on Fresh Concrete [35].

Admixture Type	Admixture Dosage*	AE Admixture Dosage*	Cement content Kg/m^3	Water content Kg/m^3	Slump mm	Air content %
Ligno	0.25	--	440	185	105	3.0
	0.25	0.0002	438	184	98	3.5
	0.25	0.0004	436	183	90	4.3
M-type SP						
	0.625	--	410	172	90	1.8
	0.625	0.0006	405	170	90	3.0
	0.625	0.0008	402	169	110	3.6
	0.625	0.0011	398	167	108	3.9
	0.625	0.0013	395	166	100	4.7
N-type SP						
	0.34	--	410	172	93	2.5
	0.34	0.0006	402	169	110	3.6

* Percentage by weight of cement, based on dissolved solids
** AE admixture - sodium resin acid

workability was produced and slump loss was comparable. The response of flowing concrete containing N-type superplasticizer to vibration was smaller than that of high-slump concrete produced with lignosulphonate. At the same placement rates, the pressure on form work was also similar. Compressive strengths of concrete were comparable in case of both the types of admixtures [50].

In an investigation on the use of N-type superplasticizer in precast concrete subjected to steam curing, it was found that when a lignin-based water-reducing agent was used as an auxiliary admixture, the curing temperature required to obtain the specified early age strength (35 MPa) decreased from approximately 50°C (with superplasticizer alone) to approximately 37°C [39]. Looking from another angle, incorporation of the lignin-based admixtures in isolation increased the early age strength at 18 hours for curing temperatures of 22, 44 and 66°C; when used along with N-type superplasticizer, the compressive strength was greater than the above and also greater than that obtained with the superplasticizer alone (Table 10) [39]. These investigations indicate that lignin-based water reducers can serve as a good auxiliary admixture to superplasticizers either for control of slump loss [18], for producing flowing concrete of improved placeability [50], or aiding steam curing in precast units [39].

Table 10. Effect of Lignin Admixture and N-type Superplasticizer on Early Age Strength of Steam-Cured Concrete [39].

Dosage, Solid/Cement %		W/C	Slump	18hr Compressive Strength, MPa, at		
Lignin-based water reducer	N-type SP		mm	22°C	44°C	66°C
-	-	0.48	57	14.8	26.5	28.6
0.06	-	0.47	51	19.0	27.4	29.3
0.12	-	0.45	51	21.3	29.3	32.7
0.18	-	0.46	44	19.5	29.4	30.4
-	0.40	0.44	51	26.7	34.3	35.7
0.06	0.40	0.41	44	27.6	37.5	37.4
0.12	0.40	0.41	44	26.4	35.9	37.9
0.18	0.40	0.40	51	24.6	33.9	36.0

Cement - ASTM Type III

Applications in Concrete Containing Industrial Wastes

Use of industrial wastes such as fly ash, granulated blast furnace slag or condensed silica fume as components of Portland cement concrete, either in the form of blended cement, or as another ingredient (mineral admixture) of concrete are well known. Additives like fly ash and silica fume impart advantages when used in concrete, by dint of their high specific surface and pozzolanic reactivity. These act as fillers, where the fine particles form a part of pore water solution, thereby improving the pore size distribution in hydrated cement paste. As a result, a more dense, stronger and less permeable concrete is obtained. In addition, these act as pozzolana, reacting with the calcium hydroxide liberated in the process of hydration of cement and forming further cementitious hydrates. However, addition of such fine powders increase the total specific surface area of the binder and the water requirement in concrete increase. In order to harness the advantage of addition of the fly ash or condensed silica fume without having to increase the water-cement ratio, use is made of water-reducing agents and superplasticizers. Typical results shown in Figure 31 indicate that by addition of a viable water reducing agent, the water content to maintain a given workability of concrete containing silica fume can be held unaltered or even lower than in the mixes without the additives [49].

Bilodeau et al., have described the performance of concrete containing up to 30% low calcium fly ash and its performance in salt scaling resistance tests [51]. The concrete contained an air-entraining admixture (neutralized vinsol resin type) and a lignin-based water reducing agent. Typical results for the three concrete mixes of water-cement and fly ash [w/(c + FA)] ratios of 0.35, 0.45 and 0.55 are shown in Table 11. The results indicate that for comparable slump and air content, replacing the cement with fly ash additions of up to 30% required increased dosages of air-entraining admixture, whereas the requirement of water-reducing agent remained the same. The compressive strength in fly ash mixes were somewhat lower up

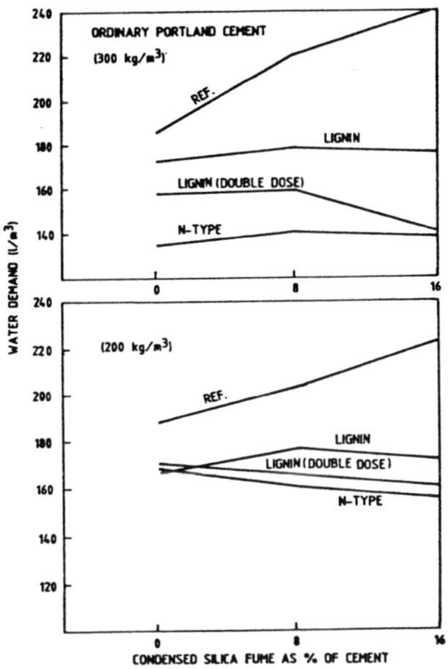

Figure 31. Effect of admixtures in controlling water demand of concrete mixes containing condensed silica fume [49].

Table 11. Properties of Concrete containing Fly Ash and Lignin Admixture [51].

W/(C+F)	Fly Ash %	AEA*	WR*	Slump mm	Air Cont. %	Compressive strength, MPa				
						3d	7d	28d	91d	365d
0.35	0	105	2130	90	7	36	41	47	51	54
0.35	30	205	2120	120	7	24	31	39	50	58
0.45	0	65	1645	90	6.2	27	32	36	40	42
0.45	30	160	1645	85	6.6	16	20	30	42	43
0.55	0	55	1325	70	6.0	21	26	33	34	38
0.55	30	155	1325	100	6.6	10	15	23	31	37

* Dosage in ml/m^3

AEA = air entraining agent; WR = water reducer

to 28 days but, were comparable after one year. Their performance in salt scaling tests will be discussed later.

Artigues et al., have reported results of investigations on the effect of addition of 10% condensed silica fume in concrete with two types of cement and a range of water-cement ratios for different workabilities [40]. Results for two series of concrete mixes with two different cements, one having 80 mm slump and another a slump of about 200 mm, are shown in Table 12. Incorporation of 10% condensed silica fume increased the resultant water-cement ratio required for the same consistency (e.g., mixes 1 and 2, 47 and 48 in Table 12); yet the incorporation of condensed silica fume increased the compressive strength after 90 days, both because of filler action and pozzolanic reaction. Addition of carboxylic polymer modified lignosulphonate water-reducing admixture (denoted as admixture A) in dosages 0.4 and 0.8 percent allowed the water-cement ratios to be lowered in comparison to the plain mixes, even with the incorporation of condensed silica fume. As a result, the compressive strength at all ages were considerably higher (e.g., mixes 1 and 3, mixes 47 and 49). Comparison is also made on the performance of two superplasticizers, one M-type (denoted as C) and another N-type (E). From the results of Table 12, it can be seen that the lignin-based water reducing admixture performed better than the superplasticizers [40].

In another investigation on the performance of concrete mixes with different types of cement incorporating 8% and 16% condensed silica fume, it was noted that the water demand increased with the content of condensed silica fume when no water-reducing agent was used [49]. Incorporation of lignosulphonate type water reducing admixture (denoted as P) in two dosages and N-type superplasticizer (denoted as LD) reduced the water requirement. These admixtures performed better in concrete mixes containing silica fume than in normal concrete. Typical results reproduced in Figure 31 indicate that lignosulphonate-based water reducing admixture was as effective as the superplasticizers, when the dose of former was doubled [49].

Table 12. Effect of Condensed Silica Fume and Admixtures on Compressive Strength of Concrete [40].

Cement Type	Mix No.	Admix. % *	CSF, %	W/C	Compressive Strength, MPa, at days 1	7	28	90
II-Z/35A (Slump-Low)	1	-	-	0.47	9.4	18.4	22.8	24.6
	2	-	10	0.52	9.4	22.1	29.6	32.6
	3	A,0.4	10	0.44	13.0	28.1	39.0	39.8
	5	C,1.4	10	0.46	10.8	25.1	33.7	37.6
	7	E,0.7	10	0.46	9.7	22.3	34.1	34.5
I/45A (Slump-High)	47	-	-	0.57	15.0	22.4	25.9	31.6
	48	-	10	0.65	15.4	28.6	35.3	38.4
	49	A,0.8	10	0.49	20.5	37.0	44.9	47.5
	51	C,2.4	10	0.58	17.8	31.2	38.6	40.3
	53	E,1.2	10	0.57	16.9	31.7	36.9	41.4

* Admixture A - Lignin based
C - M-type Superplasticizer
E - N-type Superplasticizer

EFFECTS ON DURABILITY OF CONCRETE

Performance of concrete in service is dependent not only on the loads, or external and internal stresses and strains which it is subjected to, but also on the service environment to which it is exposed, or to physico-chemical changes due to reasons residing within the concrete. As a result, performance of concrete is being expressed increasingly in terms of its durability so as to ensure satisfactory service life of the structure. Among the various reasons which can cause loss of durability of concrete, corrosion of reinforcement, carbonation and chemical degradation in aggressive environment are more prominent [52]. Frost damage caused by alternate freezing and thawing are important in colder climates.

Since the presence of moisture and ingress of harmful chemicals from the outside are important causes in many modes of degradation, the primary requirement of durable concrete is to make it impermeable, by adopting as low a water-cement ratio as possible [52]. When lignosulphonate-based water-reducing agents or other superplasticizers are used to reduce the w/c ratio of concrete, a denser microstructure with lower capillary porosity in the cement paste is obtained, which results in greater durability of concrete [8,15,53].

Air-entrainment of concrete is necessary for frost resistance. Air-entrainment caused by some lignin-based admixtures could be beneficial, provided the amount and dispersion of air entrained is adequate [4]. Alternately, lignin-based plasticizers can be used in conjunction with an air-entraining admixture to obtain appropriate air-void system in the cement paste [35].

Taflan and Facaoaru have reported the performance of sodium lignosulphonate mixed with sodium alkali-sulphonate based air-entraining water-reducing agent on resistance of concrete to artificial sea-water [54]. Two kind of cements; a sulphate resistant cement containing less than 6% C_3A with addition of slag, and a high early strength cement with higher C_3A content were used in the investigations. Water permeability tests on 200 mm concrete cube specimens under the pressure of 4 Bars showed that the admixture was quite efficient in reducing water penetration, more in case of sulphate resistant cement. Improvement in resistance to sea water due to addition of the admixture, as measured by length change under immersion in artificial sea water, showed improvement in case of both the cements. Results of freeze-thaw tests, the thawing being carried out in sea water conditions, indicated that the resistance of concrete to freezing and thawing and resistance to sea water increased when the admixture was used, both with sulphate resistant cement and rapid hardening cement [54].

Corrosion of steel in reinforced concrete is an electro-chemical process. Adequate protection to the reinforced concrete against corrosion is accorded by the alkaline medium around the reinforcing bars obtaining in good quality concrete [52] . This protection against corrosion can be destroyed if the alkalinity of the surrounding medium is lowered e.g. by occurrence of carbonation or presence of chloride ions beyond permissible limits. The carbonation of concrete depends upon the w/c ratio and increases with it (Figure 32) [53]. Any step, which allow the w/c ratio to be lowered for example, by use of plasticizers, or the porosity of the concrete to be reduced by addition of active pozzolana like fly ash, hydraulic slag or condensed silica fume (CSF), will go to reduce the extent of carbonation and increase protection against reinforcement corrosion. Collepardi et al held that any admixture, which allows reduction in concrete porosity and permeability, can be considered a corrosion-reducer admixture [53]. Such a contention is borne by the results presented in Figure 33, which shows that the amount of chloride ion penetration in concrete after 20 years' exposure to sea air decreased, as the water cement ratio in concrete was lowered. The concrete mixes contained a calcium lignosulphonate type water-reducing admixture [55]. This investigation was concerned with the effects of a number of variables such as, cement type, cement content and w/c ratio, concrete cover thickness, crack width and type of reinforcing steel etc. on the corrosion resistance of concrete beams exposed to sea water. As such, the results shown in Figure 33 should be viewed as qualitative trends.

Results of investigation on the impermeability, resistance to carbonation and resistance against exposure to various aggressive solutions (HCl 1%, acetic acid 5%, lactic acid 1%) have been reported [40,41)]. The concrete was made with two types of cement; one high early strength Portland cement and other blended Portland cement. Condensed silica fume (CSF) at 10% of mass of cement was added with carboxylic derivative modified lignosulphonate-based

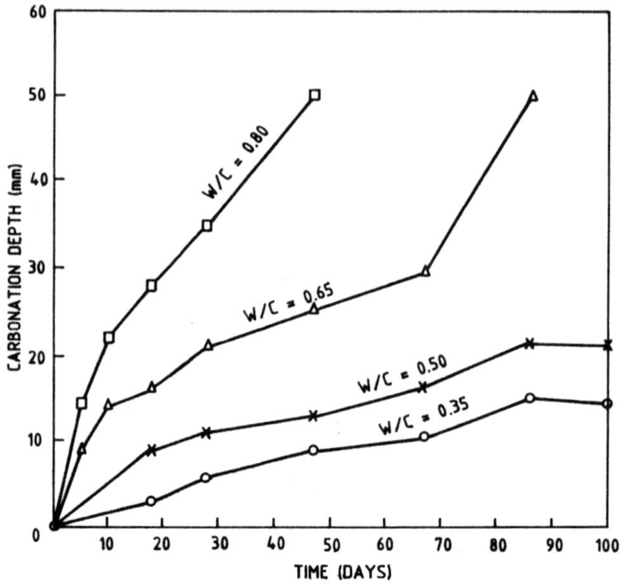

Figure 32. Influence of w/c ratio on carbonation in concrete [53].

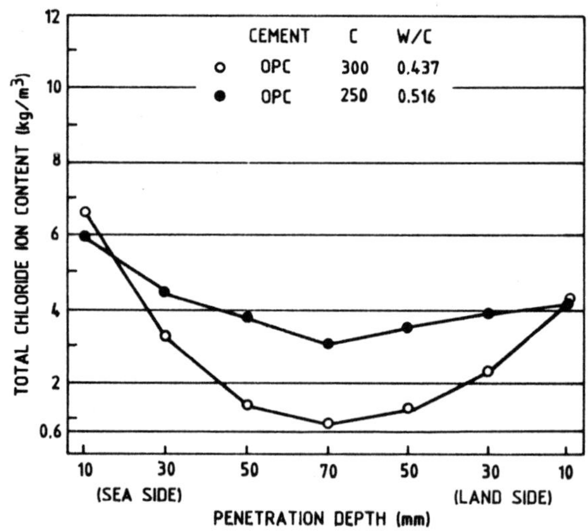

Figure 33. Effect of w/c ratio on chloride ion penetration after 20 years' exposure [55].

water-reducing admixture. Comparison was made with concrete containing other admixtures and superplasticizer. Test for water resistance was carried out on 150x50 mm thick discs made from concrete which were submerged in water-tight containers. Progressive water absorption was measured up to 14 days. Carbonation resistance was measured as per the RILEM CPC-18 method, in which 150x50 mm test pieces are placed in a hermetically sealed reactor, subjected to CO_2 pressures of 4 kg/cm^2 for a period of three months [41]. The resistance against acids was measured on 150x50 mm disc samples exposed to different aggressive solutions for up to 30 days.

Typical results are summarized in Figure 34. Additions of CSF resulted in improvement in durability of concrete and with water-reducing admixtures there was further improvement due to lowering of water cement ratio. Hydrochloric acid was found to be more aggressive to concrete than acetic acid, followed by lactic acid. In general, the modified lignosulphonate admixture performed as well as the other admixtures and superplasticizers [40,41].

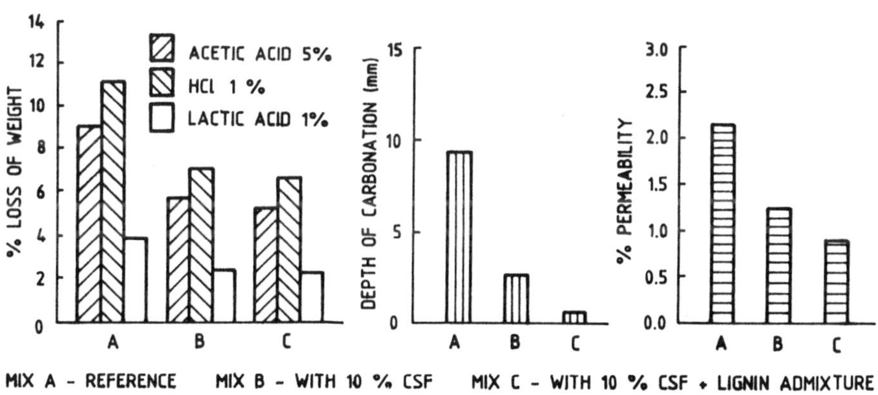

Figure 34. Influence of condensed silica fume and lignin admixture on the durability of concrete [40,41].

Freeze-thaw resistance of concrete containing calcium-lignosulphonate type water-reducing admixture along with an air-entraining admixture (sodium resin acid) was investigated [35]. Comparison was made with the performance of M- and N-type superplasticizers. Satisfactory air entrainment and air-void spacing factor was obtained with the use of lignosulphonate admixture (Figure 35). Air void spacing factor decreased with air content in concrete. Superplasticizers in general entrained less air than lignosulphonate. For a given water-cement ratio of about 0.40, the air content was required to be greater by about one percentage point when superplasticizers were used than in case of lignosulphonate, so as to result in comparable durability factor [35].

The performance of a lignosulphonate based water-reducing agent along with an air-entraining agent in concrete mixes of w/c = 0.65 was discussed earlier [34]. Satisfactory freeze-thaw resistance of such a high w/c mix was obtained in that the relative dynamic modulus was greater than 95% up to 300 cycles of tests as per ASTM C 512. The performance of lignosulphonate admixture was comparable to those of a modified superplasticizer, which was used in mixes of lower water-cement ratio [34].

Mailvaganam and his co-workers reported the performance of a calcium lignosulphonate based water-reducing admixture on the sulphate resistance of mortar mixes when subjected to accelerated soaking in 5 percent sodium sulphate solution [56]. An air-entraining admixture (neutralized vinsol resin) was added to result in 6.8 to 8 % entrained air. Comparison was made with performance of an M-type superplasticizer. ASTM Type-I and type-V cements were used. In the former case, 30 percent cement was replaced by fly ash or slag. Use of water-reducing agent or superplasticizer enabled water-cement ratios to be kept at 0.35 to 0.42. It was found that incorporation of fly ash or slag improved sulphate resistance of Type-I cement mortar to be comparable to Type-V cements. The improvement in sulphate resistance was greater in case of superplasticizer then the ligno-sulphonate water-reducer [56].

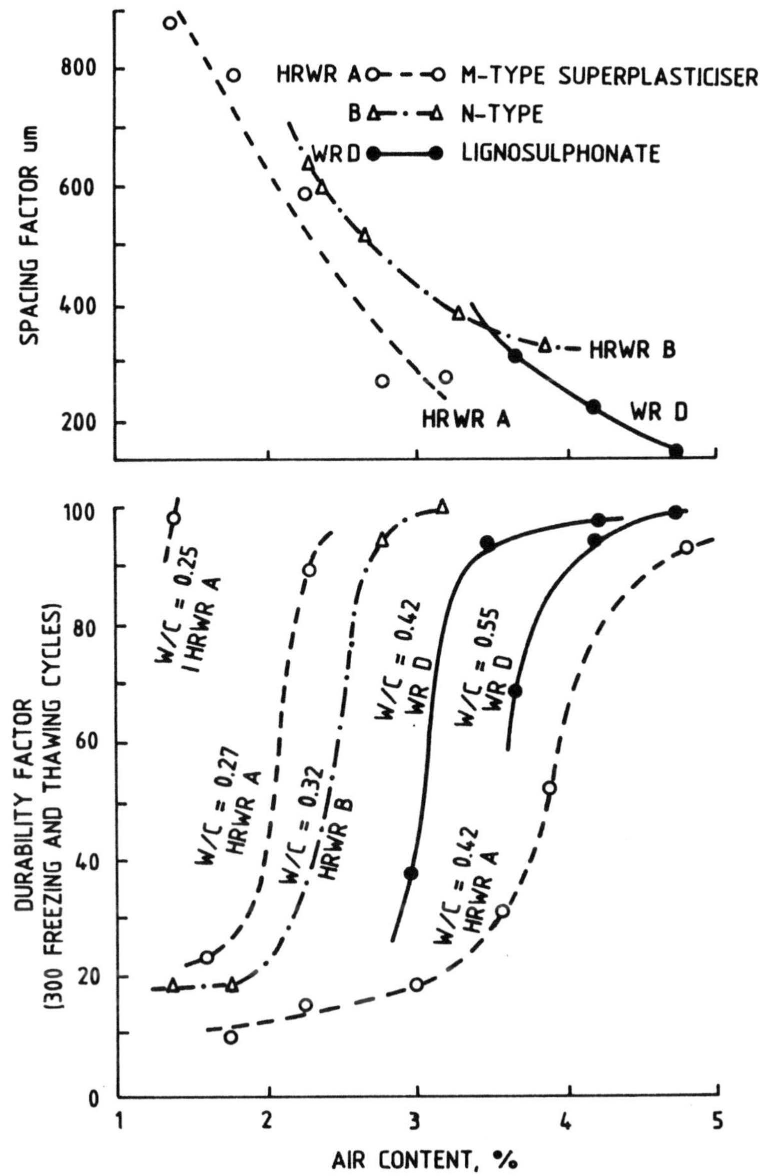

Figure 35. Relationship between air content of concrete, air-void spacing factor and relative durability factor [35].

Bilodeau et al., [51] reported the salt scaling resistance of fly ash concretes which contained lignin-based water-reducing agent and a neutralized vinsol resin type air entraining agent. The tests showed that adequate durability, in terms of air-void parameters, visual rating, as well as mass of scaling residue at the end of scaling test, was obtained, even in mixes with 30% fly ash and at the highest [w/(c + FA)] ratio employed. However, the scaling resistance was greater, the lower the fly ash addition, and particularly, lower the [w/(c + FA)] ratio [51].

In these investigations, the role of lignosulphonate based water-reducing agents in improving the concrete durability was through lowering the water/binder ratio and enabling binders like fly ash [51] or slag [56] or CSF [41] to be incorporated in replacement of cement. This, in summary, is the mechanism through which lignosulphonate based admixtures improve the durability of concrete.

CONCLUDING REMARKS

Spent liquors from wood pulping process give commercial lignins as byproducts. This waste material from the paper and pulp industry finds, among others, application in concrete manufacture, as a source of lignosulphonate-based admixtures. For nearly five decades, lignin have been widely used as the principal base of water-reducing agents. The economic and engineering advantages of an admixture which can improve workability of concrete, effect economy in cement consumption and control set properties, are obvious. With the advent of superplasticizers, the role of lignin based water-reducing agents have been somewhat restricted. Higher dosages can result in greater workability as synthetic (M- and N-type) superplasticizers do, but control over air entrainment and setting times become difficult. Much of the development has, therefore, to be directed towards modifications and purifications of lignins, removal of sugars and carbohydrates and tailoring the molecular structure, so that lignosulphonate products can continue to occupy a major share in the field of concrete admixture.

Application of lignosulphonate-based admixture in concrete has to be properly planned, keeping in mind the particular property that must be improved. Use of such admixtures has to be decided after proper evaluation with the cement, aggregate and other admixtures, if any, the mix proportions and field conditions. Concrete containing lignin admixtures require careful preparation, proper workmanship and quality control as with any other concrete and these admixtures should not be used to compensate for careless field practices.

ACKNOWLEDGMENTS

The author is thankful to Mr. M. S. Uppal, his colleague at the National Council for Cement and Building Materials (NCB) for his assistance in preparing this manuscript and to the Director General NCB, for permission to publish the paper. Dr. V. S. Ramachandran of National Research Council, Canada also provided useful information.

REFERENCES

1. ASTM-C-494-86, Standard Specification for Chemical Admixtures for Concrete, in, *Annual Book of ASTM Standards,* Vol. 04.02, pp. 325-335, ASTM, Philadelphia (1987).
2. IS 9103, *Specifications for Admixtures for Concrete,* 12 p., Bureau of Indian Standards, New Delhi, (1979).
3. ACI Committee 212, *Chemical Admixtures for Concrete, Concrete International,* 15(10): 48-53 (1993).

4. Rixom, M.R. and Waddicor, J., Role of Lignosulphonates as Superplasticizers, in: *Developments in the Use of Superplasticizers,* ACI SP-68, (V.M. Malhotra, ed.) pp. 359-379, Am. Conc. Inst; Detroit (1981).
5. The Cement Admixtures Association, *Concrete Admixtures: Use and Applications,* (M.R. Rixom, ed.) 87 p., The Construction Press, Lancaster (1977).
6. Goheen, D.W. and Hoyt, C.H., Lignin, in: *Kirk-Othmer Encyclopedia of Chemical Technology,* Third Edition, Vol. 14, pp. 294-312, Wiley - Interscience, New York (1981).
7. Hewlett, P.C., An Introduction to and a Classification of Cement Admixtures, in: *Concrete Admixtures: Use and Applications* (M.R. Rixom, ed.) pp. 9-22, The Construction Press, Lancaster (1977).
8. Rixom, M. R., Water Reducing Admixtures for Concrete, ibid, pp. 23-35.
9. Private Communciation, SIV - Magcobar Lignins Limited, Coimbatore, India (1994).
10. Henning, O. and Goretzki, L., Effect of Surfactants on the Hydration Behaviour of Cement Pastes, in: *Proc. 7th International Congress on the Chemistry of Cement,* Vol. IV, pp. 835-840, Paris (1980).
11. Ramachandran, V.S., Use of Superplasticizers in Concrete, *Il Cemento,* 84(3): 273-298 (1987).
12. Ista, E. and Verhasselt, A., Chemical Characterization of Plasticizers and Superplasticizers, in: *Superplasticizers and other Chemical Admixtures in Concrete,* ACI SP-119 (V.M. Malhotra, ed.) pp. 99-116, Am. Conc. Inst., Detroit (1989).
13. Malhotra, V.M., Superplasticizers: A Global Review with Emphasis on Durability and Innovative *Concretes*, ibid, pp. 1-17.
14. Mielenz, Richard C., *Chemical and Air-Entraining Admixtures for Concrete, Cement, Lime and Gravel,* 45(10): 251-256 (1970).

15. Collepardi, Mario and Ramachandran, V.S., Effect of Admixtures, in: *Proc. 9th International Congress on the Chemistry of Cement,* Vol. I, pp. 529-568, New Delhi (1992).
16. Mullick, A.K., Subba Rao, V.V., Narang, K.C. and Maiti, S.C., Evaluation of a Polyoxyethlene Ether-type Concrete Admixture - A Case Study, in: *Proc., National Symp. on Binder Economy and Alternate Binders in Road and Building Constructions,* Part I, 15 p., Central Road Research Institute, New Delhi (1981).
17. Legrand, C. and Wirquin, E., Study of the Strength of Very Young Concrete as a Function of the Amount of Hydrates Formed - Influence of Superplasticizer, Materials and Structures, Res. and Testing, RILEM, 27(166): 106-109 (1994).
18. Ramachandran, V.S., Effect of Retardars/Water Reducers on Slump Loss in Superplasticized Concrete, in: *Developments in the Use of Superplasticizers,* ACI SP-68 (V.M. Malhotra ed.) pp. 393-407, Am. Conc. Inst., Detroit (1981).
19. Collepardi, M., Corradi, M. and Valente, M., Influence of Polymerization of Sulphonated Naphthalene Condensate and its Interaction with Cement, ibid, pp. 485-498.
20. Massazza, Franco, Costa, Umberto and Barrila, Alessandro, Adsorption of Superplasticizers on Calcium Aluminate Monosulphate Hydrates, ibid, pp. 499-514.
21. Nagele, E. and Schneider, U., The Zeta-Potential of Cement Pt. V: *Effect of Surfactants, Cement and Concrete Research,* 18(2): 257-264 (1988).
22. Daimon, M. and Roy, D.M., Rheological Properties of Cement Mixes: II. Zeta Potential and Preliminary Viscosity Studies, *Cement and Concrete Research,* 9(1): 103-110 (1979).
23. Costa, Umberto and Massazza, Franco, Adsorption of Superplasticizers on β-C_2S - Changes in Zeta Potential of Particles and the Rheology of Pastes, *Il Cemento*, 81(3): 127-140 (1984).

24. Ramachandran, V.S., Elucidation of the Role of Calcium Lignosulphonate in the Hydration of C_3A, in: *Proc., 7th International Congress on the Chemistry of Cement,* Vol. IV, pp. 535-540, Paris (1980).
25. Massazza, F. and Costa, U., Effect of Superplastcizers on the C_3A Hydration, ibid, pp. 529-534.
26. Collepardi, M., Corradi, M., Baldini, G. and Pauri, M., Hydration of C_3A in the Presence of Lignosulphonate - Carbonate System or Sulphonated Naphthalene Polymer, ibid, pp. 524-528.
27. Kishitani, K., Tomosawa, F. and Oka, S., *Rheological Study of Superplasticized Concrete, in: Developments in the Use of Superplasticizers,* ACI SP-68 (V.M. Malhotra, ed.) pp. 423-439, Am. Conc. Inst., Detroit (1981).
28. Nishibayashi, Shinzo, Yamura, Kiyoshi and Inoue, Shoichi, Rheological Properties of Superplasticized Concrete, ibid, pp. 441-466.
29. Tattersall, G.H. and Bloomer, S.J., Further Development of the Two-Point Test for Workability and Extension of its Range, *Mag. Conc. Res.,* 31(109): 202-210 (1979).
30. Diamond, S., Cement Pastes: Rheology and Evolution of Properties and Structures, in: *Proc., 7th International Congress on the Chemistry of Cement,* Vol. IV, pp. 113-123, Paris (1980).
31. Khalil, S.M. and Ward, M.A., Influence of a Lignin Based Admixture on the Hydration of Portland Cements, *Cem. and Conc. Research,* 3(6):677-688 (1973).
32. Berke, N.S., Scali, M.J., Regan, J.C. and Shea, D.F., Long-Term Corrosion Resistance of Steel in Silica Fume and/or Fly Ash Containing Concretes, in: *Durability of Concrete,* ACI SP-126, Vol. I, (V.M. Malhotra, ed.) pp. 393-422, Am. Conc. Inst., Detroit (1991).

33. El-Rayyes, M.S., Remedies to Rapid Setting in Hot-Weather Concreting, in: *Proc. Int. RILEM Symp. on Admixtures for Concrete Improvement of Properties* (E. Vazquez, ed.) pp. 120-134, Chapman and Hall, London (1990).
34. Fukuda, M., Mizunuma, T., Izumi, T. and Iizuka, M., Slump Control and Properties of Concrete with a New Superplasticizer, I: Laboratory Studies and Test Methods, ibid, pp. 10-19.
35. Kobayashi, M., Nakakuro, E., Kodama, K. and Negami, S., Frost Resistance of Super-plasticized Concrete, in: *Developments in the Use of Superplasticizers,* ACI SP-68, (V.M. Malhotra, ed.), pp. 269-282, Am. Conc. Inst., Detroit (1981).
36. Ispas, Traian and Ionescu, Ion, Aspects of Cement and Energy Economy in Precasting Plants by the Use of Superplastcizers, ibid, pp. 319-335.
37. Fletcher, K.E. and Roberts, M.H., Test Methods to Assess the Performance of Admixtures in Concrete, *Concrete*, 5(5): 142-148 (1971).
38. Hampton, James S., Extended Workability of Concrete Containing High-Range Water-Reducing Admixtures in Hot Weather, in: *Developments in the Use of Superplasticizers,* ACI SP-68 (V.M. Malhotra, ed.) pp. 409-422, Am. Conc. Inst., Detroit (1981).
39. Roberts, Lawrence R., Cure Temperature Reduction by the Use of High-Range Water-Reducing Admixture, ibid, pp. 337-358.
40. Artigues, J.C., Curado, J. and Iglesias, E., Study of the Effectiveness of Water-Reducing Additives on Concrete with Microsilica, in: *Proc., Int. RILEM Symp. on Admixtures for Concrete Improvement of Properties* (E. Vazquez, ed.) pp. 156-167, Chapman and Hall, London (1990).
41. Idem, Impermeability and Resistance to Carbonation of Concrete with Microsilica and Water-reducing Agents, ibid, pp. 229-240.

42. Polivka, Milos and Klein, Alexander, Effect of Water-Reducing and Set-Retarding Admixtures as Influenced by Portland Cement Composition, in: *Symp. on Effect of Water-Reducing and Set-Retarding Admixtures on Properties of Concrete,* STP 266, pp. 124-139, ASTM, Philadelphia (1959).
43. Tam, C.T., Influence of Cement Composition on Setting Time of Concrete Containing a Set-Retarding Admixture, in: *Proc., 7th International Congress on the Chemistry of Cement,* Vol. IV, pp. 829-834, Paris (1980).
44. Rollet, M., Levy, C. and Cavailles, R., Evaluation of Compatible Superplasticizer for the Production of High-Strength Concrete, in: *Proc., 9th International Congress on the Chemistry of Cement,* Vol. V, pp. 115-121, New Delhi (1992).
45. Guanlun, Zhang and Ping, Ye, Mechanism of and Solutions to Rapid Setting Caused by Addition of Calcium Lignosulphonate Water Reducer to Cement with Fluorogypsum as Retarder, in: *Proc., Int. RILEM Symp. on Admixture for Concrete Improvement of Properties* (E. Vazquez, ed.) pp. 135-141, Chapman and Hall, London (1990).
46. Quon, D.H.H. and Malhotra, V.M., Effect of Superplasticizers on Slump, Strength and Degree of Conversion of High-Alumina Cement Concrete, in: *Developments in the Use of Superplasticizers,* ACI SP-68, (V.M. Malhotra, ed.), pp. 173-187, Am. Conc. Inst., Detroit (1981).
47. Tagnit-Hamou, A., Baalbaki, M. and Aitcin, P.C., Calcium Sulphate Optimisation in Low Water/Cement Ratio Concretes for Rheological Purposes, in: *Proc., 9th International Congress on the Chemistry of Cement,* Vol. V, pp. 21-25, New Delhi (1992).
48. Bialski, A.M. and Nayak, K.V., Quoted in Reference 15 (Part II).

49. Sellevold, E.J. and Radjy, F.F., Condensed Silica Fume (Microsilica) in *Concrete: Water Demand and Strength Development, in: Fly Ash, Silica Fume, Slag and Other Mineral By-Products in Concrete,* ACI SP-79, Vol.II (V.M. Malhotra, ed.) pp. 677-694, Am. Conc. Inst., Detroit (1983).
50. Kishitani, K., Kasami, H., Iizuka, M., Ikeda, T., Kazama, Y. and Hattori, K., Engineering Properties of Superplasticized Concrete, in: *Developments in the Use of Superplasticisers,* ACI SP-68, (V.M. Malhotra, ed.) pp.233-252, Am. Conc. Inst., Detroit (1981).
51. Bilodeau, A., Carette, G.G., Malhotra, V.M. and Langley, W.S., Influence of Curing and Drying on Salt-Scaling Resistance of Fly Ash Concrete, in: *Durability of Concrete,* ACI SP-126, Vol.I, (V.M. Malhotra, ed.) pp. 201-228, Am. Conc. Inst., Detroit (1991).
52. Mullick, A.K. and Chandra Satish, Physico-Chemical, Biological and Thermal Degradation including Environmental Effects, in: *Proc., 9th International Congress on the Chemistry of Cement,* Vol. I, pp 695-734, New Delhi (1992).
53. Collepardi, M., Fratesi, R., Moriconi, G. and Biagini, S., The Use of Superplasticizers as Steel Corrosion Reducers in Reinforced Concrete, in: *Proc., Int. RILEM Symp. on Admixtures for Concrete Improvement of Properties* (E. Vazquez, ed.) pp. 269-278, Chapman and Hall, London (1990).
54. Taflan, F. and Facaoaru, I., Admixtures for Concrete under Sea-Water Action, ibid, pp 360-374.
55. Ohta, T., Corrosion of Reinforcing Steel in Concrete Exposed to Sea Air, in: *Durability of Concrete,* ACI SP-126, Vol. I, (V.M. Malhotra, ed.) pp. 459-477, Am. Conc. Inst., Detroit (1991).
56. Mailvaganam, N.P., Bhagrath, R.S. and Shaw, K.L., Sulphate Resistance of Mortars Containing Superplasticizers with Fly Ash or Blast Furnace Slag, in: *Developments in the Use of Superplasticizers,* ACI SP-68, (V.M. Malhotra, ed.), pp 535-556, Am. Conc. Inst., Detroit (1981).

8

RECYCLING OF WASTE AS AN ALTERNATIVE RAW MATERIAL AND FUEL IN CEMENT MANUFACTURING

Hiroshi Uchikawa and Shunsuke Hanehara

INTRODUCTION

With the advancement of industrial activity and improvement of living standards, the amount of waste generated by society has continuously increased and the treatment and disposal of waste is a serious social problem.

Although large amounts of industrial wastes such as blast furnace slag, fly ash, silica fume, rice husk ash, red mud, by-product gypsum and nonferrous metal slag have been utilized for many years in a large amounts as raw materials and blending components in the cement industry, there are many other unused industrial wastes. For example, combustible wastes such as used tires, oils, sludges, residual dust after combustion, rubber, plastics, wood, agricultural by-products, and incombustible waste such as, coal and aluminum ash, sludge and waste molding sand, and general waste including sewage sludge and incineration ash from urban refuse. Because of the scarcity of land for disposal, the complexity of treatment to make wastes harmless and the high cost of treatment and disposal of wastes, much research is necessary for industry and manufacturing concerns to find uses for the various waste materials.

The cement industry has continuously made great efforts to develop and accumulate the technology for conserving resources and energy and to prevent environmental pollution since it is typical resource and energy intensive industry. Based on positive achievements, the cement industry is one industry capable of diversifying the use of wastes, consuming large amounts of waste and playing a role as an industrial hub for environmental concerns.

In this chapter, the types of wastes will be discussed, as will their character, discharge amounts, current treatment and disposal. The influence of the waste used as alternative raw materials and fuels for cement manufacturing processes will be discusses as will clinker formation. The character and quality of cement are explained and the required equipment investments and management concerns of cement manufacturing are described. Manufacturing of new cements other than conventional Portland cement using large amounts of wastes as raw material and the limit of the mixing ratio of wastes containing harmful elements as components for blended cement are also described.

WASTE

Classification of Wastes

Waste is defined as unused discharge products generated from human life, social and industrial activity. In Japan, wastes are classified as shown in Figure 1 according to Wastes Disposal and Public Cleaning Law. Radioactive industrial waste is regulated separately. The industrial wastes are roughly classified into residential wastes and business wastes. Residential wastes are wastes discharged from human activities and consist of refuse and human

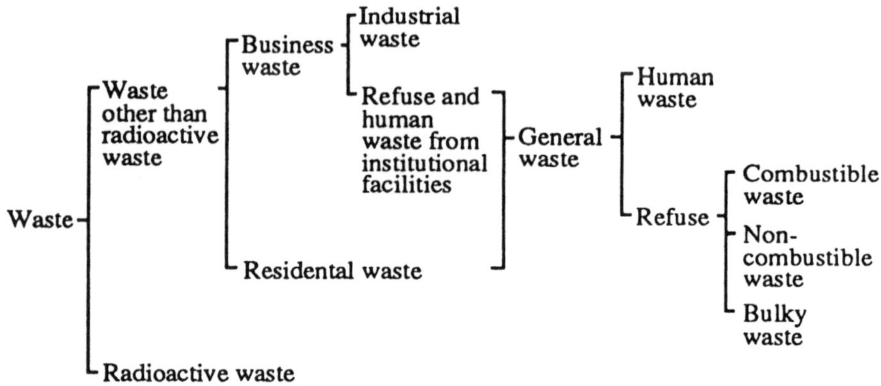

Figure 1. Classification of waste according to Wastes Disposal and Public Cleaning Law in Japan.

waste, and is referred to as general waste. It also includes refuse and human waste generated from institutional facilities. General waste other than human waste is classified as shown in Table 1. Separate collection has been applied for the convenience of treatment and recycling of wastes. Business wastes are wastes discharged from business activities such as agriculture, forest and fishery industries as well as manufacturing processes.

Table 1. Examples of General Wastes Other than Human Waste.

Combustible waste	Paper, Garbage, Fiber, Wood, Plastic, Rubber, Leather etc.
Non-combustible waste	Glass, Earthware and Stone, Metal etc.
Bulky waste	Furniture, Lumber etc.
Recyclable waste	Paper, Rag, Metal, Cullet, Waste bottle, Waste can etc.
Harmful waste	Dry battery, Fluorescent lamp etc.

Types and Amounts of Wastes

In 1990, Japan generated about 50 million tons of human waste. Approximately 60% of the waste was incinerated. Among general wastes, 40 to 50% of steel cans, aluminum cans and glass bottles (cullet) were recycled. The percentage of goods recycled tends to increase yearly due to the popularization of separate collection. Approximately 5 million tons of incineration ash were land filled with unburned general wastes. In Japan there are nineteen industrial waste types. In 1990, 400 million tons of industrial waste were discharged in Japan. Industrial waste consisted of 171.45 million tons (43.4% of the total industrial waste) of sludge discharged from sewage treatment plants and manufacturing factories, followed by 77.21 million tons (19.6%) of manure from animals, 54.80 million tons (13.9%) of construction waste and 42.50 million tons (10.7%) of slag as shown in Table 2. These four types of waste account for 88% of the total industrial waste.

Amounts of industrial wastes discharged are listed by industry category in Table 3. Discharge from manufacturing industry accounts

Table 2. Amount of Industrial Wastes Discharged and Used in the Cement Industry in Japan (1990).

Kind of industrial waste	Amount (thousand ton)	Amount used in cement industry (thousand ton)	Recycling percentage to cement industry(%)
Cinder	2,678		
Sludge	171,450	309	0.18
Waste oil	3,471	141	4.1
Waste alkali	2,674	41 (Waste alkali and Waste acid combined)	0.97
Waste acid	1,547		
Waste plastics	3,544	0.1	0.003
Waste paper	1,193		
Wood waste	6,573	7	0.11
Waste fiber	99		
Remainder of animal origin	3,543	0.2	0.006
Waste rubber	884		
Waste tire	(790)*2	(101)*2	12.8
Waste metal	8,533		
Cullet and waste earthware	5,295	876	16.5
Slag	42,507	16,023	37.7
Blastfurnace slag	(23,985)*2	(12,791)*2	53.3
Molding sand		(169)*2	
Construction waste	54,798	6	0.01
Manure of animals	77,208		
Body of animals	28	0.1	0.36
Dust	7,491	2,498	39.6
Coal ash	(5,828)*2	(2,021)*2	34.7
Others	1,220	120	9.8
Total	394,736	21,546*1	5.5

*1: Containing 1,424,000 tons of desulfurization gypsum, *2: Numerical value in parenthesis indicates the items of an account

Table 3. Amounts of Industrial Wastes Discharged (Categorized by Industry) (1990).

Category of industry	Amount (thousand ton)	Ratio (%)
Agriculture and livestock industry	77,390	19.6
Forest industry	0	0
Fishery industry	61	0
Mining industry	34,000	8.6
Construction industry	71,139	18.0
Manufacturing industry	150,388	38.1
Food industry	11,862	3.0
Paper industry, Pulp industry	27,502	7.0
Chemical industry	12,624	3.2
Glass industry, Ceramic industry	17,492	4.4
Steel industry	48,561	12.3
Utility industry (Electric power, gas and water supply)	54,983	13.9
Transportation and Communication industry	512	0.1
Wholesale trade, Retail sale trade and Food dispensing business	4,054	1.0
Service industry	2,198	0.6
Official business	12	0.0
Total	394,736	100

for approximately 150 million tons (38.1% of the discharged industrial waste). Specifically, the agriculture and livestock industries account for 77.39 million tons, construction industry generates 71.14 million tons, utility industry (electric, gas and water) 54.98 million tons and the steel industry produces 48.56 million tons. These industries are the four major discharge sources.

Character of Wastes

An example of the chemical and mineralogical composition of wastes used as raw materials and fuel for cement manufacturing, other than blast furnace slag, fly ash, silica fume, rice husk ash, lignin wastes, red mud, by-product gypsum and squeezed sugar cane waste, which are described in another chapter of this book, are shown in Tables 4 and 5. Some of them are used as alternative raw material and others are used as alternative fuels.

Waste Alternative to Raw Materials. Fly ash is roughly classified as ash accumulated by dust collectors from coal combustion gases and bottom ash deposited in furnaces. Since the utilization of fly ash as a component for blended cement has been established, new utilizations for ash as an alternative raw material is now focused on bottom ash. The main source is coal-fired electric power plants. Bottom ash consists of 40-50% SiO_2, 11-29% Al_2O_3, 2-25% Fe_2O_3, 0.6-2% CaO and 1% MgO. The ash contains 15-40% crystalline materials including quartz and mullite and about 60% glass. Although the reactivity with limestone at high temperature is generally better than clay and silicious stone, the grindability is often inferior to that of clay. The amount used is determined based on the SiO_2/Al_2O_3 ratio and reactivity. Lime modified sewage sludge is further dewatered and sterilized sewage sludge cake by adding calcined lime. It consists of 55% CaO, 8% water and 30% organic substances. The heat value is approximately 700 kcal/kg. The constituent mineral is mainly $Ca(OH)_2$. It contains slightly larger amounts of P and Cl among minor components and among trace components, respectively.

The composition of waste-molding sand varies depending on the manufacturing process of foundry mold since binders used are different in each process. In general, waste-molding sand consists of 80% or more SiO_2, 1-10% Al_2O_3 and minor components of CaO, MgO, Na_2O and K_2O. Lead and CN are found in slightly larger amounts in the trace components. The major mineral is hard quartz.

Table 4. Chemical Composition of Wastes Used as Alternatives to Raw Materials for Cement Manufacturing.

	Unit	Coal ash	Lime modified sewage sludge	Waste molding sand	Incineration ash of urban refuse	Aluminum sludge
Amount of waste*	kg/t-clinker	ca 2	< 0.01	ca 8		ca 3
Moisture		16.9-51.8	2.5-16.1	1.0-15.4		15.1-67.8
Ig.loss		14.3-21.5	27.9-41.3	0.2-16.5		19.0-44.3
SiO_2		40.8-64.9	0.2-1.8	61.9-97.4	13.7-29.1	0.4-7.8
Al_2O_3		11.3-29.0	1.4-14.5	0.8-9.2	16.4-24.9	42.1-54.0
Fe_2O_3		2.2-24.5	< 2.6	0.6-10.1	4.0-6.8	0.3-1.1
CaO		0.6-2.1	40.2-58.7	0.1-2.1	24.3-39.8	0.1-20.1
MgO	%	0.5-1.1	0.6-1.0	0.4-2.7	2.2-12.1	
S						
SO_3		0.02	0.1		1.4-2.4	
P			0.5			
P_2O_5		0.1-0.5			0.88-2.3	
Na_2O		0.2-0.3	1	0.5-1.2	2.40-3.9	0.17
K_2O		0.6-0.8	0.03	0.3-2.1	2.00-2.93	0.01
Ti						
TiO_2		1.5-1.8			0.87-1.00	
Cl			1,745		5.02-10.6 %	
Zn		31			0.546-0.657 % as ZnO	900
Cr		54-182		14-87	22-2,000	2-62
Cr^{6+}		0.2-1.5		0.5-2		
Mn		170			80-2,500	
SrO						
Ba						
BaO						
F	ppm	3.6				
Cd		0.05-0.2	0.7	0.05-2.3	1-30	0.4-0.9
As		0.3-5.3		0.5-3.4	10-90	0.1-1.6
V					1-200	
Pb		0.2-280	60	1.5-430	100-9,200	0.5-19
Cu		51				
Hg		0.005-0.04	0.04	0.005-0.1	0.1-70	0.1-0.4
Sb						
NH_3						
CN		0.1-1.0	2.2	0.2-5.0		0.1-1.2
Industrial analysis						
Ash	wt %				84.3	
Volatile matter	wt %				29.3	
Fixed carbon	wt %				< 0.01	
Heat value	cal/g				720	
Compound or mineral	—	Mullite	$Ca(OH)_2$	Quartz	CaO, $Ca(OH)_2$, Metalic aluminum, Hauyne	$\alpha-Al_2O_3$

*: Amount of waste used at present

Table 5. Chemical Compositions and Heat Values of Wastes Used as Alternatives Fuels for Cement Manufacturing.

	Unit	Waste tire	Waste oil	Sludge*1	Dust	Waste rubber	Waste plastics*2	Waste wood
Amount of waste*3	kg/t-clinker	ca 5	ca 8	< 1	ca 8	< 0.5	< 0.01	
Moisture		0.3	0.1-0.7	55.8-75.8	10.8-53.3			
Ig.loss				3.6-85.8	34.4-90.3			
Si		160-320 ppm						
SiO_2				0.5-2.0	0.1-2.9			
Al_2O_3				0.6-1.7	0.1-2.1			
Fe		10.7						
Fe_2O_3				0.2-9.4	0.1-5.1			
CaO	%			0.7-5.2	0.1-5.2			
MgO				0.4-6.9	0.1-0.3			
S		1-2	0.2-0.3					
SO_3				0.1-7.3	0.4-8.8		0.1	
P		< 27ppm		0.7ppm	0.0003-0.08			
P_2O_5								
Na_2O				0.04-0.4	0.1-2.6			
K_2O				0.03				
TiO_2					300 ppm			
Cl				93-3,860	232-800		300	
Zn		1-1.6(%)		384-1,810	32-551			
Cr		5.6-54		16-398	4-200	1.0-4.5		
Cr^{6+}				< 1	< 1	< 0.2	< 15	< 0.1
Mn		430-640		51-433	35-340			
SrO								
Ba		8.2						
BaO								
F					65			
Cd	ppm	4.4		0.2-11	0.05-2.3	0.2-1.3	< 4	< 0.15
As		0.9		2.9-95	0.5-180	0.2-0.3	1.6	0.73
V		2.9		5-4100	122-4,500		< 1000	
Pb		29.3		7.7-937	5-250	4.2-19	< 80	< 2.0
Cu		< 54		214-485	14-734			
Hg		< 0.2		0.1-4.5	0.01-2.3	< 0.005	< 0.1	< 0.005
Sb		18						
NH_3					600-20,000			
CN				< 0.5	0.1-1	< 0.05	< 0.4	< 0.3
Volatile matter		50.5						
Carbon	%	26.8						
Ash		13.5						
Calorie	kcal/kg	8,000	10,000-11,000	50-2,350	100-2,000	8,000	9,600-11,000	5,000
H		6.4						
C		78.8		1-25	1-20			
N	%	< 0.9						
S		1.6						
Others		2.5						
Compound or mineral	—	Rhodonite, Spessartite	Hydro carbon		Quartz, Mullite, Glass	Polymer of styrene and butadiene	Poly-ethylene and poly-propylene	Lignin

*1: Residual dust after combusion collected by EP from furnace fuel gas in electric power plant, *2 : Polyethylene resin , *3: Amount of waste used at present

Incineration ash of urban refuse consists of 13-30% SiO_2, 16-25% Al_2O_3, 4-7% Fe_2O_3, 24-40% CaO, 2-12% MgO, 1-2% SO_3, 2-4% Na_2O, 2-3% K_2O, 0.8-1.0% TiO_2, 0.9-2% P_2O_5, 0.5-0.7% ZnO and 5-11% Cl. It also contains significant amounts of Cr, Mn and Pb as trace elements. As shown in Figure 2, incineration ash consists of hauyne, lime (CaO), calcium carbonate, periclase, chlorine containing dicalcium silicate (Belinite) and metallic aluminum.

Waste Alternative to Fuel. Waste tires consist of automobile steel radial tires and are used as is or, as chipped tires 10-20 cm in size. Waste tires consist mainly of styrene-butadiene-copolymer rubber, carbon and steel and contain small amounts of sulfur and zinc oxide. Heat value is approximately 8,000 kcal/kg. Iron, sulfur and zinc are contained in amounts of 10%, 1-2%, and 1%, respectively and affect clinker formation and the quality of cement. Since almost all of the zinc contained in waste tires remains as clinker, the zinc content in clinker increased from several hundred parts per million to ≥1000 ppm when a large part of the fuel is replaced by waste tires. For the case in which 5% of the required heat is replaced with waste tires in a 3000 t/day NSP kiln, the waste tires used are estimated at 6.3 t/hr which causes about 0.05% increase in iron and < tens ppm increase in sulfur and zinc in clinker. Therefore, the effect on the cement quality is small. Since large clusters of belite are formed by the mixing of metallic iron, some reduction of clinker grindability may occur.

Waste oil usually comprises various kinds of oils including lubricant waste oils such as engine oil and vegetable waste oils. The main components of waste oil are carbon and hydrogen though it contains trace amount of Cr, Fe and Ni. The heat value is approximately 10,000 kcal/kg. Such trace amount of metals does not affect the manufacturing of cement. However, attention must be paid not to contain chlorine-containing organic compound in acceptance and the viscosity must be controlled in utilization.

The EP waste sludge and oil sludge contains several percent of magnesium and sulfur, hundreds to thousands ppm of chlorine,

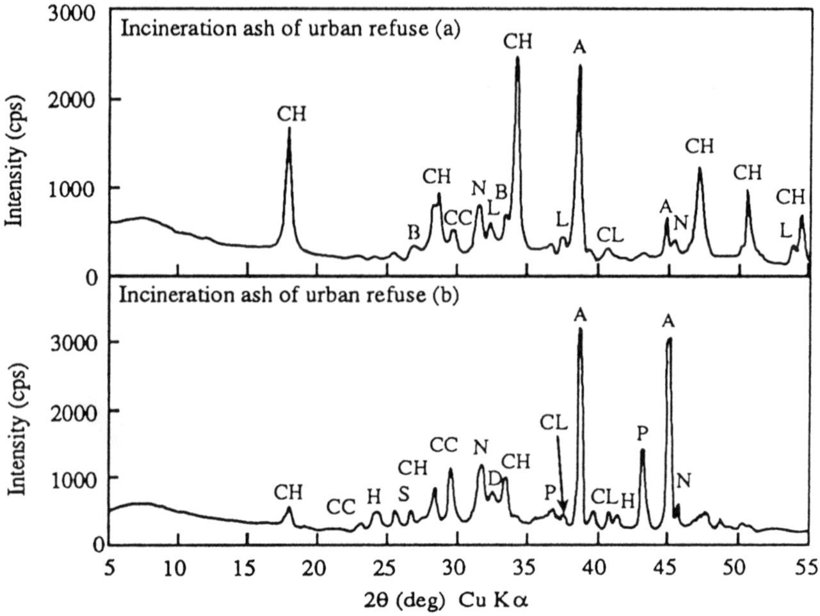

A : Metallic aluminum, L : Lime, CC : Calcium carbonate, CH : Calcium hydroxide, CL : Calcium chloride , H : Hauyne, P : Periclase, N : Sodium chloride, S: Silicon dioxide, D : Dicalcium silicate, B : Belinite

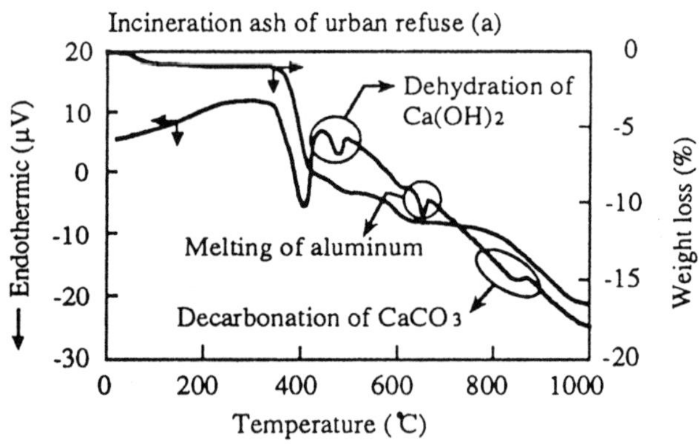

Figure 2. Powder x-ray diffraction pattern and TGA-DTA curves of incineration ash from urban refuse.

vanadium, zinc and lead. The heat value varies from 50-2350 kcal/kg. Amount to be used is decided depending on chemical composition, type of compound, reactivity and easiness of handling.

Residual dust after combustion is floating fine dust collected by an electric precipitator from furnace flue gas. It contains small amount of SiO_2, Fe_2O_3, and CaO besides large amount of carbon and the components other than carbon is similar to that of blast furnace slag. The heat value is hundred to two thousands kcal/kg. It consists of quartz, mullite and so on and about 50% of glass, and it has good reactivity with lime stone. Since ammonia content varies greatly from 600 ppm to 2% depending on whether NOx in exhaust gas is reduced with ammonia or not, attention must be paid to the generation of white plume due to formation of $NH_3(SO_3)_2$ by the reaction with SO_2 in flue gas as well as the increase in fuel NO_x in exit gas.

There are two types of rubber, one is natural rubber mainly consisting of isoprene polymer and other is syntectic rubber mainly consisting of styrene-butadiene-copolymer. Waste rubber, contains carbon, zinc oxide, iron sulfide, antimony, sulfur and chlorine other than main component. The chlorine content is checked cautiously referring the acceptance inspection standard described after. The heat value is approximately 8000 kcal/kg. Waste rubber is used in the same manner as waste tires.

Waste plastics are roughly classified as polyethylene, polypropylene, polystyrene and polyvinylchloride. They consist of 60% polyethylene and polypropylene, 23% polystyrene and 13% polyvinylchloride. Waste plastics other than polyvinylchloride can be used as alternative fuels. The main components are carbon and hydrogen. The plastics also contains small amounts of sulfur, chlorine and vanadium which originate from plasticizers and polymerization catalysts. The heat value is 9600 to 11,000 kcal/g. Large sized plastics such as polyethylene cable sheaths are sectioned into smaller pieces prior to use.

Wood waste contains small amounts of inorganic elements along with carbon, hydrogen and oxygen. The heat value is approximately 5000 kcal/g. It is used in granular or chip form.

Current Situations and Problems of Waste Treatment and Disposal

Current Situations. In Japan, the law imposes duty upon local governments including cities, towns and villages to implement waste treatment and disposal from the standpoint of public health concerns.

General waste is collected by waste collecting tracks of local governments and a portion of the waste is recycled. Combustible refuse is incinerated at facilities operated by the local government and incineration ash is disposed in landfills. Remaining non-combustible refuse is disposed of in landfills. Almost all of the incineration facilities are automated and are operated by computer control systems with dust collectors, desulfurization, denitrification and dehydrochlorination equipment. Generated heat is usually used as heat sources for district air conditioning and to heat water in community swimming pools. The amount of disposed general waste is 20 million tons which corresponds to approximately 40% of total discharged general waste. Disposal costs for general waste in 1990 was 41,000 yen/ton, 80% of which was for collection and transportation and 20% for incineration and treatment. Total cost for disposal of general waste excluding human waste treatment amounts to approximately 2 trillion yen/year.

Although a generator of industrial waste has the responsibility for treating and disposing of its waste, the discharger can select to defer waste treatment and disposal to an authorized waste management firm. Approximately 43% of the total amount of industrial waste (~ 172 million tons) is recycled. The remainder is disposed of in a landfill as is or is incinerated. The amount of disposed industrial waste is ~45 million tons/year which corresponds to approximately 11% of the discharged amount.

Type of industrial waste recycled as resource, discharging industries and material cycles are shown in Figure 3. Percentages of various slags and dust including fly ash are high, and the percentage of recycled wastes to those discharged is high, 80% for slag and 60% for dust. The industries discharging recyclable industrial waste

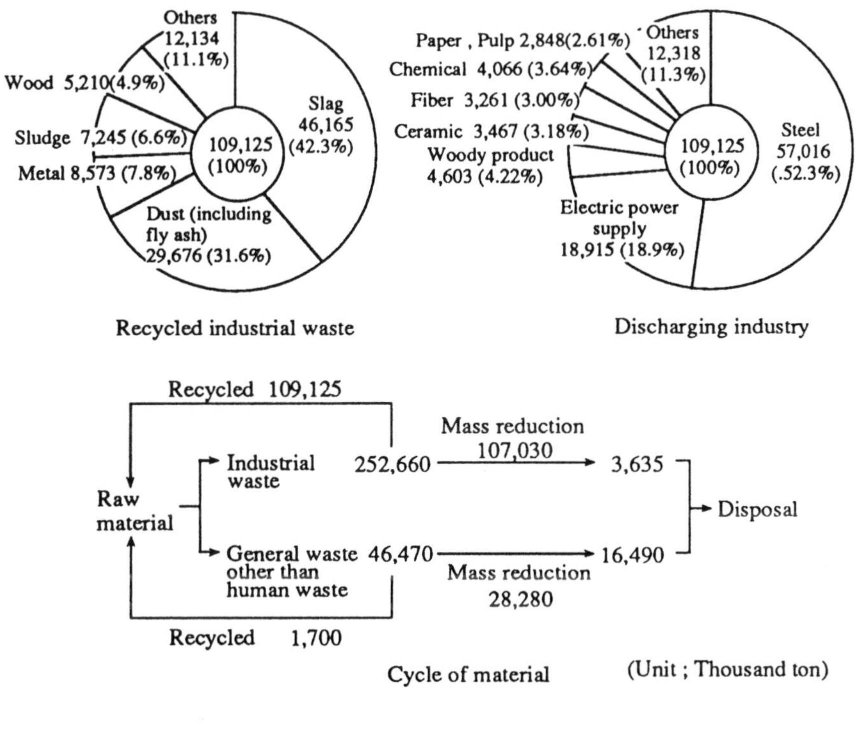

Figure 3. Types of industrial waste recycled as resource, discharging industry and cycle of material in Japan (1987).

include the steel and electric power industries, and 20 million tons of industrial wastes are annually used primarily in the cement industry. Treatment and disposal costs for industrial waste has not been determined precisely, although the purchase price in the cement industry in Japan is 5000 yen/ton for pulverized blast furnace slag and 1800 yen/ton for as-generated blast furnace slag. In contrast, electric power companies pay 2000 yen/ton to cement companies for removal of fly ash.

Problems

Security of Disposal Site. Since the population density, standards of living and industrial activity are high in developed countries, waste to be disposed of is significant despite intense efforts to reduce the amount of waste by recycling and incineration. As described earlier, approximately 20 million tons of general waste and 45 million tons of industrial waste are landfilled or disposed of in the ocean in Japan every year. However, sea disposal regulations are becoming increasingly restrictive. Therefore, land disposal is becoming the primary option for waste disposal. The environmental impact assessment and compensation for fishery rights are required in order to construct new land fill site on land or sea. Because of the difficulty in obtaining construction consent from residents and the fishing industry, as well as increased costs for construction of land fill sites, security of disposal sites is now a major social problem.

Environmental Pollution by Treatment. Sewage sludge consists of human waste is reduced in volume and made harmless by dehydration, sterilization and decomposition with active sludge. Combustible urban refuse is reduced in volume and made harmless by incineration. Generation of poisonous gases, diffusion of volatilized harmful metals into atmosphere and concentration of harmful elements into ash residue are often a result of the incineration process.

By incineration, elements such as Cl, Cd, Hg and Pb contained in waste are partially volatilized and Cl, Cr, Mn, Cu, Zn and As are concentrated in the incineration ash. The Cr^{+3} is oxidized to Cr^{+6} by high temperature heating. To dispose of the treated residue containing high amounts of these harmful elements, further insolubilization treatment is required. If the content of harmful elements exceeds certain value, numerous regulations restrict disposal even if the treated residue is subjected to insolubilization. Chlorine contained in organic waste remains not only in incineration ash as chloride but also generates chlorine gas and may generate virulently poisonous organic chlorine-containing compounds below 1000°C at incineration such as polychloro-dibenzo-p-dioxin (PCDDs), tetra-chloro-dibenzo-p-dioxin (TCDDs) and polychloro-dibenzo-furan (PCDFs) having accumulative properties as shown in Figure 4. Therefore, serious attention should be paid to prevent these compounds from diffusing into the atmosphere.

UTILIZATION OF WASTE AND ENVIRONMENTAL POLLUTION CONTROL

Pollution prevention is an important problem that must be globally tackled for the preservation of our society. The Basic Law for Environmental Pollution Control was established in 1967 in Japan. Moreover, the Air Pollution Control Law and the Water Pollution Control Law were established in 1968 and 1970, respectively. The Environment Agency was established in 1971 for administering the environmental affairs by enforcing environmental policies under a national consensus.

Active, progressive approaches to preservation of the environment have been steadily promoted in the cement industry. Close attention is paid to the reduction of emissions of air pollutants including NO_x, SO_x and CO_2 which may be the cause of global warming and the prevention of water pollution by various harmful materials and of the diffusion of dust to neighboring areas. The development of

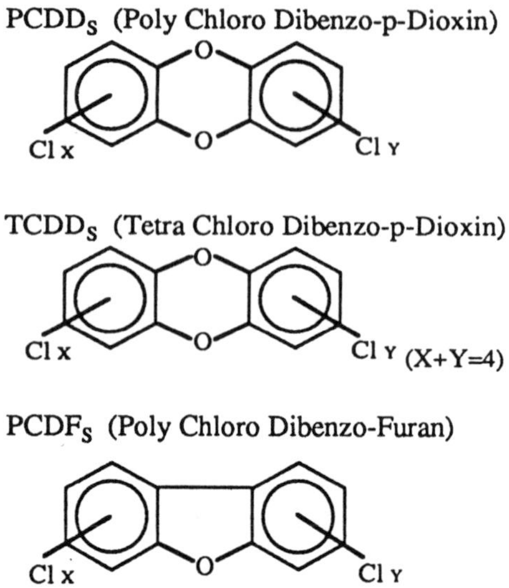

PCDDs (Dioxin) has many isomers which have various poisonousness according to the number and the position of substitutional group of chlorine.

TCDDs is tetrachloride of PCDDs and has twenty two isomers. 2,3,7,8-TCDD has a strongest poisonousness which is supposed to be ten thousand times as strong as that of DDT.

PCDFs, resembles to PCDDs chemically, is called "Pre-dioxin" and also has many isomers.

Figure 4. Chemical structure of dioxins and pre-dioxin.

technologies and the investment in the equipment have been, therefore, positively promoted for meeting the strict regulations. The environmental and emission criteria of harmful matter for the prevention of environmental pollution in Japan are listed in Table 6 [1]. Several measures for improving equipment and tightening controls by reducing or omitting the causative materials that produce air and water pollutants in the cement-manufacturing process have been introduced. Furthermore, reducing or preventing the diffusion of those pollutants by using wastes rather than raw materials and fuels have been implemented to prevent pollution in the cement industry.

Table 6. Environmental and Emission Criteria for Harmful Components in Japan [1].

Environmental criteria

Air

	SO_2	CO	Suspended dust	Oxidant	NO_2
Average*1	< 0.04ppm	< 10ppm	< 0.10mg/m³	< 0.06ppm	< 0.06ppm
Maximum	< 0.1ppm*2	< 20ppm*3	< 0.20mg/m³ *2		

*1: Average amount in a day, *2: Average amount for a hour, *3: Average amount for eight hours

Water

Cd	CN	O-P	Pb	Cr^{6+}	As	Hg	R-Hg	PCB
< 0.01ppm	nd	nd	< 0.1ppm	< 0.05ppm	< 0.05ppm	< 0.5ppb	nd	nd

nd: Not determined

Emission criteria for cement industry

Exhaust gas

- SOx SOx is regulated by K value
  - $q = K\,He^2 \times 10^{-3}$
  - q : Amount of SOx in exhaust gas (Nm^3/h)
  - He : Equivalent height of chimney (m)
  - K : Specified locally by government ranging from 1.75 to 17.5
- Soot < 0.1 g/Nm^3
  - < 0.05 g/Nm^3 (Specified area)
- NOx < 250ppm (In case that the amount of exhaust gas exceeds 100,000Nm^3/hr)
  - < 350ppm (In case that the amount of exhaust gas doesn't exceed 100,000Nm^3/hr

Dust: Regulated according to the criteria regarding the structure, operation, use and control of the equipment

General emission criteria

Waste water

Cd	CN	O-P	Pb	Cr^{6+}	As	
< 0.1ppm	<1.0ppm	< 1.0ppm	<1.0ppm	< 0.5ppm	< 0.5ppm	
Hg	**R-Hg**	**PCB**	**Cu**	**Zn**	**Mn**	**Cr**
< 0.5ppb	nd	< 3ppb	< 3ppm	< 5ppm	<10ppm	< 2ppm

nd: Not determined

Reduction of Air Pollutants

Nitrogen Oxides. Nitrogen oxides existing in the atmosphere are converted to nitrates including peroxyacetylnitrate and peroxyacylnitrate by photochemically reacting with organic compounds such as non-methane hydrocarbons by irradiation with solar radiation. Inhalation of the nitrates may be hazardous to one's health. Acidic rain formed by dissolving nitrates in water vapor in the atmosphere also damages animal and plant life.

Nitrogen contained in the fuel used for burning raw materials is combined with oxygen contained in the combustion air and the nitrogen contained in the combustion air is oxidized at high temperatures within the cement kiln, whereupon nitrogen oxides are produced. The nitrogen oxides produced in the former and latter processes are referred to as, "fuel NO_x" and "thermal NO_x", respectively. Since the combustion gas temperature generated in cement burning is generally 1800°C, the thermal NO_x dominates in the total NO_x as illustrated in Figure 5 [1].

The measures for reducing NO_x in the exhaust gas of the cement kiln have been energetically studied by developing a less NO_x-producing burner for decreasing the concentration of oxygen and reducing the maximum temperature of flame. The retention time of gas in the high-temperature region is also reduced and a fuel-dividing combustion process by partially burning the fuel in the kiln head for sintering the raw materials at a temperature as high as 1450°C or higher and partially burning it for preheating the raw materials at approximately 1000°C in the preheater installed at the tail of kiln has also been developed. An exhaust gas denitrification process has been developed that catalytically or non-catalytically reduces NO_x in the exhaust gas with ammonia.

As a result, a dry process kiln with suspension preheater and precalciner (NSP) provided with a fuel-dividing combustion system has spread because of the stable combustion, lower manufacturing cost, higher quality of product and higher manufacturing rate.

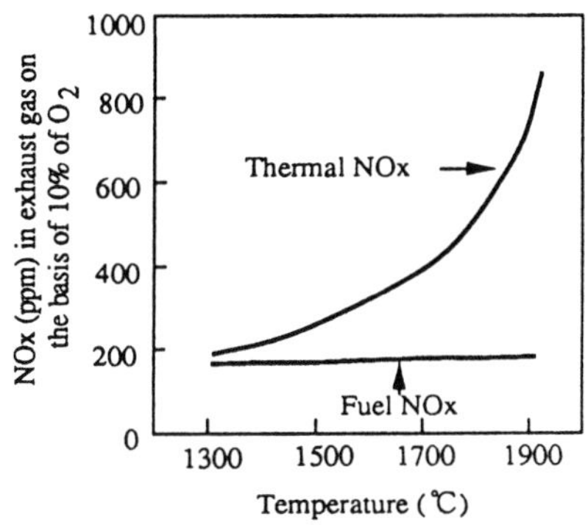

Figure 5. Relationship between burning temperature and thermal NO_x and fuel NO_x produced [1].

When waste is used as fuel, the greatest concern is the nitrogen content in the waste which is the source of fuel NO_x.

Usually, waste tire and waste plastics are fed from the kiln end and oil sludge and waste oil are fed from the precalciner. Therefore, the combustion of these wastes proceeds at relatively low temperatures which result in less generation of thermal No_x

The concentration of NO_x in the exhaust gas of kiln in Japan has been reduced from 400ppm on an average in 1970 to 200 ppm on an average in 1994 on the basis of 10% of O_2 as illustrated in Figure 6 [1].

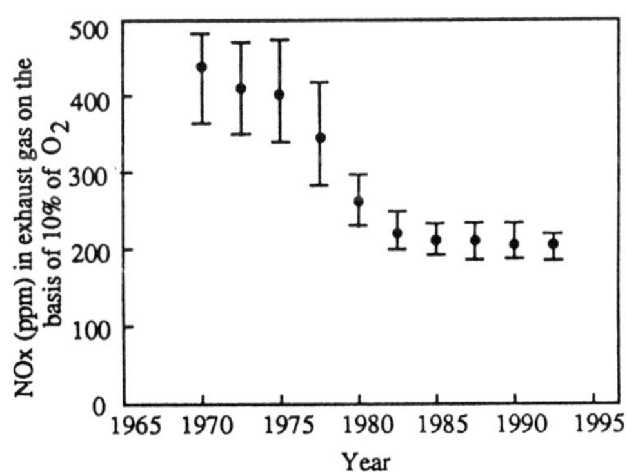

Figure 6. An example of the change of NO_x generated in a cement kiln [1].

Sulfur Oxides. A large quantity of limestone is used as one of the raw materials in cement. Because the limestone decomposes primarily into CaO which absorbs the SO_2 contained in the combustion gas as $CaSO_4$ or its complex, the SO_x, concentration in the exit gas can be maintained at levels of 20ppm or less. Even waste that contains large amounts of sulfur can be used as an alternative fuel. Thus, the cement kiln functions as a desulfurizer, thereby decreasing the concentration of SO_x in the exhaust gas to values as low as 20 ppm.

Carbon Dioxide. The manufacture of cement has intrinsic problems in that large quantities of CO_2 and NO_x are emitted because of the massive amount of limestone used and the high processing temperatures. Accordingly, CO_2 emission reduction requires a decrease in the amount of fuel and limestone consumed. A reduction in NO_x emissions requires a decrease in combustion temperature. For the reduction of fuel consumption, a decrease of the unit consumption of limestone and lowering of the burning temperature is necessary. To achieve the necessary reduction in emissions, clinker composition and an improved blended cement have been implemented and an "*environmental load-reducing cement*" [1-3] is also being developed.

Since some waste contains elements which lower the temperature of clinker mineral formation and the viscosity of clinker liquid, or contains calcium in forms other than carbonate, the suitable use of the waste is effective for reducing CO_2 generation.

Others. Fluorine, chlorine, sulfur, ammonia and heavy metals are often introduced to cement manufacturing system from sludge, waste plastics and incineration ash as alternative raw materials and fuel. However, fluorine and sulfur in exit gas are rarely exceed regulation levels since volatilized fluorine and sulfur react with CaO at high temperatures to form SiF_4 and SO_x and at low temperatures form $CaF_{2,}$ and $CaSO_4$ and sulfate complexes.

Although chlorine is converted to chlorine gas or hydrogen chloride gas in the burning process, these substances are generally captured by reacting with CaO existing abundantly in raw materials to form solid solution of clinker minerals and chlorine-containing double salts such as Na-K-Cl. In spite of this, the temporary generation of chlorine gas and hydrogen chloride gas may cause corrosion in some of the metallic manufacturing equipment. There is also a possibility that dioxins may form at temperatures $< 1000°C$ when chlorine-containing organic wastes such as plastics are used. Therefore, waste should be carefully inspected according to accepted standards.

Reduction of Dust

When dust produced in the cement manufacturing process is discharged from the system, it may cause pneumoconiosis and pollute the soil with cadmium. Dust discharge has been reduced by covering the equipment, converting to a closed-loop process and collecting the dust from the exhaust gas. The quantities of dust discharged from raw material blending, burning and finish grinding processes were significantly reduced by installing a suitable dust collection system and combining the dust collected with the appropriate cyclone, multiclone, bag filter and electrostatic precipitator dusts (according to the size of dust suspended in the gas). The dust content in the exhaust gas of kiln in Japan has been reduced from 0.05 g/Nm^3 in 1970 to 0.02 g/Nm^3 in 1994 as illustrated in Figure 7 [1].

The harmful elements contained in the raw materials are concentrated in circulating or emitted dust by repeating the evaporation and deposition in the manufacturing process. For example, the contents of Cd and Pb in the electrostatic precipitator dust are approximately ten- and three-times, respectively of those in cement. When Cr is oxidized by heating, 10-20% of trivalent Cr is converted to hexavalent Cr. Therefore, it is important to reduce dust emissions by preventing the harmful elements from diffusing into the environment.

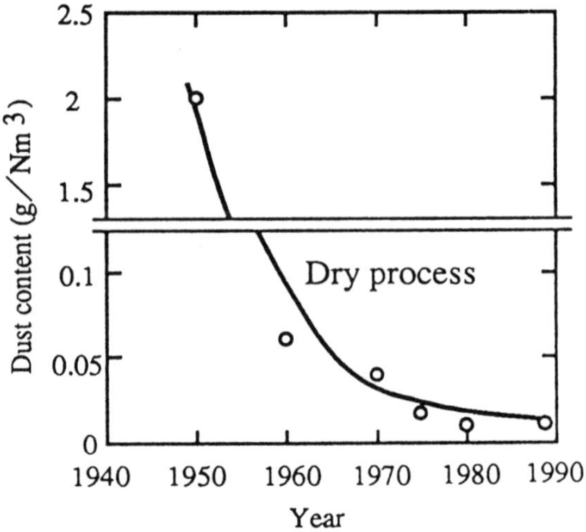

Figure 7. An example of the change in the amount of dust in kiln exhaust gas [1].

Reducing the Content of Harmful Elements

The types and contents of harmful elements contained in a cement are listed in Table 7 [1]. These elements are mainly derived from the raw materials. Figure 8 [1] reveals that Cd is mainly derived from limestone, Pb, Cu, Zn and As are derived from nonferrous metal slag and Cr and Mn are derived from various raw materials. Chromium is also derived from basic Cr-Mg-based kiln refractories.

In using waste as an alternative to raw materials and fuel, unsuitable elements inherent to the waste are brought into the cement manufacturing system. Therefore, depression of the content of such elements should be performed by strictly observing the acceptance inspection standards for waste.

Part of the harmful elements are discharged out of the plant as trace components in the cement and other portions are released

Table 7. Examples of Trace and Harmful Elements in Cement (ppm) [1].

	Mn	Cl	Cr	Cr^{6+}	Cu	Zn	As	Cd	Pb	Hg	V
Cement in 1975	851	5	72	9	125	430	11	1.1	59	45	35
Cement in 1992	1162	6	61	8	138	550	7	1.6	40	38	42

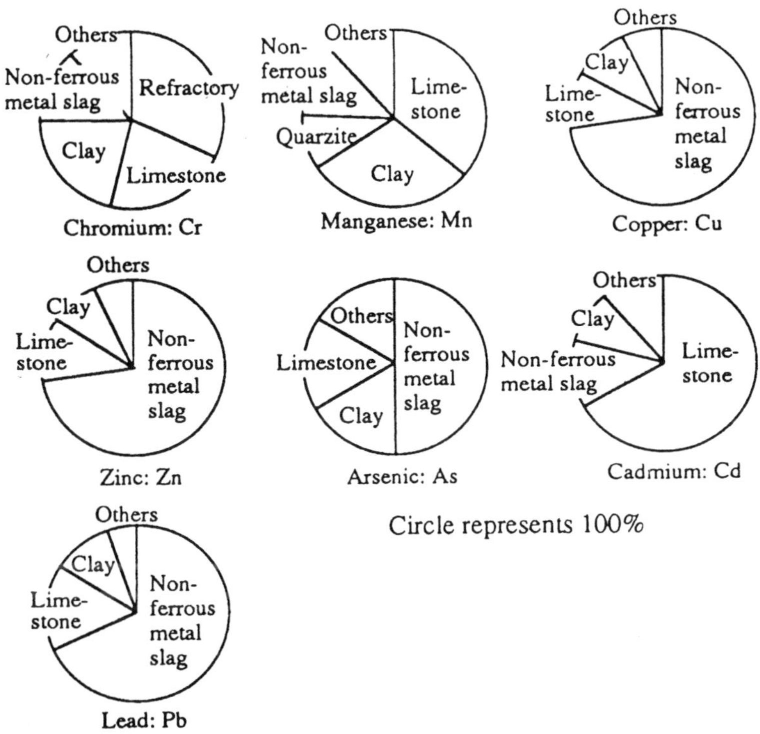

Figure 8. Sources of trace elements in cement [1].

from the plant entrained with the emissions from the manufacturing process. A great deal of effort has gone into the reduction of the harmful elements supplied to the process from the raw materials and ancillary materials by appropriate selection for reducing the content of those elements in the product. Although the range for selecting the raw materials is limited, Al-Mg-based spinel refractory has been substituted for the Cr-Mg-based refractories of burning zone of rotary kiln in order to prevent Cr from getting into the product. The content of harmful elements in the cement are kept almost constant, though industrial wastes have been increasingly used as raw materials and fuels as will be discussed later.

The trace harmful elements include Cr, Mn, Cu, Zn, Cd and Pb contained in cement are dissolved in the hydration process for preparing mortar or concrete by mixing cement, sand, gravel, blending components and admixture with water and are converted to insoluble hydroxides in an alkaline environment provided by cement hydration. Arsenic is fixed as a solid solution in calcium sulfoaluminate hydrate by substituting for SO_4^{2-} and Hg is adsorbed on the surface of C-S-H [4]. Thus, these elements are fixed in mortar and concrete, so they are rarely released into the environment. However, because of the volatility and toxicity of Hg, the introduction of Hg is strictly controlled (in particular, organic mercury). The solubility of tri-valent chromium ion (Cr^{3+}) is $5x10^{-7}$ ppm (10^{-11}M) or less and hexa-valent chromium ion (Cr^{6+}) is approximately 10,000 ppm at pH 12. Since water soluble Cr^{6+} is likely to be dissolved from hardened concrete and create health problems as a carcinogenic substance, the Cr content is strictly controlled.

Reduction of Total Discharge by Energy and Resource Saving

Reduction of total discharge is an important countermeasure not only for increasing the use of waste by-products, but also for diversification of usable wastes. It is also necessary to expand the use of wasteforms as alternative raw materials and fuels by saving energy and natural resources. The trends of the labor productivity and energy consumption in Japan are shown in Figure 9 [1,5].

Introduction of New Manufacturing Processes. Energy conservation in the cement industry has been accomplished primarily by enlarging the scale of equipment to correspond to increased rates of production required to meet increasing demands, switching to newly developed technologies, stabilizing the operation by computer control and decommissioning obsolete facilities. The burning process has changed dramatically over the last twenty years from the dry process, dry process with filter, long kiln in the wet process, shaft kiln, Lepol kiln, long kiln in the dry process, lime-calcination process and SP process to the SP with precalciner, which is the latest dominate technology, as illustrated in Figure 10 [1,2]. The cement industry has successfully overcome two oil crisis by switching fuel from oil to coal as illustrated in Figure 11 [2] and industrial waste has recently been used as an alternative to coal. As of 1992, production of cement in NSP and SP kilns in Japan accounted for 82.9% and 14.6%, respectively, of the total cement production.

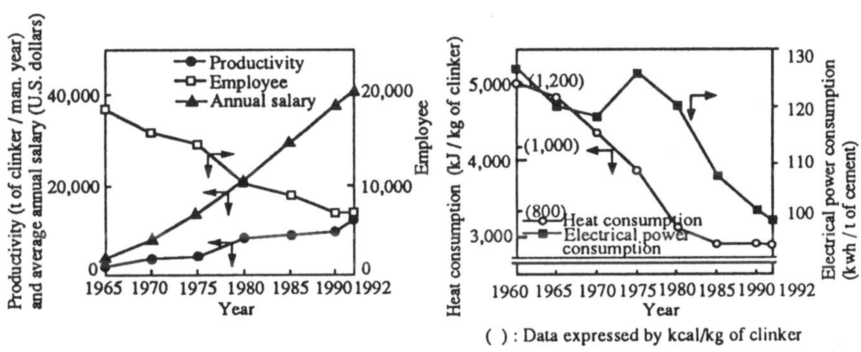

Figure 9. Productivity, number of employees, average annual employee salaries, specific heat and electrical power consumption of the Japanese cement industry [1,5].

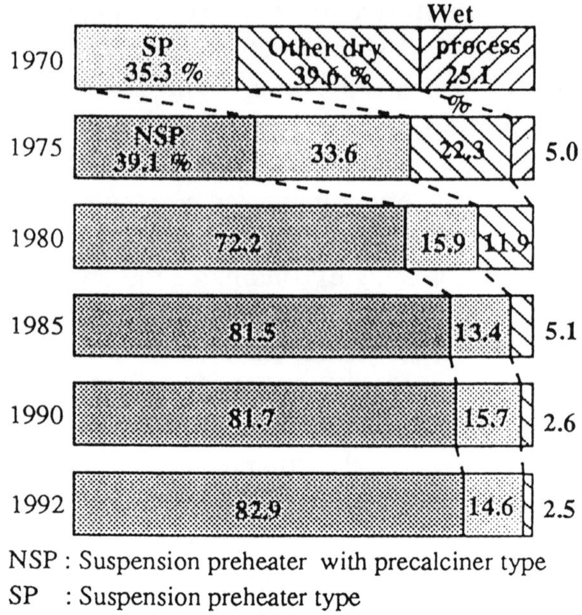

Figure 10. Kiln types in Japan [1,2].

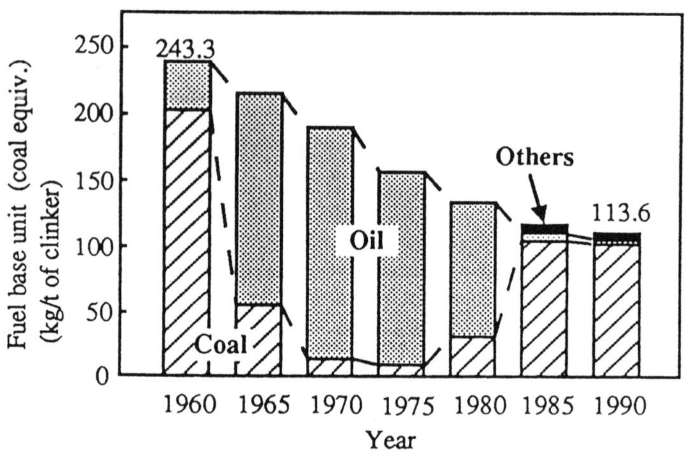

Figure 11. Change of fuel base unit (coal equivalent) [2].

The power required for grinding has been reduced mainly by introducing a hybrid grinding system which consists of a pregrinder such as a roll press or roller mill before proper grinding with a ball mill, applying large roller mills to grind raw materials and clinker and using a high-efficiency classifier [5].

Minor Improvements in the Manufacturing Process. An example of the change of specific heat and electric power consumption are illustrated in Figures 12 and 13, respectively [2]. The specific consumption of heat in Japan decreased from 3260 kJ/kg of cement in 1981 to 3010 KJ/kg of cement in 1990 and the ratio of the consumption of the industrial waste alternative to fuel to the total consumption of heat increased from 3.1% to 8.3%, while the consumption of electric power per ton of cement decreased from 110.5 to 100 kwh in the same time period.

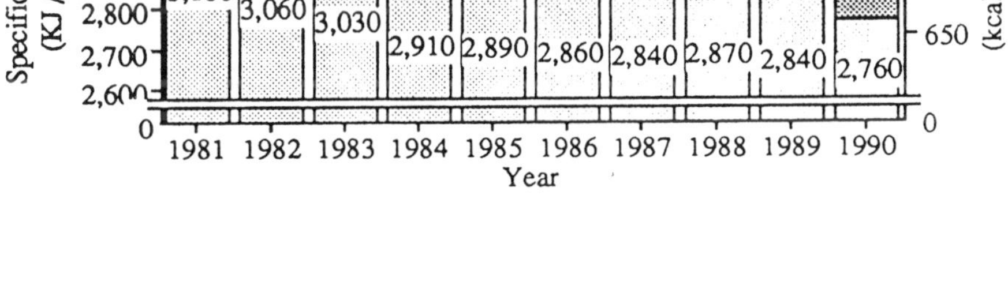

Figure 12. Change of specific heat consumption and amount of industrial waste used for cement combustion [2].

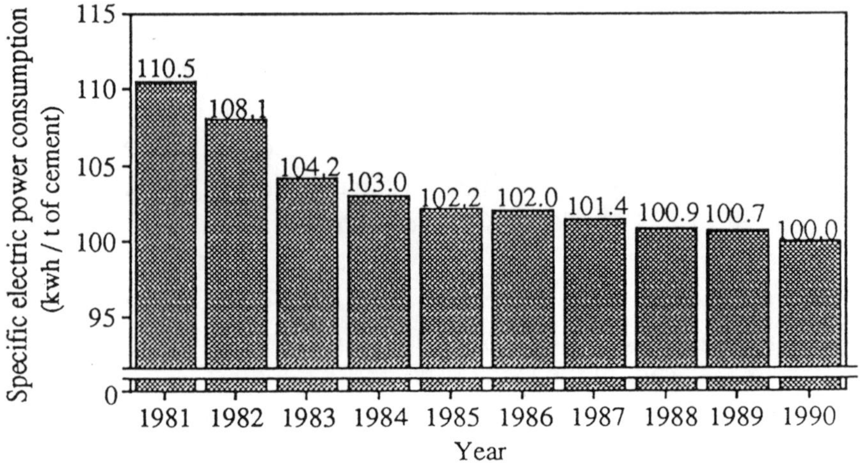

Figure 13. Change of specific electric power consumption for cement manufacturing [2].

The consumption of heat has been reduced by the cumulative results obtained from minor improvements in equipment and operations including the reduction of the carry-over latent and radiant heats of clinker and exhaust gas, the use of reactive raw materials, stabilization of the operation and the use of industrial waste. In addition, the cost has been reduced by using inexpensive oil coke, though it is less combustible.

The consumption of electric power has been reduced in the raw material, burning and finish processes.

In the ball mill system in the raw material-grinding process, energy has been saved by introducing a high performance classifier, enhancing the classification efficiency by improving the selector blade, improving grinding efficiency by introducing new liners with higher ball-classifying and lifting functions including Magotteaux-Slegten, angular spiral and duo-lift liners, by improving the air-sweeping function and by changing the material for bag filter cloth, improving the shape of opening blade ring and reducing the flow rate of air passing through the mill by optimizing the external circulation. In the roller mill system, the grinding efficiency has been improved and the service life of the mill has been lengthened by improving the materials used in the rollers and tables as well as enhancing the design. In addition to the mechanical improvements in the equipment, energy consumption has been reduced by changing the control level including a decrease in the fineness of kiln feed.

The consumption of electric power in the burning process has been reduced by, utilizing an induced fan (IDF), cooler fan and electrostatic precipitator (EP), by reducing the exhaust gas from the preheater, lowering the draft at the inlet of preheater, reducing the electric power consumed for IDF by employing the Scherbius system, optimizing the flow rate of air passing through the cooler, the capacity of the fan and flow rate of gas passing through EP, and reducing the electric power of fan.

In the finish process, the specific consumption of electric power has been reduced while maintaining the quality of cement by installation of a system to control the powder flow rate in the mill, cooling the cyclone of the separator with water and lowering the cement temperature by water spraying in the secondary compartment of the mill.

The content of coarser particles ($\geq$30 μm) in cement was reduced by using a grinding aid, developing a high-performance classifier and applying a roller mill to the grinding of clinker as shown in Figure 14 [1,2]. Since the content of particles that are 5 μm or less in size is decreased and the amount particles between 5 and 30 μm are increased, a sharp particle size distribution is obtained and the n-value in the Rosin-Rammler's formula is increased.

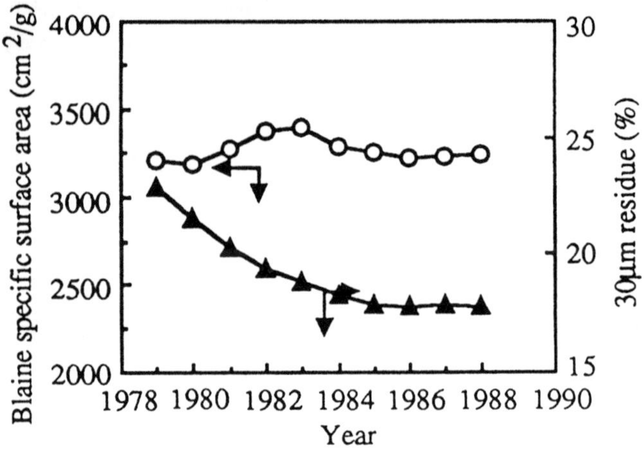

Figure 14. Blaine specific surface area and 30µm residue of cement [1,2].

Change of Control Level

In addition to mechanical equipment improvements, energy consumption has been reduced by altering control levels, including a decrease in the fineness of kiln feed. The development of large SP kilns equipped with a precalciner, a large heat load per sectional area and an increased clinker production rate has resulted in a remarkably improved feed heat transfer. Even though the feed fineness is considerably decreased, clinker sintering reactions occur with greater ease than they do in a conventional kiln. As a result, sintering temperatures can be reduced, as shown in Figure 15 [2]. The results obtained by repeating laboratory and commercial plant tests, taking the reactivity of feed into consideration, indicate that the fineness of feed can be decreased from 12 % to 30 % as the residue over the 88 μm-water sieve or from 4 % to 10 % as the residue over the 200 μm-air jet sieve without diminishing the quality of cement, as shown in Table 8 [2]. When the ball mill is operated at the same capacity as kiln, the weight of the balls can be reduced thereby decreasing the

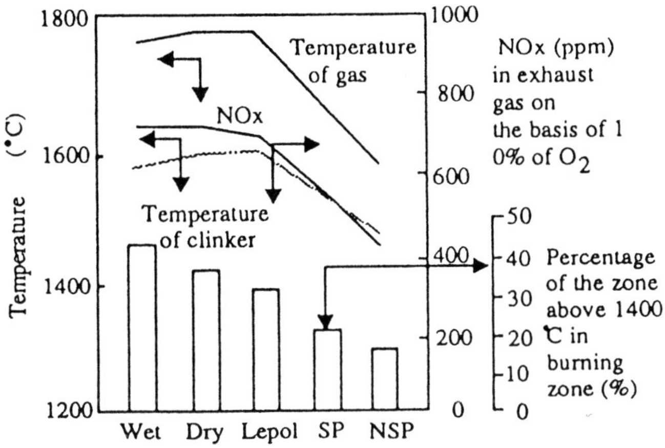

Figure 15. Temperature of clinker and gas and amounts of NO_x produced in the burning zone in various types of kilns [2].

Table 8. Amount of Free-Lime in Clinker and 28 Day Strengths of Mortar with the Change of Raw Meal Fineness [2].

Year	200 μm R*1 (%)	Free lime (%)	28 day compressive strength (N/mm²)*2
1980	4.9	0.7	39.3 (385)
1981	6.0	0.7	40.6 (398)
1982	9.6	0.5	40.8 (400)
1983	11.4	0.6	41.3 (405)
1984	14.1	0.5	41.5 (407)
1985	14.5	0.6	40.9 (401)
1986	17.8	0.3	41.3 (405)
1987	13.8	0.6	42.0 (412)
1988	13.3	0.5	42.0 (412)
1989	13.3	0.5	42.2 (414)
1990	12.6	0.5	42.5 (417)
1991	10.8	0.7	43.8 (429)

Note; *1 : 200μm residue of raw meal measured by air jet sieve, *2 : Test data of JIS mortar, ( ) : Data expressed by kgf/cm 2

*Note: *1:200μm raw meal residue measured by air jet sieve, *2: Test data of JIS mortar; ( ):* Data expressed by kgf/cm^2.

circulation ratio of ground material. The capacity of the transportation system is, therefore, optimized to reduce electric power consumption. By implementing such measures in a commercial plant, the grinding rate was increased 3-5 % reducing the consumption of power per ton of cement by 5 kwh.

Selection of Operating Condition. The volume of water required for keeping a constant fluidity increases with increasing the sharpness of the particle size distribution as shown in Figure 16 [2]. Although high-strength can be developed from cement with a sharp particle size distribution made with a constant water/cement ratio, the quantity of water required for obtaining the constant slump and fly values for practical concrete with appropriate workability and fluidity increases so much, that concrete setting is retarded and strength is decreased.

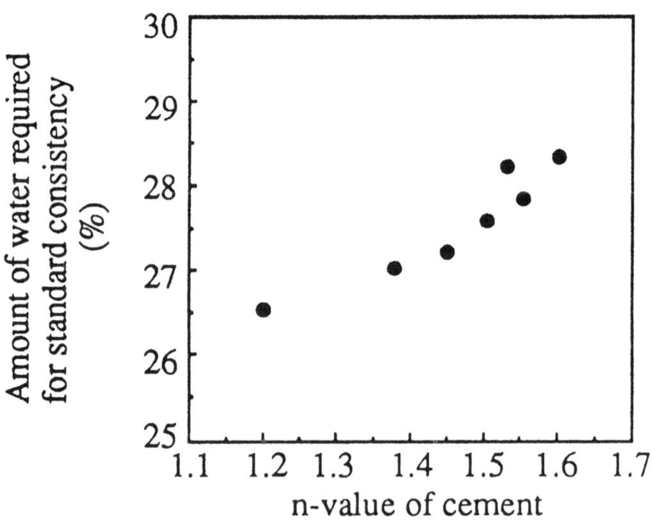

Figure 16. Relationship between n-value and amount of water required for standard consistency [2].

In practical operations, the power consumed in grinding and classification processes required for producing cement particles with the constant specific surface area is not minimized. To achieve a cement with an appropriate particle size distribution of approximately 1.2 of n, the power consumption must be increased to some extent. The fineness and the particle size distribution of a cement that meets the necessary requirements can be achieved by changing the number of revolutions, controlling the air flow in the classifier and regulating the circulation ratio in the grinding and classification systems.

Since the dehydration degree of gypsum is decreased and the state is changed by decreasing the mill temperature in the grinding and classification systems using high-performance classifiers and roller mills, the workability and fluidity of concrete are also affected. Recently, a grinding operation carried out a higher temperatures has been developed.

INFLUENCE OF USING WASTE ALTERNATIVES TO REPLACE RAW MATERIALS AND FUEL ON CLINKER FORMATION AND THE CHARACTER OF CEMENT CLINKER

Minor and trace elements introduced into cement manufacturing system from the wastes not only affect the temperature at which the main melt begins to form, the viscosity of clinker liquid phase and phase equilibrium at the burning temperature, but also are incorporated into clinker minerals to change their composition and crystal structure, or form new compounds. By using the wastes as alternative raw materials and fuel for cement manufacturing, it is important to understand the influence these minor and trace elements can have on clinker formation, character, type and amount of clinker minerals formed.

Influence on Clinker Formation

Magnesium lowers the temperature at which the main melt begins to form, increases the amount and slightly reduces the viscosity of the clinker liquid, and accelerates the formation of clinker minerals. Sulfur stabilizes belite and lime at the burning temperature, increases the size of the alite and reduces the amount of alite. In a reducing atmosphere, sulfur is converted easily to S^{2-} and exists as CaS [6].

Sodium and potassium lower the temperature at which the main melt begins to form and reduce the viscosity. However, viscosity of clinker liquid is increased when the S does not co-exist.

Titanium lowers the temperature at which the main melt begins to form and reduces its surface tension and viscosity. The temperature at which the main melt begins to form decreases from 50 to 100°C per 1% of TiO_2 [7]. As a result, the forming temperature for clinker minerals is also lowered. Titanium can function both as a flux and mineralizer and promotes the formation of alite to a greater extent than does CaF_2 [8].

Small amounts of phosphorus promotes reaction of acidic elements with CaO in the solid phase and crystallization of clinker minerals from clinker liquid]. However, the existence of large amounts of P ($\geq$ 2.5% as P_2O_5) suppresses the formation of alite and several percent more of P as P_2O_5 extinguishes the primary crystallization area of C_3S. The permissible amount of P in the raw mix is increased by the coexistence of F [10].

Fluorine accelerates the decarbonation of limestone [11-15], lowers the temperature at which the main melt begins to form and reduces the viscosity and surface tension of clinker liquid [16]. It also lowers the temperature of alite formation and enlarges the primary crystallization area of C_3S [17]. Although introduced fluorine remains mostly in the clinker as a solid solution below 1400°C, it begins to form $C_{11}A_7{\bullet}CaF_2$ in place of tricalcium aluminate solid solution when the amount exceeds 0.6% [18] and no C_3A is recognized when the amount exceeds 2.5% [19-21]. The effect of fluorine as flux and mineralizer varies depending on the element that bonds with F according to the following order: MgF_2, LiF_2 > NaF, CaF_2, SrF_2 > BaF_2, KF_2 [22] The permissible amount of F in the raw mix is approximately 0.5-0.6%.

Chorine as well as other volatile halogen elements including fluorine, bromine and iodine tend to form alkali halide gases in the kiln and circulate in the combustion system together with combustion gas. Deposited alkali halide adheres on the wall of preheaters, duct surfaces and cooler walls at low temperatures and forms low-melting temperature compounds such as KCl and alkali-bearing complexes. Chlorine promotes the formation of belite acting as a flux and mineralizer [23-25]. Usually less than 0.03% of chlorine is contained in clinker [26]. The amounts of alite and belite in clinker are reduced and tricalcium aluminate solid solution is not observed when large amounts of chlorine are introduced from the raw materials and fuel because alinite ($C_3S{\bullet}CaCl_2$), belinite ($C_2S{\bullet}CaCl_2$) and $C_{11}A_7{\bullet}CaF_2$ are easily formed when of large amounts of chlorine are present.

Chromium tends to be concentrated in clinker so that the amount in scattered dust and circulating dust is less than that found in the clinker [27]. The Cr content in cement is usually less than 100 ppm as Cr_2O_3. Although the formation of clinker minerals are promoted due to reduced viscosity and surface tension of clinker liquid [28,29] when the Cr content is as high as 1% (as Cr_2O_3), clinker formation is reduced with the Cr content is > 1.5% [30].

Manganese promotes the decomposition of limestone [31], the reduction of free lime and the formation of clinker mineral. It acts as a mineralizer below 1% as Mn_2O_3 [32].

Zinc reduces the viscosity of clinker liquid [33-35].

Strontium lowers the temperature of clinker mineral formation and promotes the formation of alite. However, since Sr enlarges the compositional region of free CaO existing at the clinker forming temperature, additional amounts of Sr accelerates the decomposition of alite. The permissible amount of Sr in raw mix is considered to be 3% as SrO [36].

Barium, as well as Sr, acts as mineralizer which promotes the formation of clinker minerals [37].

There are several other trace components which are harmful to human health. Although such elements are usually contained in cement in amounts of 200 ppm or less, there is some possibility for these elements to be introduced in considerable amount to cement manufacturing system when waste is used as an alternative raw material and fuel. Since V, Cd, Hg, and Pb are volatile elements, through the repetition of volatilization and condensation in cement manufacturing process, a major portion is concentrated in dust. However, a portion may be discharged to the environment. Therefore, it is important to fully understand the behavior of these elements.

Vanadium is a volatile element and supplied mainly from fuel and reduces the viscosity of clinker liquid. Usually it is contained in cement in amounts of 20-30 ppm, 200 ppm at a maximum. Vanadium is found in belite in the clinker.

Copper is supplied from iron raw materials including nonferrous metal slag. Copper affects the formation of alite even in small amounts. About 0.5% of CuO has the equivalent effect of 1-2% of ZnO and MnO_2 [38,39]. The content of Cu as CuO in cement is usually 200 ppm or less and most is contained in ferrite solid solution and alite. Attention must be paid when wastes containing large amount of Cu including EP dust from power stations is used as alternative raw materials and fuel.

Elements such as, As, Cd, Sb, Hg and Pb have little affect on clinker formation in the amounts contained in the raw materials.

Influence on the Character of Clinker

Composition. Various elements introduced in cement manufacturing system from waste materials are incorporated into four main clinker minerals as solid solutions during the clinker formation process. The results are changes in the crystal structure or the formation of new mineral coexisting with or replacement of one or more of the major minerals when their content exceeds a certain value. Thus, the waste composition affects the properties of the cement.

Solubility Limit of Minor and Trace Elements in Clinker Minerals. Minerals constituting the cement clinker including alite, belite, tricalcium aluminate solid solution, ferrite solid solution, periclase (MgO), CaO, K_2SO_4 and Na_2SO_4 etc. are formed through the reaction process at high temperatures. Among these minerals, alite, belite, tricalcium aluminate solid solution and ferrite solid solution are referred to as the four major minerals, and the latter two minerals are referred to as interstitial materials. The interstitial materials are further defined as dark and bright interstitial materials, depending on their optical properties. The approximate compositions of these four minerals are $3CaO{\cdot}SiO_2$ (C_3S), $2CaO{\cdot}SiO_2$ (C_2S), $3CaO{\cdot}Al_2O_3$ (C_3A) and $4CaO{\cdot}Al_2O_3{\cdot}Fe_2O_3$ (C_4AF), respectively. Since the composition of ferrite solid solution varies continuously in Al_2O_3/Fe_2O_3 ratio depending on the composition of raw mix and burning condition, it is also expressed as $C_2A_pF_{1-p}$ ($0 \leq p \leq 0.7$). Elements other than those described previously that exist in raw materials are also incorporated into the four major clinker minerals as constituent elements that construct the crystal structure. The solubility limit of the elements in the four major clinker minerals under clinker forming conditions is defined as the maximum amount of an element other than main constituent elements that can be incorporated without decomposing the major clinker minerals and without forming new minerals other than the eight minerals described earlier.

When minor and trace elements exist in the raw mix individually, the measured solubility limit of each of the four major clinker minerals and solubility limit of synthesized clinker calculated from the solubility limit of each clinker mineral are shown in Table 9.

Table 9. Individual Solubility Limits of Minor and Trace Elements in the Four Major Minerals and Calculated Solubility Limits of Synthesized Clinker.

	Clinker mineral				Clinker	
	C_3S ss*1	C_2S ss*1	C_3A ss*1	C_4AF ss*1	Calculated value *2	Measured value
MgO %	2.0	7.2	2.5	3	3.2	2.0
SO_3 %	2.9	4.7	—	—	(2.7)	—
Na_2O %	1.4	2.5	5.7	—	(2.2)	—
K_2O %	1.4	1.2	—	—	(1.1)	—
TiO_2 %	4.5	0.8	1.3	18	4.8	3.8
P_2O_5 %	—	—	—	—	—	—
F	—	—	—	—	—	—
Cl	—	—	—	—	—	—
Cr_2O_3 %	1.7	8.6	—	—	(2.7)	2.0
Mn_2O_3 %	0.7	1.5	0.8	32.6	0.9	—
ZnO %	1.0	0.3	5	6	1.6	—
SrO %	2.3	11.4	—	—	(3.7)	—
BaO %	3.0	11.6	2.5	—	(4.4)	—

*Note: *1: Solid solution, *2: This value is calculated assuming that the contents of C_3S. C_3A and C_4AF solid solution are 60, 20, 10 and 10%, respectively., Figure in parenthesis is a minimum value because solubility limit of C_3A and C_4AF solid solution are not available.*

Apportionment of the Minor and Trace Elements to the Four Major Minerals by the Formation of Solid Solution

Elements which are incorporated into the four major minerals forming solid solutions exhibit specific behavior depending on their ionic radius, electric charge and type of solid solution (substitution or penetration). The minor and trace elements are incorporated into minerals which have crystal structure suitable for the character of element. It is for this reason that the solubility limit and apportionment of minor and trace elements varies depending on each clinker mineral and the temperature and atmosphere of clinker burning. The apportionment ratio and amount of minor and trace elements to the four major minerals in various commercial clinker, chemical composition and mineralogical composition calculated from Bogue's equation , and chemical composition of the four major minerals in various commercial clinker analyzed by EPMA are shown in Tables 10, 11 and 12, respectively. In Table 12, the chemical composition of the four major minerals calculated from apportionment ratio shown in Table 10 and mineralogical composition of commercial clinker shown in Table 11 are listed for comparison. From the results shown in Table 12, the apportionment ratio of minor trace elements shown in Table 10 is considered to be substantially reasonable.

Table 10. Apportionment Ratio and Calculated Amount of Minor and Trace Elements to Four Major Minerals in Commercial Clinker.

	Alite	Belite	Aluminate*1	Ferrite*2	R_2SO_4
Mg	0.65	0.05	0.05	0.25	—
MgO %	1.6	0.4	0.8	3.9	—
S	0.05	0.2	—	—	0.75
SO_3 %	0.04	0.5	0.0	0.0	47
Na	0.25	0.3	0.15	0.1	0.2
Na_2O %	0.08	0.3	0.31	0.21	5
K	0.05	0.15	0.15	0.05	0.6
K_2O %	0.05	0.95	1.0	0.32	46
Ti	0.3	0.1	0.1	0.5	—
TiO_2 %	0.15	0.15	0.31	1.58	—
P	0.6	0.25	0.05	0.1	—
P_2O_5 %	0.11	0.14	0.06	0.12	—
F	—	—	—	—	—
Cl	—	—	—	—	—
Cr	0.45	0.2	0.05	0.3	—
Cr_2O_3 ppm*3	100	100	50	300	—
Mn	0.05	0.05	<0.01	0.9	—
MnO_2 ppm*3	30	90	<40	3320	—
Zn	0.25	0.05	0.15	0.55	—
ZnO ppm*3	250	150	920	3420	—
Sr	—	—	—	—	—
Ba	—	—	—	—	—
V	0.25	0.7	<0.01	0.05	—
V_2O_5 ppm*3	15	120	< 3	20	—
Cu	0.25	0.1	0.15	0.5	—
CuO ppm	75	90	280	950	—

Note;
Upper column : Apportionment ratio, Lower column : Calculated apportionment amount ,
*1 : Tricalcium aluminate solid solution, *2 : Ferrite solid solution,
*3 : The content of Cr_2O_3, MnO_2, ZnO, V_2O_5 and CuO are 100, 350, 590, 33 and 180 ppm, respectively

Table 11. Chemical and Mineralogical Composition of Commercial Clinker.

Kind of portland cement	Chemical composition (%)												Mineralogical composition (%)			
	Ig.loss	SiO_2	Al_2O_3	Fe_2O_3	CaO	MgO	SO_3	Na_2O	K_2O	TiO_2	P_2O_5	Total	Alite	Belite	Tricalcium aluminate solid solution	Ferrite solid solution
Ordinary	0.1	22.3	5.5	3.0	65.6	1.5	0.5	0.20	0.61	0.30	0.11	99.7	60.2	19.9	9.6	9.5
High early strength	0.1	21.4	5.3	2.5	67.9	1.1	0.4	0.19	0.42	0.25	0.12	99.7	73.7	7.8	9.9	7.8
Moderate heat	0.1	23.7	4.0	4.2	65.0	1.3	0.5	0.08	0.52	0.20	0.07	99.7	52.4	30.0	3.6	13.2
White	0.1	24.2	4.4	0.2	69.1	0.8	0.3	0.34	0.09	0.10	0.05	99.7	65.0	22.8	10.8	0.6

Table 12. Chemical Composition of Four Major Minerals in Commercial Clinker Determined by EMPA (a), and Calculated Composition (b).

Kind of portland cement		Alite (%)						Belite (%)						Aluminate *1 (%)						Ferrite *2 (%)					
		MgO	SO_3	Na_2O	K_2O	TiO_2	P_2O_5	MgO	SO_3	Na_2O	K_2O	TiO_2	P_2O_5	MgO	SO_3	Na_2O	K_2O	TiO_2	P_2O_5	MgO	SO_3	Na_2O	K_2O	TiO_2	P_2O_5
Ordinary	a)	1.2	0.06	0.07	0.05	0.25	0.27	0.5	0.47	0.29	0.47	0.32	0.61	1.2	0.04	0.30	0.2	—	—	3.7	0.03	0.33	0.19	1.33	0.06
	b)	1.6	0.04	0.08	0.05	0.15	0.11	0.4	0.50	0.30	0.95	0.15	0.14	0.8	0.0	0.31	1.0	0.31	0.06	3.9	0.0	0.21	0.32	1.58	0.12
High early strength	a)	1.0	0.07	0.04	0.01	0.21	0.21	0.8	0.94	0.48	0.54	0.23	0.19	0.7	0.04	0.50	1.2	—	—	3.5	0.04	0.24	0.20	1.58	0.05
	b)	1.0	0.03	0.06	0.03	0.10	0.10	0.7	1.03	0.73	0.64	0.32	0.38	0.6	0.0	0.29	0.4	0.25	0.12	3.5	0.0	0.24	0.27	1.60	0.15
Moderate heat	a)	0.9	0.06	0.03	0.06	0.07	0.08	0.4	0.38	0.12	0.40	0.13	0.11	0.9	0.03	0.60	1.2	—	—	3.3	0.01	0.07	0.00	0.65	0.01
	b)	1.6	0.05	0.04	0.05	0.11	0.08	0.2	0.33	0.08	0.26	0.07	0.06	1.8	0.0	0.33	2.2	0.56	0.19	2.5	0.0	0.06	0.20	0.76	0.05
White	a)	0.7	0.08	0.28	0.02	0.05	0.04	0.3	0.35	0.62	0.12	0.07	0.03	0.7	0.04	0.67	0.1	0.33	0.02	—	—	—	—	—	—
	b)	0.8	0.02	0.13	0.01	0.05	0.05	0.2	0.26	0.45	0.06	0.04	0.05	0.4	0.0	0.47	0.1	0.09	0.05	—	—	—	—	—	—

*Note: a) Composition determined by EMPA, b) Composition calculated from apportionment ratio of incorporated minor and trace elements in the four major minerals and chemical composition of commercial clinker shown in Tables 10 and 11. *1: Tricalcium aluminate solid solution, *2: Ferrite solid solution.*

Magnesium dissolves into alite and belite substitutes for Ca to form solid solution [40,41]. The Mg substitutes not only for Ca^{2+} but also for Fe^{3+} in ferrite solid solution [42]. The incorporated amount of Mg into the four major minerals is proportional to the amount of MgO in clinker. Of the total amount of Mg, 0.65 is apportioned to alite, 0.05 to belite, 0.05 to tricalcium aluminate solid solution and 0.25 to ferrite solid solution when the MgO content in clinker is less than 2%. The apportioned amount to tricalcium aluminate solid solution is small, corresponding to about 1/5 of ferrite solid solution. When MgO in clinker exceeds 2%, residual MgO exists as periclase (MgO) independently.

Sulfur combines with alkalies to form alkali sulfates including Na_2SO_4 and K_2SO_4 and exists in clinker as alkali sulfates and partially as calcium langnite $Ca_2K_2(SO_4)_3$ when $SO_3/(Na_2O+K_2O)$ ratio is 1.0 or lower. Sulfur begins to be incorporated into clinker minerals when the $SO_3/(Na_2O+K_2O)$ ratio exceeds 1.0 and excess S forms $CaSO_4$ existing independently in clinker. Sulfur is incorporated easily into belite by substituting for Si [43]. The amount of S in belite depends on the amount of alkalies in clinker, that is, higher in clinkers of high $SO_3/(Na_2O+K_2O)$ ratio. Seventy-five percent of S in clinker is apportioned to alkali sulfates, 5% to alite and 2% to belite.

Sodium is incorporated selectively into calcium aluminate [44] when S does not exist. To maintain electrical neutrality, half of the Na atoms are incorporated into the center of one of the Al_6O_{18} rings, and remaining half substitutes for Ca(III)(IV) to form orthorhombic tricalcium aluminate solid solution. Of the Na in the clinker, 0.25 is apportioned to alite, 0.3 to belite, 0.15 to tricalcium aluminate solid solution, 0.1 to ferrite solid solution and 0.2 to alkali sulfates.

Potassium is incorporated into clinker minerals as well. About 0.05 of K in clinker is apportioned to alite, 0.15 to belite, 0.15 to tricalcium aluminate solid solution, 0.05 to ferrite solid solution and 0.6 to alkali sulfates when S coexists.

Titanium is incorporated into alite by substituting for Si as $(SiO_4)^{4-}$ $(TiO_4)^{4-}$ [45]. Because the ionic radius of Ti^{4+} is nearly equal to that of Fe^{3+} and the bonding energy of Ti-0 bond is very large, Ti substitutes for Fe^{3+} in $(FeO_4)^{5-}$ tetrahedron. In calcium aluminoferrite and calcium aluminate, Ti substitutes for Al^{3+}, as well [46,47]. About 0.3 of the total Ti in clinker is apportioned to alite, 0.1 to belite, 0.1 to tricalcium aluminate solid solution and 0.5 to ferrite solid solution.

Almost all of the P contained in raw mix remains in clinker regardless of the atmosphere or burning temperature as shown in Figure 17 [48]. Phosphorous tends to be concentrated in calcium silicate rather than in interstitial materials and is apportioned more to alite than to belite. About 0.6 of the total P in clinker is apportioned to alite, 0.25 to belite, 0.05 to tricalcium aluminate solid solution and 0.1 to ferrite solid solution.

The Cr contained in raw mix remains mostly in clinker regardless of burning atmosphere and burning temperature. Chromium substitutes for Ca or $(SiO_4)^{4-}$ in alite as $(CrO_4)^{3-}$ and $(CrO_4)^{4-}$ as described below [49,50]:

$$2Si^{4+} \rightarrow 2Cr^{5+} + Ca\square^{2-},$$

$$Si^{4+} \rightarrow Cr^{4+}$$

where $Ca\square^{2-}$ denotes a vacancy of Ca in the lattice. On the other hand, Cr^{6+} substitutes for Al^{3+} in C_3A [51,52] and substitutes for Fe^{3+} and Al^{3+} of octahedral site in $C_6A_pF_{1-p}$ ($0 \geq p \geq 0.7$) [46]. Since Cr is oxidized at high temperatures, 10-20% of total Cr in clinker is hexavalent and the residual is trivalent or tetravalent, the average number of oxidation of Cr in clinker is estimated to be 4.6 [53]. As shown in Figure 18, Cr is distributed in high concentration in belite and interstitial materials, especially ferrite solid solution. 0.45 of total Cr in clinker is apportioned to alite, 0.20 to belite, 0.05 to tricalcium aluminate solid solution and 0.3 to ferrite solid solution, respectively. Excessive Cr combines with Cu and deposits on tricalcium aluminate solid solution as spherical particles with sub-micrometer in size.

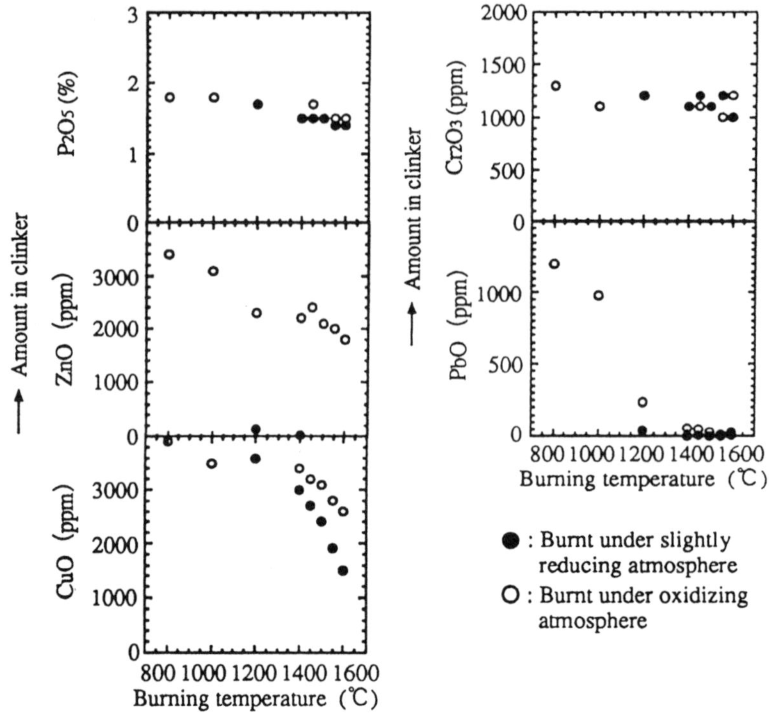

Figure 17. Change of volatilization of added minor elements with burning temperature under oxidizing and slightly reducing atmosphere [48].

Mn^{4+} substitutes for Si^{4+}, and Mn^{3+} and Mn^{4+} substitute for Al^{3+} in oxidizing atmosphere. Mn^{2+} substitutes for Ca^{2+} in reducing atmosphere [54]. Ninety percent of the total Mn in clinker exists in ferrite solid solution. Of the total Mn in clinker, 0.05 is apportioned to alite and belite, respectively, but it rarely exists in tricalcium aluminate solid.

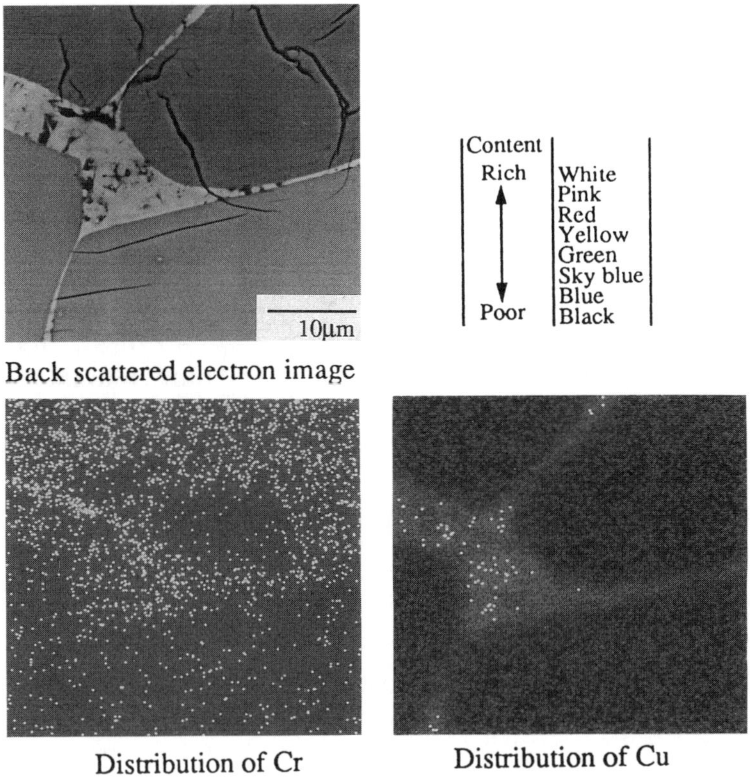

Figure 18. Back-scattered electron image and distribution of chromium and copper on the polished surface of clinker.

The amount of Zn remaining in clinker decreases with an increase in burning temperature. Approximately 60% of the Zn that initially existed in raw mix remains in clinker burnt at 1450°C but only 5% of it remains in clinker burnt in an reducing atmosphere. Zinc substitutes for Ca^{2+} in alite, Si^{4+}, Fe^{3+} and Al^{3+} in tricalcium aluminate solid solution and Fe^{3+} in ferrite solid solution [51]. Zinc exists in interstitial materials, especially in ferrite solid solutions at high concentrations. Of the total Zn in clinker, 0.25% is apportioned to alite, 0.05 to belite, 0.15 to tricalcium aluminate solid solution and 0.55 to ferrite solid solution.

Strontium substitutes for Ca in clinker minerals. For belite, Sr^{2+} substitutes for Ca^{2+} at any proportion until the composition reaches to $(Ca_{1.8}Sr_{0.2})SiO_4$. For alite, half of the Sr atoms substitute for Ca^{2+} to form solid solution. The final composition of alite is $(Ca_{2.95}Sr_{0.05})SiO_5$ with sufficient amounts of Sr [55]. Barium substitutes for Ca in clinker minerals.

Vanadium substitutes for Si^{4+} in clinker minerals. Apportionment of V to alite, belite, tricalcium aluminate solid solution and ferrite solid solution is 0.25 : 0.7 : <0.01 : 0.05.

The amount of Cu remaining in clinker decreases with an increase in burning temperature. About 80% of the Cu initially found in the raw mix remains in clinker burnt at 1450°C. The burning atmosphere has little affect on the volatilization of Cu. Copper substitutes for Ca^{2+} and exists in interstitial materials, especially in ferrite solid solution at high concentration as shown in Figure 18. The Cu in clinker is apportioned to alite, belite, tricalcium aluminate solid solution and ferrite solid solution in ratios of 0.25: 0.1: 0.15: 0.5.

Arsenic is usually contained in cement 30 ppm or less and more than half of which is supplied from nonferrous metal slag. The As is substituted easily with Al and AsO_4 is substituted easily with SO_4 and tends to exist in interstitial materials.

Cadmium is supplied mainly from limestone and its content in cement is usually 1 ppm or less. Antimony is supplied mainly from nonferrous metal slag and its content in cement is usually 40 ppm or less. The content of Hg in cement is tens ppb or less.

Lead is supplied from nonferrous metal slag and its content in cement is usually 300 ppm or less. Special care must be taken as waste oil sludge occasionally contains considerable amount of Pb which may exceed 1000 ppm. The Pb substitutes for Ca^{2+}. The volatilization of Pb is remarkable for high temperature burning in a reducing atmosphere. Almost all of Pb in the raw mix volatilizes and little Pb remains in clinker burned at 1450°C.

Amount of Clinker Minerals

Among various minor and trace elements introduced into cement manufacturing system accompanied with waste alternative to raw material and fuel, some elements changes the amount of four major minerals produced in clinker. For example, Mg changes the composition of ferrite solid solution to alumina rich composition, consequently amount of tricalcium aluminate solid solution and belite decreases and that of alite increases[(42,56-59]. When equivalent substitution of TiO_2 for SiO_2 does not take place, the amount of alite decreases and belite increases with the increase of TiO_2 amount without any change in the amount of interstitial materials [36]. Changes of belite composition to CaO rich side and compositional change in invariable point at 1338-1341°C in the system CaO-SiO Al_2O_3 -Fe_2O_3 are considered to be responsible for the charge on the clinker minerals produced with Ti [36].

Chromium decreases the amount of alite and increases the amount of belite [60]. Manganese increases the amount of ferrite solid solution [61].

Texture and Structure

Minor and trace elements introduced into cement manufacturing system accompanied with waste affect the temperature at which the main melt begin to form, viscosity and surface tension of clinker liquid and phase equilibrium of clinker mineral as described earlier, consequently the shape, grain size and texture formed by clinker minerals are also changed. The modification and crystal structure of clinker minerals stabilized at room temperature is also changed by the incorporation of minor and trace elements.

Texture of Clinker. When Mg exceeds 2% as MgO, cubic periclase crystal with a grain diameter on the order of tens of microns coexists in clinker as shown in Figure 19. When dolomitic limestone is used as a raw material and the concentration of MgO exceeds 2%

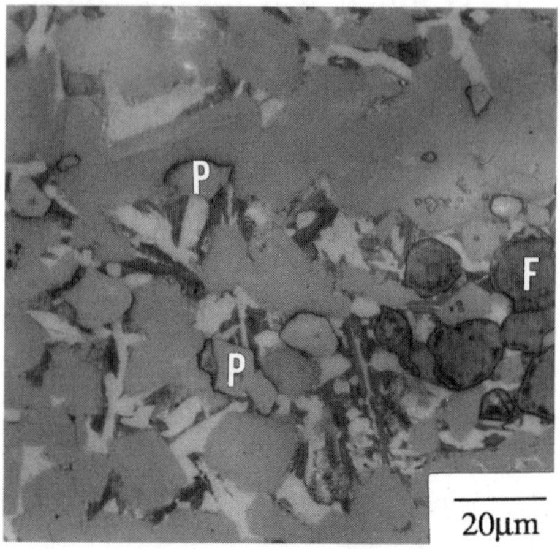

P : Periclase, F : Free CaO

Figure 19. Optical micrograph showing cubic periclase crystal in clinker (transmitted light).

locally due to heterogeneous mixing of raw material, small sized periclase with particle diameter of several microns is formed as shown in Figure 20. Increased Mg amounts promote the formation of nuclei of alite crystal and well-crystallized prismatic hexagonal alite is formed. For belite, the width of lamella corresponding to α' modification is extended under the coexistence of Mg and K. Ferrite solid solution changes its color to dark brown and shape to the dendritic form, increasing its birefringence.

Sulfur promotes the growth of alite crystals. The crystallization of alite becomes unstable [62] and various impurities, primarily large amounts of Al, are incorporated into alite crystal by co-existing with sulfur. As a result, zonal structure of alite is often

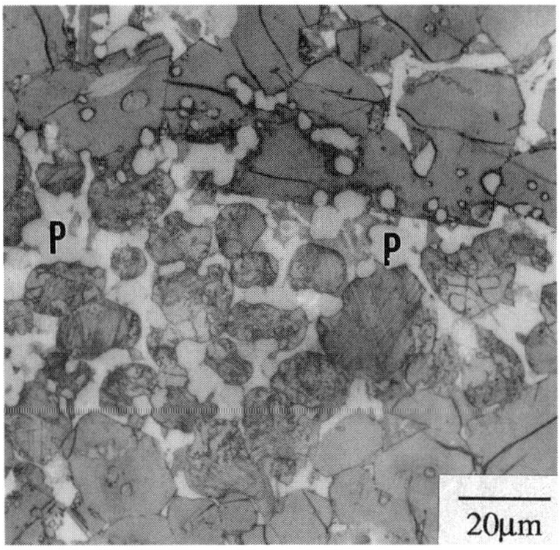

P: Periclase

Figure 20. Optical micrograph showing periclase in clinker produced by using dolomitic limestone as a raw material (reflected light).

observed when clinker is burned at high temperatures [48]. When the amount of S in raw mix exceeds several percent, the crystal size of belite becomes ≥20 microns [63] and large-sized belite coexists with free CaO. When the combustion temperature is raised to volatilize S, belite reacts with free CaO to form alite, enclosing often free CaO in alite crystal. No significant change is observed for tricalcium aluminate solid solution and ferrite solid solution.

Sodium and K promote the crystal growth of alite but edges of the crystals at the junction of the crystal planes becomes unclear. By the existence of Na, belite becomes Type I and its grain size becomes larger and the width of lamella corresponding to α modification is extended. The aspect ratio of prismatic crystal of tricalcium aluminate solid solution becomes larger and the crystals form parallel growth with ferrite solid solution. The effect of K on belite and interstitial materials is similar but not as extensive as Na.

Idiomorphic alite is formed when content of Ti in clinker is 1% as TiO_2 [64]. Belite encloses $CaTiO_3$ when content of Ti in clinker become 1.5% as TiO_2. Increased amount of Ti makes the color of belite crystal light brown and the more amount makes the color green gray.

When approximately 0.5% of P exists in clinker as P_2O_5, the formation of alite is suppressed and alite crystals become small [65]. Phosphorous is distributed within the inner part of the alite crystal and the peripheral part of the belite crystal forms a zonal structure, as shown in Figure 21 [48]. No lamella structure is observed in belite crystal, as shown in Figure 22. Color of ferrite solid solution changes from dark brown to brown by the incorporation of P.

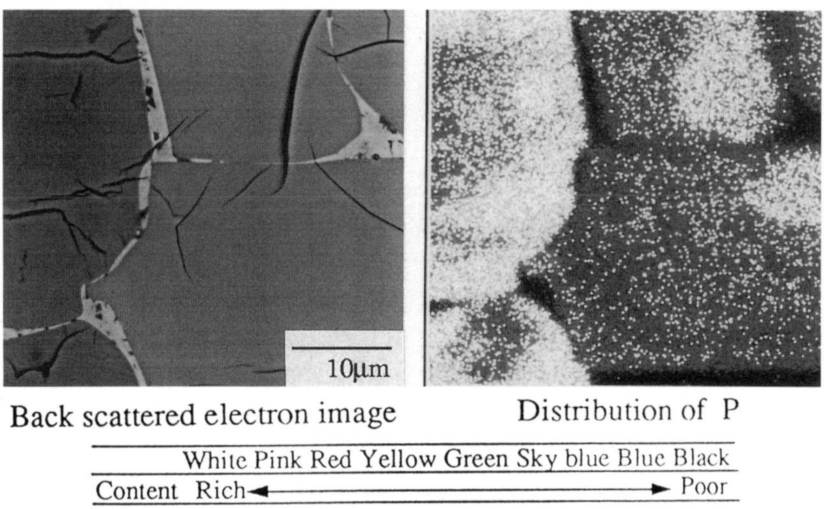

Figure 21. Back-scattered electron image and distribution of phosphorous on the polished surface of clinker.

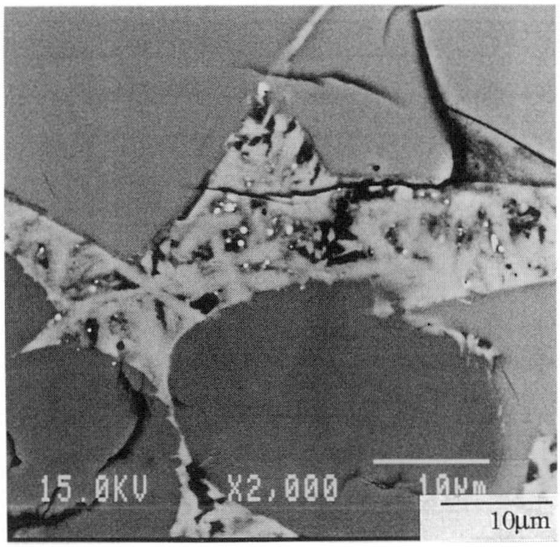

Figure 22. Back-scattered electron image of the polished surface of clinker containing phosphorous.

Fluorine affects the formation and growth of alite crystal to form step-wise pinacoid perpendicular to c-axis as shown in Figure 23. The refractive index of ferrite solid solution decreases.

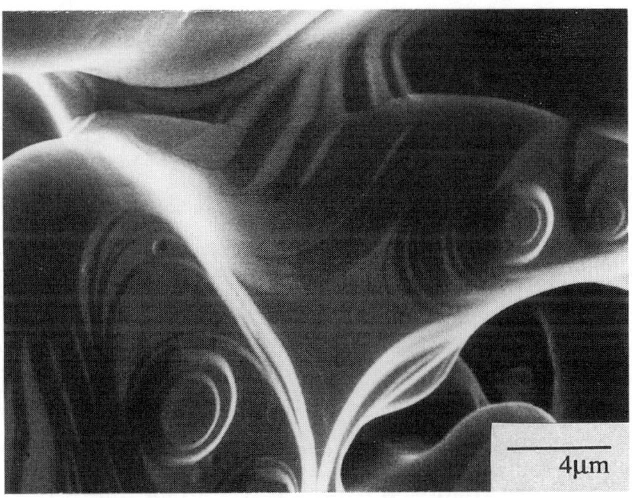

Figure 23. Crystal of alite formed in the presence of fluorine showing the step-wise growth of pinacoid perpendicular to the c-axis.

Chromium creates the color of alite and belite green. The color of the cement becomes deep green with the incorporation of Cr [9,66].

Manganese exists in the form of Mn^{4+}, Mn^{3+} or Mn^{2+} in clinker. The Mn^{2+} incorporates easily into alite to form large-sized alite with a major axis of 40 microns. When the burning atmosphere changes from reducing to oxidizing, alite decomposes and C_2F lamella substitutes with Mn forms. It is observed as a linear pattern in belite. Due to the incorporation of Mn, tricalcium aluminate solid solution in white cement is colored blue [9]. When Mn and Ti are incorporated into ferrite solid solution in maximum amounts, the brightness of clinker decreases, the hue darkens and the color becomes bluish- purple.

When Zn coexists with chlorine, Zn tends to precipitate with other trace elements such as Cr on interstitial materials, particularly with tricalcium aluminate solid solutions, as shown in Figure 24.

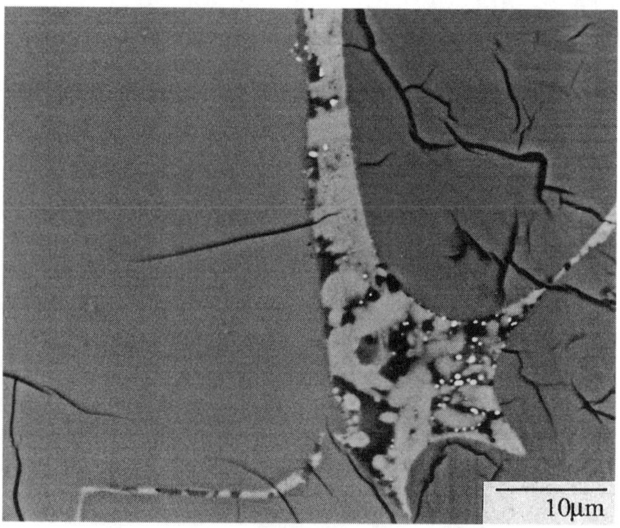

Figure 24. Back-scattered electron image showing the deposition of zinc-copper compounds on tricalcium aluminate solid solution with chlorine.

Barium does not have a significant affect on the viscosity of clinker liquid or the crystal size and shape of either alite or belite.

Approximately 0.2 -0.4% of V as V_2O_5 promotes the growth of alite crystal, but it does not affect the crystal size of belite.

Alite decomposes when the Cu content in clinker exceeds 0.5% as CuO [51].

No remarkable change in clinker texture is observed when content of P is on the order of tens of ppm.

Crystal Structure of Clinker Minerals

Since minor and trace elements introduced into the cement manufacturing system from the waste are different in ionic radius, electric charge and electronegativity from the main elements, which constitute the four major minerals, strain and defects are formed in crystal structure of clinker minerals by the incorporation of minor and trace elements. As a result, rearrangements of atoms in cooling are disturbed and transition from high temperature modifications to low temperature modifications are suppressed. The stabilization of high temperature modifications of synthesized pure C_3S and C_2S and that of alite and belite in commercial clinker at room temperature by minor and trace elements are shown in Tables 13 and Table 14, respectively.

Clinker minerals are not homogeneous and in most cases consist of some sub-grains. Therefore, the transition temperature of each modification differs slightly for individual sub-grains so that the amount of elements incorporated into stable solid solutions at temperatures for each modification have varying ranges. As a result, there are regions in which two modifications coexist.

In spite of these facts, the solubility limit of elements in the four major minerals shown in Table 9, and the amount of minor and trace elements that exist in each modification of C_3S and C_2S shown in Tables 13 and Table 14 are represented by the most frequent

Table 13. Effects of Minor and Trace Elements on the Stabilization of Each Modification in C_3S and C_2S at Room Temperature.

Element	Added component	Modification of C_3S solid solution							Modification of C_2S solid solution			
		T_1	T_2	T_3	M_1	M_2	M_3	R	γ	β	α'	α
Mg	MgO %	<0.55	0.55-1.45	—	—	—	1.45-2.0	—		<4.7	4.7<	—
S	SO_3 %	T_1 modification is stabilized							β modification is stabilized *1			
Na	Na_2O %	<0.3 (T modification)			0.3< (M modification)			—			<1	4<
K	K_2O %	T_1 modification is stabilized							β modification is stabilized			
Ti	TiO_2 %	<3.5	3.5-4.5	—	4.5<	—	—	—	γ modification is stabilized			
P		—	—	—	—	—	—	—		<2.0	2.0<	2.0<
F	F %			T_3 modification is stabilized					α' modification is stabilized			
Cl		—	—	—	—	—	—	—	—	—	—	—
Cr	Cr_2O_3 %		<1.0	—	—	1.0<	—	—	β modification is stabilized *1			
Mn	MnO_2 %	<0.4	0.4-2.2	2.2<	—	—	—	—	α' modification is stabilized			
Zn	ZnO %	<0.8	0.8-1.8	—	1.8-2.2	2.2-4.5	—	4.5-5.0	β modification is stabilized			
Sr	SrO %	<1.3	1.3-2.3	—	—	—	—	—		<11.4	21.7<	—
Ba	BaO %		T_2 modification is stabilized							<16.0	29.1-49.1	57.9<
V	V_2O_5 %	—	—	—	—	—	—	—		<4.8	9.8-16.7	26.1<
Cu		—	—	—	—	—	—	—	—	—	—	—
As		—	—	—	—	—	—	—	—	—	—	—
Cd	CdO %	T_1 modification is stabilized							—	—	—	—
Sb		—	—	—	—	—	—	—	—	—	—	—
Hg	HgO %	T_1 modification is stabilized							—	—	—	—
Pb	PbO %	T_1 modification is stabilized							—	—	—	—

Note; *1 : α' modification is stabilized if sufficient amount is added.

Table 14. Effect of Minor and Trace Elements on the Stabilization of Each Modification in Clinker Alite and Belite at Room Temperatures.

Element	Oxide	Alite	Belite
Mg	MgO %	MgO>1.2 $M_1 + M_3 \longrightarrow M_3$	Amount of β is increased with increase of MgO
S	SO3 %	SO3>1.2 $M_3 \longrightarrow M_1 + M_3$	
Na	Na2O %		α is stabilized whent the amount of Na_2O exceeds 1%

appearance value. Many studies have been carried out on pure cement compounds and synthetic clinker, since it is difficult to investigate the relationship between minor and trace elements and the structural change of clinker minerals in commercial clinker consisting of multicomponents.

Pure Cement Compounds

In pure C_3S,T_1 modification, T_2 modification, and $M_3(M_{1b})$ modifications are stabilized at room temperature by 0.55% or less, 0.55-1.45%, and 1.45-2.0% of Mg as MgO, respectively. Sodium stabilizes triclinic and monoclinic C_3S by incorporating 0.33% or less and 0.33% or more Na as Na_2O, respectively [67]. About 3.4% of K as K_2O stabilizes T_1 -C_3S at room temperature [68]. About 3.5% or less and about 3.5-4.5% of Ti as TiO_2 in C_3S stabilizes the T_1 modification and the T_2 modification at room temperature, respectively [64]. In excess of 4.5%, Ti stabilizes presumably M_1 C_3S at room temperature. For fluorine, 2.4% in C_3S stabilizes the T_3 modification at room temperatures. Large amounts of S reduce the amount of C_3S formed, though C_3S exists in stable form at room temperature as the T_1 modification even burned at 1500-1560°C with 9% of S as SO_3 [62]. At percentages of 0.4% or less, 0.4-2.2% and 2.2 % or more of Mn as MnO_2 stabilizes T_1, T_2, and T_3 modifications at room temperature, respectively [45]. Zinc as ZnO at values of 0.8% or less, 0.8-1.8%, 1.8-2.2%, 2.2-4.5% and 4.5-5.0% stabilizes T_1, T_2, M_1, M_2 and R modifications at room temperature, respectively [69]. Strontium as SrO at values of < 1.3%, 1.3 - 2.3% stabilizes T_1 and T_2 modifications at room temperature, respectively [36].

Barium stabilizes T_2 modification at room temperature [36]. Elements such as, Cd, Sb, Hg and Pb have little affect on the C_3S structure. Approximately 5% of the elements as oxides stabilize the T_1 modification at room temperature [68]. In pure C_2S, about 0.2 mol (4.7%) or less and about 0.2 mol or more of Mg as MgO stabilizes β and α' modification at room temperature as shown in Table 15 [70]. Potassium stabilizes β-C_2S at room temperature [71]. Titanium stabilizes γ-C_2S at room temperature [64]. Less than 2% and 2% or more of P as P_2O_5 stabilizes β modification, and α' and α modification of C_2S at room temperature, respectively [36,72]. The α modification stabilized by P_2O_5 rarely develops adequate strengths [72]. When F coexists, C_2S forms $(C_2S)_2 \cdot CaF_2$ at 950°C and decomposes to C_2S and CaF_2 at 1040°C. The α'-C_2S exists in a stable form at room temperature. The β-C_2S is stabilized by small amounts of Cr and α'-C_2S is stabilized by large amounts of Cr at room temperature [73]. Manganese stabilizes α'-C_2S at room temperature [74]. Zinc stabilizes β -C_2S at room temperature [3,36]. About 0.1 mol (11.4%) or less and 0.2 mol (21.7%) or more of Sr as SrO stabilizes β and α' modification at room temperature, respectively [70]. Barium as BaO with values of 0.1 mol (16.0%), 0.2 (29.1%) to 0.4 mol (49.1%) and 0.5 mol (57.9%) or more stabilizes β, α' and α modification of C_2S at room temperature, respectively [70]. About 0.05 mol (4.8%) or less, 0.1 (9.8%) to 0.2 mol (16.7%), and 0.3 mol (26.1%) or more of V as V_2O_5 stabilizes β, α' and α modification, respectively [75]. The C_2S containing V rarely develops adequate

Table 15. Modification of Pure C_2S Stabilized at Room Temperature in the Presence of Magnesium [70].

Composition				Condition of burning and cooling					
CaO	MgO	SiO2		1600℃ for 30min, Quenched in water	1500℃ for 4hr, Quenched in water	1500℃ for 30min, Quenched in water	1400℃ for 30min, Quenched in water	1350℃ for 30min, Quenched in water	1500℃ for 30min, Cooled slowly
1.9 62.4	0.1 2.4	1.0 35.2	(mol) (wt%)	$\gamma+\beta$	$\gamma+\beta$	$\gamma+\beta$	$\gamma+\beta$	γ	γ
1.8 59.7	0.2 4.7	1.0 35.6	(mol) (wt%)	β	β	$\gamma+\beta$	$\gamma+\beta$	γ	γ
1.7 56.9	0.3 7.2	1.0 35.9	(mol) (wt%)	$\alpha'+\beta$	$\alpha'+\beta$	$\alpha'+\beta$	$\gamma+\beta$	$\gamma+\beta$	$\gamma+\beta$

strengths [75]. It is considered that the addition of a large amounts of element (> 10%) to stabilize high-temperature modifications of C_2S at room temperature is likely to partially change C_2S to other compounds.

In pure C_3A, the lattice constant is shortened by substitution of Mg for Ca [76]. Cubic and orthorhombic C_3A are formed with 1.6% or less and 1.6% or more of Na as Na_2O, respectively [77]. The lattice constant of C_3A is shortened by substitution of Na for Ca. Cubic C_3A is formed with 1.4% or more of K as K_2O, and the lattice constant is made slightly longer from 15.265-15.277 Å. Small amounts of crystals similar to orthorhombic C_3A are formed by slow-cooling [78]. Titanium slightly increases the lattice constant of C_3A [76]. Zinc shortens the lattice constant of C_3A [79].

In pure C_4AF, Cr incorporated into C_4AF as Cr^{3+} makes c-axis longer slightly and b-axis shorter (80). Manganese incorporated into C_4AF as Mn^{3+} substituting for Fe^{3+} decreases the lengths of the a- and c-axes and increases the length of the b-axis significantly [81]. Chlorine increases the length of the a-, b- and c-axes of C_4AF [82].

Synthetic Clinker Minerals

In the system coexisting with Mg, Al and Fe, the crystal structure of alite changes from T_1 to T_2 and M_1 as the sum of Al_2O_3 and Fe_2O_3 amount increases when the Mg content is as low as 0.3% as MgO as shown in Figure 25 [69]. In synthetic alite in which C_3S is saturated with Al and Fe, 0.65% or less and 0.65% or more of Mg as MgO stabilizes M_1 and M_2 modifications at room temperature, respectively as shown in Figure 26 [83]. The lattice constant changes continuously with Mg content in M_3, this fact indicates the change in crystallinity of synthetic alite. About 1-4% of S as SO_3 stabilizes M_3 modifications at room temperature along with Mg, Al and Fe [84]. Approximately 2% of Na or K as Na_2O or K_2O stabilizes the M_3 modification at room temperature along with Mg, Al and Fe. About 5% of Ti as TiO_2 stabilizes the R modification at room temperature under the coexistence of Al and Fe [68]. Fluorine stabilizes R modification at room temperature under the coexistence of Mg, Al and Fe [68]. Less than 1.0% and 1.0% or more of Cr as Cr_2O_3 stabilizes the T_2 and M_2 modifications, respectively under the coexistence of Al and Fe, respectively [85]. About 5% of Mn as MnO_2 stabilizes the R modification at room temperature under the coexistence of Al [68]. Values of 3.1% or less and 4.5% or more or Zn as ZnO (solubility limit of Zn is about 3.6%) stabilizes M_3 and R modification under the coexistence of Mg, Al and Fe [86]. Barium stabilizes M_3 modification under the coexistence of 2% of more Mg as MgO [36]. Elements such as, Cd, Hg and Pb stabilize M_2 modification at room temperatures under coexistence of Al and Fe [68].

In the case of belite, small amount of S stabilizes β modification under coexistence of Al, but larger amounts of S stabilizes α' modification [62]. The β-C_2S is stabilized at room

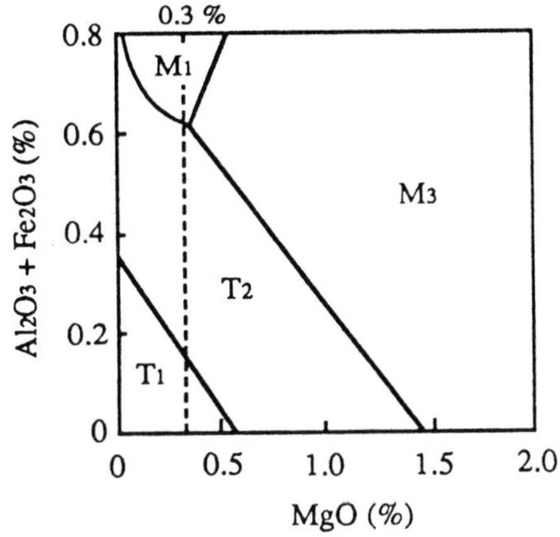

Figure 25. Relationship between amounts of magnesium and modification of alite in the coexistence of aluminum and iron [69].

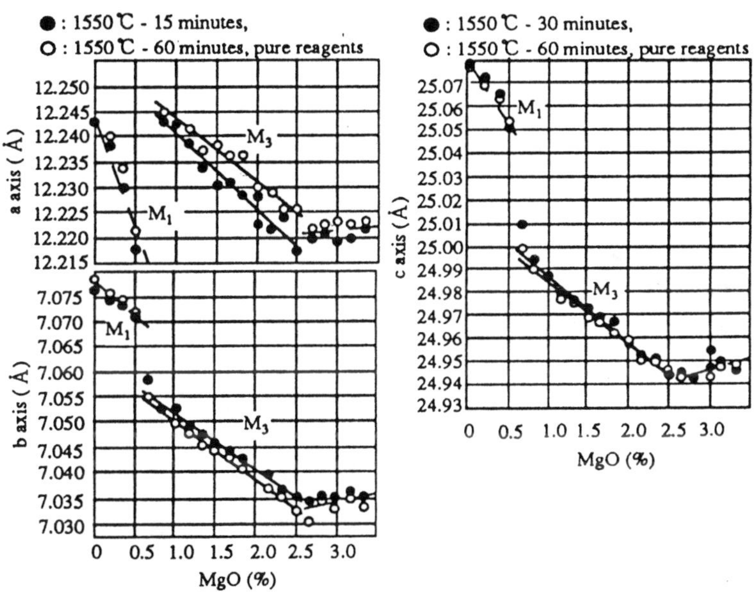

Figure 26. Relationship between the amount of MgO and modification of synthesized tricalcium silicate solid solution under saturation of Al and Fe [83].

temperature under coexistence of Na [87]. Sodium stabilizes different modifications depending on the type and amount of coexisting element, as shown in Figure 27 [87]. For example, α-C_2S is stabilized at room temperature when the sum of Al_2O_3 and Fe_2O_3 amount is 4% and the amount of Na_2O is about 4% and α'-C_2S is stabilized at room temperature when sum of Al_2O_3 and Fe_2O_3 amount is about 0.5% and amount of Na_2O is about 0.5% [88,89]. Stable modification of K-containing synthetic belite at room temperature differs depending on Al content under the coexistence of Al and Fe. For example, 0.2 mol (20.1%) of K as K_2O stabilizes β, α' and α modification under various Al contents of 0.15 mol (8.6%) or less, 0.15 to 0.25 mol (14.0%), and 0.25 mol or more, respectively [90]. Under coexistence of 2% Mg, α'-C_2S is stabilized at room

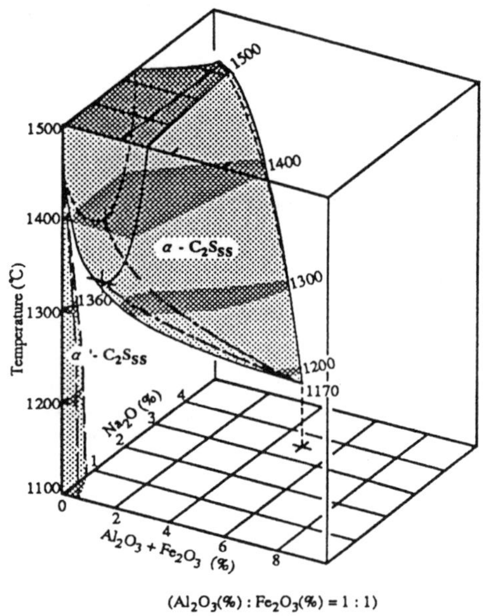

Figure 27. Relationship between the amount of Na_2O, $Al_2O_3 + Fe_2O_3$ and modification of synthesized C_2S in the presence of Mg, Al and Fe [87].

temperature with about 3% of K as K_2O [89]. Titanium stabilizes β-C_2S at room temperature under coexistence of other components [74]. Chromium stabilizes α-C_2S at room temperature under coexistence of Al [72].

Minerals in Commercial Clinker

About 0.8% or less, 0.8-1.2% and 1.2% or more of Mg as MgO in commercial clinker stabilizes M_1, M_1 and M_3, and M_3 modifications of clinker alite at room temperatures, respectively, as shown in Figure 28 [91]. Commercial clinker contains usually about 1.5% of Mg as MgO. The SO_3 contained in commercial clinker is usually about 0.5%. Amounts of 1.2% or less, 1.2-1.6% and 1.6% or more of S as SO_3 stabilizes M_3, M_1 and M_3, and M_1 modifications at room temperatures, respectively [91]. The lattice constant of clinker

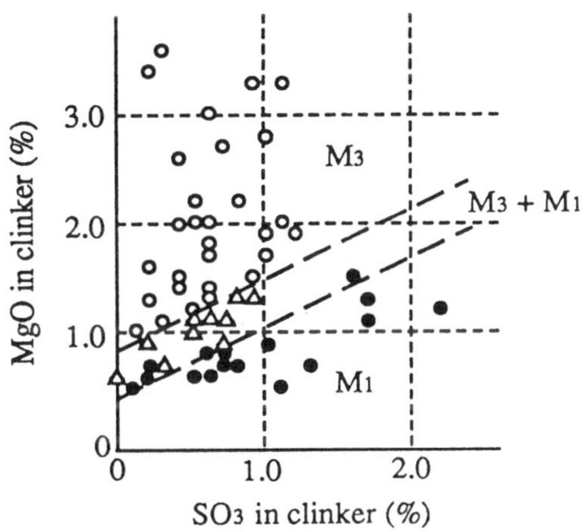

Figure 28. Relationship between amount of incorporated MgO and SO_3 and modification of clinker alite [91].

alite shows approximately constant value regardless of the type and amount of minor and trace elements.

In the case of clinker belite, the increased incorporation of Mg stabilizes the β modification at room temperature [92]. The reason for this is presumably because the activation energy required for transition of belite from α to α' modification is reduced due to incorporation of Mg in clinker belite. Sodium contained in commercial clinker is usually 0.5% or less as Na_2O. Less than 0.5% and 1% or more of Na as Na_2O in commercial clinker stabilizes β with partially α' modification, and α modification at room temperature, respectively [92].

INFLUENCE OF WASTE ALTERNATIVES ON CEMENT QUALITY

Using wastes as alternatives to raw materials and fuel in cement manufacturing causes changes in the character of cement clinker. For example, as mentioned before, minor and trace elements contained in waste have an influence on clinker formation and crystal structure through incorporation into the four major minerals, formation of other compounds and shifting of phase equilibrium at the burning temperature to produce new compounds that replace the major minerals. They, in turn, have an influence on the hydration of cement and change the structure of fresh and hardened cement paste. As a result, the properties of concrete including workability, setting, strength and durability are affected by utilization of waste as an alternative to raw materials and fuel to a considerable extent. The relationship between modification of alite stabilized with various elements at room temperature and developed strength is shown in Figure 29.

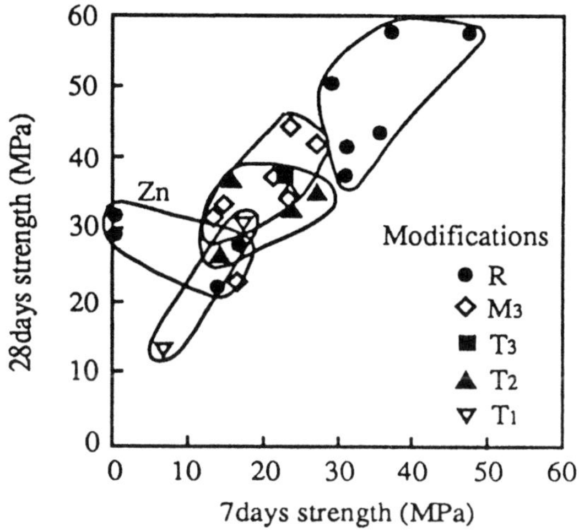

Figure 29. Relationship between modification of alite stabilized with various elements at room temperature and strength after 28 days of aging.

In this section, the influences of minor and trace elements contained in waste (as shown in Table 16) on the quality of cement from the viewpoint of hydration reactivity, microstructure of hardened cement paste, setting and strength, and durability are described.

The type of waste and the elements contained in it are diversified. The contents of minor and trace element in commercially available types of cement in Japan are shown in Table 16. The minor elements are defined as the elements other than the four major elements and trace elements are defined as the elements contained in amounts as low as 0.1% or 1000ppm or less. The effects of those elements introduced in the cement-manufacturing process as alternative raw materials and fuel on the operation and quality are also listed. Although some effects on the quality vary according to

Table 16. Influence of Minor and Trace Elements on Burnability of Raw Meal, Manufacturing Processes and Quality of Cement.

Element	Content	Volatile or non-volatile	Origin	Influence on human body	Regulation*1 (ppm) W*2	T*3	Influence on manufacturing process: Burnability	Others	Influence on mineral composition	Influence on hydraulic reactivity	Influence on quality of cement and concrete: Setting	Strength Early	Strength Late	Durability	Special remarks
Minor MgO	0.8-2.5%	non-v *4	limestone	harmless	—	—	· 1.0-1.5% ↑ · Max.=2.0% · >2.0% ↓	· Gr. *5	C_3S ↑ C_2S ↓ C_3A ↓ C_4AF ↑	in ad. amnt. *6 C_3S ↑ (<1%) C_2S ↓	accel.	up (<2%) down	down(>2%)	· problem in soundness (>4%)	Pericalse causes expansion
SO_3	0.2-0.9%	volatile	fuel, limestone	harmless, SO_x: harmful	—	—	the less, the better (<0.5%)	· Gr. · coating in Pr *7 and Kn *8 ↑	C_3S ↓ C_2S ↑	C_3S ↓ C_2S ↑ C_3A ↓	*9 accel.	up	down(>0.5%)	· caution for A.A.R. *10	Alkali sulfate is easily formed
P_2O_5	0.11-0.34%	non-v	limestone	harmless	—	—	· 0.1-0.3% ↑ · Max.<0.5% · >0.5% ↓	· coating in Pr ↓		C_3S ↓ C_2S ↑	remarkably ret. *11	up down	up (<0.5%) down(>0.5%)	if soluble (take care of p-gypsum *12)	
Na_2O K_2O (R_2O)	0.1-0.5% 0.4-1.2% 0.4-1.0%	volatile volatile	clay clay	harmless harmless	— —	— —	· 0.2-0.4% ↑ · Max.<1.0% · >1.0% ↓	· coating in Pr ↑		in ad. amnt. C_3S ↑ C_3A ↓ C_2S ↑ C_4AF ↓	accel. accel. (<0.2%: ret)	up up	down down	· A.A.R. >0.6% of R_2O	Alkali sulfate is easily formed
TiO_2	0.2-0.6%	non-v	clay	harmless			· <1.0% ↑ · lowers MF *16		C_3S ↓ C_2S ↑	in ad. amnt. C_3S ↓ C_4AF ↓	ret. (>0.2%)	up	down	many contradictory results on strength	
Trace F	310-440ppm	l-vol*14)					↑, Max.=2% · $C_3A \rightarrow C_{11}A_7$ (>0.3%)	· coating in P and Kn ↑		C_3S ↑	ret.	up down	up(<0.5%) down(>0.5%)		Cost merit is ambiguous by using F as flux
Cl	30-80 ppm	volatile	coal	harmful HCl:harmful Cl_2:harmful			↑ · coating in P ↑ (>100ppm) · metal-corrosion ↑ · brick damage ↑			C_3S ↑ C_3A ↑	accel.	up	undefinite	· metal-corrosion (<300 g/m³ - concrete)	
Cr	24-100ppm	non-v	brick, clay, metal slag	Cr^{6+}: harmful	2		↑ decreases M *15	· coating in Kn ↑	C_3S ↓ C_2S ↑	C_3S ↑	accel.	up down	(<0.5%) down(>0.5%)		Cr^{6+} content become high in high temperature burning, cement color change to green
Mn	170-350ppm	non-v	metal slag limestone		10		↑ flux effect	accel. of decarbonation		C_3S ↑ C_3A ↑ C_2S ↑ C_4AF ↓		up down	down(<1/2 subst. *13) down(>1/2 subst. *13)		Cement color changes to green or blue, Strength at later age increases by 1% addition
Zn	280-590ppm	volatile	metal slag	harmless	5		↑ · max.=1.5%	· coating in P and Kn ↑		C_3S ↓	ret.	down	down(>0.5%) (optimum: 700ppm)		Zn stablizes R-C_3S easily but hydraulic reactivity is low
SrO BaO	900-10,000ppm 10-500ppm	non-v	limestone limestone				↑ ↑ flux effect	(Max 3%)		C_2S ↑ C_3S ↓ C_3A ↑		down down	down down	Lattice defect is decreased in Sr bearing C_3S, Lattice distortion is decreased in Ba bearing C_3S High in amount of substitution	
V	24-33ppm	l-vol	coal	harmful			↑ decreases M *15			C_2S ↓					
Cu	70-180ppm	non-v	metal slag	harmful	3.0		no remarkable effect in the amount existed in cement			down	ret.		down		
As	17-25ppm	l-vol	metal slag	harmful	0.5	10					ret. (>0.1%)	down	↑		
Cd	1-28ppm	volatile	limestone	harmful	0.1	2.0				no effect	ret. (>1%)	no effect	down		
Sb	24-36ppm	non-v	metal slag	harmful						down					
Hg	119-148ppb	volatile	metal slag	harmful	0.5ppb	500ppb				no effect					
Pb	81-174ppm	volatile	metal slag	harmful	1.0	100				C_2S ↓ C_3A ↓	ret. (>0.1%)	down	no effect		
NH_3				harmful											
CN				harmful	1.0	3.0									

note; *1: Regulation for water pollution, *2: Water soluble amount, *3: Total amount, *4: Non-volatile, *5: Easy to be ground, *6: In adequate amount, *7: Preheater, *8: Kiln, *9: Accelerated, *10: Alkali-aggregate reaction, *11: Retarded, *12: Phosophogypsum, *13: Substitution for Fe_2O_3 in C_4AF, *14: A little volatile, *15: decreases M: decreases the viscosity of clinker melt, *16: lowers MF: lowers the temperature at which the main melt begins to form   References in this table are separately listed at the end of the paper expressed by the number with prefix T.

various researchers, those which the authors considered to be reasonable are listed.

Secondary electron images and back-scattered electron images observed by SEM, element distribution image by EPMA, results of element point analysis by EPMA and pore structure measured by mercury porosymetry of hardened cement paste containing these elements are shown in Figures 30, 31, 32, Table 17 and Figure 33, respectively.

Magnesium

Hydration Reactivity. Hydration reactivity of alite increases with an increase of Mg content up to 1% as MgO and then remains unchanged up to 2% [92]. Magnesium accelerates the transition of C_2S from α to β and reduces the hydration reactivity of belite after 7 days [92]. In the clinker containing a level of 2% of Mg as MgO, though the content of tricalcium aluminate is lowered by about 0.5, the composition of ferrite solid solution is close to C_6A_2F and is slightly higher in hydration reactivity than C_4AF. Cement that contains Mg (> 2% as MgO) shows accelerated early hydration up to 3 days and shows suppressed hydration at mid-term and long ages, because an increase in Mg content causes an increase in LSD leading to increased formation of free CaO. This changes the hydration reactivity of clinker minerals. Since Mg existing in clinker minerals in a form of solid solution or periclase dissolves slowly in water during hydration and shows low solubility, the Mg ion in liquid phase provides no remarkable effect on cement hydration.

Microstructure of Hardened Cement Paste. The pore structure of hardened cement paste varies little with variation of Mg content. In the hardened cement paste, Mg contained originally in cement is found concentrated in outer C-S-H on alite as well as in AFm formed in the vicinity of ferrite solid solution. Magnesium is not found in $Ca(OH)_2$. In case of blast-furnace slag cement paste, Mg contained originally in the slag is found partly incorporated into

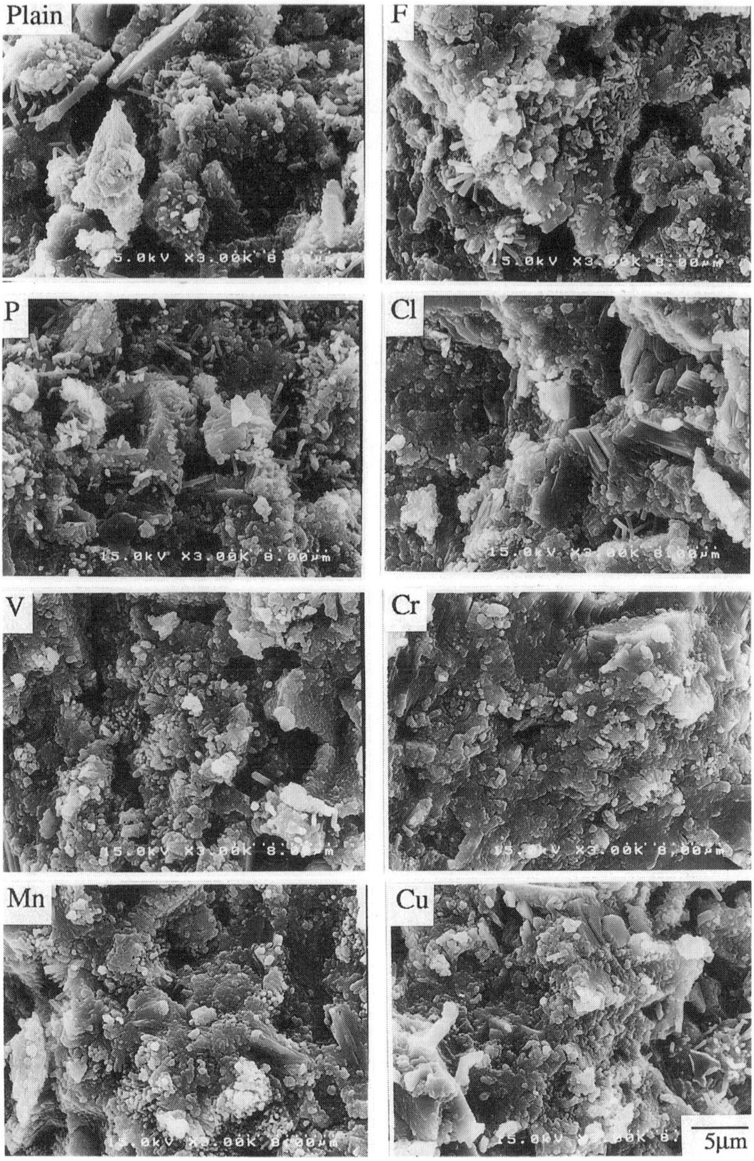

Figure 30-1. SEM micrographs of hardened paste prepared from cement containing various minor and trace elements (W/C = 0.5, 20°C, 14 days).

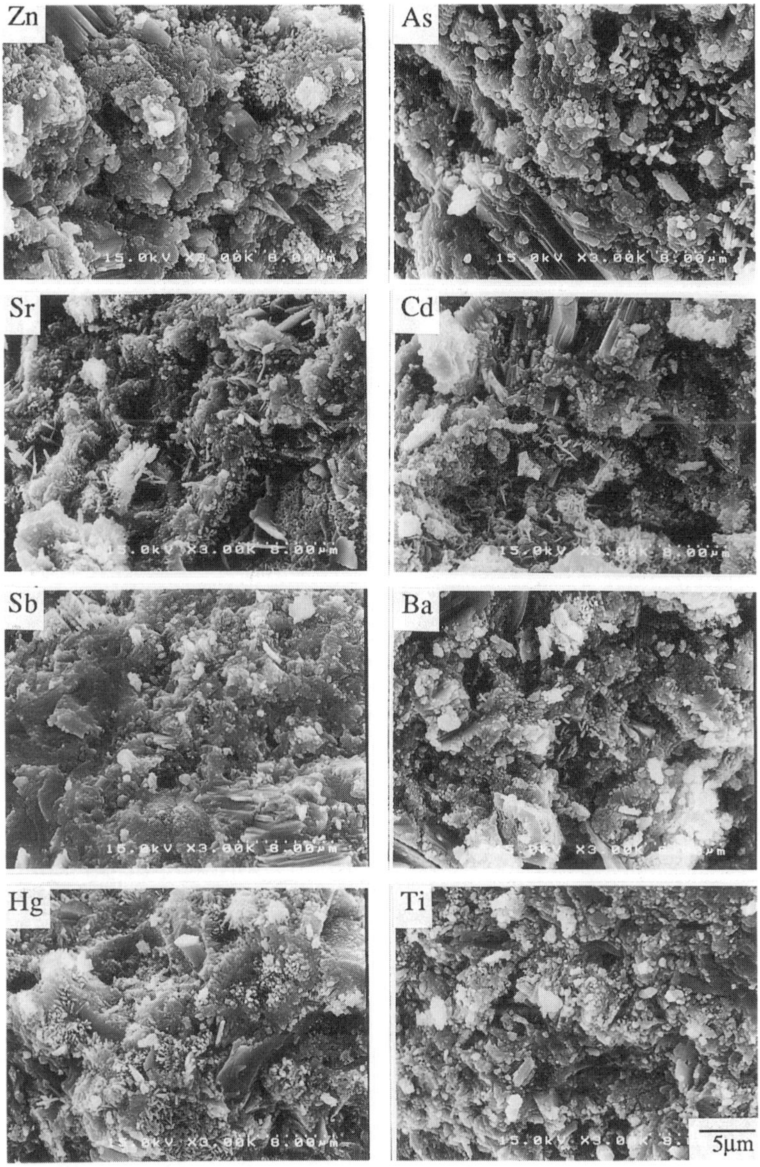

Figure 30-2. SEM micrographs of hardened paste prepared from cement containing various minor and trace elements (W/C = 0.5, 20°C, 14 days).

Figure 30-3. SEM micrographs of hardened paste prepared from cement containing various minor and trace elements (W/C = 0.5, 20°C, 14 days).

hydration products, but most remains unchanged in unhydrated slag particles.

Setting and Strength. With an increase in Mg content, setting time becomes short, early strength increases slightly, but 7 days' and 28 days' strength decreases [42,92]. When Mg content exceeds 2% as MgO, early strength decreases in spite of the increase in the amount of alite because ferrite solid solution changes in composition and the amount of tricalcium aluminate solid solution decreases.

Durability. Excess Mg over the solubility limit of clinker minerals exists as periclase (MgO) in clinker independently. Periclase hydrates gradually to change itself into brucite $Mg(OH)_2$, but this transformation proceeds long-time after contacting with water. As a result, the volume expansion due to the hydration of periclase causes destruction of already formed structure of hardened cement paste when Mg content exceeds a definite level. For this reason, permissible amount of Mg in cement is specified in most

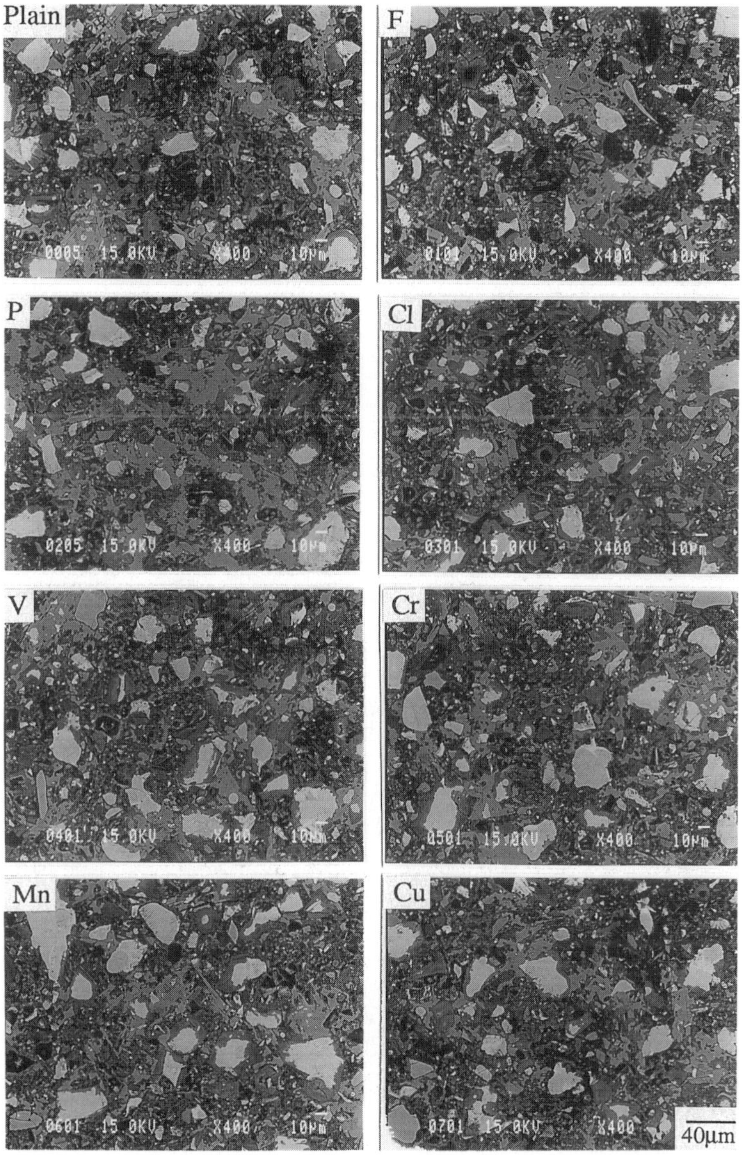

Figure 31-1. Back-scattered electron images of hardened paste prepared from cement containing various minor and trace elements (W/C = 0.5, 20°C, 14 days).

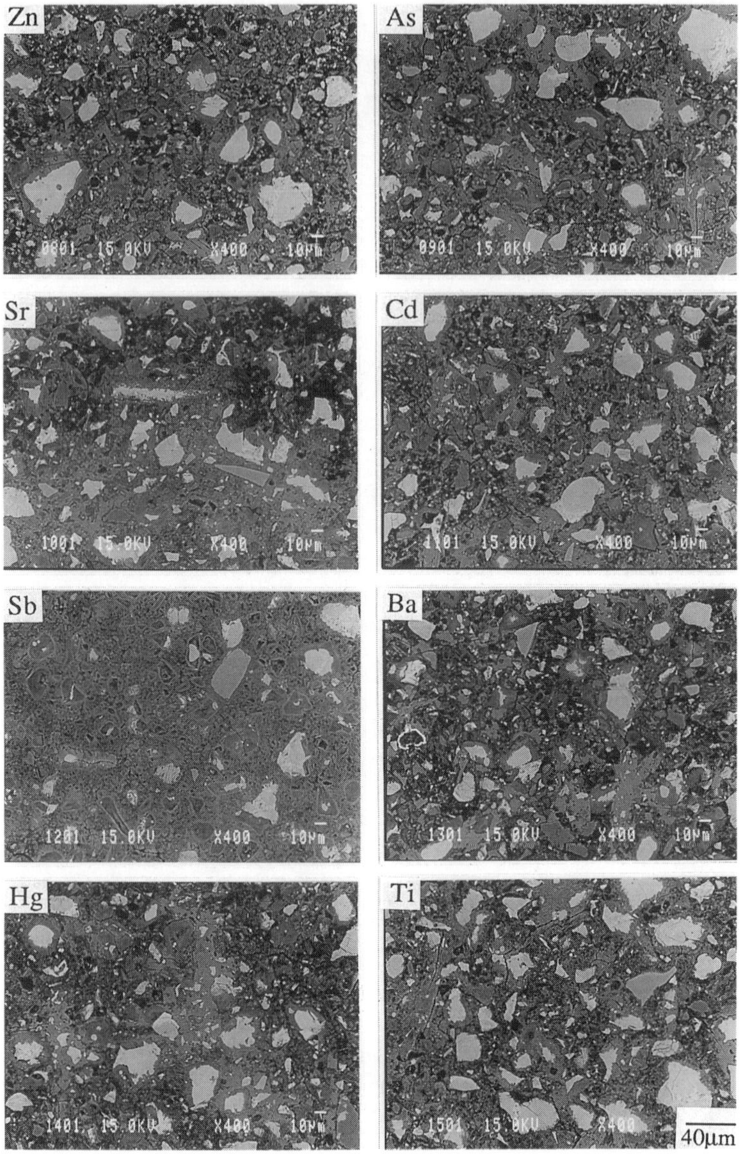

Figure 31-2. Back-scattered electron images of hardened paste prepared from cement containing various minor and trace elements (W/C = 0.5, 20°C, 14 days).

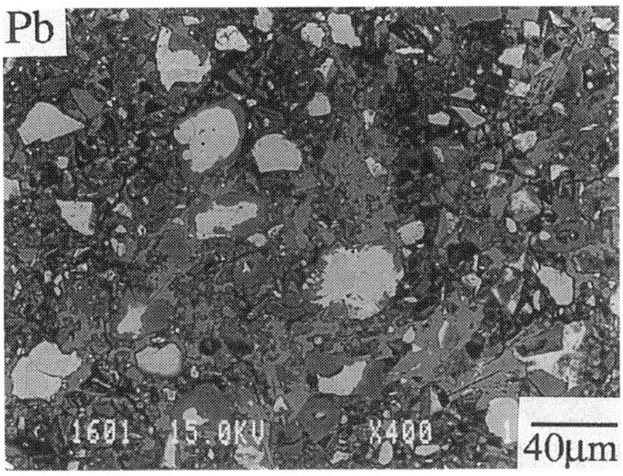

Figure 31-3. Back-scattered electron image of hardened paste prepared from cement containing various minor and trace elements (W/C = 0.5, 20°C, 14 days).

countries as follows: 4% for UK, 5% for Japan and Germany and 6% for USA as MgO.

Sulfur

Hydration Reactivity. Sulfur decreases the rate of crystal growth of alite and increases the crystal size as described earlier[62]. As a result, alite thus formed is low in the rate of hydration. Sulfur inhibits the transformation of belite into the γ-modification which results in the maintenance of its hydration reactivity. Sulfur allows increased formation of water-soluble alkali sulfates and free lime in clinker and increases early rate of cement hydration [93]. When there is a high S content in cement, the concentrations of SO_4^{2-} as well as OH^- ion in the liquid phase of the initial hydration stage become high and the induction period prior to active hydration of alite becomes short.

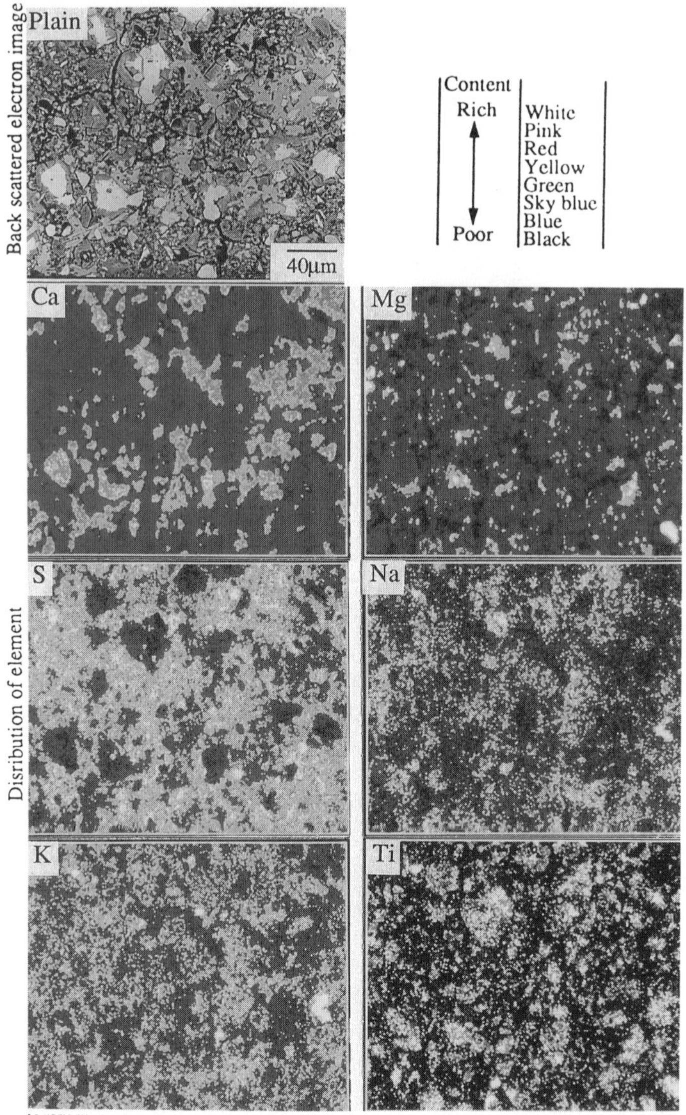

Figure 32-1. Back-scattered electron images and distribution of elements in hardened paste prepared from cement containing various minor and trace elements by EMPA (W/C = 0.5, 20°C, 14 days).

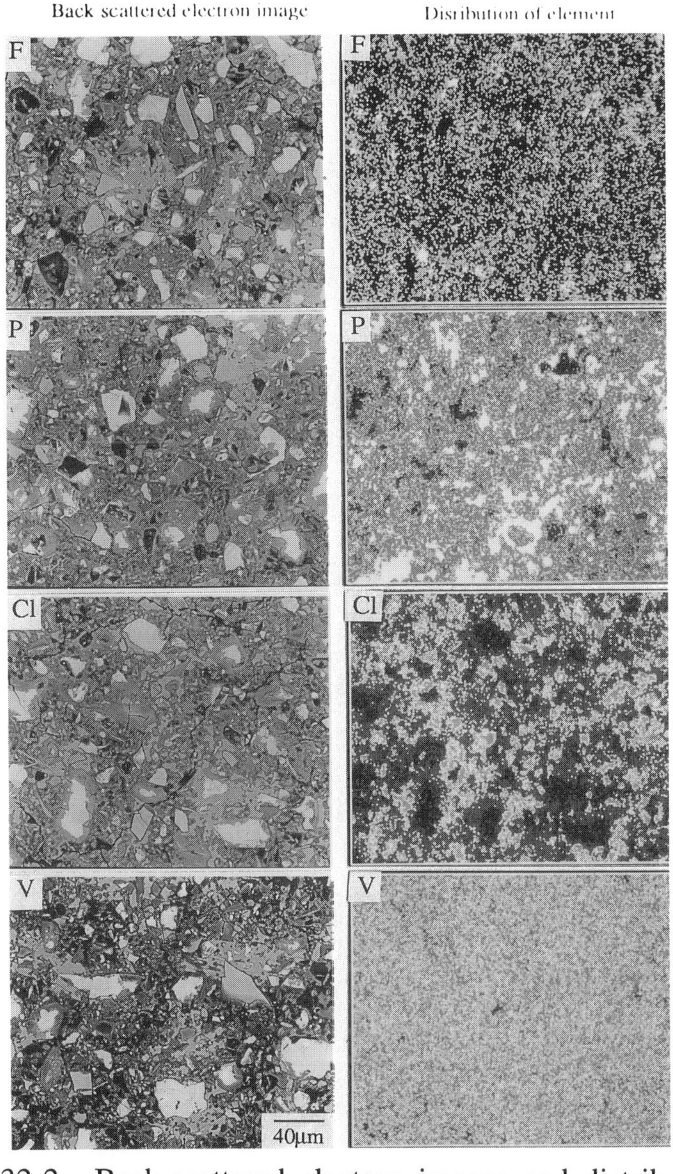

Figure 32-2. Back-scattered electron images and distribution of elements in hardened paste prepared from cement containing various minor and trace elements by EMPA (W/C = 0.5, 20°C, 14 days).

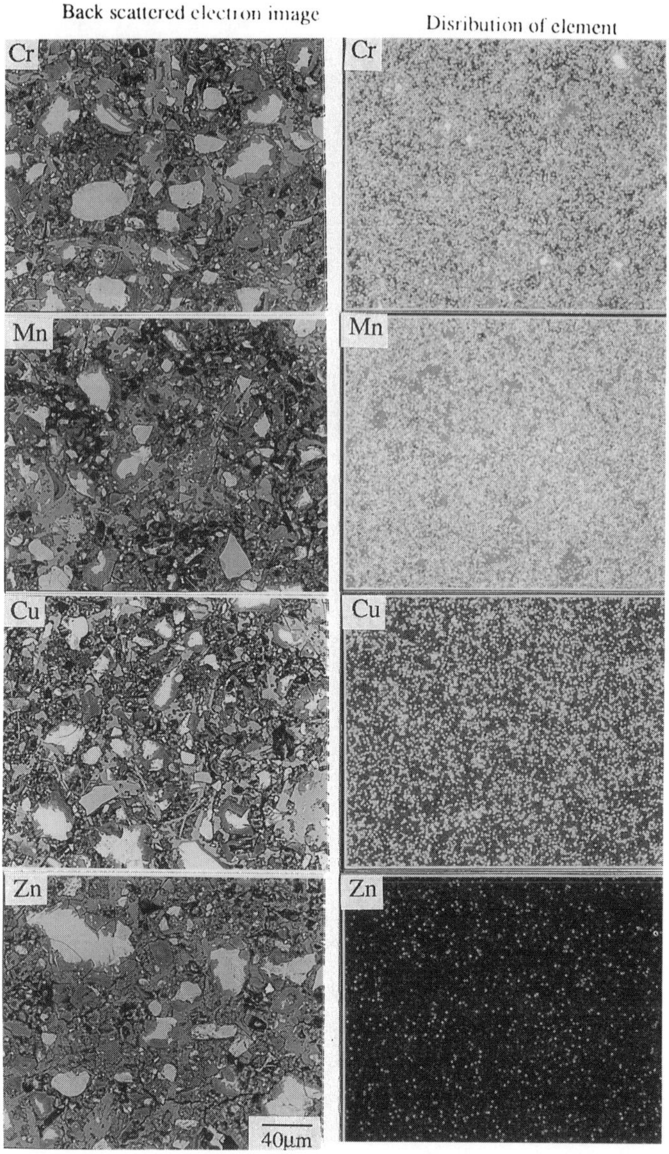

Figure 32-3. Back-scattered electron images and distribution of elements in hardened paste prepared from cement containing various minor and trace elements by EMPA (W/C = 0.5, 20°C, 14 days).

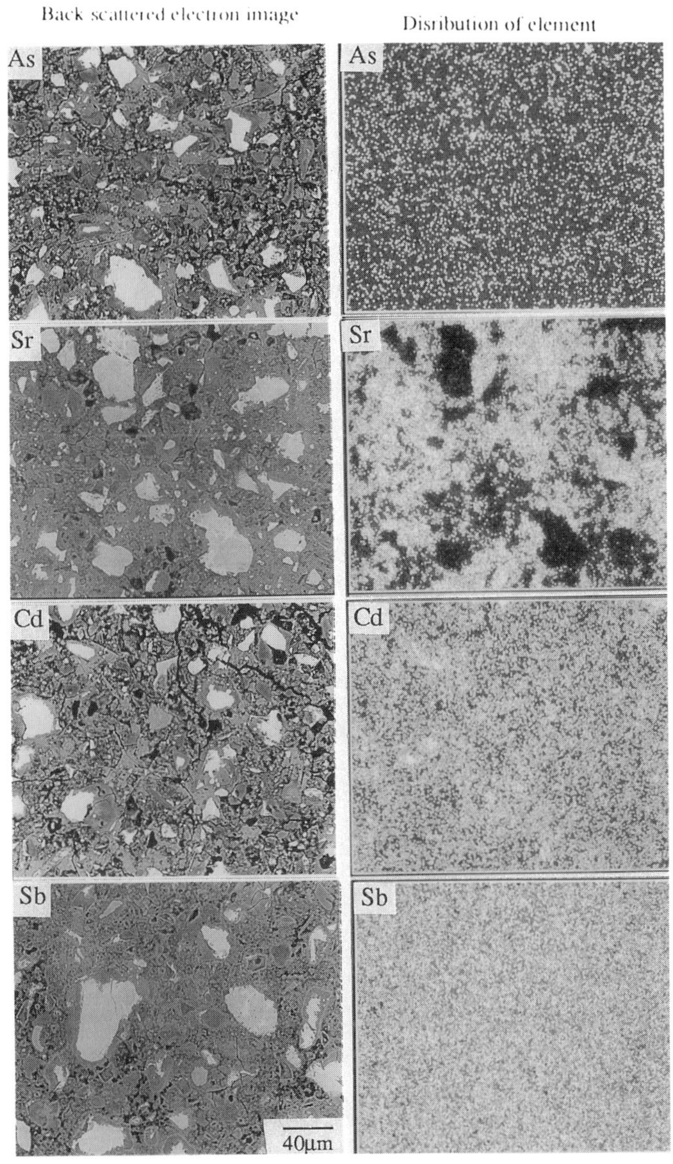

Figure 32-4. Back-scattered electron images and distribution of elements in hardened paste prepared from cement containing various minor and trace elements by EMPA (W/C = 0.5, 20°C, 14 days).

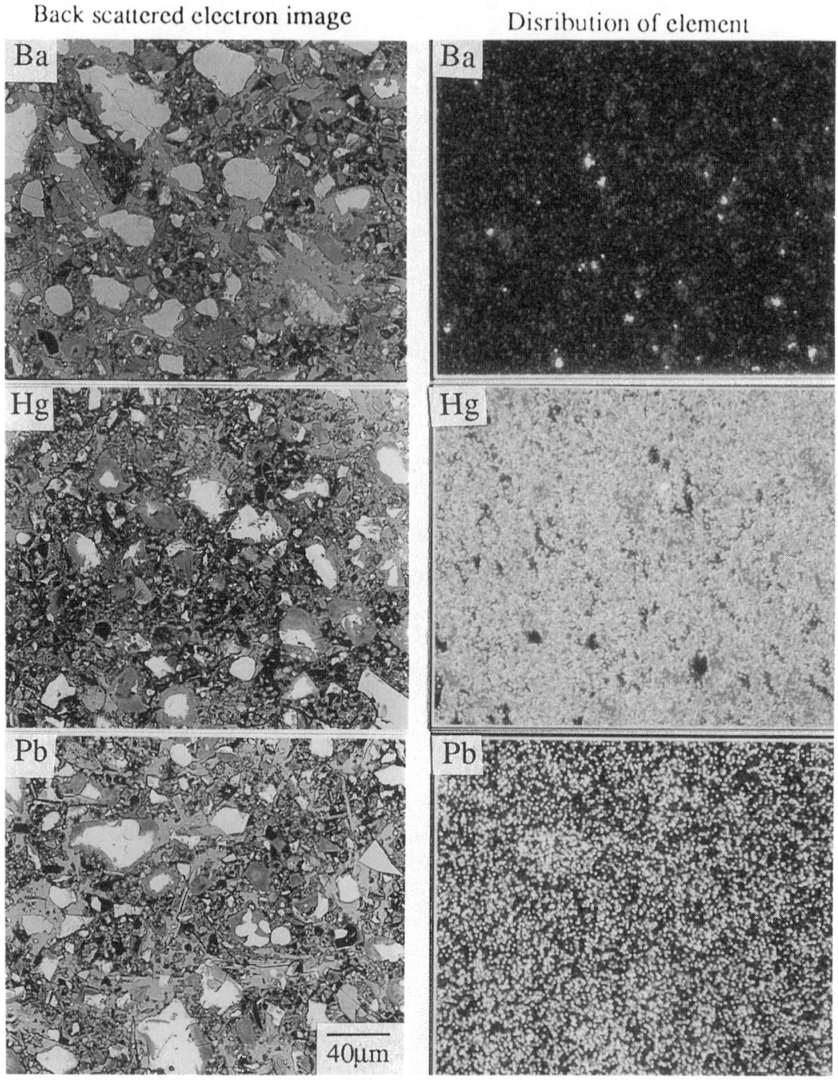

Figure 32-5. Back-scattered electron images and distribution of elements in hardened paste prepared from cement containing various minor and trace elements by EMPA (W/C = 0.5, 20°C, 14 days).

Table 17. Result of Point Analysis of Elements in Hardened Paste Prepared from Cement Containing Various Minor and Trace Elements Determined by EMPA.

Element	Amount of element in cement (%)		Concentration of element in hydrate determined by EPMA (%) C-S-H	AFm	Ca(OH) 2	
Mg	1.5	(as MgO)	0.32 -1.56	0.24 - 0.66	0.11 - 0.21	(as MgO)
S	2.0	(as SO_3)	0.4 - 1.1	4.6 - 5.5	0.2 - 0.5	(as SO_3)
Na	0.30	(as Na_2O)	0.25 - 0.45	0.08 - 0.12	0.03 - 0.07	(as Na_2O)
K	0.40	(as K_2O)	0.30 - 0.70	0.15 - 0.32	0.08 - 0.15	(as K_2O)
Ti	0.20	(as TiO_2)	0.14 - 0.32	0.22 - 0.35	0.05 - 0.09	(as TiO_2)
P *1	0.7	(as P_2O_5)	2.26 - 6.10	0.32 - 0.80	0.20 - 0.35	(as P_2O_5)
F *1	0.5		0.54 - 0.82	0.05 - 0.24	0.00 - 0.04	
Cl *1	2.0		0.16 - 0.38	1.31 - 2.88	0.01 - 0.10	
Cr *1	0.1	(as Cr_2O_3)	0.04 - 0.10	0.08 - 0.11	0.00 - 0.04	(as Cr_2O_3)
Mn *1	0.1	(as MnO)	0.07 - 0.11	0.03 - 0.05	0.00 - 0.03	(as MnO)
Zn *1	0.2	(as ZnO)	0.08 - 0.14	0.07 - 0.11	0.00 - 0.02	(as ZnO)
Sr *1	2.0	(as SrO)	2.03 - 3.10	0.81 - 1.03	0.10 - 0.32	(as SrO)
Ba *1	0.2	(as BaO)	0.09 - 0.14	0.09 - 0.12	0.00 - 0.01	(as BaO)
Cu *1	0.1	(as CuO)	0.06 - 0.11	0.06 - 0.10	0.00 - 0.02	(as CuO)
Pb *1	0.1	(as PbO)	0.15 - 0.21	0.04 - 0.07	0.00 - 0.04	(as PbO)

*1: Elements other than Mg, S, Na, K and Ti are added individually to the cement

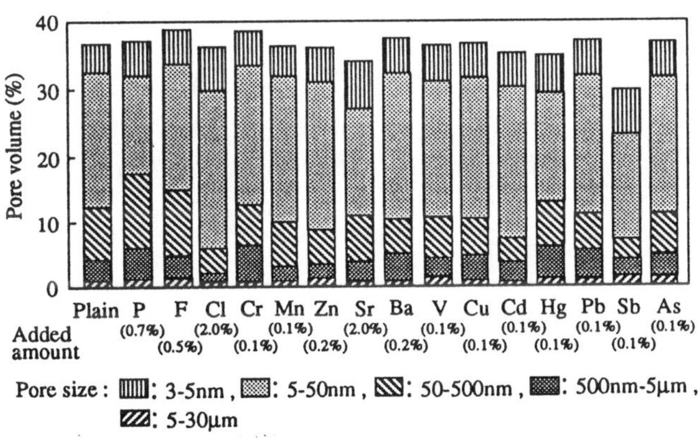

Figure 33. Pore structure of hardened Portland cement paste containing various minor and trace elements measured by mercury porosymetry.

Microstructure of Hardened Cement Paste. When the SO_4^{2-} concentration in the liquid phase in hydration is high, large amounts of smaller size ettringite (AFt) are formed at early stages of hydration [94]. Conversion of ettringite (AFt) into monosulfate hydrate (AFm) is retarded. Sulfur is found predominantly in ettringite (AFt) and monosulfate hydrate as one of their constitutive elements and partly in C-S-H. There is a greater amount of sulfur content found in C-S-H formed at a higher temperatures or at a lower C/S ratios [95]. Sulfur is not found in $Ca(OH)_2$.

Setting and Strength. Increase in S content in clinker results in short setting time and increased early strength but decreased long-term strength after 3 days. When 1% or more of S as SO_3 is contained in clinker, the amount of gypsum to be added to clinker must be reduced, but should not be lower than 1%. The upper limit of S content in clinker is considered to be 1% as SO_3 from the standpoint of cement quality.

Durability. Increases in S content result in increased formation of ettringite (AFt), leading to increased volume expansion of hardened cement paste. The JIS R-5210, 5211, 5212 and 5213 specify S content as SO_3 in cements from the viewpoint of soundness as follows: less than 3% for ordinary Portland cement, and less than 4.5% for type C blast-furnace slag cement and super high early strength Portland cements. The amount of gypsum to be added in grinding process is decided considering S content in clinker.

Sodium and Potassium

Hydration Reactivity. Sodium is incorporated into alite and K is incorporated in belite as solid solution. Solubility limits of alkalies in alite and belite crystal is in the range of 0.2 to 0.3%. These alkalies increase the hydration reactivity of both the alite and belite. Hydration reactivity of interstitial materials is enhanced by the formation of orthorhombic alkali-bearing tricalcium aluminate solid solution. But it deteriorates the rheological properties of fresh cement paste.

Microstructure of hardened cement Paste. Dissolved Na and K from clinker minerals in mixing water are found on the surface of C-S-H with large specific surface areas and may also exist in pore solution as ions. Their contents are highest in C-S-H followed by AFt and AFm. The $Ca(OH)_2$ does not contain alkalies. The amount of Na and K adsorbed by C-S-H increases with a decrease in the C/S ratio. They depress hydration of cement at later ages and increase the porosity of hardened cement paste.

Setting and Strength. Since clinker containing 0.2 to 0.4% of alkalies as R_2O usually contains less free CaO, setting is retarded slightly and higher strengths are developed at middle and later ages. In clinker containing 0.5% or more of alkalies as R_2O, the amount of alkali sulfate and free CaO increases resulting shorter setting times. Early strength develops (3 days), but the strength after 28 days is greatly reduced [96,97].

Durability. Such concrete as prepared from cement containing 0.6% or more of alkalies as R_2O and aggregate containing reactive silica and/or carbonates produces an expansive product by alkali-aggregate reaction in the presence of water which often causes the collapse of concrete structures. When the alkali content in cement is high, the hardened cement paste becomes porous. Consequently, air permeability and water penetrability become large, resulting in low durability and high susceptibility to the attack of aggressive chemical substances and weathering. Sodium and K in concrete dissolve in pore water, move through the hardened body by the drying and wetting cycle, migrate to the surface, and finally react with carbon dioxide in air to form Na_2CO_3 and K_2CO_3 which are precursors of efflorescence [98].

Titanium

Hydration Reactivity. Alite incorporating Ti is a triclinic crystal and shows a lower hydration reactivity than monoclinic [M_1] alite. Ferrite solid solution incorporating Ti has perovskite structure and relatively low in hydration reactivity. As a result, the hydration reactivity of cement [99] is reduced.

Microstructure of Hardened Cement Paste. Titanium dissolves from clinker minerals gradually during the process of cement hydration and is found concentrated in C-S-H and AFm formed in the vicinity of the original clinker minerals. Titanium is not found in $Ca(OH)_2$. The Ti has little effect on the pore-size distribution of hardened cement paste.

Setting and Strength. Titanium retards setting time [6,45,100] and provides increased strength after 3 days [45,64]. On the contrary, there is a report that addition of Ti at a level of 0.2% as TiO_2 gives increased 3 days' strength, but decreases later strength regardless of the cooling conditions [101]. These results suggest the optimum Ti content in terms of setting and strength.

Durability Less than 0.5% of Ti as TiO_2 has little effect on the durability of hardened concrete.

Phosphorus

Hydration Reactivity. Phosphorus decreases hydration reactivity of alite while stabilizes α-modification belite at room temperature and increases its hydration reactivity. Phosphorus contained in clinker dissolves slowly and, as a result, has less effect on hydration at initial stage than water-soluble P contained in by-product gypsum.

Water-soluble P reacts with Ca ion which has dissolved in mixing water during cement hydration to form a hydrate of calcium phosphate-apatite. As a result, increases in Ca ion concentration in liquid phase are suppressed and hydration of alite is also repressed.

The calcium phosphate produced precipitates on the surface of clinker minerals to form protecting film, resulting in suppression of their hydration.

Microstructure of Hardened Cement Paste. In the hardened cement paste, P is found concentrated in C-S-H and relatively little in AFm and $Ca(OH)_2$. The remaining P precipitates in the form of fine crystals of calcium phosphate. In the space among unhydrated cement particles, large crystals of $Ca(OH)_2$ are distributed. No remarkable difference is observed on the total pore volume in hardened cement paste with or without P, but it is observed that there is a slight increase in the percentage of large pores presumably built up by large crystals of $Ca(OH)_2$.

Setting and Strength. Phosphorus contained in clinker exists as a solid solution and dissolves so slowly during hydration that it has little effect on setting time. In general, P decreases the 1 day strength but has relatively little effect on later strengths.

Water-soluble P contained in by-product gypsum retards setting by more than 1 hr even when presents at a level as low as 0.1% as P_2O_5 [102]. Retardation of setting by water-soluble P varies depending on the types of phosphate. The effect of P on the retardation of alite hydration is $PO_4 = H_2PO_4 < P_2O_7 < H_2P_2O_7$ in increasing order. The retardation effect varies depending on the types of counter-cations in increasing order of H < Ca < Na < K and condensation degrees of phosphate anions.

The strength of hardened cement paste appears to be at the highest level when P content in clinker is 0.5% as P_2O_5 regardless of curing time. When the amounts exceeds 0.5%, the strength decreases with an increase in P content and the strength does not develop at levels equal to or greater than 5% [66]. High 28 days' strength can be obtained even in the case of high P content as long as the alkali content is high [23].

Durability. Phosphorus has little effect on durability.

Fluorine

Hydration Reactivity. Fluorine increases hydration reactivity of alite. When the F content in raw materials is high and formation of $C_{19}S_7 \cdot 2CaF_2$ and $C_{11}A_7 \cdot CaF_2$ takes place in the clinkering process, hydration reactivity of F-containing calcium silicate is reduced, while, that of F-containing calcium aluminate is accelerated, if sufficient SO_4^{2-} is supplied.

Microstructure of Hardened Cement Paste. Fluorine, which dissolves from clinker minerals during hydration, precipitates as fine crystals of CaF_2 and is found predominantly in C-S-H, a small amount in AFt and AFm but is not found in $Ca(OH)_2$. Fluorine causes the hardened cement paste to develop a porous structure with increased total pore volume and an increased volume of large capillary pores from 50 to 100 nm in size.

Setting and Strength. Fluorine contained in cement retards cement setting. The extent of retardation is by 40 min per 0.1% for a clinker with a 0.05-0.4% fluorine[103]. Strength increases with an increase in F content up to 0.5%, and then decreases [36]. Early strength reaches a maximum at about 0.3% of F. Various explanations for the retardation of setting by F have been proposed as follows: formation of $C_{19}S_7 \cdot 2CaF_2$ in the cooling process of clinker having low hydration reactivity [104]; formation of water-insoluble protecting film of CaF_2 on the surface of clinker minerals suppressing the water penetration and of movement of ions [105]; and formation of complexes of Ca with SiF_6^- which is formed by reacting silicate ions with F dissolved from the cement. Since water-soluble F retards the setting of cement, insolubilization treatment of F is usually carried out to prevent retardation of setting.

Durability. Air-permeability and water-penetrability of hardened cement paste is increased by the water soluble F which leads to decreased durability.

Chlorine

Hydration Reactivity. Chlorine increases hydration reactivity of alite and tri-calcium aluminate solid solution. Chlorine content as high as 0.25% accelerate the middle-stage hydration of alite.

Microstructure of Hardened Cement Paste. Chlorine, which dissolves from the cement during hydration, substitutes SO_4^{2-} in ettringite (AFt) and monosulfate hydrate (AFm) predominantly so that it is found mostly as Cl-substituted ettringite and Friedel's salt and partly in C-S-H. Friedel's salt is found in a similar morphology to that of monosulfate hydrate at later ages. The chlorine found in C-S-H may be fixed to C-S-H as a solid solution or adsorbed on its surface. Little Cl is contained in $Ca(OH)_2$. Chlorine has little effect on the pore structure of hardened cement paste within the amount normally found in clinker. However, the amount of finer capillary pores (less than 50 nm) increase and the amount of larger capillary pores decrease when large amounts of Cl are found (0.25%).

Setting and Strength. Chlorine dissolves in mixing water easily during initial stages of hydration, accelerates setting and increases early strength, but gives little effect on ultimate strength.

Durability. Chlorine improves the durability of concrete in early and middle ages by accelerating the hydration of alite and densifying the pore structure of hardened cement paste (Figure 34). It is noted, however, that Cl decreases the resistance to freezing and thawing of concrete at later ages [105]. Chlorine moves easily to the surface of iron reinforcements in concrete together with pore solution and corrodes the reinforcements. For this reason, JAS 5 and JIS A5308 specifies the content of Cl in concrete should not to exceed 300 g/m^3. Assuming that Cl content in cement is 0.03% and that the amount of cement per unit volume of concrete is 300 kg/m^3, the Cl content in concrete is 90 g/m^3 unless additional Cl is brought into the concrete from aggregate and chemical admixture used. However, special care must be exercised when waste is used as alternative raw

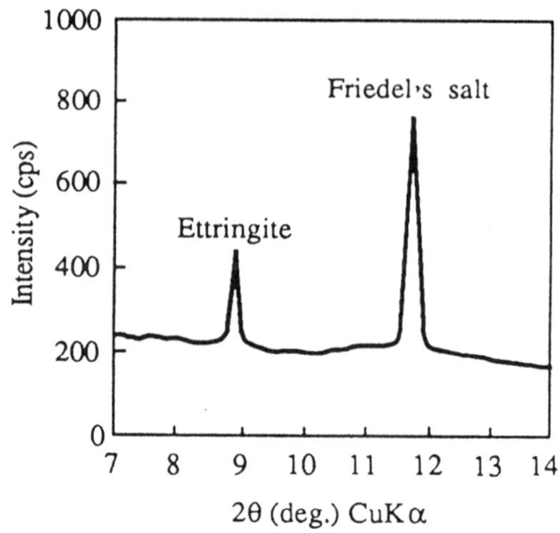

Figure 34. Powder x-ray diffraction pattern of hardened past prepared from cement containing chlorine.

material and fuel because there is some possibility of introducing Cl into cement from the waste.

Chromium

Hydration Reactivity. Hydration reactivity of alite is increased by incorporation of Cr as a solid solution. This is considered to be caused by the presence of chromium as Cr^{4+} increases the number of lattice defect of alite and makes it easy for Ca^{2+} to dissolve in water during hydration [53]. Chromium has little effect on hydration reactivity of belite.

Microstructure of Hardened Cement Paste. Chromium atoms are distributed in crystals as well as on their surfaces of C-S-H, $Ca(OH)_2$, AFt and AFm. Chromium on the surface of the crystal most likely precipitated as $Cr(OH)_3$ from the liquid phase. The Cr^{3+} in ettringite results from its substitution of Al^{3+} and that found in C-S-H results from its adsorption on the surface or in the interlayer space of the crystal [106]. About 0.1 mol/l of Cr suppresses hydration of ettringite and leads to form hydrate composed of Ca-Cr-OH [106]. Trivalent Cr rarely dissolves from hardened cement paste.

The Cr^{6+} incorporates into ettringite (AFt) as a solid solution through substitution of SO_4^{2-} as well as into the interlayer space of C-S-H in the form of CrO_4^{2-}. Presence of Cr at about 100 ppm as Cr_2O_3, normally found in cement, has little effect on the structure of hardened cement paste. When the Cr content exceeds 1%, the number of capillary pores (100 nm to 2 μm in size) in the hardened cement paste increase.

Setting and Strength. Chromium accelerates setting and increases the initial strength of cement. This takes place at about 0.05% of Cr as Cr_2O_3 and reaches maximum at about 0.5% [66]. When amount of Cr exceeds 0.5% as Cr_2O_3, strength decreases with an increase in Cr content regardless of curing age [66].

Durability. The amount of dissolved Cr^{6+} from hardened cement paste prepared from cement containing 100 ppm of Cr with 0.5 of W/C was 0.48 ppm (mg/l). The effect of Cr^{3+} was not determined [107]. Close attention should be paid to the possibility of leaching Cr^{6+} from the hardened body when the Cr content is increased by the use of waste as alternative raw materials and fuel.

Manganese

Hydration Reactivity. The presence of less than 1% of Mn increases the hydration reactivity of alite as well as that of belite, but decreases that of ferrite solid solution [31].

Microstructure of Hardened Cement Paste. Manganese is distributed predominantly in unhydrated clinker minerals and on their surfaces in hardened cement paste. Precipitated Mn on the surfaces of clinker minerals is $Mn(OH)_3$ and comes mainly from ferrite solid solution through dissolution in mixing water. The Mn does not dissolve readily from hardened cement paste [107]. The Mn^{2+} and Mn^{3+} ions are found in small amounts in C-S-H, ettringite (AFt) and monosulfate hydrate (AFm) where Mn^{2+} substitutes for Ca^{2+} and Mn^{3+} for Fe^{3+}. Manganese has little effect on pore structure of hardened cement paste when contained in amounts less than 0.1%.

Setting and Strength. Less than 0.1% of Mn as MnO_2 retards setting while it accelerates setting to a small extent exceeding above mentioned value. When coexisting with Ba, Mn is reportedly said to accelerate setting and increases early strength [37]. Development of strength is related to the substitution ratio of Fe by Mn as follows [108]: substitution ratio up to 50% shows higher strength than the case without Mn. The highest strength is developed regardless of age when the ratio equals 30% which corresponds to a Mn content of about 1% as MnO_2. However, the strength value obtained at this level is not very different from that obtained from cements containing no Mn.

Durability. The presence of Mn up to 0.1% as Mn_2O_3 does not give have an effect on air-permeability and water-penetrability.

Zinc

Hydration Reactivity. Zinc is an amphoteric element similar to Al, Cr and Pb. As shown in Figure 35 [4], Zn in clinker dissolves in water under high pH conditions and produces OH^-. The OH^- thus produced combines with Ca^{2+} to reduce the Ca^{2+} concentration in the liquid phase and retards hydration of alite. It is reported that the presence of Zn leads to precipitation of water-insoluble Ca-Zn-H_2O-CO_2 compounds on the surface of cement particles as a film, and thus to retardation of cement hydration [102]. Zinc has little effect on hydration reactivity of C_2S and interstitial materials.

Microstructure of Hardened Cement Paste. Zinc is found predominantly in C-S-H in hardened cement paste. Zinc exists in C-S-H through substitution for Ca^{2+} or formation of, for instance, calcium-bearing zinc oxide reacting with Ca on C-S-H surface. The Zn^{2+} and ZnO_4^- substitutes for Al^{3+} and SO_4^{2-}, respectively in ettringite (AFt).

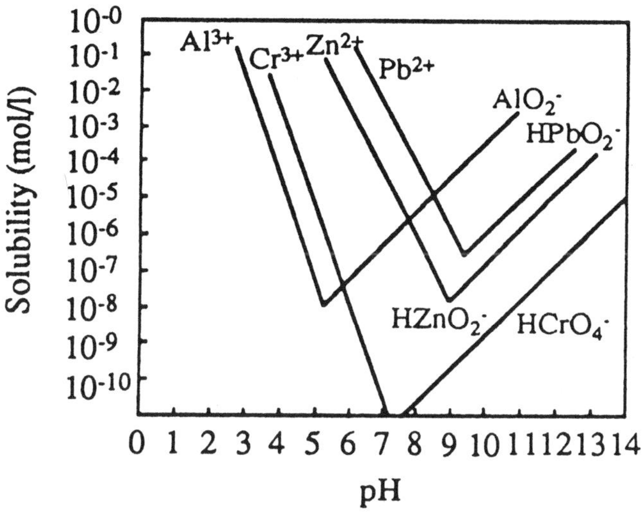

Figure 35. Relationship between pH and water solubility of zinc, aluminum, chromium and lead [4].

When Zn is contained at about 5% as ZnO, dissolved Zn from hardened cement paste is less than 1 ppm, which indicates that almost all of Zn is fixed to hardened cement paste. Presence of Zn is reported to lead the pore structure of hardened cement paste [109], but it does not differ greatly from the case without Zn when the content is less than 0.1%.

Setting and Strength. Zinc retards setting and decreases strength regardless of counter anions forming Zn compounds [110]. Both the initial and final setting is retarded with the amount of Zn in cement increases by about 45 minutes per 0.1% [111]. The effect of minor elements on setting retardation follows the following decreasing order: Zn > Pb > Cu > Sn > Cd [112]. Decreases in strength with increases in Zn content are not affected by the curing age and is unremarkable even if the amount of Zn is ≤ 0.5%.

When using waste as an alternative raw material and fuel, it is desirable to limit the amount of Zn in the clinker to an amount less than 0.1%.

Durability. Air-permeability and water-penetrability is increased when the content of Zn as ZnO exceeds 0.5% because the structure of hardened paste becomes more porous [109].

Strontium

Hydration Reactivity. Strontium decreases the size and number of lattice defects in alite. This leads to decreased hydration reactivity of alite [113]. The hydration reactivity of belite is improved by the incorporation of Sr [114].

Microstructure of Hardened Cement Paste. Strontium easily substitutes for Ca^{2+} in C-S-H. Strontium is also incorporated in AFt and AFm at a low levels, but not in $Ca(OH)_2$. The number of fine capillary pores (<50 nm) decreases with the amount of Sr (up to 4%) in cement, but Sr has little effect on the number of large capillary pores.

Setting and Strength. Strontium has a tendency to accelerate setting, but decreases the strength over the entire range of curing times.

Durability. Strontium has little effect on the durability of hardened concrete with the levels normally found in cement.

Barium

Hydration Reactivity. The lattice distortion of alite is reduced by the substitution of Ba for Ca^{2+} which causes a lower hydration reactivity [113]. Barium stabilizes α, α' and β modifications of belite at room temperature and increases its hydration reactivity [115]. Furthermore, Ba increases the hydration reactivity of tricalcium aluminate solid solution.

Microstructure of Hardened Cement Paste. Similar to Sr, Ba contained in hardened cement paste is found predominantly in C-S-H, in small amounts in AFm and AFt, but is not found in $Ca(OH)_2$. Barium slightly increases the total pore volume and increases the number of fine capillary pores smaller than 50 nm and decreases capillary pores larger than 50 nm.

Setting and Strength. Barium does not have a noticeable effect on setting and strength. Addition of BaO in combination with Mn_2O_3 results in accelerated setting and increased strength [37].

Durability. Similar to Sr, Ba is considered to have little effect on durability within the amount usually contained in cement.

Other Trace Elements

Hydration Reactivity. Vanadium decreases the hydration reactivity of belite [16]. Copper decreases the hydration reactivity of cement [116]. Cadmium has little effect on the hydration reactivity of cement in amounts less than a few ppm. Mercury has little effect on cement properties including hydration reactivity, setting and strength, and the microstructure of hardened paste. Lead is found as a nitrate ($Pb(NO_3)_2$), in amounts as low as 0.15% and retards hydration of alite and tricalcium aluminate solid solution. Antimony, in amounts as low as 0.05%, greatly accelerates hydration of cement.

Microstructure of Hardened Cement Paste. Copper is found in combination of Cr and reduces the rate of formation of ettringite (AFt), resulting in automorphic large-sized ettringite (AFt) crystals with a large axial ratio [116]. Similar to V, Cd, As and Pb, Cu is found predominantly in C-S-H, AFt and AFm and in small amounts in $Ca(OH)_2$. Small amounts of Pb have no effect on the microstructure of hardened cement paste, but act to reduce the largest pores as well as total pore volume when its content exceeds 2% [117]. An increase in specific surface area is due to the formation of $Pb(OH)_2$. Large amounts of Pb increases the density of hardened cement paste, but contributes little to the development of strength [117]. Antimony acts to reduce total pore volume as well as capillary pore volume and to increase gel pore volume. This effect is notable even when Sb content is as low as 0.05%.

Setting and Strength. Copper acts to retard setting and suppress the development of strength. The effect of Cu on setting retardation is greater for its oxide and chloride forms than it is for the sulfate and sulfide forms. Arsenic acts to retard setting and suppress the development of early strength when its content exceeds 0.1%. Cadmium acts to retard setting and suppress the development of later strength. Lead retards the initial and final setting by more than 1 hr even when its content is as low as 0.1% [102]. The following mechanisms have been proposed to explain the retardation of setting by Pb: formation of water-insoluble $Pb(OH)_2$ in liquid phase during hydration to cover the surface of cement particle [118] and formation of ions such as PbO_2^{2-} that have an effect on the concentration of Ca ions. Lead acts to reduce initial strength greatly, but has no effect on later strength after 28 days [102].

Durability. In hardened cement paste, As and Pb are found in the form of water-insoluble hydroxide. Lead is also found in hydrates where it substitutes for Ca^{2+}. Since Cd forms a solid solution substituting for Ca^{2+} in hydrates as well as $Cd(OH)_2$ and $CdCO_3$ during the hydration process, dissolution of Cd from hardened cement paste is, if found, as low as less than 0.1 ppm. Although Hg^{2+} may dissolve from hardened cement paste easier than Pb and As, it forms water-insoluble hydroxide under basic conditions and is fixed in the interlayer spaces of C-S-H by adsorption.

PROCEDURE TO USE WASTE AS ALTERNATIVE RAW MATERIALS AND FUELS

Principles of Waste Recycling

The principles to be observed for using waste as part of the raw material or fuel for cement manufacturing are as follows:

(1) The industrial waste should be more economical than conventional raw materials or fuels taking the favored treatment in industrial policy discussed later into consideration

(2) The industrial waste does not detrimentally affect the quality of cement.

(3) The use of industrial waste does not cause environmental pollution.

(4) The industrial waste is stably supplied.

In order to maintain these principles, it is necessary to closely consider the technical and economical studies of existing technical power, the investment required for new equipment for treating the industrial waste and the environmental and emission regulations of harmful materials established by local governments. The types and quantities of the main industrial waste used in cement industry of Japan are listed in Table 2.

Establishment of Acceptance Inspection Standards of Waste and Application for Use to Administrative Agency

The acceptance inspection is made by analyzing the following items of the industrial waste in the research laboratory within the company or an authorized inspection office including a certified environmental measurement office and by verifying the data submitted by the supplier of industrial waste:

(1) Harmful elements including Hg, R-Hg, Cd, Pb, Cr^{+6}, As, CN^{-} and R-P and harmful substances including PCB and dioxin,

(2) Materials with odors, including H_2S, mercaptan, NH_3 and CH_3CHO,

(3) Elements or substances detrimentally affecting the quality of cement including the element listed in Table 16,

(4) Elements that act as alternatives to the main elements including Si, Al, Fe and Ca and component alternative to heat value, and

(5) Moisture problems in the operation such as raw materials or products adhering to conveyors and chutes and consuming energy for drying, halogen elements including chlorine clogging the cyclone, chute and duct attached to the suspension preheater by producing low-melting point compounds at 800-1000°C due to concentration during circulation in the gas, and extraneous matter including metal pieces which may break the facilities for manufacturing cement.

Then, the data including the composition of exhaust gas, the contents of heavy metals and water-soluble harmful components contained in the raw material, clinker and dust and odors are collected by practically applying the industrial waste to the commercial plant.

It is indispensable to establish the acceptance inspection standards of industrial waste by comprehensively investigating whether or not the acceptance, transportation, weighing and feeding of the industrial waste can be efficiently carried out and if additional facilities are required carry out the operation. An example of the acceptance inspection standards is shown in Table 18 [1,2]. Since industrial waste contains more or less harmful matter, the standards have to provide for delivery using a closed container including container bag and covered truck to prevent the industrial waste from being released into the environment. The environmental and emission criteria of harmful matter for the prevention of environmental pollution in Japan are shown in Table 6 [1].

The environmental data collected by using industrial waste in a commercial plant have to be submitted to local administrative offices in order to obtain approval for use of industrial wastes.

Development of Applied Technologies and Investment in Facilities

Although the industrial waste that meets the requirements of the acceptance inspection standards is used in the processes shown in Figure 36 [1,2], some investment in equipment was required to make the process continue in an efficient manner.

Industrial wastes containing moisture and those free of moisture are fed into the clay preparation system before the clay dryer and into the process before the raw material grinding mill, respectively. Increases in Mg and S in the raw material by utilization of waste alternatives causes insufficient formation of clinker nodules in the kiln and forms a powdery small-size clinker, which not only reduces the heat transfer in the kiln, but also lowers the heat exchange of the cooler and reduces the combustion rate of fuel by lowering secondary air temperature. It also causes unstable operation of the kiln system due to release of the powdery clinker from the cooler grate. Since alkali, chlorine and sulfur in raw materials volatilize in burning zone and condense in the kiln end and preheater, waste containing significant amounts of these elements causes the formation of low melting point compounds including alkali sulfate and spulite etc. on the refractory surface of kiln end, preheater ducts and cyclones which often cloggs up the stream of gas and materials. Since the heat

Table 18. An Example of the Acceptance Inspection Standard of Industrial Waste [1,2].

Name of industrial waste	Acceptance inspection standard of industrial waste	
	Quality and property	Type of packing and method of transportation to the plant
Waste tire	Car tire (any tire is accepted) Cut tire or tire without cutting	Bulk
Waste oil	Liquid or mud , Low volatile matter, Chlorine content < 1000ppm, Heat generation > 12.5 MJ/kg, Water < 20%	Tank lorry or Drum (Tank lorry is preferable)
Petroleum oil, Heavyoil, Naphtha, Unburntdust	Water < 25% Heat generation > 12.5MJ/kg	Bulk vehicle with pressurized air unloader or container bag (Bulk vehicle is preferable)
Sludge	Water < 25% R_2O < 1%, Chlorine < 100ppm	Bulk dump truck (Moisture control is essential to dusty waste)
Waste sand from cast iron process	SiO_2 > 75%, R_2O < 1%, Water < 10%, Without any foreign matter	
Waste wood, Waste plastics	Chlorine < 1000ppm, Water < 10%, Size < 250x250x250mm	Bulk dump truck or container bag
Coal ash, Evacuated waste from furnace	Chlorine < 100ppm, R_2O< 1%	Bulk vehicle with pressurized air unloader
Aluminum sludge or aluminum ash	Al_2O_3 or Al > 60%, Chlorine <100ppm, Water < 70%	Bulk vehicle with pressurized air unloader , Bulk dump truck (Moisture control is essential to dusty waste)
Materials containing iron	Fe_2O_3 or Fe > 30%, Chlorine < 100ppm, Only small amount of metal such as Cr, etc. Powder, cotton-type or grindable material	Bulk dump truck or container bag

Other conditions : No bad odor, Content of harmful component (Hg, R-Hg, Cd, Pb, Cr^{6+}, As, CN, R-P or PCB) should be under the legal limit, No foreign matter such as metal, etc.

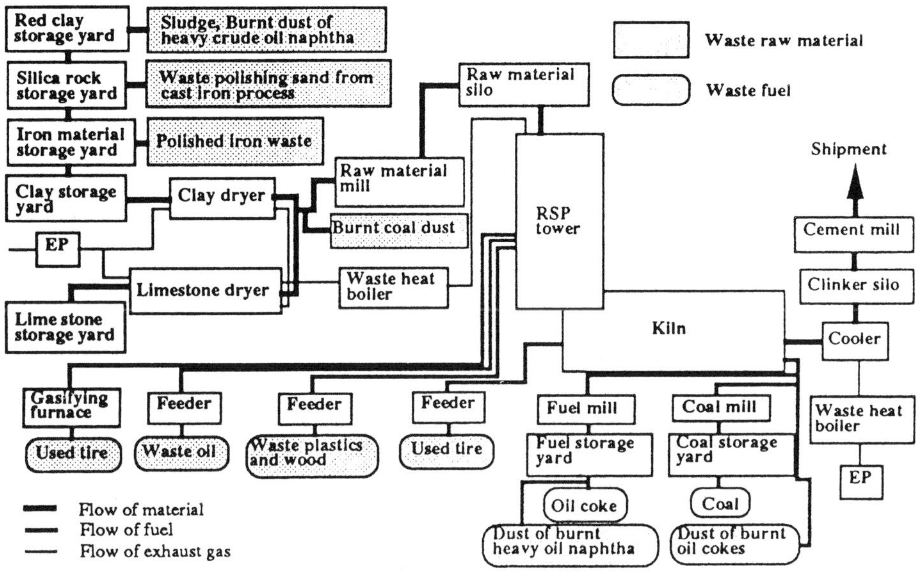

Figure 36. Flowchart of industrial waste recycling and utilization processes [1,2].

of decomposition of $Ca(OH)_2$ is less than that of limestone, the use of quicklime-modified sewage sludge as alternative raw material tends to cause unstable kiln operation. Metallic iron from steel wire of waste tires forms clusters of belite crystal in clinker and reduces the grindability of clinker. Therefore, storage facilities and industrial scales for the various types of industrial waste must be installed in order to precisely control the composition of the waste. On-line, real-time continuous analytical and control technologies are required for properly preparing different types of industrial waste with diversified compositions and properties. When waste molding sand composed of hard quartz is introduced, an additional facility for fine grinding would be necessary to ensure reactivity.

The industrial waste as alternative fuel is fed into a kiln through the burner or from the end of kiln, into the precalciner for the suspension preheater directly or through the burner and into the process after gasification. The industrial waste commonly used as raw material and fuel is mixed with the natural raw material. Fine, high combustion rate wastes are fed from the kiln head at the same time as the feed of finely ground coal. Meanwhile, coarse, relatively fire-retardant particles including EP dusts which contain unburned carbon are stored in a stockyard with oil coke and are ground together with coal and feed containing finely ground coal. Combustion residue dusts often contain harmful metals including lead and arsenic in high concentrations, a continuous, low-rate feed is necessary when such waste is used. Oil sludge and waste oil are fed into the precalciner because their offensive odor and measures to reduce their moisture content should be implemented. Although waste tires are usually fed into the kiln from the end, they are sometimes gasified in a separate gasifier and the gas produced is blown into the precalciner. Oil waste is injected into the precalciner for the suspension preheater for mixed combustion with finely ground coal. Smaller combustible solid wastes (< 1 in^3) including plastic waste and scrap wood are fed into the suspension preheater from the bottom through a drum mixer and the constant feed pusher for suspension firing.

Development of combustion technologies and installation of equipment suitable for treating each type of industrial waste are required for implementation. For example, a conveyor, air-tight constant feeder and gasifier for waste tires must be developed, and the performances of the constant feeder, combustion rate controller, extraneous matter remover and the abrasion resistance of equipment has to be improved for using industrial wastes as alternative fuels. Technical developments for preventing reactions of NH_3 with SO_3 is required for using EP dust discharged from thermal electric power plants where NO_x exhaust gas is reduced with ammonia. Since the entire plant has to be closed so that the industrial waste is not released, investments for reconstruction of the plant are required. The technical development and the investment in equipment have to be precisely investigated from the standpoint of balancing the profits obtained by the reduction in the raw materials and fuel costs and the benefits obtained from the reciprocal trade of industrial waste producers.

Expansion of Application Scope

Utilization as Mineralizer and Flux. Many types of industrial waste contain components usable as mineralizers and flux for cement as well as main components. Active research and development are also required for using such industrial waste.

For example, molybdenum contained in the residue produced after treating iron ore decreases the viscosity of the liquid phase produced in the clinker-burning process and accelerates the production of C_3S, thereby improving the early strength of cement [119]. Since the production of belite is accelerated by using Ba-containing waste produced during the manufacture of BaS, the burning temperature of cement can be decreased and energy can be saved [120].

Utilization of Sewage Sludge. Domestic waste water is sent to a sewage treatment plant through sewer pipes. Organic matter contained in the domestic waste water is decomposed by the activated sludge process using bacteria to reduce BOD and COD. After the content of harmful elements including BOD, COD and SS are reduced below the respective standards in the plant, water separated and purified in the plant is discharged to the natural environment. The annual production of sludge in the sewage treatment plants in Japan was 5 million tons in 1992 and will be approximately 7.8 million tons by the year 2000. Approximately 80% of the sludge is disposed of in landfills and sea disposal. Since the sewage treatment plants located in inland districts suffer from a shortage of disposal land, part of the sludge is used as compost. Demand for compost is, however, limited, so the development of a new sludge disposal process is expected.

Dewatered sludge cake contains bacteria as well as a large quantity of water and emits an offensive odor. As shown in Table 19 [3], it contains 80 to 90% of water and approximately 80% of the solid part is composed of organic matter and the remaining 20% or so is ash, depending upon the treated water. The heat value is approximately 5000cal/g for dried sludge cake. The contents of

Table 19. An Example of Chemical Composition of Sewage Sludge and Dewatered Cake by Blending with Equal Amounts of Lime [3].

	Chemical composition	Sweage sluge			Dewatered cake of sweage sludge by blending lime		
		Average	Range		Average	Range	
	Amount of water (wt%)	84.0	80.7	87.0	7.9	2.5	16.1
Harmful element	Cadmium (mg/kg)	0.72	<0.05	1.1	1.55	1.0	1.9
	Cyanogen (mg/kg)	0.7	ND	<1	<1	<1	<1
	Organic phosphorus (mg/kg)	0.7	ND	<1	<1	<1	<1
	Lead (mg/kg)	13.4	<0.2	30	27.9	18.7	37
	Total chromium (mg/kg)	153.3	12	810	7.52	4.78	11
	Chromium^{+6} (mg/kg)	—	—	—	0.83	0.83	0.83
	Arsenic (mg/kg)	3.9	1.9	8.9	3.5	2.7	5.2
	Total mercury (mg/kg)	0.775	0.047	1.9	0.067	<0.005	0.15
	Alkyl mercury (mg/kg)	0.003	ND	<0.005	<0.005	<0.005	<0.005
	PCB (mg/kg)	0.068	ND	0.20	<0.005	<0.005	<0.005
Other element	Total phosphorus (wt%)	1.97	1.2	3.5(P_2O_5)	0.225	0.19	0.26
	Total sulfur (wt%)	0.97	0.73	1.2	0.24	0.16	0.29
	Total nitrogen (wt%)	5.50	4.6	6.9	0.45	0.16	0.53
	Ammonium nitrogen (wt%)	1.71	1.26	2.1	0.02	0.02	0.03
	Kjeldahl nitrogen (wt%)	5.05	4.75	5.34	0.53	0.53	0.53
	Organic nitrogen (wt%)	3.55	3.05	4.04	0.51	0.51	0.51
	Nitric nitrogen (wt%)	0.05	<0.01	0.09	<0.01	<0.01	<0.01
	Nitrous nitrogen (wt%)	0.01	ND	<0.01	<0.01	<0.01	<0.01
	Chlorine (wt%)	0.239	0.054	0.52	0.1046	0.0083	0.20
	Fluorine (mg/kg)	191.0	26.4	480	90.5	11	210
	Sodium (wt%)	0.07	0.05	0.08	0.02	0.02	0.02
	Calcium (wt%)	0.69	0.59	0.78	39.0	34.2	43.7
	Carbonic ion (wt%)	1.01	0.51	1.5	1.5	1.5	1.5
	Sulfuric ion (wt%)	0.75	0.40	1.1	0.52	0.52	0.52
	Total organic carbon (wt%)	41.9	41.9	41.9	3.6	3.6	3.6
Industrial analysis	Ash (wt%)	21.1	17.5	25.4	84.3	70.7	91.9
	Volatile matter (wt%)	77.06	72.41	81.7	29.3	29.3	29.3
	Fixed carbon (wt%)	5.05	<0.01	10.09	<0.01	<0.01	<0.01
	Heat value (cal/g)	4935	4710	5160	720	720	720

harmful heavy metals are below permissible levels. The dewatered sludge cake can be used as a raw material or fuel for cement manufacturing by further dewatering, sterilizing and deodorizing. At present, the further dewatering of dewatered sludge cake are being studied using the slaking reaction of lime and sterilizing and deodorizing of the cake are being studied using the slaking heat and alkalizing. A mixture of lime and sludge (50/50) shown in Table 19, consists of grains and powders. The content of particles 0.15mm or more in diameter is approximately 50%. The water content in those particles is as low as few percent and heat value is approximately 700kcal/g. Since the mixture of lime and sludge is easily weighed and transported, it can be used as an alternative raw material and fuel for cement manufacturing. Ammonia-nitrogen in the sludge is converted to ammonia gas by adding lime and odor is emitted. The gas produced during grinding and preparation processes of the lime/sludge mixture must be treated by absorption with water or oxidative destruction. Ammonia is, however, thermally decomposed in the burning process, so no odor is emitted. The addition of the sewage sludge treated with lime to the raw mix of cement is as low as 3%, at present.

Utilization of Incineration Ash of Urban Refuse as Raw Material. The quantity of urban refuse continues to increase as the population in urban settings increases. The total quantity of urban refuse in Japan increased from 41.5 million tons in 1985 to 48.6 million tons in 1989. Although urban refuse was disposed of by landfilling or sea disposal in the past, incineration is now in widespread use to reduce the volume of refuse because disposal sites are increasingly difficult to locate. The volume of refuse is reduced by incineration to approximately 3%. According to 1989 statistics, the annual quantity of urban refuse incinerated and that of incineration ash were 35.93 million tons and 3.23 million tons, respectively. The latter is approximately 9% of the former.

The data in Table 20 [3] reveals that the contents of main components in incinerated urban refuse ash are suitable for use as

Table 20. Chemical Composition of Incineration Ash of Urban Refuse [3].

Major and minor element (%)

	SiO_2	Al_2O_3	Fe_2O_3	CaO	MgO	SO_3	Na_2O	K_2O	TiO_2	P_2O_5	ZnO	Cl
Average	22.9	19.7	5.6	30.4	4.8	2.1	3.34	2.55	0.93	1.80	0.60	8.47
Range	13.7 - 29.1	16.4 - 24.9	4.0 - 6.8	24.3 - 39.8	2.2 - 12.1	1.4 - 2.4	2.40 - 3.9	2.00 - 2.93	0.87 - 1.00	0.88 - 2.3	0.546 - 0.657	5.02 - 10.6

Trace element (ppm)

	Cd	Pb	Cr	As	Hg	Mn	Co	Ni	V
Average	11	311	538	55	3.5	438	33	345	67
Range	1-30	100-9,200	22-2,000	10-90	0.1-70	80-2,500	1-400	20-500	1-200

raw materials for cement, however, it contains considerable quantities of heavy metals and other harmful elements such as chlorine, phosphorus and zinc. The most serious problem is chlorine contained in quantities as low as 10%. Only alinite cement can make use of incineration ash containing high amounts of chlorine. There are still many problems awaiting solution for manufacturing as will be discusses later and the uses are limited due to insufficient performance and quality. It is now uncertain whether or not the incineration ash of urban refuse can be used in bulk as a raw material for cement. Its use depends upon the progress of research and development.

Improvement of Reactivity by Pre-Treatment One of the useful methods for using variously reactive industrial waste containing diversified components as raw material is to improve the reactivity by pretreatment. An application example of the sol-gel method for the sintering of advanced ceramics may be applied as a pre-treatment of cement raw materials from the wastes mentioned above [121]. However, attention must be paid to the cost effectiveness.

Strict Control of Operation and Quality

In order to use those industrial wastes as alternatives for raw materials and fuel for cement manufacturing, the wastes must be appropriately put to practical use and the quality of cement must be carefully controlled by considering the various influences the waste materials may have on the cement.

Future Problems

In order to meet the requirements of the four principles for using industrial waste as an alternative to fuel and raw materials and expedite its use, cement manufacturers and industrial waste producers should implement the following suggestions:

(1) Reduce the amounts of harmful elements that are detrimental to the environment, that effect the quality of cement and processes for cement manufacturing to their permissible limits or less,

(2) Development of application technologies for industrial waste; Development of new types of cement using harmful elements as useful elements; Development of efficient dehydration and handling techniques for industrial waste with high contents of water; Development of techniques for homogenizing industrial waste with diversified contents and properties; Acceleration of labor saving and unmanned operations in processes using industrial waste.

(3) Apportionment of costs caused by using industrial waste between the cement manufacturer and industrial waste producer.

It is estimated, however, that the use of industrial waste will be increased in the future for meeting social demands, as illustrated in Figure 37 [1,2]. Even industrial waste which cannot be used at present due to the variable of quality will be used in the near future, whereupon the technical development costs and the investment in equipment will be increased to provide for using such industrial wastes. The heat required for a process is not necessarily reduced by using industrial waste alternatives for fuel [122]. It is estimated that the cost, including the interest and depreciation expenses for the

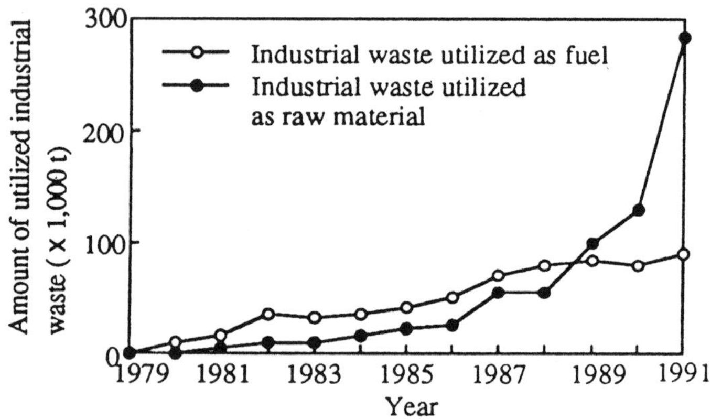

Figure 37. An example of the amount of industrial waste used in cement plants, excluding blending components for blended cements [1,2].

investment in equipment will be higher than that without using industrial waste in the near future. The following incentives for expediting the use of industrial waste should be considered:

(1) Administrative guidance in the indication of chemical composition of industrial waste.

(2) Tax credits for technical developments and equipment investments to provide for using industrial waste.

(3) Financing the firms with low-interest funds.

(4) Financial assistance including subsidies

The final disposal of urban refuse and waste building materials as well as industrial waste is an important social problem. Since the treatment and disposal of such massive amounts of refuse and waste causes increasing financial and land-use difficulties, incentives provided to deal with the waste problems will eventually pay for themselves in the end.

OTHER TYPES OF CEMENT-UTILIZING WASTE

Several investigations have been planned to utilize waste containing large amounts of harmful elements, not only for alternative raw materials and fuel for Portland cement manufacturing, but also as alternative raw materials or blending components for other types of cement. For example, incineration ash of urban refuse containing considerable amount of chlorine and metallic aluminum, phosphors and zinc is now being tried to use as raw materials of alinite cement, and as blending component of blended cement by previously treating the ash hydrothermally.

Alinite Cement

Alinite cement was developed in the Soviet Union in 1977 [123]. The cement is produced by substituting alinite ($Ca_{9.9}Mg_{0.8}\square_{0.3}(SiO_4)_{3.4}(AlO_4)_{0.6}O_{1.9}Cl$, where $\square$ = vacancy) for alite, $C_{11}A_7{\cdot}CaCl_2$ for C_3A, and adding several percent of gypsum dihydrate to clinker which contains 60-80% alinite, 10-30% belite, 5-10% $C_{11}A_7{\cdot}CaCl_2$ and 2-10% ferrite. The clinker contains 1.0-2.9% chlorine. The temperature range producing each mineral component is shown in Figure 38 [1,2]. The figure reveals that the burning temperature is as low as 1000-1200°C with flux effect of $CaCl_2$. Calcium chloride silicates produced in addition to alinite include $C_2S·CaCl_2$. The hydraulic reactivity is high and the later strength is fairly low, though early strength is higher than that of ordinary Portland cement. Although the early strength is increased by increasing the amount of $C_{11}A_7·CaCl_2$ and gypsum, stability is decreased. A portion of the Cl is fixed as a hydrate in the hydration process and the fixed chlorine is released again by the conversion and decomposition of hydrate. Since reinforcements and prestressing wires are remarkably corroded by the released chlorine, the alinite cement is used for unreinforced concrete and foundation-improving materials required for neither high-strength nor durability. Also in this case, close attention has to be paid to the dissolution of harmful heavy metal elements and Cl^- in the environment. Although more than ten years have passed since alinite cement was developed, it is produced only in a 15t/d mini-plant located at the Research Industrial Association Institute in the Republic of Kazaf.

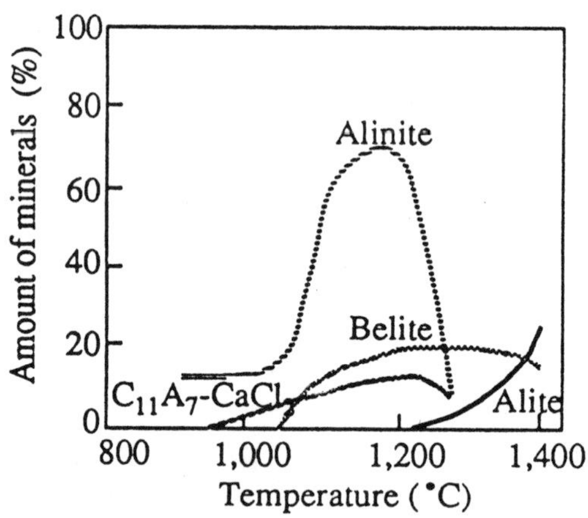

Figure 38. Temperature-phase relationship in the formation of alinite cement clinker [1,2].

Since the burning temperature of alinite cement is low and chlorine-containing wastes, including incinerated ash of urban refuse may be used as raw materials in the cement, new attention has been focused on this cement. There are, however, still many problems awaiting solution prior to commercial production and marketing including the quality, corrosion of manufacturing equipment by chlorine, coating problems, recovery of chlorine from the exhaust gas, emission prevention, accumulation of harmful elements contained in the incinerated ash of urban refuse in clinker and diffusion of harmful elements into environment when the cement is actually utilized by dissolution of hydrated cement, mortar and concrete even though the content of chlorine is as low as 2.5% or less, taking the quality of alinite cement into consideration.

Blended Cement with Incineration Ash of Urban Refuse as Blending Component

The amount of incineration ash of urban refuse amounts to approximately 6 million tons in Japan, and almost all of the ash was disposed of in landfills. In Japan, local governments treat incineration ash using one of several methods: glassification, solidification with cement and chemical treatment or acid extraction before land filling to prevent the harmful elements, which include heavy metals, from leaching out into the environment.

Incineration ash of urban refuse consists of hauyne, calcium hydroxide, quick lime, periclase, belinite i.e., chlorine-containing dicalcium silicate, sodium chloride, calcium chloride and metallic aluminum. It is a powder with a Blaine specific surface area of 3000-4000 cm^2/g, an average particle diameter of 15-20μm and a specific gravity of 2.4-2.5. It emits heat by the hydration of constituent compounds including quicklime, hauyne and belinite after contacting with water.

Metallic aluminum decomposes water in the presence of alkalies and generates hydrogen gas. Therefore, the mixture of cement and incineration ash generates heat and foams profusely immediately after mixing with water. As a result, the fluidity of fresh paste prepared from cement blended with incineration ash is reduced, the hardened body becomes porous and the paste continues expansion after hardening. If one's sole purpose is to prevent the dissolution of harmful elements, the dimensional stability as well as strength of waste solidified by cement to be land filled is not important. Therefore, most incineration ash is now disposed using a cost-effective cement solidification method that involves solidification by adding 10-20% Portland cement.

Recently attempts to use incineration ash of urban refuse as a resource have been investigated and the utilization of ash as a blending component of cement is being studied. Using the ash in this way requires that, when mixing with water, there is no vigorous generation of heat, no swelling and no profuse foaming, and no expansion after hardening. For these reasons, incineration ash is hydrated in advance so that the metallic aluminum and quicklime may change into stable compounds such as $Ca(OH)_2$, $AlO(OH)$ and $Al(OH)_3$. It takes at least seven days at 20°C, three days at 50°C and one day at 95°C for hydration of quicklime and metallic aluminum under atmospheric pressure. Because incinerators of urban refuse provide a plentiful supply of heat, it is an effective means for heating. Hydration and drying of incineration ash would allow one to make incineration ash a resource without increased costs.

The hydrates formed during the early stage of hydration of incineration ash blended cement are ettringite, Friedel's salt and $Ca(OH)_2$. Considerable amount of C-S-H are also observed in the later ages. Although the total pore volume of hardened paste of incineration ash blended cement is higher than that of blast furnace slag cement and fly ash cement as shown in Figure 39, dissolved elements from hardened paste of blended cement containing 25% of

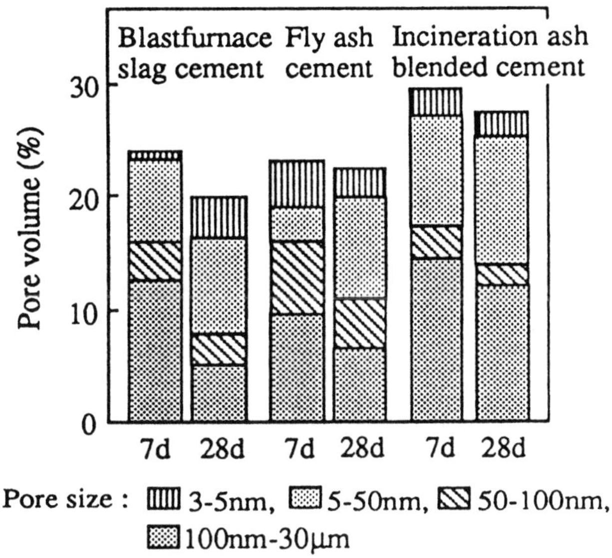

Figure 39. Pore structure of hardened mortar of blended cement (W/C = 0.65, S/C = 2.0).

incineration ash measured according to Environment Agency Notification No.13 is below the limits prescribed by law as shown in Table 21. The setting is retarded but still remains within acceptable limits and early strength is comparable to blast furnace slag cement and fly ash cement at low blending ratios as shown in Table 22. From these results, it is considered that incineration ash of urban refuse modified by hydrothermal treatment is adequate for use as blending component for blended cement.

Table 21. Dissolved Amount of Harmful Elements from Hardened Paste of Incineration Ash Blended Cement Measured in Accordance with Notification No. 13 of Environment Agency in Japan.

	R-Hg	T-Hg	Cd	Pb	O-P (mg/l)	Cr^{6+}	As	CN	PCB
Measured value	ND	ND	ND	0.1	ND	ND	0.045	<0.01	ND
Criteria	ND	0.005	0.3	3	1	1.5	1.5	1	0.003

R-Hg : Alkyl mercury compound , T-Hg : Total mercury , O-P Organic phosphorus , ND : Not determined

Table 22. Setting Time and Mortar Strength of Blended Cements.

	Blending ratio (%)				Setting time (JIS) (hr-min)		Compressive strength (MPa) (JIS)		
	OPC	I	S	F*1	Initial	Final	3d	7d	28d
OPC	100	0	0	0	2-40	3-30	16.7	26.5	41.2
Incineration ash blended cement	90	10	0	0	3-30	4-50	15.2	22.5	31.9
	75	25	0	0	3-40	5-25	7.4	10.3	15.2
	60	40	0	0	4-50	7-05	3.9	5.4	7.2
Blastfurnace slag cement	90	0	10	0	2-00	2-50	15.2	24.5	43.1
	75	0	25	0	2-15	3-05	14.7	23.5	42.6
	60	0	40	0	3-10	4-15	12.7	22.1	42.2
Fly ash cement	90	0	0	10	3-05	4-00	14.2	23.0	38.2
	75	0	0	25	3-30	4-35	12.3	19.1	30.9
	60	0	0	40	4-05	5-10	9.3	15.7	28.9

*1: OPC: Ordinary portland cement , I : Incineration ash of urban refuse, S : Blastfurnace slag , F : Fly ash

SUMMARY

In order to help the cement industry coexist with the natural environment and contribute to the welfare and prosperity of mankind, it is necessary to not only control the emission pollutants from the manufacturing process and prevent the diffusion of such pollutants to the natural environment, it is also necessary to preserve natural resources by utilizing waste materials.

Maintenance of manufacturing equipment is essential as is improving the efficiency of the operating control systems. Development of a waste recycling system with an environment-cleaning function to effectively use industrial wastes and urban refuse as raw materials and fuel is critical.

In this chapter, problems for treatment and disposal as well as type, amount and character of industrial waste and general waste which are currently used for cement manufacturing or may be used in the near future, excluding blast furnace slag, fly ash, silica fume, rice husk ash, red mud, by-product gypsum and squeezed sugar cane wastes which were described separately in other chapters of this book, were described.

Also, the influence of the use of waste as an alternative raw material and fuel on the process and operation of cement manufacturing was described. The required measures to decrease the generation of pollutants from manufacturing processes and reduce their discharge to environment was discussed.

Minor and trace elements introduced into cement manufacturing system from the wastes and their affect on clinker formation such as changes in composition, texture, and structure were discussed. The minor and trace elements have an effect on the properties of hydrated cement and concrete including viscoelastic properties, setting, strength and durability. Influences of six typical minor elements and fourteen typical trace elements were described in detail. It was also pointed out that attention should be paid to maintaining the quality of cement when using wastes as alternative raw materials and fuel.

In using waste as an alternative raw material and fuel in a plant, it is indispensable to establish acceptance inspection standards for preventing the pollutants contained in the waste from diffusing to environment, to develop technology and to invest in equipment for homogenizing raw material, improving the ease of handling, stabilizing kiln operations, expanding the utilization of waste, and to obtain the approval of local administrative offices for using waste.

Investigation from a different point of view is also necessary to utilize wastes not only as alternative raw materials and fuel for Portland cement manufacturing but also as raw materials or blending components for other type of cements. This subject was also discussed in a later section of this chapter.

It is expected that this description is useful for the diversification and expansion of the utilization of waste as raw materials and fuel for cement manufacturing.

REFERENCES

1. Uchikawa, H., Cement Industry as Environmentally Compatible and Waste Recycling System, Invited paper of *International Symposium on Environmental Issues of Ceramics* organized by The Ceramic Society of Japan (Under printing), Sapporo (1994).
2. Uchikawa, H., Present Problem in Cement Manufacturing, *9th International Congress on the Chemistry of Cement*, New Delhi, Special Lecture, (Invited Paper), Vol.VII (under printing) (1992): *Journal of Research of the Onoda Cement Company*, 45(128): 7-44 (1994).

3. Uchikawa, H., Management Strategy of Cement Industry in Technology Toward the Next Century, Engineering Foundation Conference; *Advances in Cement and Concrete* (M.W. Grutzeck and S.L. Sarkar eds.) held at New England Center, the University of New Hampshire, Durham, NH, pp.524-561, Amer. Soc.Civ. En., (1994): Management Strategy in Cement Technology for the Next Century, *World Cement*, 25(9): 66-78, 25(10): 49-56, 25(11): 47-50, 25(12): 58-63 (1994).
4. Uchikawa, H., Fixation of Harmful Element in Industrial Waste and Sludge with Cement, *Ceramics Japan* 12: 103-117 (1977).
5. Uchikawa, H., The Future for Cement/Concrete in Japan - The Way to the Prosperity of Japanese Cement Industry through Self-Renewal-, *Onoda Pacific Conference*, Sydney, Vol.2: Uchikawa pp.1-77 (1990)/ *Journal of Research of the Onoda Cement Company*, 42 (122): 1-62 (1990).
6. Uchikawa, H., Hanehara, S., Shirasaka, T. and Hashimoto, S., Influence of Burning Atmosphere on Distribution of Minor and Trace Element in Clinker and Formation of Brown Color Clinker, *JCA Proceeding of Cement and Concrete* 46: pp.32- 37 (1992).
7. Knofel, D., Beeinflussung einiger Eigenschaften der Portlandzementklinkers und des Portlandzements durch TiO_2, *Zement-Kalk-Gips* 30: 191-196 (1977).
8. Royak, S.M., Astreyeva, O.M. and Lukina, M.N., Effect of Titanium Dioxide on Clinker Formation Processes, *Tsement* 21 (3):12 (1955).
9. Akatsu, K., Maeda, K. and Ikeda, I., The Effect of Cr_2O_3 and P_2O_5 on the Strength and Color of Portland Cement Clinker, *CAJ Review of the 24th General Meeting* 24: pp.20-23 (1970).

10. Timashev, V.V., Galperina, T.YA., Bykova, S. N., Sevostianova, L.I. and Yankova, N.F., Effect of Phosphorous on Phase Composition and Properties of Cement Produced of High- alkali Raw Materials, *Tsement* 45(3): 6-7 (1979).
11. Vogel, E., Die Wirkung von Fluβspat und Kiesabbrand auf den Reaktionsverlauf von Zementrohmehlen unterhalb 1100, *Silikat-technik* 10: 415-419 (1959).
12. Luginina, I.G., Influence of Mineralizers on the Clinker Formation by Rapid Firing, *Tsement* 27(4): 20-22 (1961).
13. Gutt, W., Manufacture of Portland Cement from Phosphatic Raw Materials, *5th International Symposium on the Chemistry of Cement*, Tokyo, vol.I: pp.93-105 (1968).
14. Schlegel, E, Die Einwirkung von Flussspat auf die Entsauerung von Magnesit - Calcit und Dolomit, *Zement-Kalk-Gips* 21: 386-389 (1968).
15. Klemm, W.A., Holub, K.J. and Skalny, J., The Effects of Fluxes and Mineralizers in Lowering Cement Kiln Temperatures, *Progress Report No1, July 1977 (Prepared for Div. Adv. Energy and Resources)*, Research and Technology National Science Foundation, Washington (1977).
16. Timashev, V.V., The Kinetics of Clinker Formation - The Structure and Composition of Clinker and its Phases, *7th International Congress on the Chemistry of Cement* Paris, Vol.1: pp. I-3/1-20 (1980).
17. Johansen, V. and Christensen, N.H., Rate of Formation of C_3S in the System CaO-SiO_2-Al_2O_3-Fe_2O_3-MgO with Addition of CaF_2, *Cem. Concr. Res.* 9: 1-5 (1979).
18. Imlach, J.A., The Influence of Heating Conditions on the Production of Fluorine Containing Portland Cement Clinker, *Cement Technology* 5: 403-406 (1974).
19. Brisi, C. and Rolando, P., Sul Meccanismo di Decomposizione dell Alluminato Tricalcico in Presenza di Fluoruro e Cloruro di Calcio, *Annali di Chimica* 56: 224-231 (1966).

20. Massazza, F. and Pezzuoli, M., Equilibres a l'etat solide dans le systeme quaternaire CaO-Al_2O_3-SiO_2-CaF_2, *Rev. Mater. Constr.*, (642): 81-86 (1969).
21. Uchikawa, H. and Kohono, K., Ultra-rapid Hardening Cement (Jet Cement), *New Concrete materials*, (R.N. Swamy eds.) Surry University Press vol.1: pp.70-136 (1973).
22. Courtault, B., Etude des reactions a l'etat solide jusqu'a 1600°C au moyen de l'analyse thermique differentielle, *Rev. Mater. Constr.* (569): 37-47, (570): 67-78, (571): 110-124, (572): 143-156, (573): 190-203 (1963).
23. Moir, G.K. and Glasser, F.P., Mineralisers, Modifiers and Activators in the Clinkering Process, *9th International Congress on the Chemistry of Cement*, New Delhi, vol.I: pp.125-152 (1992).
24. Nudelman, B., Bikbaou, M., Sventsitski, A. and Ilukhine, V., Structure and Properties of Alinite and Alinite Cements, *7th International Conference on the Chemistry of Cement* Paris, Vol.III: pp.169-174 (1980).
25. Fitkos, C., Georgiades, A. and Philippou, T., Preparation and Hydration Study of Alinite Cement, *Cem. Concr. Res.,* 21: 1129-1136 (1991).
26. Agarwal, R.K., Paralkar, S.V. and Chatterjee, A.K., Chloride Salts as Reaction Medium for Low Temperature Clinkerisation - A Probe into Alinite Technology, *8th International Conference on the Chemistry of Cement*, Rio de Janeiro, vol.II: pp.327-333 (1986).
27. Krishnan, N.A., Gore, V.K., Khadllkar, S.A., Hargave, R.V., Page, C.H. and Chatterjee, A.K., Trace Element Balances in Kiln System of a Few Indian Cement Plants, *9th International Congress on the Chemistry of Cement*, New Delhi, vol.II: pp. 67-73 (1992).

28. Butt, Y.M., Timashev, V.V. and Malozohn, L.I., The Crystallization of Compounds in the Presence of Cr_2O_3, P_2O_5, or SO_3 and the Properties of the Resultant Cement, *5th International Symposium on the Chemistry of Cement*,Tokyo, Vol. I: pp.340-347 (1968).
29. Butt, Y.M., Timashev, V.V. and Osokin, A.P., The Mechanism of Clinker Formation, Processes and the Modification of Its Structure, *6th International Congress on the Chemistry of Cement*, Moscow, pp.I-3/1-33 (1974).
30. Sichov, M.M., Problem of Admixtures, *5th International Symposium on the Chemistry of Cement*, Tokyo, vol.I: pp.157- 162 (1968).
31. Parkash, R., Ahluwalia, S.C., Kathal, N.K. and Mathur, V.K., Role of Manganese on the Kinetics of Formation and Hydration of C_3S Phase, *9th International Congress on the Chemistry of Cement*, New Delhi, vol. I: pp.420-426 (1992).
32. Valkov, V., Sitchiov, M.M., Gigova, L. and Nikiforov, Yu.V., Common Influence of BaO and Mn_2O_3 on the Obtaining and the Properties of the Portland Cement, *7th International Conference on the Chemistry of Cement* , Paris vol. II: pp.I-95 - I-98 (1980).
33. Xu, G., Huang, W., Lu, Z. and Qian, G., The Effect of ZnO on Burning of Portland Cement Clinker in Rapid Heating-up Burning, *9th International Congress on the Chemistry of Cement*, New Delhi, vol.I: pp.372-378 (1992)
34. Tsuboi, T., Ito, T., Hokinoue, Y. and Matsuzaki, Y., Die Einflusse von MgO-SO_3 und ZnO auf die Sinterung von Portlandklinker, *Zement-Kalk-Gips* , 25: 426-431 (1972).
35. Knofel, D., Beeinflussung einiger Eigenschaften des Portlandzementklinkers und des Portlandzements durch ZnO und ZnS, *Zement-Kalk-Gips* , 31: 157-161 (1978).
36. Bucchi, R., Influence of the Nature and Preparation of Raw materials on the Reactivity of Raw Mix,*7th International*

Conference on the Chemistry of Cement, Paris, vol. I: pp.I-3- I-43 (1980).

37. Akatsu, K.and Monna, I., Effect of Mn on the Portland Cement Clinker and Cement, *CAJ Review of the 20th General Meeting* 20: pp.41-45 (1966).
38. Odler, I. and Abdul-Maula, S., Effect of Mineralisers on the Burning Process of Portland Cement Clinker, Part 1: Kinetics of the Process, *Zement-Kalk-Gips*, 33: 132-136 (1980).
39. Odler, I. and Abdul-Maula, S., Effect of Mineralisers on the Burning of Portland Cement Clinker, Part 2: Mode of Action of the Mineralisers, *Zement-Kalk-Gips*, 33: 278-282 (1980).
40. Nurse, R.W., Phase Equilibria and Formation of Portland Cement Minerals, *5th International Symposium on the Chemistry of Cement*, Tokyo, vol.I: pp.77-89 (1968).
41. Regourd, M. and Guinier, A., The Crystal Chemistry of the Constituents of Portland Cement Clinker, *6th International Congress on the Chemistry of Cement*, Moscow, pp.I-2/1-62 (1974).
42. Nikiforov, Yu. V., Zozulya, R.A. and Ivanova, N.M., Role Played by Magnesia in Clinker and Cement Technology,*6th International Congress on the Chemistry of Cement*, Moscow, pp.I-3/14 1-12 (1974).
43. Budnikov, P.P. and Kuznetsova, I.P., Investigation of the Influence of Gypsum of Mineral Formation in Cement Clinker, *Journal of Applied Chemistry, (USSR)*, 35(5): 913-916 (1962).
44. Gies, A. and Knoefel, D., Influence of Alkalis or the Composition of Belite-rich Cement Clinker and the Technological Properties of the Resulting Cements, *Cem. Concr. Res.*, 16(3): 411-422 (1986).
45. Kondo, R. and Yoshida, K., Miscibilities of Special Elements in Tricalcium Silicate and Alite and the hydration Properties of C_3S Solid Solutions, *5th International Symposium on the Chemistry of Cement*, Tokyo, vol. I: pp.262-274 (1968).

46. Sakurai, I. and Sato, T., Written Discussion on "Analysis of Portland Cement Clinker," Yamaguchi G. and Takagi S., *5th International Symposium on the Chemistry of Cement*, Tokyo, Vol.I: pp.221-225 (1968).
47. Fierens, P., Thauvoye, M. and Verhaegen, J.P., Effects of Iron (III) and Titanium (IV) on the Tricalcium Aluminate Structure and Reactivity, *Il Cemento* , 71: 3-10 (1974).
48. Shirasaka, T., Kobayashi, K. and Uchikawa, H., Influence of Specified Elements in Wastes on the Texture and Elemental Distribution in Four Major Clinker Minerals, *JCA Proceedings of Cement & Concrete*, 45: (under printing) (1995).
49. Johansen, V., Solid Solution of Chromium in Ca_3SiO_5, *Cem. Concr. Res.*, 2: 33-42 (1972).
50. Johansen, V., Absorption Spectra of CrO_4^{4-} in Ca_3SiO_5 and $Ca_3Al_2O_6$, *Cem. Concr. Res.*, 8: 245-250 (1978).
51. Hornain, H., Sur la repartition des elements de transition et leur influence sur quelques proprietes du clinker et du ciment, *Rev. Mater. Constr.* , (671-672): 203-218 (1971).
52. Kurdowski, W. and Szummzr, A., L'influence des composants mineurs sur les proprietes hydrauliques du clinker portland, *Silicates Ind.*, 33: 183-191(1968) .
53. Sakurai, T., Sato, T. and Yoshinaga, A., The Effect of Minor Components on the Early Hydraulic Activity of the Major Phases of Portland Cement Clinker, *5th International Symposium on the Chemistry of Cement*, Tokyo, vol.I: pp.300- 321 (1968).
54. Puertas, F., Glasser, F.P., Blanco-Varela, M.T and Vaquez, T., Influence of the Kiln Atomosphere on Manganese Solid Solution in C_3S and C_2S, *Cem. Concr. Res.,* 18: 783-788 (1988).
55. Brisi, C. and Appendido, P, Equilibri allo stato solidio nel sistema $CaO\text{-}SrO\text{-}SiO_2$, *Annali di chimica,* 65: 1213- (1965).

56. Cirilli, V. and Brisi, C., Influenza del modulo calcareo sila composizione della fase ferrica del clinker di portland, *Ind. Ital. Cem.*, 25: 5-8 (1955).
57. Kramer, H. and Zur Strassen, H., Discussion. Phase Equilibria and Constitution of Portland Cement Clinker (R.W.Nurse), *4th International Symposium of the Chemistry of Cement* Washington, Vol.1: pp.32-33 (1960).
58. Miyazawa, K. and Tomita, K., Der Einfluss von MgO auf die Eigenschaften des Portlandzements, *Zement-Kalk-Gips* 19: 82-85 (1966).
59. Knofel, D. and Spohn, E., Der quantitative Phasengehalt in Portlandzementklinkern, *Zement-Kalk-Gips* 22: 471-476 (1969).
60. Teoreanu, I., Balasoiu, H., Radovici, C. and Ciomartan, D.,The Influence of Cr_2O_3 Micro Additions on the Chemical and Mineralogical Composition of Portland Cement Clinkers, *7th International Conferencen the Chemistry of Cement* Paris, Vol.II: pp.I/90-94 (1980).
61. Knofel, D., Strunge, J. and Bambauer, H.U., Effect on Manganese of the Properties of Portland Cement Clinker and Portland Cement, *Zement-Kalk-Gips* 36: 402-408 (1983).
62. Gutt, W. and Smith, M.A., Studies of the Role of Calcium Sulphate in the Manufacture of Portland Cement Clinker, *Trans. Brit. Ceram. Soc.*, 67: 487-509 (1968).
63. Kondo, R., Effects of Special Components on the Mineral Compositions of Portland Cement Clinker, *JCEA Review of the 17th General Meeting* 17: pp.31-32 (1963).
64. Knofel, D., Der Einbau von TiO_2 in die Phasen des Portlandzementklinkers, *Zement-Kalk-Gips.*, 32(1), 35-40 (1979).
65. Halicz, L. and Nathan, Y., The Influence of P_2O_5 on Clinker Reactions, *Cem. Concr. Res.*, 14: 11-18 (1983).
66. Uchikawa, H., Uchida, S., Ogawa, K. and Hanehara, S., Influence of the Amount, State and Distribution of Minor

Constituents in Clinker on the Color of White Cement, *Il Cemento*, 83: 153-168 (1986).

67. Yamaguchi, G. and Uchikawa, H., Untersuchungen über die Mischkristalle im System $3CaO{\cdot}SiO_2$-Na_2O, *Zement-Kalk-Gips*, 13: 497-504 (1961).
68. Ono, Y., Uno, T. and Kanai, Y., Synthesis of Five Polymorphic Modifications of C_3S, *CAJ Review of the 19th General Meeting* , 20: pp.36-41 (1965).
69. Woermann, E., Eysel, W. and Hahn, Th., Chemische und strukturelle Untersuchungen der Mischkristallbildung von Tricalciumsilikat, *Zement-Kalk-Gips* , 22: 414-422 (1969).
70. Suzuki, K., Morita, M. and Sugiyama, H., α'-Ca_2SiO_4 Stabilized by Ba, Sr and Mg, *CAJ Review of the 21st General Meeting*, 21: pp.43-45 (1967).
71. Udagawa, S., Cement Chemistry of Dicalcium Silicate, *CAJ Proceeding of the 34th General Meeting* (in Japanese) 34: pp.31-39 (1980).
72. Welch, J.H. and Gutt, W., The Effect of Minor Components on the Hydraulicity of the Calcium Silicates, *4th International Symposium on the Chemistry of Cement* Washington, vol.1: pp.59-68 (1960).
73. Kondo, R. and Goto, S., A Study of the System CaO-SiO_2-Cr_2O_5- Al_2O_3, *CAJ Review of the 19th General Meeting*, 19: pp.42-44 (1965).
74. Tilley, C.E. and Vincent, H.G., *Mineralogical Magazine and Journal of the Mineralogical Society*, 28: 255 (1948).
75. Suzuki, K., Shoono, E. and Sakai, R., $\alpha'-Ca_2SiO_4$ Stabilized by B, V and P, *CAJ Review of the 22nd General Meeting*, 22: pp.24- 25 (1968).
76. Nagashima, M and Saito, M, Some Aspects of the Hydration of $3CaO{\cdot}Al_2O_3$ Solid Solutions, *CAJ Review of the 31st General Meeting*, 31: pp.37-38 (1977).
77. Maki, I., Processing Conditions of Portland Cement Clinker as viewed from the Fine Textures of the Conctituent Minerals,

Cement Technology (E.M. Garter and H.Uchikawa eds.), Ceramic Transactions 40: pp.3-17 (1994).
78. Maki, I., Nature of the Prismatic Dark Interstitial Material in Portland Cement Clinker, *Cem. Conc. Res.*, 3: 295-313 (1973).
79. Ono, M., Nakamura, T. and Takeuchi, H., Effect of Na_2O on the Composition of Ferrite Phase, *CAJ Review of the 30th General Meeting*, 30: pp.23-25 (1976).
80. Sakurai, T., Sato, K. and Yoshinaga, A., On the Influence of Chromium Oxide on $2CaO{\cdot}(Al, Fe)_2O_3$, *CAJ Proceeding of the 30th General Meeting*, 21: pp.53-56 (1967).
81. Sakurai, T., The Behaviour of Manganese in Cement Clinker Minerals, *JCEA Review of The 14th General Meeting* 14: pp.23-24 (1960).
82. Matsuda, Y., Kotani, H., Joe, Y. and Tanaka, H., Influence of Chlorine Ion and Sulfate Ion on the Phase Equilibrium of Ferrite, *CAJ Review of the 38th General Meeting* 38: pp.32-35 (1984).
83. Ono, Y., Kawamura, S. and Yamamoto, M., Lattice Constants of Alite in Laboratory Clinker, *CAJ Review of the 39th General Meeting*, 39: pp.28-31 (1985).
84. Ono, Y., Change of Lattice Constants of Alite by Alkalies, SO_3 and Cooling Condition, *CAJ Review of the 40th General Meeting*, 40: pp.32-35 (1986).
85. Sato, T. and Fukunaga, M., Lattice Strain in $3CaO{\cdot}SiO_2$ Solid Solution, *CAJ Review of the 24th General Meeting*, 24: pp.1-4 (1970).
86. Tanaka, M., Uchida,I. and Kuroki, Y., Influence of Zinc Oxide on Property of Alite, *CAJ Review of the 38th General Meeting*, 38: pp.36-39 (1984).
87. Hirai, M. and Yanagi, M., Formation of Belite Burned in the Presence of Al_2O_3 and Na_2SO_4, *CAJ Review of the 22nd General Meeting* 22: pp.25-26 (1968).

88. Yamaguchi, G., Ono, Y., Kawamura, S. and Kanai, Y., The Solid Solution Region of the High Temperature Modifications of Ca_2SiO_4 for Al_2O_3, Fe_2O_3 and Na_2O, *JCEA Review of the 17th General Meeting* , 17: pp.28-30 (1963).
89. Yamaguchi,G., Ono,Y., Kawamura, S. and Sooda, Y., Strength of Mortars Made from the Modifications of Ca_2SiO_4, *JCEA Review of the 17th General Meeting*, 17: pp.39-43 (1963).
90. Suzuki, K., Shibata, S., Fujii, N. and Ito, S., Hydration of α, β and β Dicalcium Silicate with Na-, K-Fe or Na-, K-Al Components, *CAJ Review of the 33rd General Meeting*, 33: pp. 45-45 (1979).
91. Maki, I and Goto, K., Factors Influencing the Phase Constitution of Alite in Portland Cement Clinker, *Cem. Concr. Res.*, 12: 301-308 (1982).
92. Soda, Y., Mizukami, K. and Shirasaka, M., On the Decline in the Development of the Strength of Portland Cement by the Increase of Magnesia, *CAJ Review of the 22rd General Meeting*: 22: pp.48-51 (1968).
93. Uchikawa, H., Ogawa, K. and Uchida, S., Influence of Character of Clinker on the Early Hydration Process and Rheological Property of Cement Paste, *Cem. Concr. Res.*, 15: 561-572 (1985).
94. Uchikawa, H., Uchida, S., Ogawa, K. and Hanehara, S., Influence of $CaSO_4{\cdot}2H_2O$, $CaSO_4{\cdot}1/2H_2O$ and $CaSO_4$ on the Initial Hydration of Clinker Having Different Burning Degree, *Cem. Concr. Res.*, 14: 645-656 (1984).
95. Uchikawa, H., Hanehara, S. and Sawaki, D., Structural Change of Hardened Mortar by Drying, *Proceeding of 3rd NCB International Seminar*, New Delhi, vol.4: pp.VIII/1-12 (1991).
96. Osbaeck, B., Der Einfluss von Alkalien auf die Festigkeitseigenschaften von Portlandzement, *Zement-Kalk-Gips*. 32: 72-77 (1979).

97. Gebauer, J. and Kristmann, M., The Influence of the Composition of Industrial Clinker on Cement and Concrete Properties, *World Cem. Technol.*, 10: 46-51 (1979).
98. Takemoto, K., Takahashi, H., Tokairin, T. and Makino, H., Basic Research on Production of the Eflorescence of Concrete, *Cement and Concrete* (191): 15-25 (1963).
99. Knofel, D., Relationships between Chemistry, Phase Content and Strength in Portland Cements, *Zement-Kalk-Gips*, 32: 448- 454 (1979).
100. Fierens, P., Thauvoye, M. and Verhaegen, J. P., The Influence of Iron (III) and Titanium (IV) on the Structure and Reactivity of Tricalcium Silicate, *Il Cemento*, 69: 211-220 (1972).
101. Rio, A and Saini, A., Through on X-ray Spectrometry of Cement Material by Mineralogical Material, *Il Cemento* 69: 151-162 (1972).
102. Lieber, W., The Influence of Lead and Zinc Compounds on the Hydration of Portland Cement, *5th International Symposium on the Chemistry of Cement*, Tokyo, vol.2: pp.444-454 (1968).
103. Moir , G.K., Improvements in the Early Strength Properties of Portland Cement, *Phil. Trans. Roy, Soc.*, A 310: 127-138 (1983).
104. Tanaka, M., Sudoh, G. and Akaiwa, S., New Compound $Ca_{12}Si_4O_{19}F_2$ in the system CaO-SiO_2-CaF_2 and the Role of CaF_2 in the Burning of Cement Clinker, *5th International Symposium on the Chemistry of Cement* Tokyo, vol.1: pp.122- 135 (1968).
105. Taylar, H.F.W, *The Cement Chemistry*, pp.386, Academic Press Limitied (London) (1990).
106. Ivey, D.G., Heimann, R.B., Neuwirth, M., Shumborski, S, Conrad, D., Mikula and R.J. Lam, W.W, "Electron Microscopy of Heavy Metal Waste in Cement Matrices, *Journal of Materials Science*, 25: 5055-5062 (1990).

107. Takahashi, H., Shinkado, M., Hirakida, H., Shibazaki, M., Tanaka, M., Fundamental Study on Solidification of Hazardous Industrial Wastes Containing Heavy Metals with Portland Cement, *Journal of Research of the Onoda Cement Company*, 25 (3): 1-10 (1973).
108. Yamamoto, J. The Effect of Minor Components and Bourning Atmosphere on the Color of Cement, *CAJ Review of the 20th General Meeting* 20: pp. 45-46 (1966).
109. Poon, C.S., Clark, A.I., Perry, R., Barker, A.P. and Barnes, P., Permeability Study on the Cement Based Solidification Process for the Disposal of Hazardous Wastes, *Cem. Concr. Res.*, 17: 161-172 (1986).
110. Bonen, D. and Sarkar, S.L., The Present State-of-the-art of Immobilization of Hazardous Heavy Metals in Cement-based Materials, Engineering Foundation Conference, *Advances in Cement and Concrete* (M.W. Grutzeck and S.L. Sarkar eds.) held at New England Center, the University of New Hampshire, Durham, NH, pp. 481-498, Amer. Soci. Civ. En., (1994).
111. Akatsu, K., Ikawa, K. and Maeda, K., Some Effects of Zinc Oxide in the Raw Mix on the Portland Cement and Clinker, *CAJ Review of the 22nd General Meeting*, 22: pp. 40-43 (1968).
112. Lieber, W., Effect of Inorganic Admixture on the Setting and Hardening of Portland Cement, *Zement-Kalk-Gips*, 26: 75- (1973).
113. Butt, Y.M., Timashev, V.V. and Kaushanski, V.E., Properties of Tricalcium Silicate and Its Solutions, *9th Conf. Silicate Ind. (Siliconf 1967)*, Budapest, vol. 1: 59- (1968).
114. Suzuki, K., Shibata, S. and Ito, S., The Hydration and the Strengths of High Temperature Phase of $2CaO{\cdot}SiO_2$, *CAJ Review of the 32nd General Meeting*, 32: pp. 37-38 (1978).

115. Suzuki, K., Morita, M. and Sugiyama, H., α-Ca_2SiO_4 Stabilized by Ba, Sr and Mg, *CAJ Review of the 21st General Meeting*, 21: pp.43-45 (1967).
116. Tashiro, C and Oba, J, The Effect of Cr_2O_3, $Ca(OH)_2$, ZnO and PbO on the Compressive Strength and Hydration of C_3A Paste, *Cem. Concr. Res.*, 9: 253-258 (1979).
117. Alford, N. McN., Rahman, A.A. and Salhi, N., The Effect of Lead Nitrate on the Physical Properties of Cement Pastes, *Cem. Concr. Res.*, 11: 235-245 (1981).
118. Thomas, N.L., Jameson, D.A. and Double, D.D., The Effect of Lead Nitrate on the Early Hydration of Portland Cement, *Cem. Concr. Res.*, 11: 143-153 (1981).
119. Pan Yi Zhou; Studies of Producing Portland Cement with Tailings of Molybdenum-Iron Ore, *9th International Congress on the Chemistry of Cement*, New Delhi, vol. II: pp. 271-277 (1992).
120. Rajczyk, K. and Nocun-Wczelik, W., ,Studies of Belite Cement from Ba-containing By-Product, *9th International Congress on the Chemistry of Cement*, New Delhi, vol. II: pp. 250-254 (1992).
121. Varadarajan, V., Thombare, C.H., Borkar, S.A., Page, C.H., and Chatterjee, A.K., Experiences in Use of Industrial By-Product Sludge for Cement Manufacture by Sol-Gel Technology, *9th International Congress on the Chemistry of Cement*, New Delhi, vol. II: pp. 278-284 (1992).
122. Sprung, S., Delort; M., "Effect of Energy Conservation and Enviromental Control Measures on Clinker Properties,*9th International Congress on the Chemistry of Cement*, New Delhi, vol. II: pp. 155-174 (1992).
123. Nudelman, B.I., U.S.S.R. Patent SU-2397380 (1977).

9

USE OF SILICA FUME IN CONCRETE

Satish Chandra and Leif Berntsson

INTRODUCTION

The interest in the use of silica fume started with the strict enforcement of air pollution controls in many countries, which implied that the industry had to stop releasing silica fume along with the other flue gasses into the atmosphere. This required the use of special devices for accumulation of fine materials. The issue remaining was what to do with the material collected. In an effort to find a solution to this problem, studies were initiated to find a suitable use for the silica fume. After some investigation, it was found that the silica fume could be used as a very reactive pozzolanic material in concrete.

ORIGIN

Silica fume, SF, is an industrial by-product mainly from ferro-silicon producing companies. It is produced during the reduction of high purity quartz with coal or coke and wood chips in an electrical arc furnace during the reduction of silicon metal or ferro silicon alloy. The SiO_2 content of silica fume is highly dependent on the type of product produced. The higher the productivity of the silicon alloys, the smaller the amount of silica fume. For example, in the case of silicon metal and ferro-silicon alloy, 550 kg and 350 kg of silica fume are produced per ton, respectively. It is summarized approximately as follows:

Alloy type	SiO_2-content of silica fume
50% ferrosilicium	61-64%
75% ferrosilicium	84-91%
98% silicon metal	87-98%

The majority of published data and field-use of silica fume has been from alloys of 75% ferrosilicium or higher. The production and use of silica fume worldwide is given in Table 1 according to the RILEM Committee 73 SBC Report 1988 [1].

Table 1. Annual Production and Utilization Rates of Silica Fume (tons), RILEM reports 73 SBC, 1988 [1].

Country	Condensed silica x 10^3	
	Production	Utilization
Australia	60	20
Canada	23	11
China	None	None
Denmark	None	None
France	60	None
Germany	25	None
India	None	None
Japan	25	None
Netherlands	None	None
Norway	140	40
South Africa	43	None
Sweden	10	1
United Kingdom	None	None
United States	100	2

Silica fume is a very fine amorphous material and is used in concrete as a mineral admixture. However, it is relatively new as a mineral admixture compared to natural pozzolan, fly ash and blast furnace slag. According to the RILEM TC 73 classification, silica fume is a highly active pozzolan. The earliest studies have been conducted in the North European countries including Norway, one of the primary countries producing silica fume. Silica fume has been used as a cement replacement material in the beginning. The earliest example is the use of 15% silica fume to replace cement in the construction of a tunnel in Oslo in 1952 [2]. Later, additional research was done in Scandinavia. Work representative of these studies has been reported by the following: Fascia et al., 1971 [3], Traetteberg 1977 [4], Asgeirsson & Gudmundsson 1979 [5], Jahr 1981 [6] Loland 1981 [7], and Gjørv and Loland 1982 [8].

SILICA FUMES AND SYNTHETIC SILICA

Silica Fumes

Silica fume comes in three forms; powder, condensed and slurry. A short description of these silica fumes which are commercially available is as follows.

Powder Silica Fume. This is also referred to as-produced silica fume. Silica fume as collected is in very fine powder form. Its loose bulk density is 130 to 430 kg/m^3 compared to 1000 kg/m^3 for Portland cement. Because of its very low loose bulk density and extreme fine particles, there can be handling problems. It can also create other problems such as clogging of the pneumatic transport equipment, stickiness, and bridging in storage silos. Bagged as-produced silica fume has been used by discharging the material directly into the truck mixer. But it creates a lot of dust and is expensive. Due to these difficulties, as-produced silica fume has not been extensively used.

Slurried Silica Fume. The difficulties encountered with powder silica fume are solved by water-based silica fume slurry.

These slurries can have a density of approximately 1320 kg/m^3. It also depends upon the solid content of silica fume. The solid content in the slurried silica fume can vary between 42 to 50 percent by mass. Considering the amount of water in the slurried silica fume, its transportation is much more practical than that of powder silica fume. The slurries are available with and without chemical admixtures such as high range water reducers, HRWR (Superplasticizers) and retarders. The amount and type of chemical admixture depends upon the supplier. Generally, the amount of chemical admixtures are calculated so as the extra water demand by the use of silica fume will be balanced. Therefore, it is necessary to follow specifically the instructions given by the supplier for the use of slurried silica fume. It is supplied in 208 L drums, and 19 L pails.

Densified (Compacted) Silica Fume. The problems associated with the as-produced silica fume are also overcome by densified silica fume. Its loose bulk density ranges between 30 to 45 lb/ft^3 (~500-700 kg/m^3). These are heavy enough to be transported economically. One method to produce densified (compacted) silica fume is by blowing compressed air from the bottom of the silos filled with as-produced silica powder. By doing so, the particles tumble and agglomerate. The heavier agglomerates fall to the bottom of the silo and are periodically removed. This type of silica fume is frequently used. Another method to produce densified silica fume is to compress the silica fume powder mechanically. Mechanically densified silica fume is available in United States. The densified silica fume is available with and without chemical admixtures. These can be supplied in bulk, in bulk bags (bags ~2000 lb [907 kg] and in small bags weighing about 50 lb [23 kg]).

Pelletized Silica Fume. Silica powder as-produced can be pelletized by mixing with water. The pelletization process is not a reversible process. These pellets are too hard to break during the concrete production. This hinders its use as mineral admixture in concrete production. However, the pellets can be interground with cement clinker.

Properties of Silica Fume

The properties of silica fume depends very much upon the type of the product produced and upon the process used for its manufacture. Silica fume generally contains more than 90% SiO_2. However, the content of SiO_2 and the degree of the amorphous nature may vary considerably depending upon the method of production. The chemical composition and physical properties of some of the silica fumes are given in Table 2 [9].

Color. The color of silica fumes varies from light to dark grey. This color is due to contaminates as pure SiO_2 is colorless. These contaminations are mainly carbon and iron oxide. The greater the carbon content, the darker is the color of silica fume. The color depends upon the manufacturing process, i.e., wood chip composition, ratio of wood chip to the coal used, furnace temperature, exhaust temperature, and the type of metal produced. The color can also be influenced by the compaction process.

Uniformity. Silica fume like blast furnace slag and fly ash may vary from source-to-source and also within the source. The variation is similar to that of ground granulated blast furnace slag, but smaller than that found in the case of fly ash. Not much data is published on the variability of silica fume. The widths of variation are shown in Table 2 [9]. One way to minimize the variation is to mix the silica fume from different sources or to mix several days of produced silica fume.

Synthetic Silica

Some other amorphous silica products are also available. These are synthetically made. In this way their physical characteristics and chemical composition do not change. Whereas in the case of silica fume there is no guarantee on its physical properties and chemical composition, it depends upon the conditions under which it is produced. Main important factors are the amount of coal and the other impurities which can play an important role in the rheology and

Table 2. Characteristics of Silica Fume [9].

Type/Company	Si		FeSi-75 %			FeSi-50 %		Scancem	Elkem	Iceland	Hanna
	*	**	*	**	***	**	***		G980-960	Alloy	Canada
Chemical composition											
ig. loss	0.8-1.5	2.5	2.0-4.0	2.7	1.7-9.2	3.6	1.7	2-4	1.0-3.0	–	–
SiO_2	94-98	94	86-90	89	69.8-91.6	83	91.0	86-96	95-97	92-94	92-94
Al_2O_3	0.1-0.4	0.06	0.2-0.6	0.4	0.2-1.8	2.5	0.8	0.2-2.2	0.3-0.7	0.61	0.20-0.30
Fe_2O_3	0.02-0.15	0.03	0.3-1.0	0.6	0.6-4.1	2.5	0.6	0.3-2.2	0.2-0.3	0.94	0.10-0.30
CaO	0.08-0.3	0.5	0.2-0.6	0.2	0.1-2.9	0.8	0.2	0.1-0.6	0.3	0.39	0.10-0.50
MgO	0.3-0.9	1.1	1.0-3.5	1.7	0.3-3.9	3.0	1.4	0.3-3.5	0.2-0.5	1.58	0.10-0.15
Na_2O	0.1-0.4	0.04	0.8-1.8	0.2	0.20-2.33	0.3	0.92	0.8-1.8	0.1-0.3	–	0.10-0.20
K_2O	0.2-0.7	0.05	1.5-3.5	1.2	0.80-5.52	2.0	3.04	1.5-2.5	0.5-0.8	–	0.10(Na)
C	0.2-1.3	1.0	0.8-2.3	1.4	0.4-3.3	1.8	0.6	0.5-2.5	0.7-2.5	–	0.10(K)
SiC	0.2-1.0	–	0.1-0.4	–	–	–	–	–	–	–	3-5
SO_3	–	–	–	–	0.0-1.3	–	0.0	0.4-1.5	0.1-0.4	–	0.10(s)
TiO_2	–	–	–	–	0.00-0.03	–	0.00	–	–	–	–
MnO	–	–	–	0.06	0.01-0.51	0.2	0.15	–	–	–	–
Cl	–	–	–	–	–	–	–	–	0.05-0.1	–	–
P_2O_5	–	–	–	–	–	–	–	–	0.1-0.23	–	–
others	0.1-0.5	–	0.5-0.9	–	– –	–	–	–	–	–	–
Physical properties											
Specific gravity	–	2.23	–	2.21	2.18-2.66	2.30	2.59	2.3	2.2	2.10	2.02
Surface area (m^2/g)	–	20(BET)	–	15(BET)	14.9-24.2	13.5(BET)	14.3	15-20	20	15-20	–
Ave.Diameter(μm)	–	0.18	–	0.26	–	0.21	–	0.15	–	–	0.1
Pozzolanic activity index (%)	–	100	–	97	–	56	–	–	–	–	–

* Traetteberg A.: Silica fume as a pozzolanic material, CEMENT, Vol.75,No3, p. 369-376,1978
** Aitcin P.C. et al.: Physical and chemical characterization of condensed silica fumes. CERAMIC BULLETIN, Vol.63,No.12, p. 1487-1491, 1984.2
*** Silica fume in concrete, State-of-the art report, Cement Association of Japan, March, 1986,11

durability properties of concrete. These have potential of performing good in concrete. Synthetic silicas may be expensive but are more safe and the reproducibility of concrete with their addition can be guaranteed, which is not the case with silica fume or any other mineral admixture which is an industrial by-product.

Synthetic silicas are produced from sodium silicate. Examples of synthetic silica are; fumed silica, precipitated silica, colloidal silica and silica gel. Examples of some of the companies producing different types of synthetic silica are given below. The information is taken from the Chemical Economic Handbook-SRI International [10].

Degussa:	Fumed silica, precipitated silica
Tokuyama Soda:	Fumed silica, precipitated silica
W.R.Grace:	Silica gel, zeolite catalysts
Crosfield:	Precipitated silica, silica gel
Akzo:	Precipitated silica, zeolite catalysts
Wacker:	Fumed silica, precipitated silica
PQ:EKa Nobel	Colloidal silica, silica gel, precipitated silica, and catalyst zeolites

The consumption of specialty silicas in different parts of the world are shown in Table 3.

Precipitated Silica. Precipitated silica is a particulate, amorphous silica, generally prepared by precipitation from sodium silicate solutions. Particle sizes initially formed in the slurry are in the range of 20 nanometers, but the particles aggregate upon drying. Particle diameters in the finished products vary from approximately 1-100μm. Smaller diameters less than 50μm are produced by milling the dried aggregate. Precipitated silica slurry does not have the rigid three-dimensional network characteristic of silica gel. The primary markets are as a specialty fillers and carriers.

Table 3. Consumption of Specialty Silicas by Major Region 1988 (Thousands of metric tones, 100% SiO_2) [10].

	United States	Canada	Western Europe	Japan
Precipitated silica	101-112	2.5-3	250-280	55
Silica gel	22-23	0.7-1	10-15	16
Silica colloid	12.7	1-1.5	na	15
Fumed silica	17-18	0.5-0.6	35-40	6
Fused silica	34-36	na	na	38

Silica Gel. Silica gels are produced by acidification of sodium silicate solutions with a mineral acid or carbon dioxide. The gel is allowed to set to a gelatinous mass called a hydrogel. It is broken into pieces and washed to remove the acid and salts, and may be further treated to modify surface area, pore volume and particle size. The gel is then slowly dried. It may also be spry dried, ground or pelletized. Gels can also be produced from colloidal silica or by hydrolysis of silicon compounds such as silicon tetrachloride or ethyl silicate. In the later case, pure, dense gels can be formed.

Silica gel is an ordered three-dimensional network of silica particles. In a hydrogel, water remains in the pores of the network. In aerogels or xerogels the water has been removed. In xerogels the gel shrinks and crumbles upon drying, whereas in an aerogel shrinkage is minimized and the final structure closely resembles that of the wet gels.

Colloidal Silica (Silica Sol). Colloidal silica is produced from a solution of sodium silicate. The solution is passed through an ion exchange to partially remove the sodium ions, forming a silica suspension. The pH is adjusted to control the particle size. Another counter ion such as ammonium ion may be introduced.

Colloidal silica or silica sol is a suspension of the fine silica particles in an aqueous medium and is derived from sodium silicate. Its major applications are, binders in investment castings, paper processing and coating, silicon wafer polishing, and as binders in refractories and building materials.

Fumed Silica. Fumed silica is produced by the combustion of a chlorosilane in a hydrogen/oxygen flame at approximately 1000°C. When the particles cool, they form branched, three-dimensional aggregates.

$$SiCl_4 + 2H_2 + O_2 \rightarrow SiO_2 + 4HCl$$

Silicon tetrachloride is produced from the reaction of silicon metal with hydrochloric acid. In this way hydrochloric acid and hydrogen can be recycled.

$$Si + 4HCl \rightarrow SiCl_4 + 2H_2$$

The fumed silica produced is hydrophilic, with silanol linkages on the surface of the particles. However, for use in some applications, particularly in silicone formations, a hydrophobic surface is desirable. It is obtained by treating the particle surface with silane such as dimethyl-dichloro-silane or a siloxane.

Fumed silica, also known as pyrogenic silica, is a fine, particulate, amorphous silica consisting of spherical particles with diameters in the range of 7-20 nanometers. These particles, consequently, have very high specific surface area i.e., ratio of surface area to volume. Metallic contaminations are very low. Fumed silica has a high electrical resistivity.

Fused Silica. Fused silica is produced by fusion of a pure silica sand in an electrical arc furnace. The fused product is cooled and ground to the desired particle size. It is used as a special filler in refractory (high temperature concrete) and electronics applications.

It is generally produced as powder or as grains approximately 8 μm in diameter.

EVOLUTION OF HYDROGEN

Some research done on mixes of silica fume and lime resulted in hydrogen gas evolution [11]. The reaction involves free silicon which may be present in some of the silica fumes in small quantities. This reaction is similar to the one that occurs when mixing aluminum for producing gas concrete. However, Warren 1987 [12] reports that since hydrogen gas is produced instead of atomic hydrogen, there is no risk of hydrogen embrittlement in the prestressing steel.

In a Swedish precast concrete factory that produces prestressed hole core slabs, silica fume was used as a mineral admixture. Some years ago there was quite a serious explosion during cutting of the hardened concrete elements with a diamond blade. An investigation revealed that the explosion was most likely caused by hydrogen gas which concentrated inside the holes of the concrete elements during heat curing. Hydrogen and oxygen created a detonating gas mixture which was brought to ignition by the formation of sparks during the cutting operation. Earlier, on other occasions, a small banging noise was noted during cutting of the same type of concrete elements, but no measures had been taken to investigate the reason. After excluding silica fume from concrete mixtures, no further incidents have been reported.

POZZOLANIC REACTIVITY, HYDRATION AND MICROSTRUCTURE

Pozzolanic Reactivity and Hydration

Condensed silica fumes perform partly as fine filler materials and as pozzolanic additives. In this way, the strength of concrete with the addition increases with time. Pore structure becomes denser and permeability decreases. This helps to improve the durability of concrete. The pozzolanic reactivity of condensed silica fume (CSF) has been studied by many researchers around the world. It has been reviewed by Hjorth [13], Gjørv [14] and Regourd [15]. Results generally show high pozzolanic reactivity [15-19], although some researchers have reported medium [20] or low reactivity [21]. The pozzolanic reactivity has been calculated by measuring the amount of calcium hydroxide at different times in the pastes with varying dosages of CSF. The amount of calcium hydroxide in the mature cement paste made with the use of white Portland cement is shown in Figure 1. The water to cement ratio is kept constant at 0.50, and various amounts of CSF are added. Upon extrapolation of the curve, it is seen that 24% of CSF is required to eliminate all the calcium hydroxide [17]. The CSF has been found to have an accelerating affect on the hydration of white cement. It is similar to the effect of fine calcium carbonate filler. The pozzolanic reaction does not occur immediately after mixing. It has been shown that the calcium hydroxide content in the CSF is higher initially, but after two days it becomes substantially lower [17]. This is presumably due to the pozzolanic reaction.

In another study, Cheng-Yi and Feldman [18] reported that the addition of CSF and ground quartz sand accelerates the hydration of cement paste and mortars. After 14 days, calcium hydroxide content was eliminated in pastes containing 30% CSF. Mixes with 10% CSF addition have shown an 8% reduction in calcium hydroxide content by weight of cement.

Pozzolanic reaction is generally accelerated at higher temperatures. Tenoutasse and Marion [22] studied the pastes of Portland cement blended with CSF at 20 and 40°C. The pozzolanic behavior was investigated by determining the $Ca(OH)_2$ content. The CSF content was 5, 10, 20 and 30 %. The study was done after hydration periods ranging from 8 hours to 60 days. The results are reproduced in Tables 4 and 5, respectively.

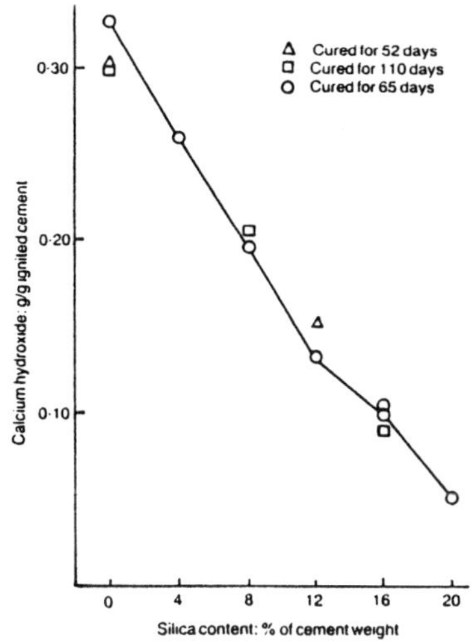

Figure 1. Calcium hydroxide contents (measured by thermal gravimetric analysis (TGA)) of mature pastes made from Portland cement. The water to cement ratio was 0.50 [17].

Here a parameter $\Delta Ca(OH)_2$,(ΔCH), was introduced which is the difference between the amount of CH measured by the classical wet chemical analysis according to the Frankies method and the amount of CH calculated.

It can be seen from the data in Table 4 that during the initial hydration period, from 8 hours to 7 days at temperatures of 20 °C, the ΔCH values were positive for all the samples tested, even for the low silica fume blended cement (5%). It was postulated that during this period, silica fume works purely as fine filler that consist in a catalytic action of very small particles with very large surface areas. After 7 days of hydration, some negative values were observed due to the pozzolanic reaction of silica fume. The quantity of CH fixed was related to the amount of SF in the cement paste. The more significant was exhibited by the 30% SF blended cement. The pozzolanic reaction continued for 60 days. Further data are not

Table 4. The $\Delta Ca(OH)_2$ (%) in OPC + SF Blended Pastes Hydrated at 20°C. SF is The Amount in Percent Blended with OPC [22].

SF	8hrs	24hrs	2days	5days	7days	9days	12days	14days	28days	60days
5%	+0.04	+0.66	+0.50	+1.05	+1.48	+0.12	-0.25	-0.40	0.77	-0.60
10%	+0.54	+0.79	+0.56	+1.10	+2.00	-0.35	-0.30	-0.80	-1.74	-1.97
30%	+0.80	+0.89	+0.98	+0.50	+0.40	-3.05	-3.20	-3.40	-5.22	-6.99

Table 5. $\Delta Ca(OH)_2$ (%) in OPC + SF Blended Pastes Hydrated at 40°C. SF is The Amount Blended with OPC in Percent [22]

SF	8hrs	24hrs	2days	5days	7days	12days	14days	28days	60days
5%	+1.43	-0.57	---	-1.90	-2.48	-3.13	-3.51	-3.53	-3.61
10%	+1.00	-0.41	-2.52	-3.97	-4.11	-4.45	-5.21	-5.42	-5.52
20%	+1.00	-1.39	-3.57	-7.68	-7.81	-8.03	-9.62	-10.46	-10.44
30%	+0.71	-2.72	-4.98	-7.97	-8.00	-8.10	-9.45	-9.51	-9.52

available at this time, but it is speculated that the pozzolanic reaction would continue with time.

The behavior was quite different when the samples were hydrated at 40°C, as shown in Table 5. It was observed that only after a hydration period of 24 hours, ΔCH attained negative value, even for the samples that contained 5% CSF, and it appeared that the pozzolanic reaction was complete after 14 days of hydration. There was a good correlation between the ΔCH and CSF values. However, after 5 days of hydration no substantial difference in the ΔCH value between the 20% and 30% silica fume blended cement was observed.

Kurbush et al., [23] cured mixes of calcium hydroxide and CSF at a water-to-solid ratio of 0.50-0.25. The paste was cured at 50 and 90°C. It was observed that at 90°C, 68-95% of added lime had reacted after 2.5 hours, while at 55°C only 25-55% had reacted in 2.5 hours. At 20°C the reaction was found to be very slow [17] for a mix with C/S = 1.0 (C/S = CaO/SiO_2) but after 110 days all the calcium hydroxide was consumed. A consequence of low C/S in CSH leads to an increased capacity to incorporate more external ions such as alkalies and aluminum. This is further substantiated by the analysis of pore water pressed from hardened cement-CSF paste [24,25] (see Figure 2). Because of the increased capacity to incorporate external ions, the concretes made with CSF exhibit a better resistance to aggressive chemicals and alkali silica reactions.

Microstructure

Pozzolanic reaction of silica fume is very well illustrated by Sarkar [26] through microstructural studies, done on a concrete made with 6% SF and 0.24 water-cement ratio. Microphotographs are very illustrative and show distinct morphology of silica fume particles during pozzolanic reaction, Figures 3-7. The rate of dissolution was followed from 1-28 days by transmission electron microscope. The morphology and size of silica particles are shown in Figure 3. The pozzolanic reaction was not advanced in the early stage, Figure 4,

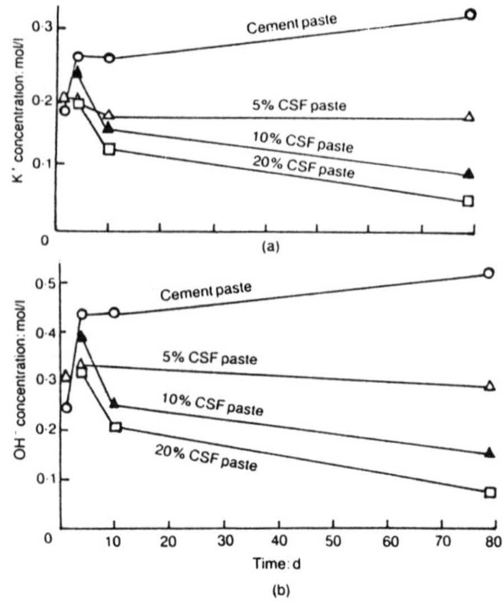

Figure 2. Concentration of a) K^+ and b) OH^- in pore solutions expressed at times indicated from cement paste and CSF bearing pastes. Water binder ratio = 0.50 [24].

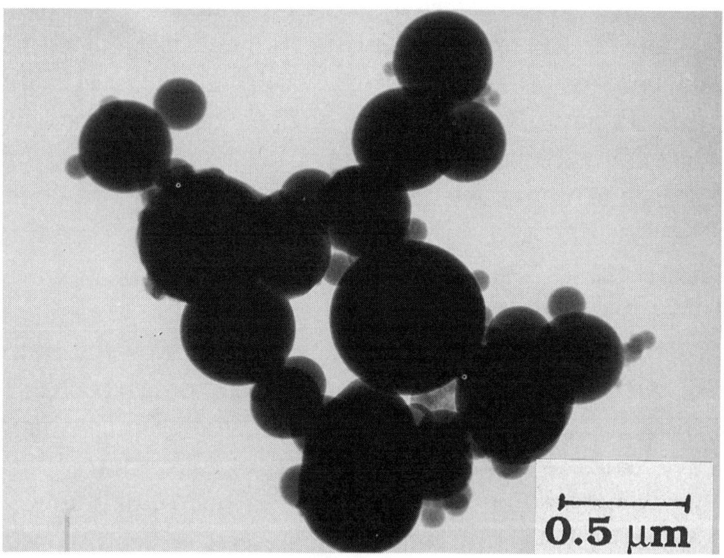

Figure 3. TEM micrograph of silica fume particles [26].

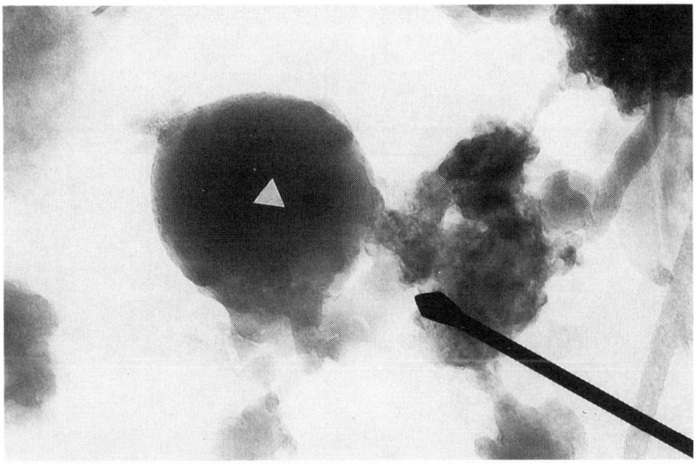

Figure 4. TEM micrograph of a slightly reacted SF particle after 1 day [26].

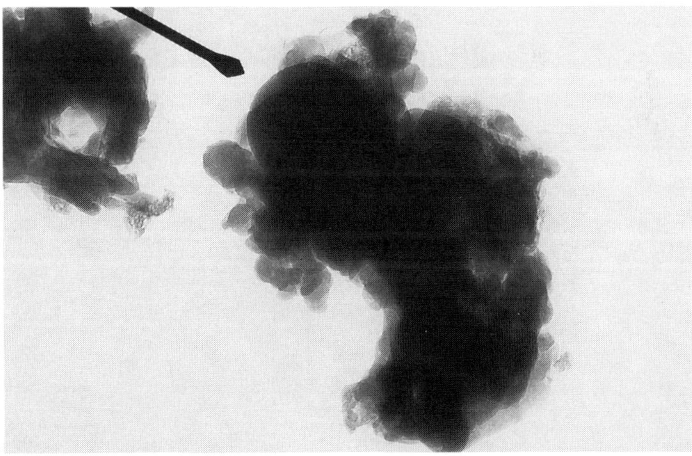

Figure 5. TEM micrograph of unreacted SF particles after 1 day [26].

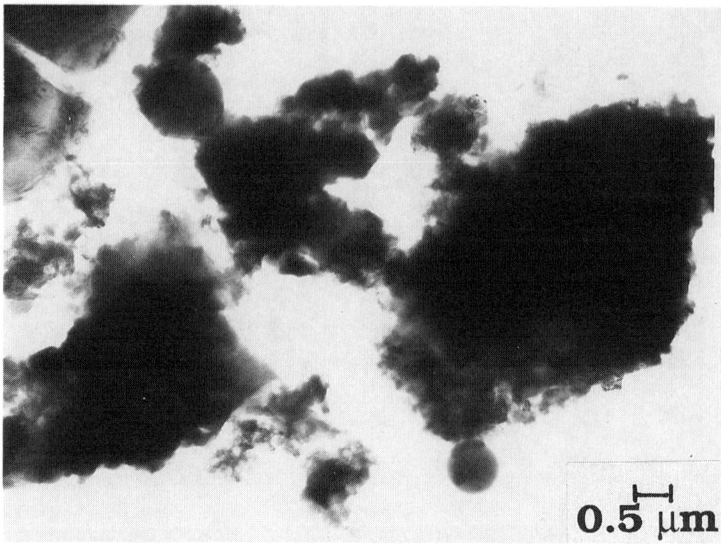

Figure 6. TEM micrograph illustrating gel formation in SF particles after 7 days [26].

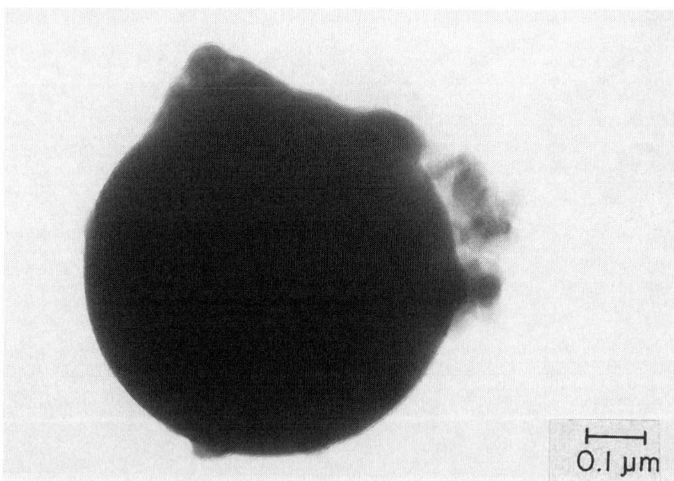

Figure 7. TEM micrograph of a partially reacted SF particle after 28 days [26].

and a number of unreacted silica fume particles were seen, Figure 5. During this period they acted merely as fillers. After 7 days the reaction was advanced, the particles lose their spherical morphology, and change to a featureless mass without any geometrical configuration, as shown in Figure 6. It was observed that even with a silica fume addition as small as 6%, there were still unreacted particles after 28 days as shown in Figure 7. This may be due to the unavailability of water, reducing the dissolution rate of the SF particles [27].

The concrete microstructure studied did not show any ettringite crystals ($C_6AS_3H_{32}$) during any stage even after 91 days, even though these are formed in normal concrete at quite early stages. Ettringite formation and pozzolanic reaction of silica fume both require a large amount of water. Furthermore silica fume particles are sites for nucleation [18]. When these siliceous particles come in contact with cement solution, they adsorb OH^-, Ca^{2+} and alkalies. Sarkar noted that along with the silica fume particles also adsorb SO_4^{2+} and Al^{3+} ions on its surface, see Figures 8-10. It was shown that the peak ratio of Ca/Si increases from 0.98 (1 day) to 2.1 (28 days). It was further shown that there was a gradual formation of C-S-H with the advancement of pozzolanic reaction between silica fume and calcium hydroxide. The concrete after 28 days became very dense and compact, as shown in Figure 11, which is not the case with conventional concrete.

Transition Zone. The transition zone is the interphase between the aggregate and the hydrated cement paste. It is very important both from the view point of mechanical strength as well as durability. Carles-Gilbergues et al., [28] investigated the influence of water-cement ratio and pozzolanic admixtures on the two important parameters of the transition phase; namely its thickness and the degree of orientation of calcium hydroxide crystals. It is reported that with the increase in water-cement ratio, both the thickness of the transition zone and the degree of orientation of calcium hydroxide crystals is increased. This is thought to be due to the increase in the internal bleeding.

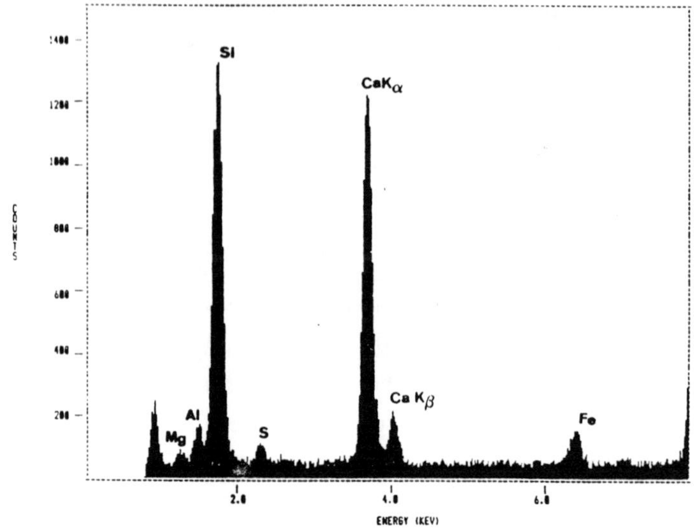

Figure 8. The EDXRA spectrum of a partially reacted SF particle after 1 day [26].

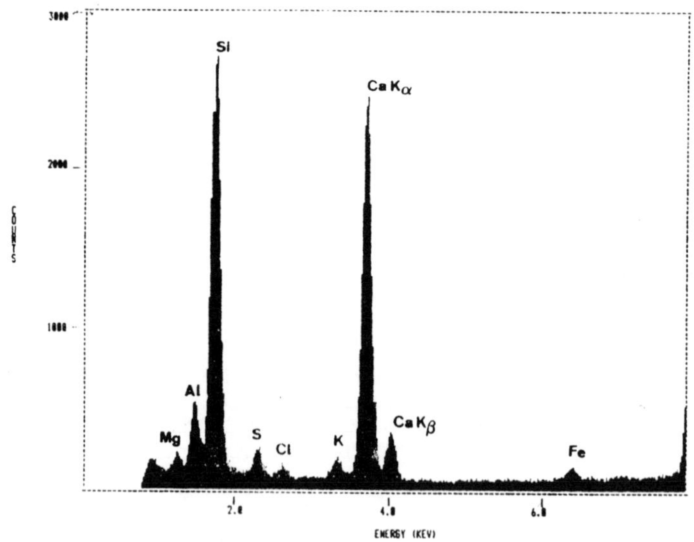

Figure 9. The EDXRA spectrum of a silica-rich gel after 7 days [26].

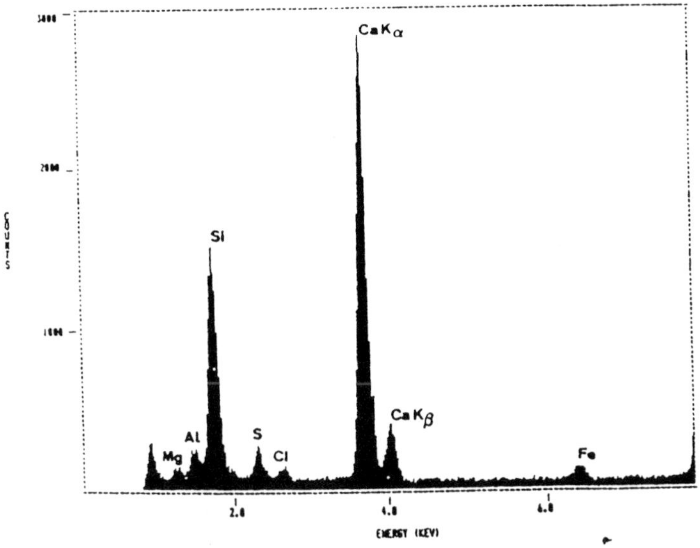

Figure 10. The EDXRA spectrum of C-S-H after 28 days [26].

Figure 11. Microstructure of dense concrete with SF after 28 days [26].

Water is generally surrounded around the aggregates. Thus, a substantial part of calcium hydroxide is formed in this area with preferentially oriented crystals. Otherwise, well-crystallized large tabular crystals appear in the hydrated cement paste, as shown in Figure 12 [26]. Silica fume significantly reduces internal bleeding, consequently decreasing the thickness of the transition zone. Furthermore, due to the pozzolanic reactivity of silica fume, calcium hydroxide is consumed and additional C-S-H phase is formed. It works as a binder and provides stronger bonding between the aggregate and paste. This subsequently reduces the amount of free calcium hydroxide and densifies the transition zone. The dense structure of transition zone is shown in Figure 13 [26] and Figure 14 [27].

Pore Structure. Pore structure governs to a great extent, the permeability and consequently, the durability properties of cement-based products. There is a relationship between pore structure, permeability and durability for blended cements [29]. In order to evaluate the pore structure, it is essential to study the porosity and pore size distribution. Total porosity may be the same, but and pore

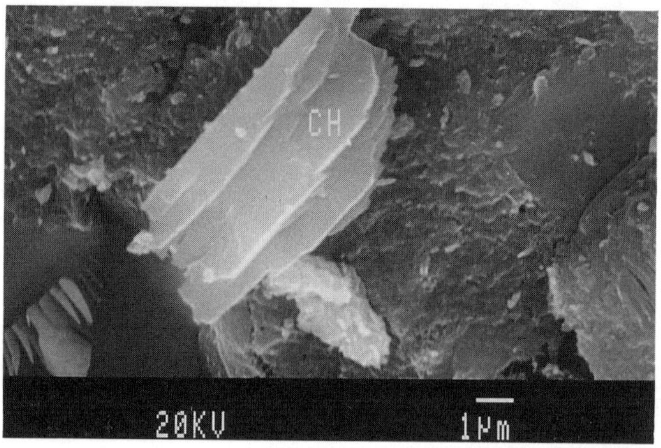

Figure 12. A large, well-crystallized CH crystal on the left and C-S-H gel on the right in normal strength concrete (conventional), W/C = 0.5 [26].

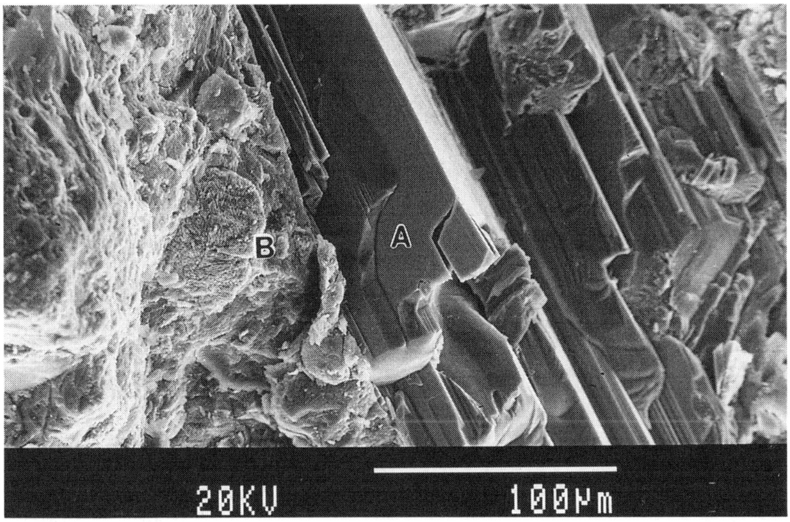

Figure 13. The strong paste-aggregate bonding after 1-day hydration in high strength concrete, paste on the left and limestone aggregates on the right. [26].

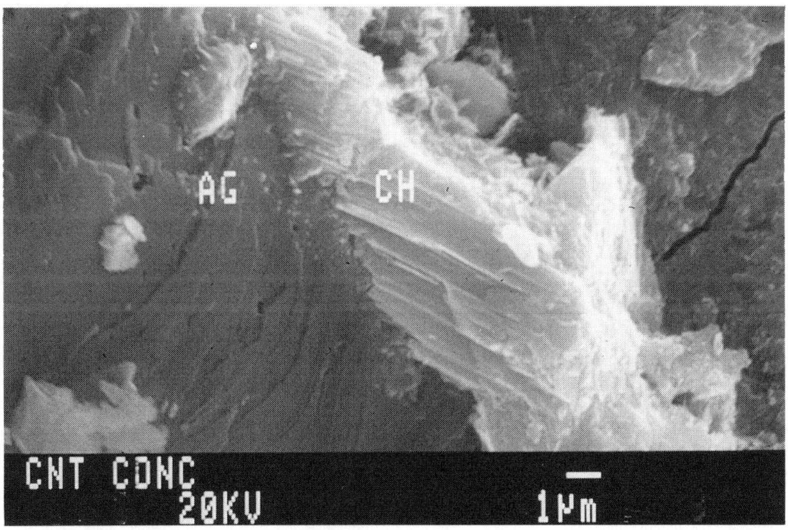

Figure 14. Oriented $Ca(OH)_2$ crystals at the interfacial transition zone, AG = aggregate; CH = $Ca(OH)_2$ crystals [27].

size distribution can be different with the addition of silica fume, as it refines the pores. Total porosity and pore size distribution has been investigated by many researchers. Tenoutasse and Marion [22] studied the total porosity, pore size distribution and the hydration process with the addition of condensed silica fume at different temperatures and times. Their results are cited and discussed here.

Total Porosity. Total porosity blended with 5 to 30 % condensed silica fume and hydrated at 20 and 40°C are shown in the Tables 6 and 7, respectively as was measured by mercury intrusion porosimetry [22]. It is seen from the data in the tables that the total porosity of OPC and blended pastes decreases with increase in hydration time. At 20°C, total porosity values after 8 hours of hydration shows that there is increase in the total substantial increase in the porosity with the addition of silica fume, as shown in Table 6. This may be due to the high specific surface area of silica fume and its water adsorption at the initial period of hydration. From 8 hours to 7 days, the total porosity of blended cement is larger than that of the OPC paste. This difference becomes lower after 14 days, and after 28 days it becomes lower than that of the reference. On the other hand, at 40°C, no significant difference is observed from 8 hours to 5 days between the total porosity of OPC and blended paste, as shown in Table 7.

Table 6. Total Porosity (cm^3/g) of Cement Paste Blended with Different Amounts of Silica Fume at 20°C[22].

SF	8hrs	24hrs	2days	5days	7days	14days	28days
0%	0.334	0.265	0.237	0.201	0.183	0.165	0.160
5%	0.369	0.294	0.285	0.236	0.214	0.123	0.177
10%	0.420	0.312	0.308	0.242	0.219	0.188	0.169
30%	0.428	0.333	0.312	0.243	0.226	0.176	0.143

Table 7. Total Porosity (cm^3/g) of Cement Paste Blended with Different Amounts of Silica Fume at 40°C [22].

SF	8hrs	24hrs	2days	5days	7days	14days	28days
0%	0.342	0.275	0.231	0.205	0.225	0.201	0.191
5%	0.354	0.288	0.254	0.233	0.215	0.174	0.187
10%	0.355	0.285	0.225	0.188	0.187	0.184	0.140
30%	0.347	--	0.348	0.221	0.178	0.165	0.151

These results are in agreement with the results cited before on the CH determination showing the pozzolanic activity of silica fume after 9 days. At higher temperature the pozzolanic activity is higher as is shown by the lower total porosity of for all the blended paste in comparison to the OPC only after 7 days of hydration.

Pore Size Distribution. The pore size distribution of OPC and blended pastes from 8 hours to 60 days at different temperatures is illustrated in Tables 8 - 15 [22]. It is seen that after 8 hours of hydration, at 20°C (Table 8), the cement blended with 5-10% silica fume presents the same microporosity between 60 and 500 Å as the OPC. However, at 40°C silica fume addition significantly increases the percentage of the pores in this area as well as between 1100 and 500 Å. Further, it is interesting to note that for the plain cement, 60% of pores exist between 0.57 and 2μm, whereas in 30% blended paste there are no pores in this region. This can be attributed to the pozzolanic activity of silica fume. A decrease in macropores is also noted in relation to the presence of silica fume. This is due to the "filler effect."

After 24 hours, significant increase in the pores < 0.57 microns is noted as shown in Table 9. Transformation to the smaller pores is related to temperature and the amount of silica fume in the paste. This tendency follows with the hydration period. From 2 to 7

Table 8. Pore Size Distribution of OPC With Different Amounts of SF Hydrated at 20 and 40°C, 8 Hours After Hydration [22].

Pore in microns	100% OPC		OPC + 5% SF		OPC+10%SF		OPC+20%SF		OPC+30%SF	
	20°C	40°C	20°C	40°C	20°C	40°C	20°C	40°C	20°C	40°C
2.00	54	11	27	2	12	1	-	0	0	0
2.00-0.57	28	39	42	31	52	16	-	1	1	0
0.57-0.25	5	17	7	25	12	33	-	12	7	2
0.25-0.11	4	13	7	16	11	19	-	48	61	44
0.11-500Å	4	9	7	11	9	13	-	19	19	31
500-60Å	4	11	5	13	4	15	-	19	9	21

Table 9. Pore Size Distribution of OPC With Different Amounts of SF Hydrated at 20 and 40°C, 24 Hours After Hydration [22].

Pore in microns	100% OPC		OPC + 5% SF		OPC+10%SF		OPC+20%SF		OPC+30%SF	
	20°C	40°C	20°C	40°C	20°C	40°C	20°C	40°C	20°C	40°C
2.00	3	12	2	2	0	0	-	0	0	0
2.00-0.57	16	16	10	9	5	6	-	1	0	2
0.57-0.25	27	13	30	19	27	15	-	11	0	5
0.25-0.11	18	17	21	23	25	29	-	25	14	13
0.11-500Å	13	17	13	16	16	18	-	21	55	14
500-60Å	21	24	20	27	23	30	-	41	27	65

Table 10. Pore Size Distribution of OPC With Different Amounts of SF Hydrated at 20 and 40°C, 2 Days After Hydration [22].

Pore in microns	100% OPC		OPC + 5% SF		OPC+10%SF		OPC+20%SF		OPC+30%SF	
	20°C	40°C	20°C	40°C	20°C	40°C	20°C	40°C	20°C	40°C
2.00	0	1	4	1	0	0	-	0	0	3
2.00-0.57	11	11	18	16	7	5	-	1	0	3
0.57-0.25	21	14	16	13	27	18	-	10	0	8
0.25-0.11	18	16	16	15	20	17	-	17	11	14
0.11-500Å	18	21	16	18	17	19	-	19	46	9
500-60Å	27	36	25	35	27	39	-	50	40	59

Table 11. Pore Size Distribution of OPC With Different Amounts of SF Hydrated at 20 and 40°C, 5 Days After Hydration [22].

Pore in microns	100% OPC		OPC + 5% SF		OPC+10%SF		OPC+20%SF		OPC+30%SF	
	20°C	40°C	20°C	40°C	20°C	40°C	20°C	40°C	20°C	40°C
2.00	2	1	3	1	1	0	-	0	0	0
2.00-0.57	20	7	11	24	6	6	-	4	0	5
0.57-0.25	15	17	19	7	15	24	-	9	1	11
0.25-0.11	9	14	10	14	15	18	-	17	17	11
0.11-500Å	16	20	18	16	19	15	-	15	24	4
500-60Å	34	39	38	29	42	36	-	52	57	62

Table 12. Pore Size Distribution of OPC With Different Amounts of SF Hydrated at 20 and 40°C, 7 Days After Hydration [22].

Pore in microns	100% OPC		OPC + 5% SF		OPC+10%SF		OPC+20%SF		OPC+30%SF	
	20°C	40°C	20°C	40°C	20°C	40°C	20°C	40°C	20°C	40°C
2.00	4	3	1	1	0	0	-	0	0	0
2.00-0.57	21	22	7	21	6	7	-	3	0	5
0.57-0.25	17	8	19	11	11	14	-	9	11	14
0.25-0.11	8	12	12	11	14	10	-	25	23	14
0.11-500Å	10	19	14	11	20	26	-	10	17	5
500-60Å	36	35	42	21	44	40	-	51	54	67

Table 13. Pore Size Distribution of OPC With Different Amounts of SF Hydrated at 20 and 40°C, 14 days after Hydration [22].

Pore in microns	100% OPC		OPC + 5% SF		OPC+10%SF		OPC+20%SF		OPC+30%SF	
	20°C	40°C	20°C	40°C	20°C	40°C	20°C	40°C	20°C	40°C
2.00	1	0	0	0	0	0	-	0	0	1
2.00-0.57	20	17	2	1	3	6	-	2	2	3
0.57-0.25	17	13	19	15	21	13	-	24	5	13
0.25-0.11	10	16	17	28	22	17	-	18	24	19
0.11-500Å	16	19	22	12	18	29	-	7	20	6
500-60Å	34	33	36	42	33	33	-	47	48	52

Table 14. Pore Size Distribution of OPC With Different Amounts of SF Hydrated at 20 and 40°C, 28 Days After Hydration [22].

Pore in microns	100% OPC		OPC + 5% SF		OPC+10%SF		OPC+20%SF		OPC+30%SF	
	20°C	40°C	20°C	40°C	20°C	40°C	20°C	40°C	20°C	40°C
2.00	0	1	0	1	0	0	-	0	0	0
2.00-0.57	9	24	6	29	1	4	-	2	0	4
0.57-0.25	18	10	27	12	26	10	-	11	2	12
0.25-0.11	14	11	15	15	17	20	-	32	23	22
0.11-500Å	22	17	19	15	23	29	-	11	29	7
500-60Å	35	34	30	27	28	36	-	41	41	56

Table 15. Pore Size Distribution of OPC With Different Amounts of SF Hydrated at 20 and 40°C, 60 Days After Hydration [22].

Pore in microns	100% OPC		OPC + 5% SF		OPC+10%SF		OPC+20%SF		OPC+30%SF	
	20°C	40°C	20°C	40°C	20°C	40°C	20°C	40°C	20°C	40°C
2.00	0	-	0	-	0	-	-	-	0	-
2.00-0.57	7	-	3	-	3	-	-	-	3	-
0.57-0.25	24	-	22	-	31	-	-	-	9	-
0.25-0.11	13	-	16	-	21	-	-	-	45	-
0.11-500Å	19	-	20	-	16	-	-	-	14	-
500-60Å	36	-	35	-	27	-	-	-	28	-

days (Tables 10 to Table 12), formation of more fine pores < 1100 Å is noted, which increases with the addition of CSF. From 14 days, the amount of pores between 0.11 and 0.57 μm increases, and the pores between 60 and 500Å decrease. This effect is found in the samples containing 5-10% silica fume, Tables 13-15. This intrinsic change of porosity is probably due to the modification of microstructure of hydrosilicates formed by the pozzolanic reaction between silica fume and calcium hydroxide [22].

DURABILITY PROPERTIES

Deterioration of concrete can have many causes. These can be physical, chemical or a combination of both. Physical changes are connected with the microstructure, permeability etc., and chemical degradation is due to the chemical interaction of cement hydration products generated during the hydration of Portland cement like calcium hydroxide, alkalies, calcium aluminated hydrates, etc. These are the essential factors which control the chemical durability of concrete. Condensed silica fume behaves like a pozzolanic material and thus, modifies the pore structure and produces concrete of dense microstructure with fewer or no transition zones as was described earlier. The permeability of concrete is decreased. Consequently, permeation of solutions containing aggressive ions like sulphate and chloride is expected to decrease significantly. Permeability, sulphate resistance, alkali-silica or alkali-aggregate resistance, chloride ingress and freezing and thawing will be discussed.

Permeability

Sellevold et al., [17] measured the drying rate of 3 mm thick discs made of CSF paste containing from 0-20 % of CSF by weight of cement. In one series, the water-cement ratio was kept constant and CSF was used as an additive. It was observed that the relative diffusion coefficient decreased with CSF addition. In another series, the water-cement ratio was kept constant and the cement was replaced by CSF in a ratio of 3:1. It gave equal diffusion coefficients for 0.8 and 16 % CSF contents. These results indicate that with respect to water diffusion during drying, the efficiency factor, K, of CSF in cement paste is about 3, close to the value of K generally found for compressive strength. Sørensen [30] calculated the relative diffusion coefficient for concrete discs with equal amounts of CSF by weight of cement, specifically, 0.8 and 16%. The efficiency factor with respect to drying was in the range of 6-8. This was higher than K for compressive strength and showed that CSF had a more positive effect on the concrete than did the cement paste. The higher efficiency factor indicates a possible improvement in the boundary phase between aggregate and cement paste which in turn influences water permeability.

Water permeability tests done on a lean mix (water-cement ratio = 0.89) with 20% cement by weight replaced by CSF showed no water penetration under 7 atmospheres of pressure for 15 days (60 mm thick disc specimens were used) [31]. Water penetration tested on concrete using another method [32] showed that penetration was reduced by half with a 10% addition of CSF to the cement. Penetration depth was further reduced to less than half the value for concrete in which 20% CSF by weight of the cement was added.

Hustad and Loland [33] tested water permeability of 25 concretes. The cement content was between 100 and 500 kg/m^3 and CSF was 0.10 and 20% by weight of cement. Not much difference in permeability was noticed for concretes having 28 days compressive strengths of 30-40 MPa. Concretes with strengths in excess of these values were found to be 'water-tight". The permeability coefficients were in the range of 10^{-13} to 10^{-15} m/s. Sandvik [34] tested the water permeability of concrete with a cement content of 300 kg/m^3 with 5, 10 and 20 % replacements of cement by CSF. Sandvik found that the permeability coefficient was reduced from 3×10^{-11} to 6×10^{-14} m/s for 5% cement replacement, for higher cement replacements, the coefficient was less than 10^{-14} m/s and could not be measured. Skurdal [35] measured the water permeability of concrete with 10 %

CSF after curing at 20, 30 and 50 °C. The permeability coefficients are shown in the Table 16.

Table 16. Water Permeability after Curing at Various Temperatures [35].

Concrete	20 °C	30 °C	50 °C
no CSF	$7.2 \cdot 10^{-13}$	$27 \cdot 10^{-13}$	$90 \cdot 10^{-13}$
10 % CSF	$0.5 \cdot 10^{-13}$	$0.8 \cdot 10^{-13}$	$74 \cdot 10^{-13}$

It can be seen from the data in Table 16, that the permeability coefficient increased with an increase in curing temperature for both types of concrete. The CSF works as a pozzolan and reacts with calcium hydroxide and forms CSH-gel. Buck and Burkes [36] detected well-crystallized calcium silicate hydrate (CSH) after 7 days of curing at 38°C. Grutzeck et al., [37] observed a silica-rich gel on CSF surfaces shortly after mixing in calcium hydroxide solution or cement paste. Later, the solution interacts with $Ca(OH)_2$ and forms CSH. Kurbus et al., [38] cured a mixture of CSF and calcium hydroxide at 55 and 90°C. It was observed that the reaction was temperature dependent. At 90°C 68-95% of the added lime had reacted within 2.5 hours, whereas at 55°C only 25-55% had reacted during the same curing time. At 20°C, the reactions occur, but very slowly.

Maage [39] studied discs drilled from various structures both with and without CSF and found the same tendencies stated before; the water permeability decreases with CSF addition.

Mindess et al., [40] also reported a decrease in permeability while testing cement paste as well as cement aggregate composite. Actually, little is known about the role of the cement and aggregate interface in controlling the permeability of concrete. The paste aggregate interface is commonly considered to be "weak link" in concrete [41] and there is some evidence that the interfacial region is a preferred pathway for the flow of water [42,43]. However, Mindess [40] showed that the presence of interface does not have major effect on water permeability. This is in agreement with the findings of Wakeley and Roy [44] who found that the interface between some aggregate types and certain cementitious mixes does not provide a preferred pathway for water flow. However, it must be noted that because of the very small volume of the interfacial region for the specimens size used, only large differences in permeability would be detected.

Water permeability is related to the microstructure of concrete. The denser the structure, the lower the permeability. By adding CSF, the pore structure is modified. The concrete becomes very dense and amorphous as was reported by Regourd [45]. Many other researchers including Manmohan and Mehta [29], Feldman [46] and Sellevold et al., [47] have shown that compared to the Portland cement pastes, the hydrated Portland cement-pozzolan type pastes contain finer pores which are not readily permeable. Mehta and Gjørv [48] studied pore size distribution with 10% addition of pozzolan by weight of cement as shown in Figure 15. They confirmed the pore refining capacity of CSF when present in a Portland cement paste.

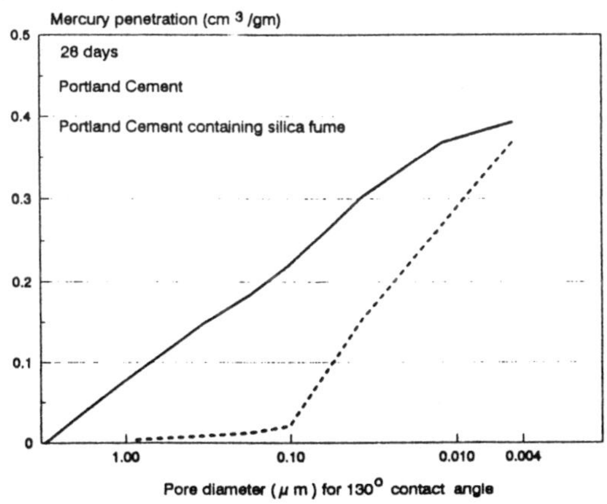

Figure 15. Influence of silica fume addition on pore size distribution of well-hydrated cement paste [48].

Unlike normal concrete in a CSF concrete, the CSF gel particles can not be "visualized as individual particles," but as a massive dense structure. Calcium hydroxide appears as small local crystals rather than the large crystals (portlandite) found in normal concrete [49]. It also decreases the microcracks that usually occur due to the large crystal formations. The dense paste structure extends to the true aggregate boundary in dense CSF concrete [50]. Thus the CSF eliminates the normal porous region of about 40-50 μm, rich in calcium hydroxide, which surrounds the aggregate grains in normal concretes. These observations are less pronounced in moderate CSF content concrete.

Sulfate Resistance

Sulfate attack occurs on concrete structures due to increases in industrial pollution. The gases, SO_2 and SO_3, subsequently interact with calcium hydroxide and calcium aluminum hydrates producing salts such as calcium sulphate and ettringite. These produce immense expansion and cause degradation of concrete.

Sulfate resistance is tested using sodium sulphate (Na_2SO_4), magnesium sulphate ($MgSO_4$) and ammonium sulphate ($(NH_4)_2SO_4$) solutions as the source of sulphate ions. Ammonium sulphate is very aggressive because of its high acidity.

Bernhardt [2] tested sulfate resistance of CSF concrete by storing the concrete specimens in a 10% sodium sulphate solution. The results indicated improvement in sulphate resistance with 10 - 15 % CSF addition. Mather [51] tested expansion of mortar specimens stored in sodium sulphate solution. Three high C_3A cements (14.6, 13.1 and 9.4 % C_3A) were used alone and with 30% replacements of different pozzolans. It was found that CSF in combination with each of the cements was better able to resist sulfate attack.

Hooton [52] tested mortar bars containing 10% condensed silica fume in sodium sulphate solution. Tests were performed according to ASTM C 1012. The results are shown in Figure 16. It was shown that the mortar bars containing silica fume showed a

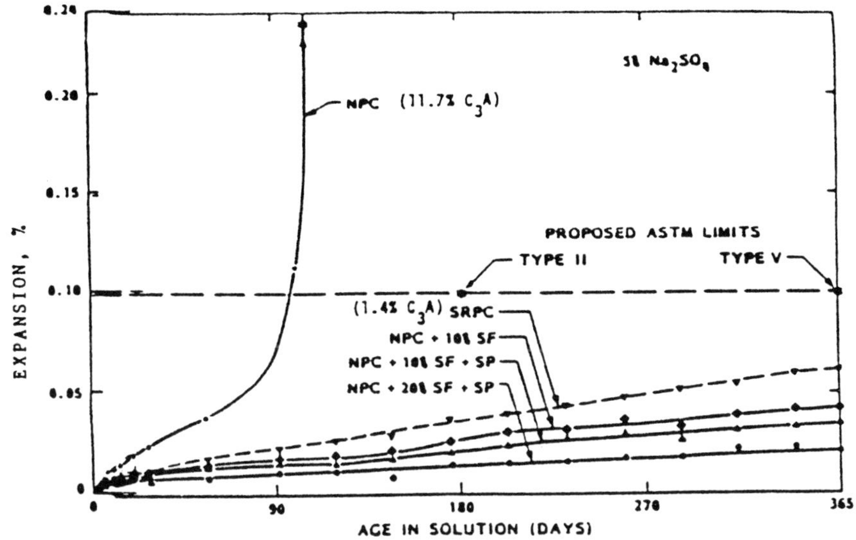

Figure 16. ASTM C 1012 sulfate resistance of silica fume mortar bars, flow 115±5 [52].

better resistance to sulfate than Portland cement. The resistance increased with the increase in the condensed silica fume (20%). It should be pointed out that the CSF is more effective with superplasticizer (SP), as it reduces the water-binder ratio and produces a denser and less permeable concrete structure. However, the condensed silica fume was used with an 11.7 % C_3A cement, which is more vulnerable to sulfate attack.

Buck [53] obtained similar results using cement containing 15 % C_3A. Yeginobali and Dilek [54] tested the sulfate resistance of mortar according to ASTM C452 and ASTM C1012. In their tests, cube specimens were also immersed separately in 10% Na_2SO_4 and 8.4% $MgSO_4$ solutions. Compressive strengths of the cubes were tested at the age of 60 weeks. It was concluded that the addition of silica fume improved the durability of mortars in sodium sulfate solution. However, it had a detrimental effect on mortars stored in magnesium sulfate solution of similar sulfate ion concentration. Blending of silica fume improved the sulfate resistance tested by both the ASTM methods. In very high concentrations of magnesium sulphate, equal to a 4.5% SO_3 solution, silica fume addition impaired the performance of the paste made with both ASTM Type I and Type V cements as reported by Cohen and Bentur [55]. Calcium silicate hydrate (C-S-H) was found to have decomposed and there was an absence of magnesium hydroxide which tends to block pores and provide protection to C-S-H from further attack. Irrespective of the cement type, addition of silica fume reduced the strength and increased mass loss by a factor of 5 to 10.

Popovic et al., [56] tested the sulfate resistance of small prisms, 25 x 25 x 160 mm, stored in ammonium sulphate solution. He concluded that 15% additions of CSF in mortar prevents the sulfate corrosion. It was interesting to note that CSF was equally effective with and without superplasticizers at different water to cement ratios. This is attributed more to the chemical effect of CSF than the reduced permeability. Mehta [57] compared the resistance to chemical aggression of concrete with OPC, latex modified concrete and concrete with 15% CSF using a water-binder ratio of 0.33. The test was performed using 5% sodium and ammonium sulfate solutions. No appreciable influence of sodium sulfate solution could be noticed where as ammonium sulfate was found to be equally destructive to CSF as to control concrete. This was due to the ability of ammonium sulfate to decompose CSH. These results are in contrast to those obtained by Popovic et al., [56], as mentioned above, showed that even in higher ammonium sulfate concentrations (10%) CSF concrete is much more resistant than the control samples. This difference may be due to many reasons for example, the type of superplasticizer, source of cement, and most important, the mixing technique. In the authors experience, the mixing technique and the sequence of adding the chemical admixtures has a significant effect on the quality of concrete. Similarly, superplasticizers from the same basic group but from different commercial sources can behave in very different ways and can produce concrete with diverse properties.

Some field testing has been conducted in Norway. In these tests, concrete samples were placed in a tunnel in Oslo's alum shale region for long periods of time. The ground water contained up to 4 mg of SO_3 per liter and the pH varied between 2.5 and 7. Fiskaa [58] has published two reports, one after 12 years and the other after 20 years of exposure. Samples were made using different cements and additives. Among others, 15% replacement of cement by CSF was also included. All of the samples had 0.50 water-cement ratio, whereas in the case of CSF, the water-cement ratio was 0.62. After 20 years of exposure, it was found that the most resistant concretes to sulfate attack were those made with Danish sulfate-resistant cement and those containing 15% CSF, as shown in Figure 17. However, this comparison with sulfate-resistant cement is not correct as it had low water-cement ratio. It is expected that the CSF concrete made with a water-binder ratio of 0.50 using superplastisizer could have shown better resistance.

These results agree with the results of Mather [51]. The sulfate resistance of CSF concrete, as mentioned earlier, is more due to the chemical effect of CSF than the lower permeability it imparts to the concrete structure. Further improvements in sulfate resistance are due to the following factors:

- Lower content of free calcium hydroxide due to its reaction with CSF.
- An increased amount of alumina can be incorporated in CSF products. Subsequently less alumina will be available to interact with sulfate, producing expansive double salts such as ettringite [13].

Alkali Silica Reaction

Influence of condensed silica fume (CSF) on alkali-aggregate reaction (AAR) has been studied for at least two decades. Pulverized

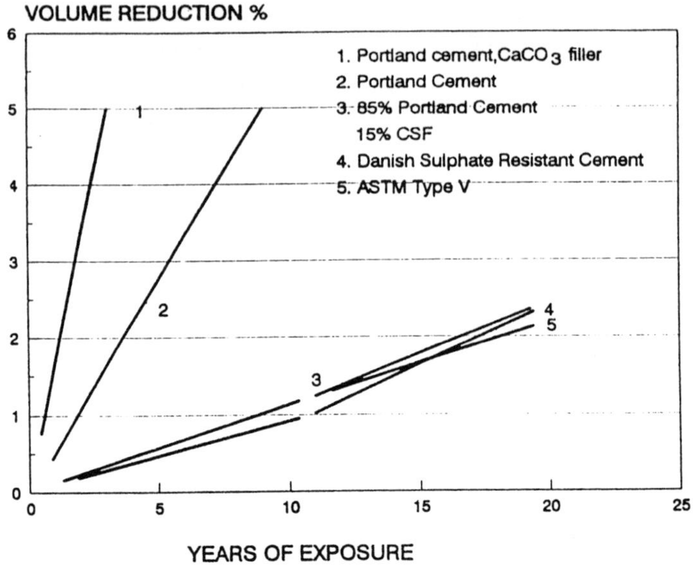

Figure 17. Volume reduction of 100x100x100 mm concrete prisms stored for 20 years in acidic solution and sulfate-rich water in Oslo alum-shale region [58].

fuel ash and ground granulated blast furnace slag act primarily as alkali dilutors when used as partial replacements of high alkali cement. When CSF is used as a partial replacement, alkali-aggregate reaction is enhanced in fresh state of concrete. This is primarily due to the highly amorphous fine silica particles. If sufficient microsilica is used, the sodium and potassium alkalies released by the cement and CSF can be largely consumed before the concrete develops strength [59]. If all or most of the metal alkalies in the pore fluid are depleted before the concrete develops requisite strength, the reactive silica present in aggregate or as aggregate will remain unreacted. In contact with alkali at later stages, silica will interact and produce expansion in the hardened concrete.

Sprung and Adadian [60] and Oberholster and Westra [61] have studied AAR with partial replacement of cement with CSF. Their results are summarized in Figures 18 and 19. According to the former investigators, 4% replacement of cement by CSF prevented expansion, whereas the latter investigators found that the replacement of cement by 15% CSF was necessary to prevent expansion (0.04 %). Perry and Gillott [62] studied the effect of CSF on alkali-aggregate reaction. Opal and dolomite limestone were used as alkali sensitive aggregates. The CSF was used in the amounts of 5, 10, 15, 20 and 40% by weight of total binder. It was found that the replacement of cement by 20% CSF is needed to suppress expansion. Replacement of cement by 10% and 15% of silica fume reduced expansion.

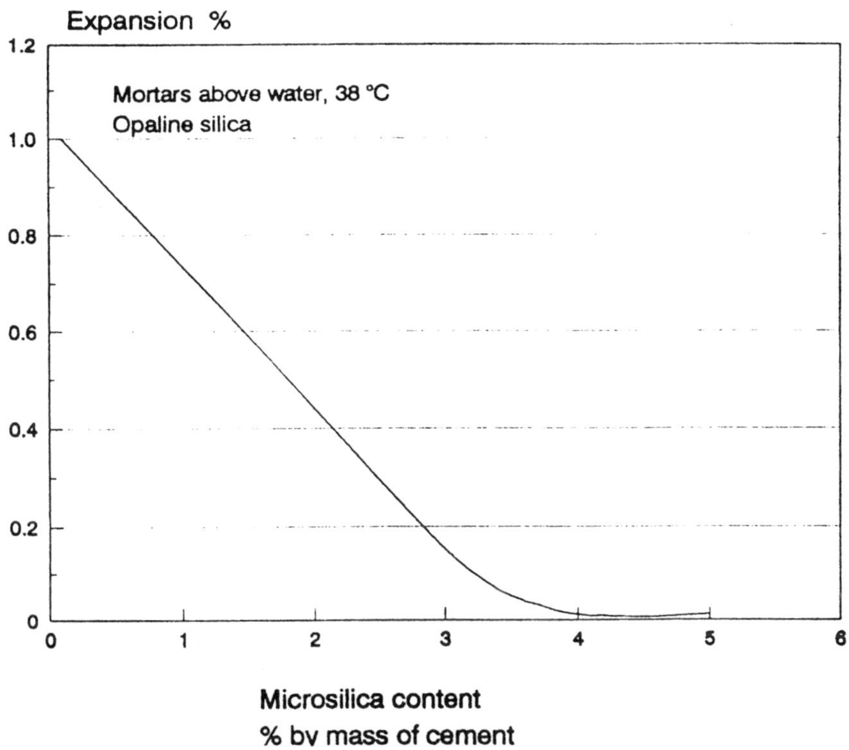

Figure 18. Influence of microsilica on expansion of mortar bars; W/C = 0.45 and aggregate/cement ratio = 2.25 and 4% by mass in aggregate type of opaline silica [60].

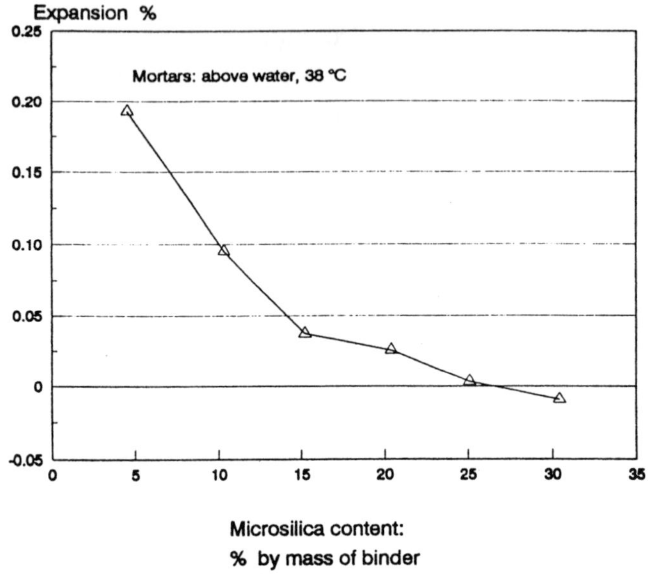

Figure 19. Influence of microsilica on expansion of mortar bars, aggregate/cement ratio = 1.5; Malmsburry hornfels aggregate [61].

However, when a smaller proportion of opal is present (~4%), the magnitude of expansion is likely to be unacceptably high. Hooton [52] reported that a 10% replacement by mass of the high alkali cement with silica fume was needed to reduce the expansion to 0.02% at 14 days.

Use of a smaller amount of CSF, 5%, resulted in increased expansion. Swamy [63] has also seen in his investigation that low amounts of OPC replacement does not control AAR. It was pointed out that even at 10% replacement by CSF, the expansion was considerably lower at initial stages. But at latter stages, the results showed expansion, see Table 17, and Figure 20.

It is further noted here that with the extra addition of alkalies at 10% replacement level, there was an increase in expansion of about 7% compared to the control concrete. However, with 30%replacement and 0.9 kg/m^3 of added alkali, the expansion was reduced by nearly 90%. These results showed that at alkali content

Table 17. Effect of Salt on ASR Expansion [63].

	Cement	Alkali content kg/m³			% expansion at 278 days	
Mix	kg/m³	in cement	added	total	fresh water	salt water
A1	300	3.0	-	3.0	0.285	-
A2	270	2.7	-	2.7	0.229	0.332
A3	210	2.1	-	2.1	0.027	0.026
B1	300	3.0	-	3.0	0.285	-
B2	270	2.7	0.3	3.0	0.306	0.482
B3	210	2.1	0.9	3.0	0.034	0.032
C1	300	3.0	3.0	6.0	1.070	-
C2	270	2.7	3.3	6.0	0.974	1.120
C3	210	2.1	3.9	6.0	0.019	0.016

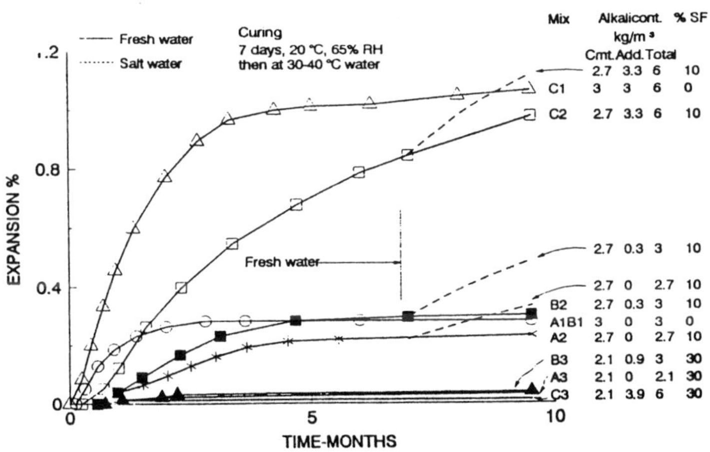

Figure 20. Effect of CSF replacement on expansion of mortar bars caused by alkali-silica reaction. Cmt = % alkali in cement, Add = added alkali [63].

levels of 2.7 to 3.0 kg/m^3, 10% CSF replacement was insufficient to control the expansion, particularly when part of the alkali is introduced from the external source.

Hobbs [64] also reported an increase in expansion when, after 220 days of aging, NaCl was added to the small quantity of water around a number of specimens to increase the equivalent Na_2O content by 2 kg/m^3. According to Hobbs, as is shown in Figures 21 and 22, 10-15% CSF was sufficient to prevent expansion caused by the extra addition of sodium chloride, in both the cases, for 0.5% as well as for 1.0% opaline silica aggregates.

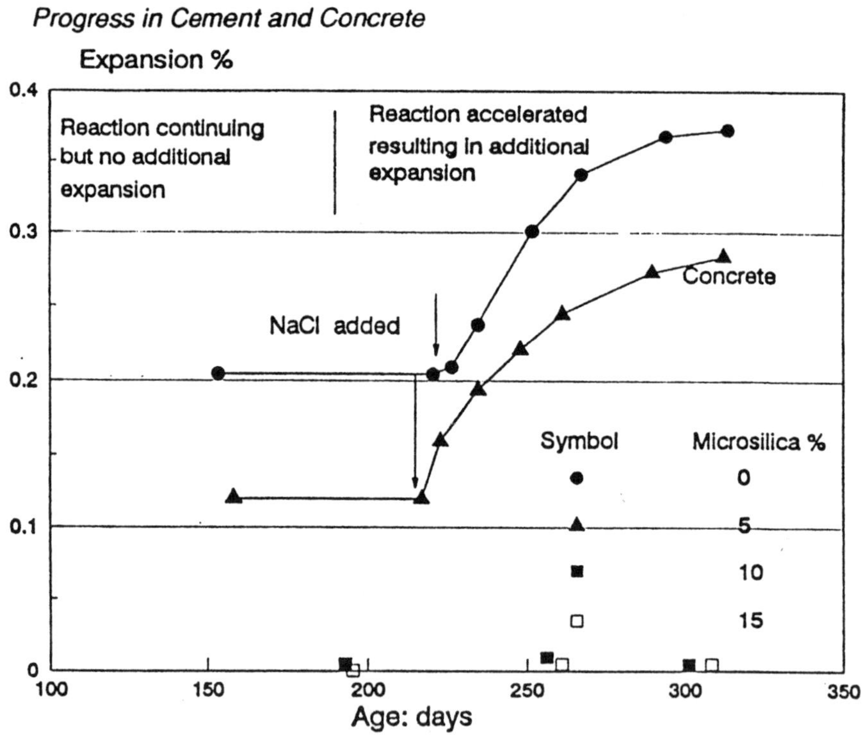

Figure 21. Influence of added sodium chloride on expansion; 0.5% opaline silica in aggregate [64].

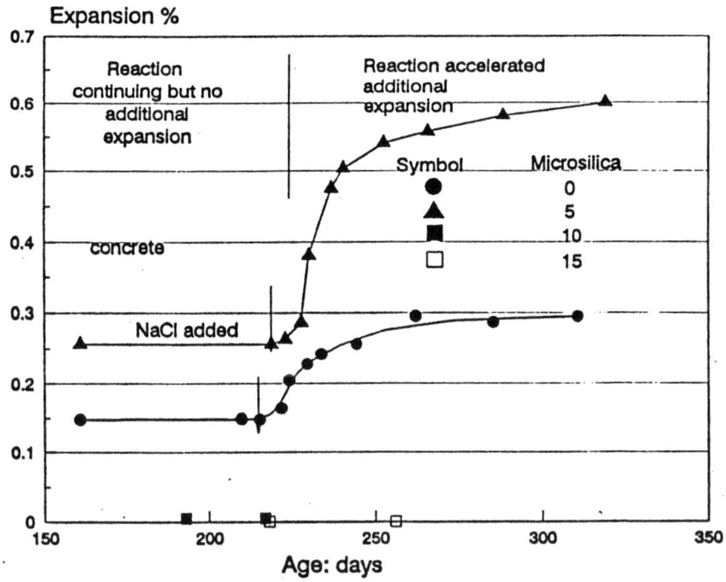

Figure 22. Influence of sodium chloride on expansion; 1.0% opaline silica in aggregate [64].

Kawamura and Takemoto [65] noted that the replacement with CSF delays expansion. They also observed higher expansion at the 10-15% CSF level. It was further reported that after 8-12 months of reaction, the expansion continued. Thus, short term data indicating the effectiveness of mineral admixture to control expansion can be misleading. In another report Kawamura et al., [66] found that the four silica fumes tested have shown a large variation in their behavior on alkali-aggregate expansion of mortars containing Beltane opal, a reactive aggregate. They further reported that the addition of small amounts of these silica fumes increased expansion while the addition of larger amounts of some of these silica fumes have completely stopped expansion of the mortars. They have also pointed out that this is not necessarily due to the pozzolanic activity of condensed silica fume.

However, with 30% replacement levels, no damage could be noticed due to AAR at any level of alkali. The reduction in damage due to AAR by CSF addition is attributed to its pozzolanic reaction. Calcium hydroxide is consumed and less alkali becomes available for reaction with reactive aggregates. This was confirmed by the analysis of pore water [67,68].

Olafsson and Thaulow [69] pointed out that the alkali-aggregate reaction was increased in the presence of salts. They used Icelandic blended cement made with CSF. Asgeirsson and Gudmundsson [5] used high alkali Icelandic cement and reactive sand. They demonstrated that cement replacements at 5 to 10% by CSF were able to control mortar bar expansion. The tests were performed according to ASTM C 227. Based upon their tests, 5-7.5% CSF is now intermilled into Islandic OPC. The ASTM C 441 Pyrex mortar test reported by Hooton [70] showed similar results. It was found that the expansion was reduced with replacement by CSF from 5 - 20 % and that 5 - 10 % replacement was adequate to satisfy various ASTM expansion limits.

The ASTM criteria is not generally been found to be satisfactory. According to ASTM C 441 for example, the criterion for performance at 14 days is met when 10% of the cement is replaced by the CSF, but specimens have undergone expansion at later stages and cracked, as shown by Swamy [63]. The ASTM C 441 criterion can thus be unreliable. Similar results were observed by Perry and Gillott [62]. Further this standard is unsuitable when alkalies enter from external source.

Thus it is necessary to modify the ASTM standards for expansion studies caused by alkali-aggregate reactions such that it is based on long-term testing. The CSF also works as an inhibitor for pop-outs. Nilsson and Petersson [71] studied the relationship between the moisture state and the pop-outs caused by alkali-aggregate reactions. They reported that 5% CSF was efficient in preventing pop-outs. Further, they found that 10-15% CSF content was needed to prevent expansion.

With increases in CSF replacement levels, water requirements also increase, due to its high specific surface area, but it is controlled by the use of some types of superplasticizer. In this connection, a very noteworthy work was done by Perry and Gillott [62]. They showed that superplasticizers dramatically increased expansions relative to the mortars having no superplasticizer (see Figures 23 and 24). From the Figure 23, it is seen that with 5% addition of CSF, expansion is substantially increased, but with 10% CSF expansion is controlled. When superplasticizer was added, expansion substantially increased in both the cases.

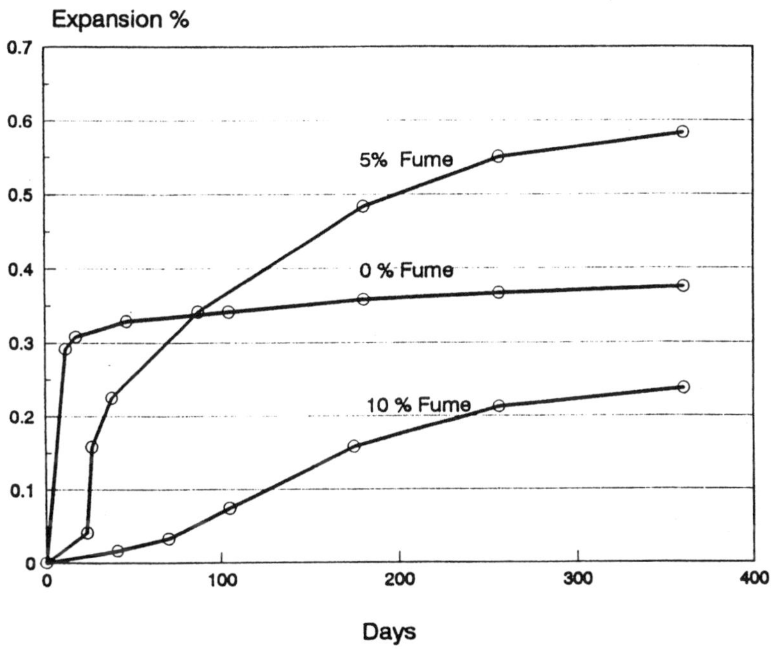

Figure 23. Expansion of mortar bars containing different amounts of silica fume without superplasticizer (sulphonated napthalene formaldehyde), aggregate contained 4% opal, T= 30°C at 100%RH [62].

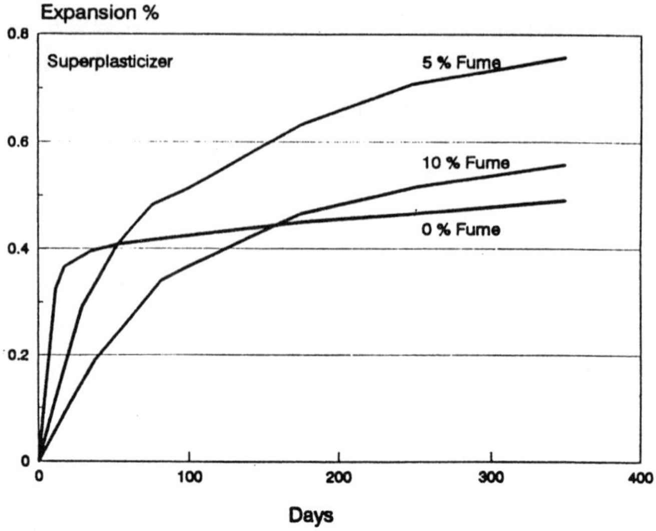

Figure 24. Expansion of mortar bars containing different amounts of silica fume without superplasticizer (sulphonated napthalene formaldehyde), aggregate contained 4% opal, T= 30°C at 100%RH [62].

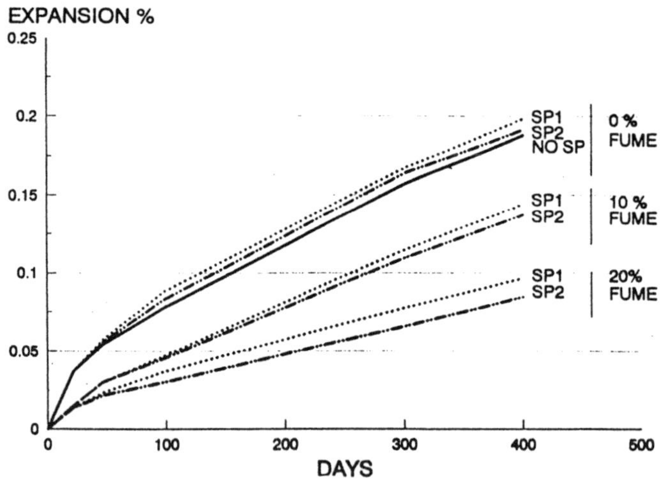

Figure 25. Expansion of concrete prisms made with reactive dolomite limestone and various amounts of silica fumes (23°C, 100%RH) SP = superplasticizer; 1 = sulphonated naphthalene formaldehyde; average of 3 determinations; ranges up to 0.01 [62].

In another test done on concrete prisms made with reactive dolomitic limestone and various amounts of silica fume and addition of two different types of superplasticizers, a decrease in expansion was observed with an increase in CSF content, as shown in Figure 25. Further it was shown that a superplasticizer based on sulphonated naphthalene formaldehyde (SNF) lead to an increase in expansion whereas a superplasticizer based on lignosuphonate decreased expansion. The relative increase in expansion with SNF-type superplasticizer was greater in specimens made with silica fume. It may be due to the difference in the dispersion of CSF which can vary with the superplasticizer.

Aitcin and Regourd [72] in a follow-up study of field tests have reported that a lean concrete with CSF contents (20-40%) showed no traces of gel formation whereas, traces were found in concretes containing 15% CSF. However, the alkali-aggregate reaction was found to be under control in all concretes.

There are some contradictory test data on the effectiveness of CSF to control alkali-aggregate reaction, based upon short term test (14 days) performance according to ASTM 441. The results can be very misleading that can lead to damage, expense and possibly loss of life. Performance under load should be based on long-term tests. The CSF concretes are made with low water-binder ratio. The CSF present in the concrete does not fully react. It is present in the reactive form as fine filler. It is available for reaction in the hardened concrete. Consequently, if alkalies enter the hardened concrete from an external source, there is a possibility that deleterious expansion may result from AAR.

Freeze-Thaw Resistance

Freeze-thaw resistance is an important property of concrete and much study has been devoted to understanding its effects on concrete. Much work is yet to be done in this area as the results obtained thus far are often controversial and there is little general agreement. Freeze-thaw tests are done both with and without deicing salt. A comprehensive review is provided here.

Condensed silica fume addition refines the pore structure of concrete as described in the section on permeability. In general, the silica fume reduces the water permeability in concrete and as a result, the freeze-thaw resistance should increase. Normally, air entraining agents are used for improving the freeze-thaw resistance of normal concrete. When the air entraining agents are used along with pozzolanic materials, the effects can be different and the mechanisms involved become more complicated. When CSF is used due to its high specific surface area, higher amounts of water are necessary to achieve the desired consistency of concrete. This necessitates the use of water-reducing admixtures (superplasticizers). At times this may have a negative influence on the concrete air-void system.

Some studies carried out on the air-void systems for concrete both with and without CSF showed that by proper adjustment of air entraining agents, it was possible to obtain the desired air content in both types of concrete [73,74]. The amount of air entraining agent needed was more for concrete containing CSF compared to the reference concrete, i.e., concrete without CSF. The difference decreased with the addition of superplasticizer. Further it was pointed out that the air content was more stable during vibration of concrete in mixes with both water-reducing admixtures and air-entraining agents. This is in agreement with the results obtained by Okkenhaug [74]. However, Aitcin and Vezina [75] reported a somewhat greater loss in air from CSF concrete during handling.

Carette and Malhotra [76] reported the need for additional air-entraining agents for a 1:1 replacement of cement by CSF for water-binder ratio of 0.40. Similar studies were done by Lehtonen [77] on a concrete of water-binder ratio of 0.45 to which 0.5, 10 and 15% of CSF had been added. He also reported that more air-entraining agent was needed with an increase in the CSF content. These results are contrary to the results published by Okkenhaug and Gjørv [73], who used 8% CSF by weight of cement, and use of 5% CSF improved both the spacing factor and the specific surface. Higher content of

CSF led to less favorable values than for the control concrete. Virtanen [78] observed that for a particular dosage of air-entraining agent, the air content was dependent on CSF content. The investigations were done on a concrete with a compressive strength of 40 -50 MPa and contained 0-16% CSF.

Another characteristic which influences the freeze-thaw resistance of concrete is water absorption. Sellevold et al., [17] studied ice formation in hardened concrete pastes with various amounts of CSF. They found that for the control cement paste with a water-cement ratio of 0.6, a large primary freezing peak appeared near 0 °C, indicating poor frost resistance. By additions of 8% CSF or more, the peak shifts to -20 °C, indicating better frost resistance. Other experiments showed that the pastes with extremely low water-cement ratios (about 0.30) and high CSF contents (> 8 %), did not provide any primary freezing peak even after drying and resaturation. This indicates that it is possible to produce CSF concrete which is frost resistant without air entraining agents. For concretes with a water-binder ratio of 0.45 and 0 and 10% CSF contents, the critical degree of saturation measured did not vary [77], but the water suction behavior appeared to be different. The rate of water absorption in concrete containing CSF was very low.

Freeze-thaw resistance was tested on concrete with a cement content of 730 kg/m^3 and W/C = 0.40, CSF addition, however no admixtures were used [2]. The CSF content varied from 10 to 30%. The test procedure consisted of freezing in air and thawing in sea water. It was concluded that the concrete with CSF performed significantly better than the control concrete. Traettenberg [79] tested the freeze-thaw resistance of 15 day old mortars with various air contents, CSF additions of 0-25 % and water-binder ratio was in the range of 0.45-0.83. The test procedure consisted of freezing in air and thawing in water. The damage was evaluated in terms of residual strength and modulus of elasticity. He found that all CSF mortar with water-binder ratio of 0.60 or less showed excellent freeze-thaw resistance. To get the same freeze-thaw resistance in control mortar, a W/C = 0.40 was needed. Traettenberg also studied the pore size distribution of mortars by mercury intrusion and attributed the improvement in frost resistance to the alteration in the pore structure.

Chang-Yi and Feldman [80,81] tested-freeze thaw resistance according to ASTM C 666-84 procedure B, freezing in air and thawing in water. Mortar had water-binder ratio of 0.45 and 0.60 containing 0, 10 and 30% CSF. It was reported that at the higher water content addition of CSF, improves the frost resistance and at lower water-binder ratios, the use of up to 10% of CSF led to improved frost resistance. With the addition of 30% CSF, the frost resistance was not good, despite the high strength. The reason for improvement was attributed to an increase in the pore volume of capillary pore sizes from 0.20 to 0.35 μm. These capillary pores are believed to act as air bubbles because of their bottle shape. It was suggested that lowest frost resistance of high strength concrete is caused by a combination of low permeability and higher CSF content than could react with the calcium hydroxide from cement hydration.

Malhotra [82] tested CSF concrete with and without air-entraining agents. Water-binder ratios ranged from 0.40 to 0.60 and CSF substitute was 0, 5, 10, 15 and 30% by weight of cement. Superplasticizer was used to compensate for the slump loss. It was concluded that non-air-entrained prisms, irrespective of the value of water-binder ratio and silica fume content, showed very low durability factors and excessive expansion when tested in accordance with ASTM C 666, procedures A and B. The samples showed greater deterioration with an increase in silica fume content. Air-entrained concrete prisms performed satisfactorily up to 15% CSF, as a partial replacement for cement. However, concrete prisms incorporating 30% CSF with a water-binder ratio of 0.42 performed very poorly, with a durability factor less than 10. This was probably due to the high spacing factor. It was not possible to entrain more than 5 % air, and this amount of air volume probably could not provide satisfactory spacing in the hardened concrete. Galeota et al., [83] reported similar results. With 20% CSF and 5% of entrained-air concrete showed poor performance of frost resistance. This poor performance was attributed to the high value of spacing factor.

The results above are contradictory to results reported by the other researchers for low CSF contents. For 30% CSF content, however, they agree with the results of Cheng-Yi and Feldman [80,81]. These results are also in contradiction to the calorimetric investigation results of Sellevold and Bager[84], that showed the disappearance of the primary freezing peaks in concrete with high CSF content and low value of water-binder ratio.

Aitcin and Vezina [75] studied frost resistance of concrete made with silica fume additions in a trial section of highway A25 in Montreal. The normal concrete used had 351 kg/m^3 cement and the water-cement ratio was 0.43. In the silica fume concrete, the cement used was 286 kg/m^3 and silica fume was 21.7 kg/m^3, i. e., 65 kg/m^3 cement was replaced by 21.7 kg/m^3 silica fume. The total binder was 308 kg/m^3. Thus, the CFS/(C+CSF) was 0.07. The missing volume was replaced by an equal volume of sand. Water-binder ratio was 0.50. The test method used was ASTM C 666-84, procedure B. The two concretes were air-entrained and had the same spacing factor. It was shown that the concrete with silica fume (less cement) had higher 28-day compressive strength and better resistance to freeze-thaw.

Yamato et al., [85] tested concretes with water-binder ratios of 0.25, 0.35, 0.45 and 0.55 and CSF contents of 0.5, 10, 20 and 30%. Mixes were air-entrained. It was found that none of the samples, including the air-entrained samples, had a satisfactory air-void system. Test results showed that all the concretes with a water-binder ratio of 0.25 had durability factors above 90. It decreased with an increase in CSF content. For higher values of water-binder ratios, none of the concretes performed satisfactorily. In general, the poorest performance was for those with 20 to 40% CSF. These results are in agreement with Cheng-Yi and Feldman [80,81] and Malhotra [82].

Pigeon et al., [86] studied high performance concrete (HPC) which is usually made with the addition of mineral admixture and very low water-binder ratio. It was reported that the type of cement has a profound influence on the freeze-thaw resistance. The ASTM Type I cements behave completely different than the ASTM III cements. Concrete samples made with 6% silica fume and Type I cement requires air entrainment even after 7 days of curing at a water-binder ratio of 0.30. Results on the spacing factor, on the other hand, are very controversial. For some of the samples, the critical value of the spacing factor was reported to be 200μm and for others, 500μm for good freeze-thaw resistance. For high strength non-air concretes, the spacing factor is usually in the 600 μm to 1.0 mm range, and the concrete exhibits freeze-thaw resistance.

In another test [86], concrete was made with Type I cement, 6% silica fume, water-binder ratio of 0.26 and a pore spacing factor of 690 μm. The concrete showed excellent freeze-thaw resistance. For Type I cement, the limiting value of water-binder ratio was about 0.25, below which no air entrainment was needed for good freeze-thaw resistance. Similar results have been reported by Foy et al.,[87] for concrete made without the addition of silica fume and cured for 7 days. Hammer and Sellevold [88] on the other hand, showed that a non-air-entrained concrete, with a water-binder ratio of 0.26 containing almost no freezable water, was severely damaged by freezing and thawing. They suggested that this may be due to the difference between the coefficient of thermal expansion of the paste and the aggregate and not to the freezing effect.

Freeze-thaw resistance of concrete is related to ice formation. Hammar and Sellevold [89] have shown using low temperature calorimetry techniques, that in HPC there is a very small amount of freezable water down to -8°C which is less than 10% of the evaporable water content. The authors further postulate that the ice formed did not propagate as a front through a continuous pore system, but was formed in isolated areas that may have been filled with water during the vacuum resaturation of the sample before the test.

Low water-to-binder ratio leads to self desiccation which aids in emptying the water-filled pores. Subsequently, the number of unsaturated voids are increased. This may influence the increase in the resistance to microcracking.

According to Fagerlund's [90] calculations, a small amount of water, 0.7%, is sufficient to exceed the tensile strength of the paste at

-20°C. Hence, a very small amount of freezable water is sufficient to damage a saturated concrete.

For high performance concretes, freeze-thaw damage does not appear as surface scaling which leads to mass loss as it does for ordinary concrete. The concrete may appear undamaged on the surface, but it may be full of microcracks. Jacobsen [91] showed that microcrack formation occurred in non-air-entrained HPC even with low amounts of freezable water. This confirms Fagerlund's theoretical calculations, while other researchers [92-95] show that the non-air concrete can be frost resistant.

Freeze-thaw resistance does not only depend upon the water-cement ratio, it also depend upon the type of cement used. For non-air-entrained HPC, controversial results have been reported. Air entrainment reduces strength and it is difficult to entrain air in concrete with a low water-binder ratio. Very high amounts of air-entraining agents are necessary. Its influence on the long time durability has yet to be tested.

The results mentioned above deal with freezing in pure water. Some investigations have also been done on freeze-thaw tests using deicing salts on CSF concrete with and without air entraining agents [96]. It was reported that air-entrainment substantially increased freeze-salt resistance, but concrete with 10% CSF and a water-binder ratio of 0.88 was found to be resistant without air entrainment. In another study done by Kukko and Matala [97], it was shown that with 5% CSF incorporation in high strength concrete, a compressive strength of more than 80 MPa was achieved and without air entrainment, the high strength concrete performed well both in freeze-thaw as well as freeze-salt resistance tests. Further it was reported that the aging procedure, (repeated wetting and drying cycles), did not adversely affect frost and frost-salt resistance.

In another test performed on air entrained concrete with 0 - 15 % of the cement replacement and with water-binder ratio of 0.45, it was found that the freeze-salt resistance improved significantly with an increase in CSF content. The rate of scaling increased after 28 cycles. After 56 cycles, the scaling of all CSF concretes were found to be much lower than that of the control concrete [77]. Similar experiments were done by Virtanen [98] on normal strength concrete with a compressive strength of 30 - 40 MPa. He concluded that the CSF addition improves the freezing and thawing resistance of concrete compared with the control concrete having the same strength and air content. In other experiments Virtanen [78] reported that scaling was reduced with an increase in CSF content, even when the air content decreased with an increase in CSF content. The samples were put in salt solution at -15 °C for 8 hours and in water at 20 °C for 16 hours. However, the scaling rate increased.

Petersson [99] tested three mixes with different values of CSF/cement: 0.11 and 0.19. The freezing and thawing was done in 3 % sodium chloride solution. It was observed that for the same water-concrete ratio (0.35), CSF concrete had excellent resistance up to 110 cycles, as shown in Figure 26. However, it was noted that at more than 110 cycles sudden loss of mass occurred and the samples collapsed in another few cycles. In sample B, water- binder ratio was high. Its scaling resistance at the beginning was similar to A without CSF. After 60 cycles it showed the same trends as was observed in sample C. In this test no air-entrainment was used. The same tendencies were observed by Virtanen [98].

Jacobsen and Sellevold [100] tested freeze-salt resistance of concrete with and without condensed silica fume according to Swedish standard SS 13 72 44 (the "Borås-method"). It was shown that the concrete made with 8% CSF lead to reduced scaling at all curing temperatures up to 60°C. The concrete was tested up to 280 cycles. There was no abrupt collapse. Based upon this work, the concrete was classified as freeze-salt resistant. This is in contradiction to the results reported by Petersson. Freeze-thaw resistance of concrete using deicing salt is to be investigated. Xu et al., [101] showed that the addition of polymer to 10% CSF concrete significantly improves freeze-salt resistance. Generally polymer dispersions introduce air bubbles into concrete. In this case, silica fume worked as defoaming agent. The combined effect of the polymer and silica fume compensated for the reduction in strength.

Combinations of polymer and silica fume resulted in a more durable material than obtained by adding only silica fume. The polymer had an influence on the water absorption and the strength between paste and surface of aggregate.

Chandra and Berntsson [102] observed that concrete with 20-25 % silica fume and a a water-binder ratio as low as 0.18 behaves very well in freeze-salt resistance tests and abrupt deterioration was not seen. In another test in which mortar was used with water-binder ratio of 0.50 and 5 % CSF addition, a significant improvement on freeze-salt resistance was observed when compared to the reference mortar samples prepared without CSF additions [103].

Chandra and Aavik [104] showed that the freeze-thaw resistance performed using different test methods, gave different results. The rate of deterioration depended upon the test conditions, which varied with the test methods used. Therefore, it is very important to have the same testing parameters when comparing results done at different laboratories. Some of the results reported are controversial. However, there is a tendency of freeze-thaw resistance improvement up to 15% replacement of cement by CSF, whereas 20% and 30% replacement decreased freeze-thaw resistance even for air-entrained concrete. This is most likely due to non-homogeneous dispersion of silica particles.

Feldman [105] postulated that CSF may create its own micro-air void network around fine aggregate particles. Normally, the calcium hydroxide liberated by the hydration of Portland cement preferentially deposits in the transitional zone around the aggregate particles. The CSF would consume the $Ca(OH)_2$ during secondary hydration leaving voids. With sufficiently high fine aggregates of CSF, these created voids stay close to each other so as to provide the required spacing for improved freeze-thaw resistance. Another explanation suggests that due to the lower permeability, the CSF concrete does not attain a critical saturation state. As concrete contains reactive CSF as filler, water which may penetrate will be

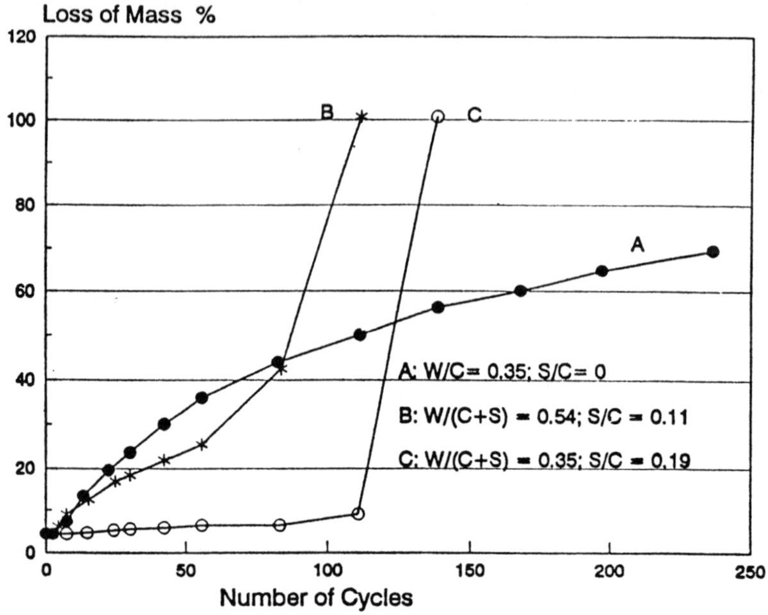

Figure 26. Scaling resistance curves (loss of mass) for concretes without (A) and with (B) and (C) silica fumes as functions of the number of freeze-thaw cycles (W = water, C = cement, S = silica) [99].

consumed by CSF for further hydration. Concrete thus becomes more impermeable. Hooton [106] has given some indirect evidences for this explanation.

The CSF is a very fine material with a very high specific surface area. Dispersion of such fine material may be problematic. Homogenous dispersions are more easily achieved with low CSF contents. Because of this, massive amounts of CSF can collect in localized areas, similar to a balling effect. In this case, some calcium hydroxide will still be produced and crystallize in areas where there are no silica surfaces with which to interact. This creates expansion. Moreover, during pozzolanic reaction which is influenced by the temperature and moisture variations, high stresses can develop in the areas with high CSF concentrations and can cause concrete deterioration. This hypothesis fits well, as CSF is found to improve freeze-thaw resistance specifically for lower CSF contents and higher water-binder ratio. The difference in freeze-thaw resistance with the same amount of CSF, in the authors' opinion, is due to the difference in dispersion of the silica fume itself. This can be influenced by the type and dosage of chemical admixtures used and also upon the mixing technique.

Freeze-salt resistance test has shown an abrupt discrepancy of concrete after 110 cycles, a feature which is not normally observed in freeze-salt tests. It indicates that the deterioration is not only due to the factors involved in normal freeze-salt tests. It may be is logical to speculate that the alkali from the deicing salt, sodium chloride, may react with CSF over time, causing the alkali-silicate reaction. This adds to the other factors involved in freeze-salt deterioration of concrete and cause a sudden collapse in the concrete.

CONCLUDING REMARKS

Condensed silica fume is a very fine, amorphous and reactive mineral admixture. It reacts readily with the calcium hydroxide which is produced during Portland cement hydration. Silica addition refines pore structure and produces concrete of improved mechanical strength. Durability properties tested on concrete gave controversial results which depend primarily on the test method used. Different test methods have different parameters and the rate of deterioration depends upon the parameters to which the concrete is exposed.

Silica fume has a very high specific surface and acts as a reactive pozzolan. Normally, silica fume is used in rather small amounts compared to other pozzolanic materials. Its homogeneous dispersion in concrete is problematic. This inhomogeneity may cause a ball bearing effect and localized problems. Non-uniformity in pore structure will directly hamper the durability properties.

Dispersing chemical admixtures such as superplasticizers are to be selected carefully, as expansion has been reported with certain superplasticizers. It is generally accepted that freeze-thaw resistance improves with air entrainment. It is reported that freeze-thaw resistance decreases with silica fume incorporation for air-entrained concrete. On the other hand, non-air entrained silica fume concrete gives better results. This may be related to the dispersion of silica fume particles, which subsequently influences the pore structure and consequently, the durability properties. Dispersion is also related to the superplasticizer.

From certain perspectives, silica fume is a very promising mineral admixture particularly for producing high-strength to ultra-high strength concrete, but care must be taken during mixing in securing proper dispersion of the particles. Failure to do so may cause serious durability problems and will the resulting concrete will prove to be inferior to ordinary concrete.

REFERENCES

1. RILEM Committee 73 SBC Report (1988).
2. Bernhardt, C.J., SiO_2-Stov som Cementtillsettning (SiO_2 Dust as an Admixture to Cement), *Betongen i dag,* V. 17, No.2, pp. 27-53 (in Norwegian) (1952).
3. Fascia, Hansen and Moum., Betong i Alunskifer, (Concrete in Alum Shale), *Norwegian Geotechnical Institute*, Publ. 86, p.32 (in Norwegian with English Summary) (1971).
4. Traetteberg, A., Waste Materials in the Construction Industry SINTEF Report No. STF65 A78006, *Norwegian Cement and Concrete Research Institute;* Trondheim, p. 13 (1977).
5. Asgeirsson, H., and Gudmundsson, G., Pozzolanic Aactivity of Silica Dust, *Cement and Concrete Research*, V.9, No. 2, pp 249-252 (1979).

6. Jahr, J., Possible Health Hazards from Different Types of Amorphous Silicas, *Symposium on Health Effects of Synthetic Silica Particulates*, STP 732, ed. D.Dunnom, ASTM, Philadelphia, pp. 199-210 (1981).
7. Loland, K.E., Silica Fume in Concrete, SINTEF Report, No. STP 65 F81011, *Norwegian Cement and Concrete Research Institute*, Trondheim, p. 32 (1981).
8. Gjørv, O. and Loland, K.E., Condensed Silica Fume in Concrete, *Proc. Nordic Research Seminar on Condensed Silica Fume in Concrete*, Report No. BML82-610, Norwegian Institute of Technology, Trondheim, p. 293 (Norwegian language with abstract in English) (1982).
9. Japan Silica Fume Technology Association, *State of the Art Report on Silica Fume-Draft II* (Japanese), August 1992. Also published by S. Nagataki in Concrete Technology, Ed. P.K.Mehta, ACI, SP144, p 447 (1994).
10. Smart, M., Leder, A., and Sakuma, Y., Silicates and Silicas, CEH Marketing and Research Report, *Chemical Economic Handbook-SRI International* (1990).
11. Buil, M., Witier, P. and Paillere, A., Etude de la Production d'Hydrogene Gazeux par les Fumees de Silice dans les Materiaux a Base de Ciment, (study of the hydrogen gas production by the silica fume in cementitious materials), *Bulletin de Liaison des Laboratorite des Ponts and Chausses* 153:123-126 (1988).
12. Warren,D., Hydrogen Effects on Steel, *Materials Performance*, pp 38-48 (1987).
13. Hjorth, L., Microsilica in Concrete, *Nordic Concrete Research*; Nordic Concrete Federation, Oslo, Norway, December,No.1, 9/1 (1982).
14. Gjørv, O.E., Durability of Concrete Containing Silica Fume, in Proceedings, *First International Conference on the Use of Fly Ash, Silica Fume, Slag and Other Mineral By-products in Concrete*; Montebello, July 31 - August 5, 1983,

Editor; V. M. Malhotra, ACI Special Publication SP 79, pp 695-708 (1983).

15. Regourd, M. et al., Hydraulic Reactivity of Various Pozzolanas, *Proc. 5th Int. Symp on Concrete Technol.*, Monterrey; Mexico, March (1981).
16. Regourd, M. et al., *Proc. 4th Int. Conf.on Cement Microscopy*, Las Vegas, March (1983).
17. Sellevold, E.J. et al., *Norwegian Institute of Technology*, Trondheim, Report BML 82.610, pp. 19-50 (1982).
18. Huang Cheng-yi and Feldman, R.F., *Cement and Concrete Research*, 15, pp. 585-592 (1985).
19. Huang Cheng-yi and Feldman, R.F., *Cement and Concrete Research*, 15, pp. 285-294 (1985).
20. Traetteberg, A., Silica Fume as a Pozzolanic Material, *Il. Cemento*, 75,No.3, pp. 369-376 (1978).
21. Scheetz, B.E. et al., *Proc. 3rd Int. Conf. in Cement Microscopy*, Houston, pp. 307-318, March (1981).
22. Tenoutasse, M. and Marion, A.,M., *Proc; CANMET/ACI Conference on Fly Ash, Slag Silica Fume and Natural Pozzolana*, Madrid (Spain), April (1986).
23. Kurbush, B. et al., *Cement and Concrete Research*, 15, pp. 134-140 (1985).
24. Diamond, S., *J. Am. Ceramic Soc.* 66, May, No.5 (1983).
25. Page, C.L. and Vennesland, Ø., *Materials and Structures*, 16, No. 91, pp. 19-25 (1983).
26. Sarkar, S.L., *Microscope*, 38, pp. 141-152 (1990).
27. Sarkar, S.L., Microstructure of Durable Concrete; Real World Concrete; *Proceedings, R.N:Swamy Symposium*, Ed. Gurdev Singh, Milwakee, USA, pp. 221-233 (1995).
28. Carles-Gibergues, A., Grandet, J. and Ollivier, J., *Proc, CANMET/ACI First Int. Cong. on Fly Ash, Silica Fume, Slag and Other Mineral By-Products in Concrete*, Montebello, Canada, SP 79, p. 20 (1983).
29. Manmohan, D. and Mehta, P. K., *Cement, Concrete and Aggregate* 3,1, pp. 63-67 (1981).

30. Sørensen, E. V., Report BML 82.610, *The Norwegian Institute of Technology*, Trondheim, Norway, pp. 189-202, Feb. (1982).
31. Markestad, S. A.,Report STF 65A 77027, FCB/SINTEF; *The Norwegian Institute of Technology*, Trondheim, Norway, June (1977).
32. Johansson, L., "Tillsats av kiseldioxidstoft i betong"; CBI Report No. 7587; *Swedish Cement and Concrete Research Institute at the Royal Institute of Technology*, Stockholm, Sweden; October (1975).
33. Hustad, T. and Loland, K. E., Report STF 65 A 81031; FCB/SINTEF; *The Norwegian Institute of Technology*, Trondheim, Norway, June (1981).
34. Sandvik,M.,"Silika betong" Herdevarme egenskapsutvickling; Report STF65 A83063; FCB/SINTEF; *The Norwegian Institute of Technology*, Trondheim, Norway (1983).
35. Skurdal, S., "Egenskapsutvickling for silika betong ved forskjellige herdetemperaturer; Report BML 82.416; *The Norwegian Institute of Technology*, Trondheim, Norway, December (1982).
36. Buck, A. D. and Burkes, J. P., Miscellaneous paper SL 81 - 13; *US Army Engineer Waterways Experimental Station, Structures Laboratory*, Vicksburg, US, July (1981).
37. Grutzeck, M. W., Atkinson, S. and Roy, D. M., Proceedings, *First International Conference on the Use of Fly Ash, Silica Fume, Slag and Other Mineral By-Products in Concrete*, July 31 - Aug 31, ACI Special Publication SP 79, Editor; V. M. Malhotra, pp. 643-664 (1983).
38. Kurbus, B., Bakula, F. and Gabrovsek, R., *Cement and Concrete Research* 15, pp. 134-140 (1985).
39. Maage, M., Report STF 65 A 84019, FCB/SINTEF; *The Norwegian Institute of Technology*, Trondheim, Norway; March (1984).

40. Mindess, S., Gray, R. J. and Skalny J. P., *Proceedings ACI/RILEM Symposium on Technology of Concrete when Pozzolans, Slag and Chemical Admixtures are Used*, Monterrey, Mexico, pp.121-141, March (1985).
41. Strubel, L., Skalny, J. P. and Mindess S.,*Cement and Concrete Research*, Vol 10, No 2, pp.277-286 (1980).
42. Valenta, O., *International Symposium on the Durabilty of Concrete*, Prague, Preliminar Report RILEM, pp. 53-87 (1961).
43. Valenta, O., *Proceedings, Fifth International Symposium on the Chemistry of Cement*, Tokyo 1968, vol 3, Cement Association of Japan, Tokyo, pp. 193-225 (1969).
44. Wakeley, L. D. and Roy, D. M., *Cement and Concrete Research*,Vol 12, No.4, pp.533-534 (1983).
45. Regourd, M., "In condensed silica fume" Editor P.C. Aitcin, *Booklet Universitet de Sherbrook*, Quebec, Canada, pp. 20-24, August (1983).
46. Feldman, R., *Proceedings, Symposium on Fly Ash, Silica Fume and Slag, Material Research Society*, Boston, pp. 124-133, November (1981).
47. Sellevold, E. J., Bager D. H., Klitgaard Jensen E. and Knudsen T., Proceedings, Nordic Mini Seminar on Silica in Concrete, *The Norwegian Institute of Technology*, Trondheim, Norway, pp. 1-32, December (1981).
48. Mehta, P. K. and Gjørv, O. E., *Cement and Concrete Research*, 12(5), pp. 587-595 (1982).
49. Diamond, S., *First Seminar-Elkem Microsilica Technology*, Elkem Chemicals, Sao Paulo, Brazil; September (1984).
50. Regourd, M., *Material Research Society, Symposia Proceedings*, 4, pp. 3-17 (1985).
51. Mather, K., *Proceedings, Seventh International Conference on Chemistry of Cements*, Paris, France, IV, pp. 580-585 (1980).

52. Hooton, R.D., Influence of Silica Fume Replacement of Cement on Physical Properties and Resistance to Sulphate Attack, Freezing and Thawing and Alkali-Silica Reactivity, *ACI Materials Journal*, vol. 90, No.2, March-April, pp 143-151 (1993).
53. Buck, A., Use of Pozzolan or Slag in Concrete to Control Alkali-Silica Reaction and Sulphate Attack, Technical Report SL-88-29, *U.S Army Engineer Waterways Experimental Station*, Vicksburg, US (1988).
54. Yeginobali, A. and Dilek, F.T., *Proceedings Fifth CANMET/ ACI International Conference on Fly Ash, Silica Fume, Slag and Natural Pozzolans in Concrete*, Ed. V.M. Malhotra, Milwaukee, SP 153-42, pp. 795-813 (1995).
55. Cohen, M.D. and Bentur,A., *ACI Materials Journal*, vol. 85, No.3, pp. 148-157 (1988).
56. Popovic, K., Ukraincik, V. and Djurekovic, A., *Durability of Building Materials* 2; Elsevier Science Publishers B.V, Amsterdam, Holland, pp. 171-186 (1984).
57. Mehta, P. K., "Chemical Attack of Low Water Cement Ratio Concretes Containing Latex or Silica Fume as Admixture," in Proc. ACI/RILEM Symposium, Tecnhology of Concrete When Pozzolans, Slag and Chemical Admixtures are Used, Monterray, Mexico. pp. 325-340 (1985).
58. Fiska, O.M., "Betong I alunskifer," Publication No. 101, *The Norwegian Geotechnical Institute*, Oslo, Norway (1973).
59. Hobbs, D.W., *Magazine of Concrete Research*, 33, pp. 208-231 (1981).
60. Sprung, S. and Adadian, M., *Symposium: Effect of Alkalies on the Properties of Concrete,* London, C. & CA Slough, pp. 125-137 (1976).
61. Oberholster, R.E. and Westra, W.B., *Proceedings, Fifth International Conference; Alkali Aggregate Reaction in Concrete*, Cape Town, 252/31 (1981).

62. Perry, C. and Gillott, J.E., *Durability of Building Materials,* Elsevier Science Publications, Amsterdam, pp. 133-146 (1985).
63. Swamy, R.N., Silica Fume and Alkali-Silica Reaction, *Proceedings, 2nd International Seminar, Durability of Concrete; Aspects of Admixture and Industrial By-Products, June 1989,* Chalmers University of Technology, Swedish Council of Building Research, D9, pp. 179-191, (1989).
64. Hobbs, D.W., *Alkali-Silica Reaction in Concrete*, Thomas Telford Ltd., London, pp. 123-130 (1988).
65. Kawamura, M. and Takemoto, K. , *Cement Association of Japan,* pp. 258-261 (1985).
66. Kawamura, M., Takemoto, K. and Hasaba, S., *Proceedings: Katherine and Bryant Mather International Conference, SP 100* (J.M. Skanlon, ed), Atlanta, ACI, pp. 1809-1819 (1987).
67. Diamond, S., *J. Am. Ceram. Soc.,* 66:5 (1983).
68. Page, C.L. and Venessland, O., *Materials and Structures,* RILEM, 16-91, pp. 19-25 (1983).
69. Olafsson, H. and Thaulow, W., *Nord Test Project 173-79, Technological Institute*, Copenhagen, Denmark (1981).
70. Hooton, R.D., *International Workshop on Condensed Silica Fume in Concrete,* Canada (1987).
71. Nilsson, L-O. and Petersson, O., "Alkali-Silica Reactions in Scania, Report TVBM-3014, *Division of Building Materials,* Lund Institute of Technology, Lund, Sweden (1983).
72. Aitcin, P.C. and Regourd, M., *Cement and Concrete Research*, 15, pp. 711-715 (1985).
73. Okkenhaug, K. and Gjørv, O.E.,Report STF 65A 82044, FCB/SINTEF, *The Norwegian Institute of Technology,* Trondheim, Norway (1982).
74. Okkenhug, K., CBI Report 2-83, *Swedish Cement and Concrete Research Institute at the Royal Institute of Technology,* Stockholm, Sweden, pp. 101-105 (1983).

75. Aitcin, P.C. and Vezina, V.M., *Cement Concrete and Aggregates,* pp. 38-42, 6:1 (1984).
76. Carette, G.G. and Malhotra, V.M., *Cement Concrete and Aggregates, pp. 3-13, 5:1 (1983).*
77. Lehtonen, V., Publication 22:85, *Dansk Betonforening*, Copenhagen, Denmark, pp. 217-230 (1985).
78. Virtanen, J., Publication 22:85, *Dansk Betonforening*, Copenhagen, Denmark, pp. 231-254 (1985).
79. Traettenberg, A., *First International Conference, Durability of Building Materials and Components* (P.J. Sereda and G.G. Litvan, eds), Ottawa (1978); *ASTM Special Technical Publication*, 691, pp. 536-548 (1980).
80. Cheng-Yi Huang and Feldman, R.E., *ACI Journal*, pp. 740-743, Sept/Oct. (1985)
81. Felman, R.E. and Cheng-Yi Huang, *Proceedings of the RILEM Seminar on The Durability of Concrete Structures Under Normal Outdoor Exposure*, Hannover, Germany, March (1984).
82. Malhotra, V.M., *Second International Conference on the Use of Fly Ash, Silica Fume, Slag and Natural Pozzolan in Concrete* (V.M. Malhotra, ed), ACI Special Publication 91, pp. 1069-1094, Madrid (1986).
83. Galeota, D., Giammatteo, M.R. and Volta, V., *Second ACI/CANMET International Conference on Durability of Concrete* (V.M. Malhotra, ed), ACI Special Publication SP 126, Montreal, Canada, pp. 249-255 (1991).
84. Sellevold, E.J. and Bager, D.H., Publication No. 22:85, *Dansk Betonforening*, Copenhagen, Denmark, pp. 47-74 (1985).
85. Yamato, T., Emoto, Y. and Soeda, M., *Second International Conference on the Use of Fly Ash, Silica Fume, Slag and Natural Pozzolans in Concrete* (V.M. Malhotra, ed), ACI SP 91, pp. 1095-1118, Madrid (1986).

86. Pigeon, M., Gagne, R., Aitcin, P.C. and Banthis, N., *Cement and Concrete Research,* Vol. 21, pp. 844-852 (1991).
87. Foy, C., Pigeon, M. and Banthia, N., *Cement and Concrete Research*, 18:4, pp. 604-614 (1988).
88. Hammer, T.A. and Sellevold, E.J., *Second International Symposium on Utilization of High Strength Concrete,* Berkeley, p. 17 (1990).
89. Hammer, T.A. and Sellevold, E.J., *ACI Special Publication*, SP-121, pp. 457-489 (1990).
90. Fagerlund, G., Report TVBM-3056, Division of Building Materials, Lund Institute of Technology, p. 38 (1993).
91. Jacobsen, S., "The frost durability of high performance concrete," Ph.D. Thesis, Norwegian Institute of Technology, Trondheim, Norway (1994).
92. Pigeon, M., Gagne, R., and Aitcin, P.C., *Cement and Concrete Research*, Vol.21, No.5, pp 844-852 (1991).
93. Kobayashi, M., Nakakuro, K., Kodama, S., and Negami, S., ACI Special Publication SP-68, pp 269-282 (1981).
94. Okada, E., Hisaka, M., Kazama, Y., and Hattori, K., ACI Special Publication SP-68, pp 215-231 (1981).
95. Mitsui,K., Kasami, H., Yoshika, Y., and Kinoshita, M., ACI Special Publication SP-119, pp 215-231 (1989).
96. Sørensson E. V., *Proceedings, First International Conference on the Use of Fly Ash, Silica Fume, Slag and Other Mineral By-Products in Concrete*, Montebello, Canada, July 31 - August 5, 1983, V. M. Malhotra, ed, ACI Special Publication SP-79, 1983, pp. 709-718 (1983).
97. Kukko H. and Matala., *Second ACI/CANMET International Conference on Durability of Concrete*, Montreal, Canada, 1991, V. M. Malhotra, ed, ACI Special Publication SP-126, pp. 229-247 (1991).

98. Virtanen J., *Proceedings, First International Conference on the Use of Fly Ash, Silica Fume, Slag and other Mineral By-Products in Concrete*, Montebello, Canada, July 31 - August 5, 1983, V. M. Malhotra, ed; ACI Special Publication SP-79, pp. 923-942 (1983).
99. Petersson P.E., *International Seminar on Durability of Concrete: Aspects of Admixtures and Industrial By-Products*, D1:1986, Swedish Council of Building Research, Stockholm, Sweden, D1, pp. 179-189 (1988).
100. Jacobsen, S. and Sellevold, E.J., Frost-Salt Scaling of Concrete, Effect of Curing Temperature and Condensed Silica Fume on Normal and High Strength Concrete, 6th CANMET/ACI Int. Conf. On Fly Ash Silica Fume and Natural Pozzolans in Concrete, V.M. Malhotra, ed (1992).
101. Xu A., Chandra S. and Aavik J., *Proceedings, 6th International Congress, Polymers in Concrete*, Tongji University, Shanghai, China, pp. 304-311 (1990).
102. Chandra S. and Berntsson L., "Freeze-salt resistance of very high strength concrete," Unpublished Work.
103. Chandra S. and Berntsson L., "Influence of silica fumes on freeze-salt resistance of mortars," Unpublished Work.
104. Chandra S. and Aavik J.,*Seminar; Freeze-Thaw Resistance of Building Materials,* Swedish National Testing and Research Institute, Borås, Sweden (1991).
105. Feldman R. F., *Second International Conference on the Use of Fly Ash, Silica Fume, Slag and Natural Pozzolan in Concrete*, Madrid, Spain, 1986, Editor, V. M. Malhotra, ACI Special Publication, SP-91, pp. 973-989 (1986).
106. Hooton R. D., *Proceedings, International Workshop on Condensed Silica Fumes in Concrete,* paper nr. 18, Montreal, Canada (1987).

10

PALM OIL SHELL AGGREGATE FOR LIGHTWEIGHT CONCRETE

A. A. Abang Abdullah

INTRODUCTION

Increased agricultural production and the development of agro-based industries in many countries of the world have brought about the production of large quantities of agricultural wastes, most of which are not adequately managed and utilized. Agricultural wastes have been used for animal feed, fertilizer and fuel for energy production, but little work has been carried out to develop utilization of these wastes in the production of building materials. The need to conserve the traditional building materials which are facing depletion, have obliged engineers to look for alternative materials. The rising cost of transportation encourages the use of materials which are readily available within the region surrounding the construction activity.

Low-cost building materials can be produced using inexpensive indigenous raw materials. Agricultural wastes which are renewable and are found in abundance in many countries, present an interesting alternative to the traditional and sometimes imported building materials, particularly for low-cost construction. The types of agricultural wastes available for use in each country need to be identified. Utilization of these materials and the development of a domestic building materials industry based on the available materials should be encouraged.

The use of agricultural wastes as aggregates in lightweight concrete can provides an alternative to conventional methods for the production of lightweight concrete. Lightweight concrete results in a reduction in the overall deadweight of a structure and provides a significant saving in the overall cost of construction. The use of less expensive alternative lightweight aggregates in the production of the concrete can provide further reductions in expense.

LIGHTWEIGHT CONCRETE

The use of lightweight aggregates dates back about 2000 years when the Romans built the Pantheon, the Aqueduct and the Colosseum in Rome, using lightweight concrete made of natural pumice aggregate. In Britain, clinker aggregate concrete was used in the construction of the British Museum in the early part of this century. S.J. Hayde's patented lightweight aggregate, "Haydite" made by the expansion of shale, came into production in the United States in 1918 [1].

Today, lightweight concrete includes aerated and no-fines concrete and is generally defined as concrete with an air-dry density not exceeding 1850 kg/m^3 (normal concrete has a density of 2300 kg/m^3). Lightweight concrete has many applications, including in reinforced and pre-stressed concrete.

Lightweight concrete is used in many structural applications as well as for insulation purposes. It provides a significant reduction in dead weight, which represents a large proportion of the total load on the structure. However, there is a close relationship between the strength, thermal insulation properties and density of lightweight concrete. High density concretes tend to be stronger and are better conductors. For insulation purposes, lightweight concretes are usually under 400 kg/m^3, while structural lightweight concretes range from 1600 kg/m^3 to 2000 kg/m^3.

Lightweight concrete can be produced using lightweight aggregates, aeration or omitting the fine aggregates to produce no-fines concrete. A combination of the above methods is also possible. The American Concrete Institute defines structural lightweight concrete as having an air-dry density of less than 1840 kg/m^3 and a compressive strength of over 17.2 N/mm^2 after 28 days. The British Code of Practice BS8110 recommends a minimum characteristic strength of 15.0 N/mm^2 at 28 days.

PALM OIL SHELL AGGREGATE

Agricultural Waste

The use of agricultural wastes as aggregate for the production of building materials has several practical and economic advantages. Apart from being a zero-value product in the current market situation, each waste type has characteristic properties which make it more suitable for specific applications. The applications of the different agricultural waste types can be grouped into three broad categories as follows.

- Production of cementitious materials
- Fibers for particle boards or sheets
- Aggregates in concrete

The palm oil industry which is important in many countries such as Malaysia, Indonesia and Nigeria yields a large amount of wastes which can be utilized for the production of building materials. Palm oil shells are produced in large quantities by the oil mills and can be used as aggregates in the production of lightweight concrete.

Palm Oil Industry

Megat Mohd Nor [2] reported that in 1989, world palm oil production reached about 9.6 million tons per year from a modest start in the early 1920s due to increasing demand for vegetable oil. Malaysia contributed 5.5 million tons per year from a palm oil cultivation of 1.85 million hectares during the same year, making it a major producer of palm oil. Oil palm is cultivated by many countries in the equatorial region.

Palm oil processing is separated into six stages: sterilization, threshing, pressing, depericarping, separation of kernel and shell, and clarification. The wastes produced during processing include empty bunches, shells, pericarp and the palm oil mill effluent (POME). The palm oil shells produced are occasionally used for paving roads on some plantations. However, most mills burn the shells along with the fruit bunch stalk and pressed fibers for energy.

Palm Oil Shells

About 1.1 tons of shells or 5.5% of the weight of the fresh fruit bunch is produced annually from each hectare cultivated. Palm oil shells have a bulk density of 620 kg/m^3 and a specific gravity of 1.25.

LIGHTWEIGHT CONCRETE USING PALM OIL SHELLS

The material properties and structural performance of lightweight concrete with palm oil shells as aggregates are similar to lightweight concrete produced using the more common lightweight aggregates such as clinker, foamed slag and expanded clay.

Lightweight Aggregates

The palm oil shells are hard and are received as crushed pieces as a result of the process used to release the oil. Manual sieving is necessary to remove the large amount of fine particles. The shells are then air-dried before use. A typical particle size distribution curve for palm oil shell in comparison with sand is shown in Figure 1 [4].

Since the palm oil shells are lighter than the cement matrix, the shells tend to segregate in wet concrete mixes. On the other hand, dry mixes using palm oil shells are relatively more difficult to work with when compared to ordinary concrete. Trial mixes are necessary to achieve a good mix design.

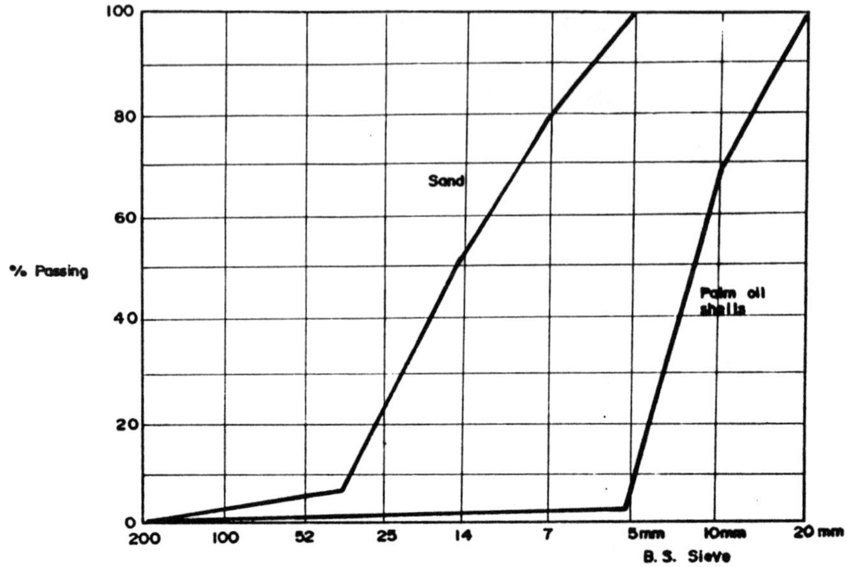

Figure 1. Particle size distribution curves for palm oil shells and sand.

Compressive Strength

The workability and compressive strength of lightweight concrete with palm oil shells as aggregates is affected by the proportion of palm oil shells and the water-to-cement ratio. The 28 day cube compressive strengths of the lightweight concrete vary between 5.0 to 19.5 N/mm^2.

Variations of workability and compressive strength values of the lightweight concrete with different proportions of palm oil shells, sand and water-to-cement ratios are shown in Figures 2a, 2b and 2c. These values are summarized in Table 1 [3].

Flexural Strength

Reinforced concrete beams made from lightweight concrete with palm oil shells exhibited satisfactory structural behavior. With a

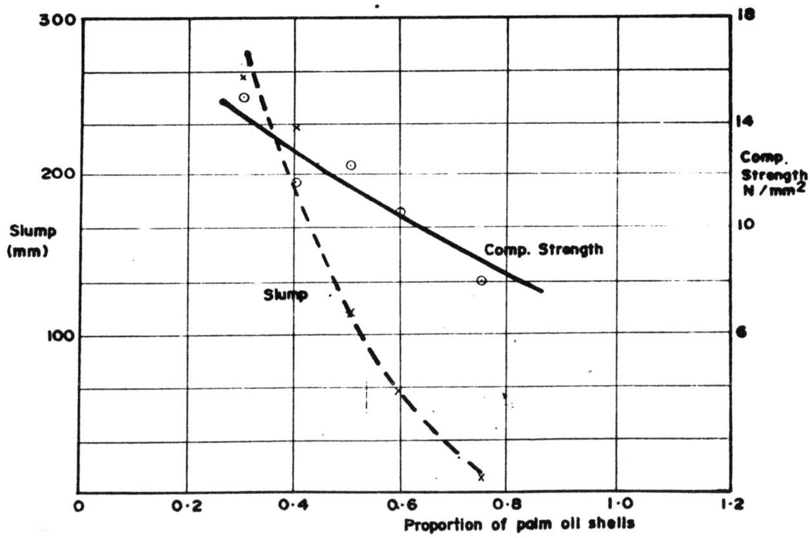

Figure 2a. Variation of slump and compressive strength with different proportions of palm oil shells.

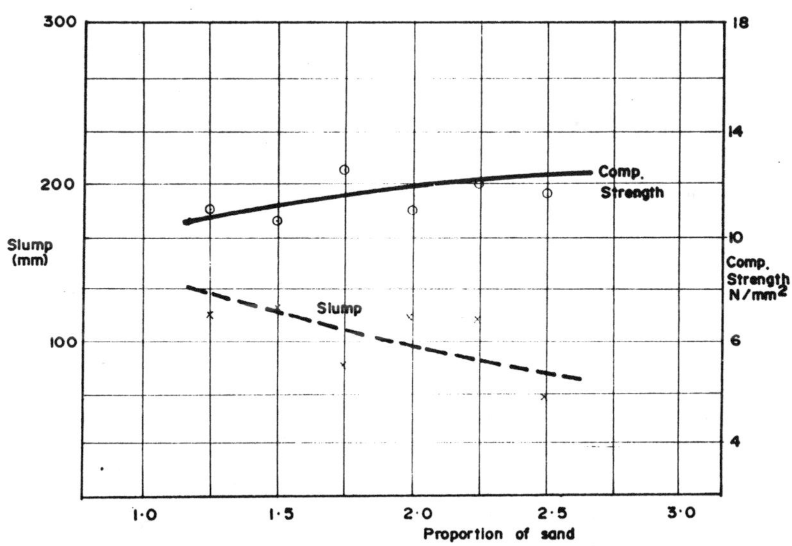

Figure 2b. Variation of slump and compressive strength with different proportions of sand.

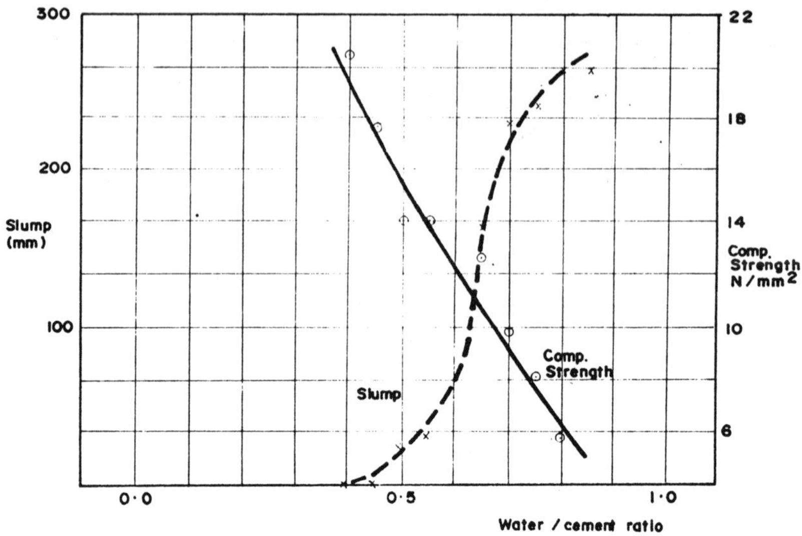

Figure 2c. Variation of slump and compressive strength with water-to- cement ratio.

Table 1. Variations of Workability and Compressive Strength With Lightweight Concrete Mix Ratios.

Mix Ratios*	Workability		Compressive Strength (N/mm^2)		Air-Dry Density (kg/m^3)
	slump (mm)	Comp. Factor	7 days	28 days	
1:1.5:0.5-1.0/0.6	200-0	0.98-0.78	10.0-2.5	15.0-5.5	1890-1725
1:2.0:0.5-1.0/0.6	230-5	0.99-0.79	8.0-2.0	11.5-5.0	1930-1780
1:2.5:0.3-0.75/0.6	260-10	0.97-0.87	9.5-5.5	15.0-8.0	1985-1895
1:1.25-2.5:0.6/0.6	115-60	0.97-0.86	7.0-8.0	11.0-11.5	1905-1935
1:2:0.6/0.40-0.85	0-260	0.72-0.97	16.5-0	20.5-0	2050-1840

* Mix ratios: cement:sand:palm oil shells/water to cement ratio.

lightweight concrete mix of 1: 1.5: 0.5/0.5 cement, sand, palm oil shells and water-to-cement ratio, a 28-day cube compressive strength for the lightweight concrete was 17.5 N/mm^2. A typical load-deflection curve for the reinforced concrete beam using palm oil shells as aggregates is shown in Figure 3 [3].

Creep

The creep test performed according to ASTM applied a load equivalent to a stress of 6.0 N/mm^2 to three 150 mm diameter cylinders made using lightweight concrete with palm oil shells as aggregates. The instantaneous strain was obtained by deducting the initial reading from the final reading immediately after loading. Readings were taken six hours later, then daily for one week, weekly for one month and monthly for nine months. Strain readings on control specimens were taken according to the same time schedule. The total strain divided by the average stress provided the total strain per unit stress. This total strain per unit stress was plotted on semi-log coordinate paper. The creep curve for lightweight concrete using palm oil shells [3] shown in Figure 4 shows a large creep compared to ordinary concrete and the creep rate for lightweight concrete using palm oil shells did not achieve a constant value after 3 months. This property may be a cause for concern when using lightweight concrete with palm oil shells as structural concrete.

Shrinkage

Specimens 50 mm square and 250 mm long prepared using lightweight concrete with palm oil shells were tested for shrinkage. The lengths of the specimens were read before and after oven drying for at least 44 hours at a humidity of 17%. This cycle was repeated until a constant length was obtained. The initial wet measurement minus the final dry measurement gave the drying shrinkage. This was expressed as a percentage of the dry length after shrinkage had taken place. The specimens were kept standing in the measurement apparatus at the prevailing relative humidity and readings were taken at specific intervals.

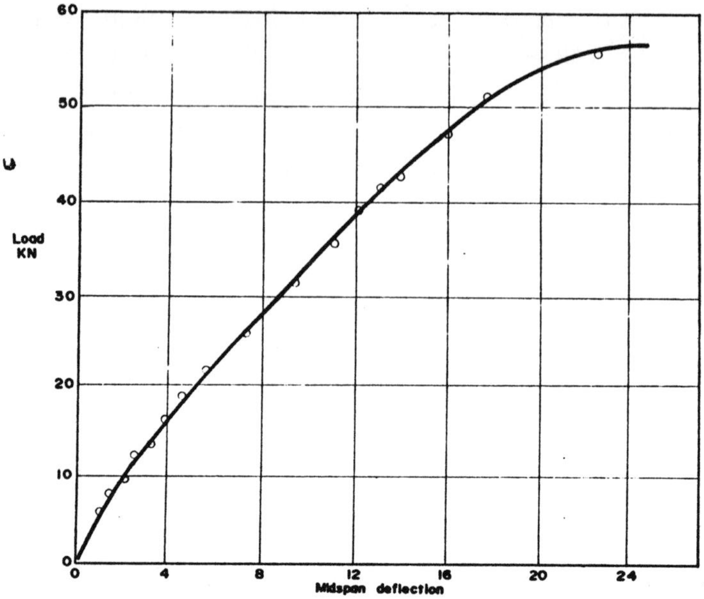

Figure 3. Typical load-deflection curve for reinforced concrete beams using palm oil shells as aggregates.

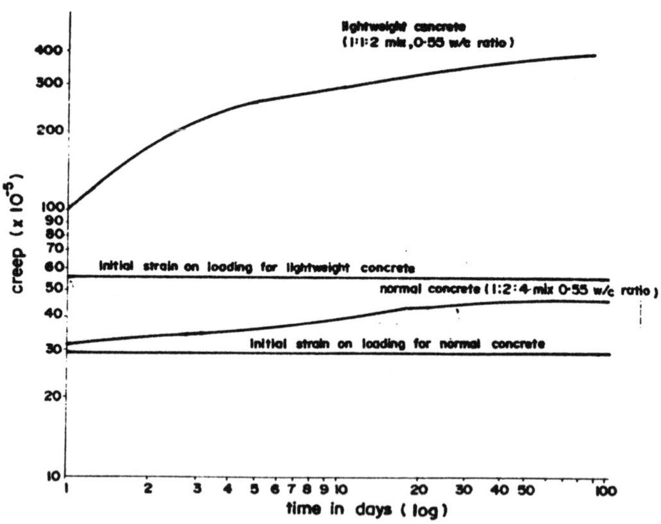

Figure 4. Creep of lightweight concrete using palm oil shell as aggregates in comparison with normal concrete.

The shrinkage strain was plotted against logarithm of time. This curve gave an indication of the progressive shrinkage and how long it took the sample to reach the dry length.

The shrinkage strain versus time curves for lightweight concrete using palm oil shells in comparison with normal concrete are shown in Figure 5 [4]. The curve shows shrinkage rate decreased after about 60 days. However the shrinkage of lightweight concrete using palm oil shells was about five times that of normal concrete. A semi-logarithmic plot of the same curve is shown in Figure 6 [4].

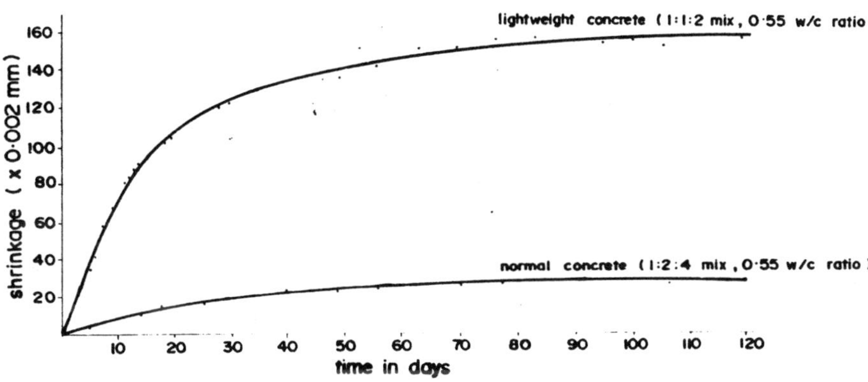

Figure 5. Shrinkage of lightweight concrete using palm oil shell as aggregates in comparison with normal concrete.

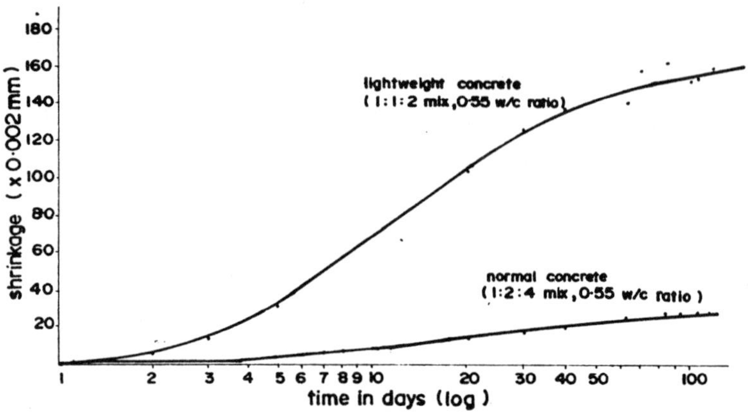

Figure 6. Semi-logarithmic plot of shrinkage of lightweight concrete using palm oil shell as aggregates in comparison with normal concrete.

COMPARISON WITH OTHER AGRICULTURAL WASTES

The strength properties of lightweight concrete with palm oil shells were compared with those of lightweight concrete using other agricultural wastes as aggregates. In this study, four types of agricultural wastes were studied and included, palm oil shells, palm oil clinkers, rice husks and coconut shells. Both the palm oil clinkers, which are by-products of the palm oil mills energy generating burners, and the coconut shells had to be broken into pieces not larger than 20 mm before use. The agricultural wastes were air-dried in the laboratory. After preliminary tests, a constant concrete mix ratio of 1:1:2 of cement, sand and agricultural wastes with a water-to-cement ratio of 0.55 was adopted. The bulk density, compressive and tensile strengths of lightweight concrete using the different types of agricultural wastes as aggregates are summarized in

Table 2 [4]. From the experimental test results, light-weight concrete using palm oil clinkers and palm oil shells were found to produce relatively higher compressive and tensile strengths values.

Table 2. Compressive and Tensile Strengths of Lightweight Concrete Using Agricultural Wastes as Aggregates.

Agricultural Wastes	Bulk density of wastes (kg/m^3)	Lightweight Concrete Strength (N/mm^2)		
		Compressive		Tensile
		7 days	28 days	28 days
Oil palm clinker	830	20.7	29.8	1.37
Oil palm shell	620	12.6	17.4	1.12
Coconut shell	445	8.9	11.6	1.15
Rice husk	136	3.4	6.3	0.82

APPLICATIONS

Lightweight concrete with palm oil shells as aggregates has many potential applications such as in the production of structural concrete, loadbearing and non-loadbearing concrete blocks as well as for insulating purposes. It can be used in precast wall panels and roof units.

ACKNOWLEDGMENTS

The author thanks former staff members and students of the Faculty of Engineering, University Pertanian Malaysia particularly to Dr. S.A.Salam, Abdul Razak Abdul Manap and Tan Kee Liong who made this study possible. This work was supported by the Malaysian Government IRPA and Ministry of Housing and Local Government grants.

REFERENCES

1. Spratt, B.H., *An Introduction to Lightweight Concrete, Cement and Concrete Association,* Slough, United Kingdom (1975).
2. Megat Mohd Nor, M.J. et. al., *Management and Utilization of Oil Palm Wastes - A Review,* Regional Information Centre on the Management and Utilization of Wastes, University Pertanian Malaysia, Serdang, Malaysia (1990).
3. Abang Abdullah, A.A., *Use of Palm Oil Shells as Aggregates in Lightweight Concrete*, RILEM-NCCL Joint Symposium on Use of Vegetable Plants and Their Fibres As Building Materials, Baghdad, October (1986).
4. Abang Abdullah, A.A., *Basic Strength Properties of Lightweight Concrete Using Agricultural Wastes as Aggregates*, Int. Conf. on Low-Cost Housing for Developing Countries, Roorkee, November (1984).

INDEX

D

E

G

H

I

K

L

M

Q

R

S

X

Z